CHAOTIC DYNAMICS

CHAOTIC DYNAMICS

From the
One-Dimensional Endomorphism
to the
Two-Dimensional Diffeomorphism

Christian Mira
Groupe d'Etude des Systèmes
Non Linéaires et Applications
INSA Toulouse
France

World Scientific
Singapore • New Jersey • Hong Kong

Published by

World Scientific Publishing Co. Pte. Ltd.
5 Toh Tuck Link, Singapore 596224
USA office: 27 Warren Street, Suite 401-402, Hackensack, NJ 07601
UK office: 57 Shelton Street, Covent Garden, London WC2H 9HE

Library of Congress Cataloging-in-Publication data is available.

British Library Cataloguing-in-Publication Data
A catalogue record for this book is available from the British Library.

CHAOTIC DYNAMICS
From the One-Dimensional Endomorphism to the Two-Dimensional Diffeomorphism

ISBN-10 9971-5-0324-7 (hardcover)
ISBN-13 978-9971-5-0324-6 (hardcover)
ISBN-13 978-981-4503-21-1 (ebook for institutions)
ISBN-13 978-981-4603-44-7 (ebook for individuals)

For any available supplementary material, please visit
https://www.worldscientific.com/worldscibooks/10.1142/0413#t=suppl

A Kraljica Mira
(Med. 25.6.86, l'Oasis)

A ma femme Nicole,

à mes enfants,

à mes petits enfants.

Petite, et dabitur vobis .

Quaerite, et invenietis .

Pulsate, et aperietur vobis .

(Mt. 7_7)

PREFACE

The subject matter concerns the study of complex motions of dynamical systems, via the basic mathematical tool of maps (equivalent denomination recurrences; a non-invertible map being an endomorphism, an invertible and differentiable one being a diffeomorphism). More precisely, the book treats the properties of a class of determinist mathematical models, the solutions of which have not the periodical regularity. Nowadays, depending on the context, such solutions are called "chaotic". Since few years, chaotic dynamics has become a favour choice for the most part of disciplines dealing with evolution phenomena. It results a considerable increase of the number of publications.

Due to its sudden and "explosive" growth, the subject matter has been developed independently of fundamental results obtained before. Then many present results are rediscoveries, or variants of older ones.

From this point of view the book will appear non-conventional, because its basic references are not those of the literature now considered as classical. Some readers will no doubt find occasions to feel irritated by it. More particularly, the assertion that so popular present notions as those of "invariant coordinates", "kneading invariant",and others , are variant of Myrberg's results will appear disappointing for some. Nevertheless one of the purposes of this publication is to give to the reader an information about generally unknown works of Hadamard, Lattès, Cigala, Myrberg, Neimark, Leonov, Pulkin, ..., and thus to restore certain anteriorities. However, it is worth of note that this choice does by no means constitute a negative judgment of the works of the literature now called classical. Generally written according to the present standards of the "abstract theory of dynamical systems", they are excellent and necessary from the point of view of formal exposition and rigour. With these criterions, such works conserve wholly their own merits, and must be considered as productions of quality.

The debate which might result from the choices of this book would not be a new one. So, in his paper, "George David Birkhoff and his mathematical work" (Bull. Amer. Math. Soc., May 1946, vol. 52, n° 5, part 1, pp. 357-391), Morse wrote :

"Birkhoff was uncompromising in his appraisal of mathematics, by the test of originality and relevance. For him the systematic organization or exposition of a mathematical theory was always secondary in importance to its discovery. I recall his remarks on a mathematical treatise that had come to his attention and eventually had a wide circulation but which he did not regard as original. Birkhoff said : "I read this book through in half hour"... Some of the current mathematical theories were regarded by Birkhoff as no more than relatively obvious elaborations of concrete exemples".

Besides before Birkhoff, the famous mathematician Halphen is also known to have often complained that non-essential generalizations are overcrowding the publication media. In connection with this, it is well known that the majority of scientists were not led to their discoveries by a process of deduction from general postulates or general principles, but rather by a thorought examination of properly chosen particular cases. The generalizations have come later, because it is far easier to generalize an established result than to discover a new line of argument |G23|.

The book may also appear non-conventional from another point of view, which incidentally is a consequence of the first one. Indeed, another possible reason of irritation for certain abstractly inclined readers might be related to the terminology adopted from the original works and the absence of the previously mentioned ars-pro-artis generalizations so popular in contemporary mathematics. The resulting formulation of problems may therefore appear primitive, sometimes even simplistic. The style may appear less formal and rigorous than that of contemporary papers about the abstract theory of dynamical systems. The answer to this eventual remark consits in stating precisely that *the book does not pretend to be a mathematical treatise on dynamical systems* (it is published in a physics collection), because it is not presented in a sufficiently elaborate and rigorous form. Its objective is far more modest and consists of :

-(a)- giving among the basic tools of dynamical systems theory, a large part of the most efficient ones for solving practical problems (cf. such problems for example in |G 35||G 18||N 5|).

-(b)- understanding dominating internal mechanisms of evolution processes, and thus preserving a phenomenological transparency through the simplest possible concrete examples (in the sense of the above mentioned Birkhoff's citation), from which perhaps "standard" mathematical theories might be elaborated. Indeed, it can be considered that the field of "*concrete dynamic systems*" is constituted by two sets of results. The first one is related to the study of problems directly suggested by practice (Physics, Engineering,...). The second one concerns the study of equations, not directly tied with practice, but having the lowest dimension, and the simplest structure, which permit to isolate in the purest form a "mathematical phenomenon", by eliminating "parasitic effects" of a more complicated structure. The bringing to light of the phenomenon of Myrberg's period doubling cascades, and their accumulations, from the very simple discrete dynamic system $x_{n+1} = x_n^2 - \lambda$ (x being a real variable, λ a real parameter, n = 0, 1, 2, ...) is of such a type, by reason of its many practical implications. The Smale's horseshoe is also of this type. Consequently, the approach which is adopted is a geometric one, based on an important number of figures.

-(c)- stimulating some ways of research from the choices of examples.

Is is worth of note that the choice of this less formal style has again the advantage of optimizing the interdisciplinary communication, even if it has to appear old-fashioned to abstractly inclined readers.

Beyond as systematic as possible a study of critical cases in the Ljapunov'sense, and of the crossing through the corresponding bifurcations, it is attempted to realize the mentioned objective by showing that *two basic fractal bifurcations structures* play a fundamental role in the understanding of the dominating internal mechanisms in a wide class of chaotic behaviours. They were called :

- *"box-within-a-box"* (or *"embedded boxes"* in |G 14|) *bifurcations structure* ("structure de bifurcations boites-emboitées" in French |G 32| |M 25|, 1975), which corresponds to an ordering of the Myrberg'spectra (often called in the contemporary literature Feigenbaum's cascades of bifurcations by period doubling).

- *"boxes in files" bifurcations structure* ("structure de bifurcation boites en files" |M 27|, 1978) which corresponds to an ordering of the Farey's sequences of fractions in their lowest terms, and which was elaborated from an adaptation of the Leonov's results |L 14| about piecewise linear maps bifurcations structure.

From the knowledge of these structures,it resulted the elaboration of a *particular symbolism* associated with their fractal properties, and the introduction of new notions such as :

. *fuzzy*, or *chaotic*, *basin boundary* ("frontière floue" |M 23|, 1975)
. *average value cycle* ("cycle en valeur moyenne" |M 22|, 1976) known now under the denomination *phenomenon of intermittency*
. *average value cyclic chaotic segment* ("segment stochastique cyclique en valeur moyenne" |M 22|, 1976) related to what it is now called *attractor in crisis*.
. *adjoint cycles*, and *self adjoint cycle* |K 3||F 7|(1985) which extend the possibilities of the symbolic dynamics, and which permits an understanding of the complex communications between sheets of a foliated box-within-a-box bifurcation structure |F 7||F 8||F 9|.
. the notions related to various non-classical global bifurcations, a part of them characterizing the transition order $\rightleftarrows$ chaos.

A non negligible part of the book constitutes a synopsis of slowly accumulated results obtained by the research group of Toulouse which works about non linear dynamical systems since 1962. The existence of this group would not be possible without my meeting I. GUMOWSKI in 1958.

From this personal event resulted my "initiation" into the famous (but in this time rather little known in western countries) works of the Andronov's school |B 14|. Since that time, a long friendly collaboration has permited a continuous and fruitfull action in this field, and many common publications using Andronov ideas. I. Gumowski is gratefully acknowledged for his invaluable contribution to the researches of the Toulouse group, which thus has profited of his interdisciplinary scientific knowledges of exceptional extent. All the researchers who worked, and who presently are working in this group, are also acknowledged. A particular mention is made to H. Kawakami, who efficiency collaborated at this task as a visiting Professor in 1984-1985. His presence gave a new and important impulsion to the studies related to the two-dimensional diffeomorphism, when he introduced the above mentioned fruitful notion of "adjoint cycles" and "self adjoint cycle".

Since 1978, R. Thom has accepted to present some papers of the group in "Comptes Rendus de l'Académie des Sciences de Paris". He is warmly thanked for his interest to the corresponding work, and for some fruitfull discussions which resulted.

Voluntarily certain subjects are not treated (renormalization, fractal dimension, applications to practical dynamical problems, ...). For a usefull complementary information on them, and others developed in this book, among the numerous available publications, it is suggested to consult :

* The Neimark's book |N 5| (1972) devoted to the method of point mappings and their applications to the theory of automatic control and the theory of dynamical systems. Scientist of the Andronov's school, Neimark has brought one of the most important contributions to this question.

* The Guckenheimer and Holmes book |G 18| (1983) dealing with non linear oscillations, dynamical systems and bifurcations, which is one of the most interesting publications recently published.

* The Gumowski and Mira's book |G 35| (1980) about chaotic dynamics, based on the point mappings method, with applications to engineering problems. The book |G 36| (1980) of the same authors completes the theoretical aspects of |G 35|.

* The book "Chaos" by Hao Bai-Lin (World Scientific Singapore, 1984), which is a selection of the classical literature on chaos with 41 papers among the most known.

* The Holmes and Whitley's paper |H 10| (1984) which constitutes the most complete study on homoclinic tangency in two-dimensional diffeomorphisms.The original study of |H11| (1985) by Holmes and Williams.

* The series of papers by Grebogi, Mc Donald, Kostelich, Ott, and Yorke |G 6|-|G 10|, |M 3||M 4| (1983-1986) representing a very important contribution to the questions of fractal and fuzzy basin boundaries,

chaotic transients.

* The book "Circuits non linéaires" by M. Hasler and J. Neirynck (Presses Polytechniques Romandes, 1985) which gives examples of electrical circuits with chaotic behaviours.

* A recentsystematic study of chaotic dynamics in an electrical circuit can also be found in the paper "The double scroll" by L.O. Chua, T. Matsumoto and M. Komuro, I.E.E.E. Trans. on Circ. and Syst., vol CAS 32, n° 8 (1985), 798-813. About such studies, it is worth of note the important contribution of Hayashi and Kawakami since 1972 (cf. some references in |K 2| and |M 48|).

* The very recent book "Difference equations and their applications" by A.N. Sharkovsky, Yu. L. Maistrenko, E. Yu. Romanenko (Kiev Naukova Dumka, 1986). Here, new results of the authors, concerning methods of qualitative study of different classes of difference equations, and differential-difference equations, are laid.

°°°

Not having a usual practice of english, the quality of the english of this book is certainly affected. The reader may excuse this fact.

A part of the research of Chapter 6 was executed under contract DRET n° 85-1303 (French Ministry of Defense).

The most part of the figures were drawn by G. Roussel, and a part of the text was typed by Mrs C. Grima. These contributions are gratefully acknowledged.

Toulouse, April 1987 Christian MIRA

CONTENTS

CHAPTER 6 : TWO-DIMENSIONAL DIFFEOMORPHISMS AND THE FOLIATED BOX-WITHIN-A-BOX BIFURCATIONS STRUCTURE.

Chapter 1
DYNAMIC SYSTEMS AND RECURRENCES. GENERALITIES

1.1. CONTINUOUS DYNAMIC SYSTEMS AND DISCRETE DYNAMIC SYSTEMS

The content of this monograph concerns evolution processes, or equivalently dynamic systems. Till now, a generally accepted, and unambiguous, definition of this notion seems not yet exist. In a concrete context, that of the experimental physics, for example, a "dynamic system" is a limited configuration of objects, or again a more or less subjective description of this configuration in a verbal, or mathematical, form. Because it is impossible to study a limited configuration when it is imbedded into the whole universe, it is necessary to separate, or isolate, this configuration via an appropriate artifice consisting at the final step in choosing a certain number of relevant variables, and parameters. This indispensable process of separation leads to what is called a constraint of separation, which can have different forms according to the considered limited configuration, the desidered mathematical formulation, and the nature of the sought solution. In an abstract context, it is supposed that the constraint of separation is known, and the expression "dynamic system" simply designates the equation of the motion |G 33|.

A first possible formulation of motion equations is that ordinary differential equations with real variables, the time t being the independant variable, in one of the two explicit forms :

(1.1) $dx_i/dt = f_i(x_1, x_2, \ldots, x_n),\ i = 1, 2, \ldots, m,\ x_i(t_o) = x_{i_o}$

(1.2) $dx_i/dt = f_i(x_1, x_2, \ldots, x_n, t),\ i = 1,2,\ldots,m,\ x_i(t_o) = x_{i_o}$

The first one is called a m-dimensional *autonomous equation* and the second one a m-dimensional *non-autonomous equation*. When the existence and uniqueness conditions are satisfied for the initial condition $x_i(t_o)$, the solution is a continuous function of t, and corresponds to what is called a *phase trajectory*, or a motion, in the *phase space* of coordinates $x_1, \dots, x_m$. The equations (1.1), (1.2) also can be represented in a vectorial form $X^\bullet = F(X)$, $X^\bullet = F(X, t)$, $X^\bullet = dX/dt$.

A second possible formulation is that of recurrences, or equivalently *point-mappings, maps, iterations* (these forms are indifferently used in this book), for which the time t is no longer a continuous independant variable but a sequence of integers, n = 0, 1, 2, ... (discrete time), in one of the two explicit vectorial forms :

(1.3) $X_{n+1} = F(X_n)$, or $X_{n+1} = T\,X_n$, $X\,(n=0) = X_o$

(1.4) $X_{n+1} = F(X_n, n)$, or $X_{n+1} = T_n\,X_n$, $X(n=0) = X_o$

X being a real vector with components $(x_1, \dots, x_p)$, F a single valued vectorial, function of X. In (1.3) (*autonomous* equation), T is the p-dimensional point-mapping which gives the point X_{n+1} from the point X_n, and which is independant of n. In (1.4), (*non autonomous* equation), the point-mapping depends on n. When existence and uniqueness conditions are satisfied for the initial condition X_o, the solution $X_n = X(n, X_o)$ is a sequence of points (*iterated sequence*, or *orbit*, or *discrete phase trajectory)* in the p-dimensional *discrete phase plane*. The recurrences (1.3), (1.4) can be *directly* the mathematical models of a class of dynamic systems which are by nature of the discrete type, i.e. the available information about their states is only obtained in a sampled form. However, the autonomous recurrence (1.3) plays a fundamental role, when it is associated with an autonomous differential equation (1.1), the dimension p of (1.3) being not higher than (m-1), or with (1.2) when the f_i are periodical functions of t with a fixed period. In this case (1.3) is the result of the application of the *Poincaré's method of section*. The reduction of the effective dimension of the differential equation, which is so obtained, makes easier its study. Details about this method will be considered in an other paragraph. It should be stressed that the inverse process, consisting in the construction of a differential equation from a recurrence (1.3), is generally not possible. The functional space of recurrences is thus much richer than that of differential equations, and this richness induces some far-reaching theoretical and practical properties |G 35| |G 36|.

1.2. BIRKHOFF CLASSIFICATION OF DYNAMIC MOTIONS. IMPREDICTIBILITY OF CHAOTIC MOTIONS

In his book "Dynamical Systems" |B 9| Birkhoff wrote : "*the final aim of the theory of dynamical systems must be directed towards the qualitative determination of all possible types of motions and the interrelations of these motions*". Going towards this purpose, by using ideas developped by Poincaré, as a first step Birkhoff proposed en 1927 a classification of dynamic motions, which has been refined by Andronov in 1933 with the diagram of fig. 1.1. In this diagram, starting from the lower part up to the top, an increase of the structural complexity of motions manifests itself as a gradual transition from orderly to chaotic (erratic, or stochastic) motions.

It is recalled that steady states of *quasi periodic* type correspond to oscillations containing a finite number M of mutually incommensurable frequencies. Steady states of almost periodic type correspond to oscillations containing infinitely many mutually incommensurable frequencies ($M \to \infty$). Let us denote by $X(t, X_o)$ a dynamical system defined in some metric space R, the trajectory of which passes through the point X_o.

The motion $X(t, X_o)$ is called *recurrent*, if for any $\varepsilon > 0$ there can be found a $\tau(\varepsilon) > 0$ such that any arc of the phase trajectory of time length τ approximates the entire trajectory with a precision to within ε. In other words, whatever may be the numbers t_1, t_2, there can be found a number t_3 such that $t_2 < t_3 < t_2+\tau$, and $\rho\left[X(t_1,X_o), X(t_3,X_o)\right] < \varepsilon$, where $\rho(\alpha,\beta)$ is the distance between α and β |N 8|.

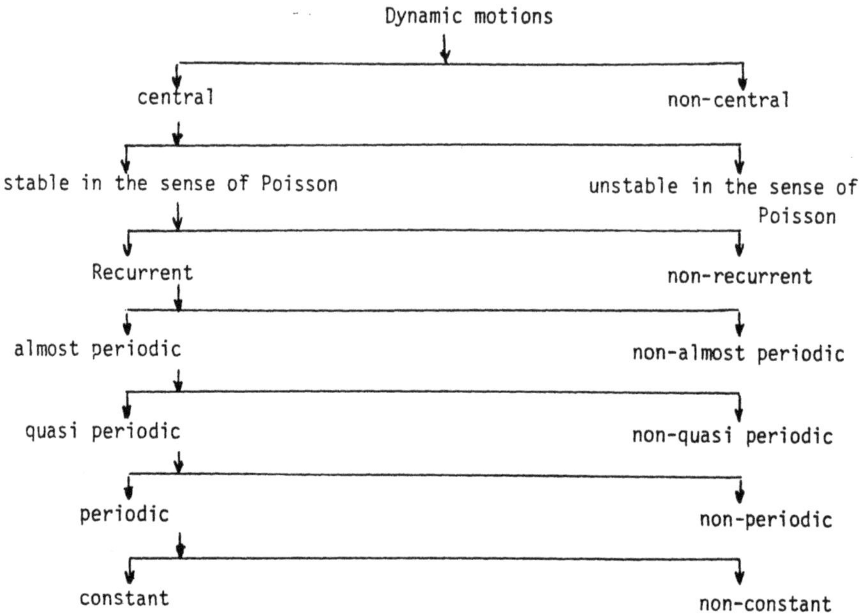

Figure 1.1

A point $\bar{X}$ is called *positively stable according to Poisson* (written stable P^+) if, for any neighbourhood U of $\bar{X}$, and for any $T > 0$ there can be found a value $t \geq T$ such that $X(t,\bar{X}) \in U$. Similarly, if there can be found a value $t \leq -T$ such that $X(t, \bar{X}) \in U$, then $\bar{X}$ is *negatively stable according to Poisson* (written stable P^-). A point stable according to Poisson as $t \to \infty$, and as $t \to -\infty$ is called *stable according to Poisson* (written stable P). A theorem states that if $\bar{X}$ is stable P^+ then every point of the phase trajectory $X(t,\bar{X})$ is also stable P^+. Similar theorem is obtained for stability P^- and stability P. Thus motions, or phase trajectories, are called positively, negatively, or simply stable according to Poisson, and every recurrent motion is stable according to Poisson | N 8 |.

A point $\bar{X}$ is *non-wandering* if, for any neighbourhood $U(\bar{X})$ containing it, there can be found arbitrarily large values of $|t|$ such that $U(\bar{X}) \cap X[t, U(\bar{X})] \neq 0$. The set Ω of points non-wandering with respect to some metric space R is a closed invariant set which may be empty. The wandering points are those which do not verify the definition of $\bar{X}$. Every point stable P^+, or P^-, is non-wandering. In the simplest form, the set of *central motions* may be defined as the greatest closed set the points of which are all non-wandering with respect to this set.

All the precedents definitions,given for the solution (or flow) $X(t,X_0)$ of (1.1), (1.2), have their analogue for the solution $X(n,X_0)$ of (1.3). Wandering points corresponds to transient behaviour, while non-wandering ones are related to asymptotic behaviour, or steady states.

It is conjectured that in general the more the dimension of the dynamic system increases, the more the structural complexity of the motion increases, i.e. the more obtained motions correspond to those located towards the top of the diagram of the fig. 1.1.

Nowadays, the numerous different abstract definitions of a *chaotic motion* correspond to non-periodic motions of the diagram. Till now, it seems that this diagram reflects, better than the mentioned abstract definitions, the "degrees" of chaos which can be obtained. Such a degree, corresponding to a given line of the left hand side of fig. 1.1, indicates the complexity of *deterministic dynamic motions*. Chaotic motions are characterized by the property of *practical impredictibility*, or *sensitivity* to initial conditions. The problem of prediction depends whether an analytical solution is exact, or only approximate, in a dynamically disorderly, or chaotic, situation. If the analytical solution is exact, and all operations on numbers can be carried out without any error (which is without any practical sense), then the prediction is possible in principle. However prediction may be difficult, or even impossible in practice, when the numbers and the operations upon them are approximate, as it is always the case with numbers of experimental origin. The simplest example is that of the quasi periodic motion, $f(t) = \sin \omega_1 t + \sin \omega_2 t$, ω_1, ω_2 being two mutually incommensurable angular frequencies. In the practical situations, the most one can expect is a qualitative, or probabilistic prediction, and not a quantitative one. This type of difficulty occurs whenever a process is intrinsically very sensitive to its initial state.

It will be shown in this monograph that the considered chaotic states are associated with the existence of infinitely many unstable periodic states, in the domain of the phase plane where they are located. Such a configuration gives rise, either to an asymptotically stable chaotic state, called frequently *strange attractor*, or to an unstable one, sometimes called *strange repeller*. The unstable case leads either to a *chaotic transient* (cf. § 2.8.3) towards a stable chaotic, or a non chaotic steady state, or to a *fuzzy*, or *chaotic, boundary* separating the domains of influence (or basins) of two, or more, stable steady states (cf. § 2.8.4) |M 23|(1975), |M 30||M 31||G 35|| G 6 |-|G 10| | M 3 || M 4|. With respect to the impredictibility due to a strange attractor, or a fuzzy boundary, the chaotic transient has the following surprising property : a short-dated prediction is very difficult, but a long-dated one is possible when the transient tends towards a stable non chaotic state.

From a practical point of view, a stable periodic motion, having a period larger than the possible duration of observations, and an irregular evolution during this period, also can be considered as chaotic. About this point, the term "chaos in a non-strict sense", or "*non strict chaos*", will be used with respect to the *strict chaos* denoting the non-periodic motions of fig. 1.1. In chapters 3 and 4, it will appear that periodic motions with an increasing period τ (when a parameter of the dynamic system varies) have a limit (when $\tau \to \infty$), which is a strict chaos , what justifies this extension of the notion of chaos.

1.3. SOME CONSIDERATIONS ABOUT RECURRENCES

Let us consider (1.3), or (1.4), the point X_{n+r}, r being a positive integer is called the *rank r consequent* of the point X_n. Similarly, X_n is called the *rank r antecedent* of X_{n+r}. Symbolically, it is written

$X_{n+r} = T^r X_n$ and $X_n = T^{-r} X_{n+r}$. The single-valuedness of F (in (1.3)) does not imply anything about the existence and uniqueness of the antecedents because the inverse recurrence (or the inverse map T^{-1})

(1.5) $$X_n = F^{-1}\,(X_n)$$

where $F\left[F^{-1}(X)\right] = F^{-1}\left[F(X)\right] = X$, may not exist, or it may be multi-valued. The uni-dimensional example, $x_{n+1} = x_n^2 - \lambda$, λ being a real parameter, illustrates this point with T^{-1} given by $x_n = \pm\sqrt{x_{n+1}+\lambda}$. The x_n are double-valued for $x_{n+1} > -\lambda$, and are not real for $x_{n+1} < -\lambda$.

When F in (1.3) is a continuous and continuously differentiable function of X, and if T^{-1} exists and is unique in the definition domain of F, the map T is called a *diffeomorphism*. If F is not continuous or not differentiable, T is called an *invertible map*. When F in (1.3) is such that T^{-1} may be multi-valued, or may not exist in a certain domain of the phase space, the map T is non invertible and then T is called an *endomorphism*. Conditions ensuring that a map T is a diffeomorphism will be given in a next paragraph.

Practical problems, or theoretical ones, leading to associate recurrences via the Poincaré method of section, exceptionally give rise to the explicit form (1.3). Considering the one-dimensional case with the variable x, the following *parametric*, or *implicit forms*, can be obtained :

(1.6) $$f(x_n) = g(v_n),\quad g(x_{n+1}) = f(v_n)$$

(1.7) $$\Phi_i(x_n,\ x_{n+1}\ ;\ v_n^{(1)}, \ldots, v_n^{(m)}) = 0 \qquad i = 1, 2, \ldots, m+1$$

(1.8) $$x_n = \Psi_1(v_n),\quad x_{n+1} = \Psi_2(v_n)$$

(1.9) $$f(x_n,\ x_{n+1}) = 0$$

where the v are considered as auxiliary parameters, changing for every iteration |G 35|.

Such auxiliary parameters must not be confounded with the

"internal" parameters λ of the recurrence, which maintains a fixed value during a whole iteration sequence. With the explicit autonomous form, the recurrence T is written :

$$(1.10) \qquad X_{n+1} = F(X_n, \ \lambda)$$

For an initial condition X_o, the iterated sequence is $X_n = X(n, X_o, \lambda)$. T can be considered as an implicit definition of the function X_n, which plays the role of the recurrence solution. Though theoretically quite satisfactory, such a definition practically is almost useless, because in general the function X_n is unknown, except for the linear case, and for very few examples in the non linear case. In all non-contrived cases X_n cannot be expressed explicitly in terms of known elementary and transcendental functions. In order to characterize the non-classical transcendental function X_n, it is possible to use the analogue of the characterization of a function of the complex variable via the knowledge of its singularities (poles, zeros and essential singularities). For (1.10), a meaningful characterization of X_n consists in the identification of its *singularities*, and the behaviour of the latter as the parameter λ varies These approaches constitute what is called *qualitative methods* of the theory of dynamic systems. Any so-obtained change of singularities, and their qualitative properties, is called a *bifurcation*. In the parameters space, the boundary separating behaviours of X_n, which are qualitatively different, is called a set of *bifurcation values* of the parameters.

The simplest singularities are zero-dimensional : *cycles of order k* (or periodic points of period k). A cycle of order k is a set of k points X_i^*, i = 1, 2, ..., k, such that $X_{i+1}^* = T\,X_i^*$, $X_i^* = T^k X_i^*$, $X_i^* \neq T^\ell\, X_i^*$, $\ell < k$. A cycle of order k = 1 is called a *fixed point* of T. Every point of a cycle of order k is a fixed point of T^k. *Invariant manifolds* by T, or T^k, of dimension d = 1, 2, ..., p-1 (p is the dimension of the vector X) constitute singularities with an increasing degree of complexity. They are solutions of *functional equations* (cf. chapter 5 for p = 2).

1.4. CONSIDERATIONS ABOUT STABILITY. SOME DEFINITIONS

Fixed points, or cycles, are characterized by their stability. So a fixed point X^* is called *attractive*, or *asymptotically stable*, if for every initial condition X_o, belonging to a sufficiently small neighbourhood U of X^*, $\lim X_n = X^*$ when $n \to \infty$. It is called *repulsive*, or *unstable*, if for an initial condition $X_o \in U$, $n > N$, X_n remains outside U. Equivalent definitions can be used for a cycle of order k. There exists situations for which the definition of the stability of a fixed point is not possible. In particular, it is the case when a fixed point is a limit point of a sequence of other fixed points. For the one-dimensional recurrences, this point appears with the example $T : x_{n+1} = x_n + x^4 \sin(\pi/x)$. Here, T has a countable set of fixed points $x = 0$, and $\pm 1/m$, $m = 1, 2, \ldots$ The linear approximation of T, in a neighbourhood of $x = \pm 1/m$, is $x_{n+1} = [1 - (\pi/m^2) \cos m\pi] x_n$. Then, when m is even (resp. odd), $x = \pm 1/m$ is attractive (resp. repulsive), and the stability of the fixed point $x = 0$ cannot be defined |G 18|.

A point X^*_ω is an *ω-limit point* of $X(n, X_o)$, $n = 1, 2, \ldots$, if there exists a sequence of points $X(n_i, X_o)$, $i = 1, 2, \ldots$, n_i = integers, given by (1.3) such that $X(n_i, X_o) \to X^*_\omega$ when $n_i \to \infty$. A point X^*_α is an *α-limit point* of $X(n, X'_o)$ if there exists a sequence $X(n_i, X'_o)$ such that $X(n_i, X'_o) \to X^*_\alpha$ when $n_i \to -\infty$. An ω(resp. α)-limit set is the set of ω (resp. α) limit points of $X(n, X_o)$. The notion of stability can be extended to more complicated steady states than those corresponding to fixed points, or cycles. So a closed invariant set A of the discrete dynamical system (1.3) is called *attracting set* if there exists some neighbourhood U of A such that $X(n, X_o) \in U$ for $n \geqslant 0$ and $X(n, X_o) \to A$ when $n \to \infty$, for all the initial conditions $X_o \in U$. Attractive cycles are particular zero-dimensional attracting sets. A *repelling set* is defined analogously replacing n by −n. A *no-escape region* of the phase space is a simply connected set R of this space such that $X(n,R) \subset R$ for $n = 1, 2, \ldots$, and an attracting set can be defined as the intersection of all the domains $X(n, R)$, $n = 0, 1, 2, \ldots$

Considering T given by (1.10), a range of parameters λ may be such that a fixed point X^* (of T or T^k) simultaneously is an ω-limit point and an α-limit point of a sequence $X(n_i, X_0)$ for $n_i \to +\infty$, $n_i \to -\infty$. The points of this sequence constitute a *homoclinic* iterated sequence (or discrete trajectory, or orbit). If this property remains for infinitely small perturbations of the equation (1.10), it is called a *transversal* (or structurally stable) *homoclinic* iterated sequence. The definition of structural stability is given just below. Using the definition of |G 18| (cf. p. 256) one can say : an *attractor* is an indecomposable closed invariant set Λ with the property that, given $\varepsilon > 0$, there is a set U of positive Lebesgue measure in the ε-neighbourhood of Λ such that $X_0 \in U$ implies that the ω-limit of $X(n, X_0)$ is contained in Λ, and $X(n, X_0)$, $n > 0$, is contained in U . This attractor is a *strange atrractor* if it contains a transversal homoclinic orbit.

In the two-dimensional general analytic case, a Birkhof theorem |B 13| states that an arbitrarily small neighbourhood of a homoclinic point contains infinitely many (repulsive) cycles and a homoclinic point. *Then a strange attractor is related to the existence of infinitely many repulsive cycles.*

The *influence domain,* or *domain of attraction,* or *basin* of an attractor Λ is the open set (D) of the points X_n such that the consequents of all $X_n \in$ (D) approach asymptotically Λ as $n \to \infty$. This last notion, as those of ω(or α)-limit, those of attracting set, homoclinic orbit, attractor, have equivalent definitions for the solutions of the differential equations (1.1), (1.2).

The differential equation (1.1), or the recurrence (1.3), is *coarse,* or *inert,* in *the sense of Andronov-Pontrjagin,* or *structurally stable* |A 2| |G 23| ,if in the phase space the topological structure of the trajectories, or of the singularities, does not change for small changes (perturbations) of the right hand sides. When these right hand sides are sufficiently smooth, a quantitative definition classically is given for perturbations with small amplitudes, and small absolute values of their partial first order derivatives |A 2| |G 18| |G 23|.

1.5. THE POINCARE'S METHOD OF SURFACE OF SECTION

A rather fundamental role is played by recurrences in the works of Poincaré, although at first sight it appears that they are introduced there merely as an artifice for the study of trajectories of dynamic systems, defined by a system of ordinary differential equations, where the independent variable represents time. The advantage obtained by means of this artifice consists in the reduction of the dimensionality of the problem by one unit in the non-conservative systems, and by two units in the conservative ones. This is obtained via a suitable surface of section, transverse to the phase trajectories (trajectories "without contact" in the Poincaré's terminology). It is with this method that firstly Poincaré proved the existence of limit cycles in the autonomous two-dimensional case, and after, in more general cases, that of periodic solutions on the subject of which he said : "what renders so precious these solutions, is that they are the only gap through it is possible to penetrate in a place which till now was inaccessible".

The existence of a "regular surface of section" is not proved in the general case. In the autonomous conservative case, with two degrees of freedom, the existence seems very probable. It is also the case for the autonomous three-dimensional equation (1.1). In order to illustrate the method, let us consider the following differential equations which are the simplest ones which can give rise to chaotic behaviours :

(1.11) $du_i/dt = f_i(u_1, u_2, u_3)$, $i = 1, 2, 3$

(1.12) $du_i/dt = g_i(u_1, u_2, \omega t)$, $i = 1, 2$

where the functions f_i, g_i, are at least continuous and differentiable with respect to their real arguments, g_i being periodic in t with the period $\tau = 2\pi/\omega$. From a dynamical point of view, it is considered that (1.11) and (1.12) have the same *effective dimension* m = 3. Indeed, adding to (1.12) the relation $du_3/dt = 1$, an equivalent form to (1.11) is obtained. The associated recurrence with (1.11), or (1.12) has either an explicit, autonomous, two-dimensional form T :

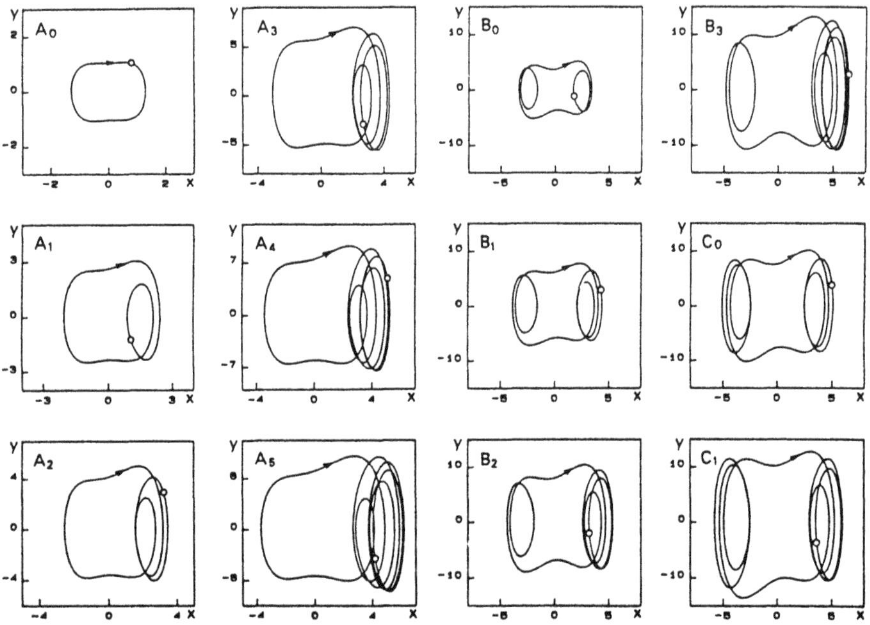

FIG.1.2 - Trajectories of periodic solutions with period 2π, corresponding to each island of fig. 1.3. Small circle and arrow indicate respectively the fixed point and increasing t. (according to $|K2|$)

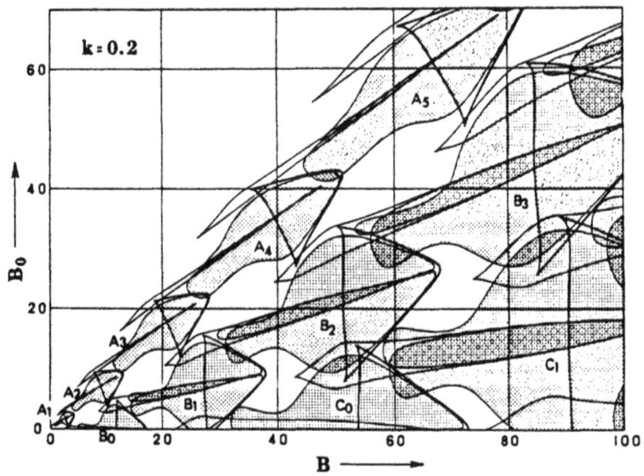

FIG.1.3 - Bifurcation diagram of Duffing's equation (1.14'). (according to $|K2|$).

(1.13) $\quad x_{n+1} = P(x_n, y_n),\ y_{n+1} = Q(x_n, y_n),\ n = 0, 1, 2, \ldots$

or an implicit one :

(1.14) $\quad F(x_{n+1}, y_{n+1}, y_n, y_n) = 0,\ G(x_{n+1}, y_{n+1}, x_n, y_n) = 0$

but in the two cases T is invertible, and is a diffeomorphism if the data are differentiable. For equation (1.11), x_n, y_n are the coordinates of a point M_n intersection of a phase trajectory (in the three-dimensional space) with a "regular" transverse surface of section (a plane in particular cases) associated with reference axes x, y. With increasing t, the next intersection, occuring in the same direction flow, is M_{n+1} (x_{n+1}, y_{n+1}). So a fixed point $(M^* = M_n = M_{n+1})$ of (1.13), or (1.14) corresponds either to an equilibrium point of (1.11), or to a periodic solution. A cycle of order k corresponds to a more complex periodic solution, giving a closed phase trajectory crossing through the surface k-times. For equation (1.12), if t = 0, $u_1(0)$, $u_2(0)$ are the initial conditions, x_n, y_n are defined by $x_n = u_1(n\tau)$, $y_n = u_2(n\tau)$. A fixed point M^* of (1.13) or (1.14) is either an equilibrium point of (1.12), or a periodic solution of period τ. A cycle of order k corresponds to a *subharmonic oscillation* of period $k\tau$.

It is worth of noting that the characteristics of a fixed point, or of a cycle, of (1.13) (1.14) do not give the whole information about the corresponding periodic solutions of (1.11) or (1.12). This point simply appears by considering the non autonomous Duffing's equation

(1.14') $\quad du_1/dt = u_2,\ du_2/dt = -ku_2-u_1^3 + B\cos t + B_o$

According to the region of the parameter space (k, B, B_o), the closed trajectory, corresponding to a periodic solution with $\tau = 2\pi$, has various shapes passing through the associated fixed point of (1.13). Fig. 1.2 shows different possible shapes corresponding to the regions A_i, i = 0, ..., 5, B_j, j = 0, ..., 3 and C_o, C_1 of the parameter plane given by fig. 1.3 obtained for k = 0.2. The boundaries of these regions, which can overlap themselves, are bifurcation curves (the regions of overlaping are noted with a darker shadow) (for more details, cf. |K 2|). The same occurs with cycles and corresponding subharmonic oscillations. Thus the shape of the closed phase trajectory passing through a fixed point, or a cycle, is necessary to complete the characterization of a periodic solution of (1.11) or (1.12).

The associated recurrence obtained via the method of section must not be confused with the recurrence obtained from a numerical simulation of (1.1), or (1.2) by using the classical methods of the numerical analysis. Here there is not a dimensionality decrease, as for the method of section. The methods with separated steps (Euler, Euler-Cauchy, Runge-Kutta) lead to recurrences having the same dimension as that the differential equations. Those with bound steps (Adams, Adams-Basforth) lead to recurrences having a higher dimension.

1.6. DIFFEOMORPHISM AND ENDOMORPHISM. THE HADAMARD'S THEOREM. THE TWO CLASSES OR ENDOMORPHISMS.

Let (1.3), $X_{n+1} = F(X_n)$, X = real vector, be the considered p-dimensional recurrence, or map T. A fundamental problem is that of the invertibility of T, or that of an answer to the question : is it possible to obtain univocally T^{-1}, i.e. X_n, from X_{n+1} ? This question has a double aspect. The first one is the existence of a real X_n. The second one is the single-valuedness, or unicity, of X_n.

The differential calculus methods lead to look for the solution of this problem in the infinitesimal properties of the functions components f_j, j = 1, ..., p, of F. Supposing that the $f_j(x_1, \ldots, x_p)$ are continuous, and continuously differentiable functions of $x_1, \ldots, x_p$, a *necessary condition of invertibility is that the Jacobian determiant* $J = D(f_1, \ldots, f_p) / D(x_1, \ldots, x_p)$ *may have a constant sign.* So contrary to widespread belief the property of non-vanishing Jacobian is not universal, because *the validity of the implicit-functions theorem is only local and not global.* Globally, this condition is not sufficient. The examination of the simple example :

$$x_{n+1} = a \exp\left[-y_n^{-2}\right], \quad y_{n+1} = (1/2)x_n y_n^3 \exp\left[y_n^{-2}\right] + b\, y_n^2$$

the Jacobian of which is $J = -a$, illustrates this point. Indeed, two points (x_n, y_n) and $(-x_n, -y_n)$ have the same consequent (x_{n+1}, y_{n+1}) (cf. |G 35 |p. 334, and |G 36 | p. 181), and the condition of unicity of T^{-1} is not satisfied.

In the one-dimensional case, $p = 1$, $x_{n+1} = f(x_n)$, if $f^{\bullet}(x) = df/dx$ is always positive, T^{-1} cannot be multi-valued. But if the two integrals $\int_{-\infty}^{\alpha} f^{\bullet}(x)dx$ and $\int_{\alpha}^{\infty} f^{\bullet}(x)dx$ have not infinite values, there exist intervals of x such that T^{-1} does not exist.

Let μ be the minor axis of the deformation ellipsoïd, i.e. the smallest value of $\left[(df_1^2 + \ldots + df_p^2) / (dx_1^2 + \ldots + dx_p^2)\right]^{1/2}$, and μ_ρ the least value of μ on the sphere $x_1^2 + \ldots + x_p^2 = \rho^2$. The quantity μ is equal to zero when $J = 0$, and is different from zero when $J \neq 0$. Hadamard |H 3 | showed that infinitesimal properties of the functions f_j are sufficient to study the invertibility of T, *if these properties are known in the whole phase space.* The corresponding theorem is :

(P1.1) *The recurrence, or map, T, given by (1.3), is invertible (T^{-1} exists and is unique), if μ (or J) is different from zero in every point of the phase space, and if the integral $\int_0^{\infty} \mu_\rho \, d\rho$ has an infinite value.*

With (P1.1), one has the necessary and sufficient condition in order that a map T be a diffeomorphism (T is sufficiently smooth). However, the hypothesis of differentiability of the functions f_j is not necessary for the invertibility of T, being then simply invertible It is sufficient to suppose that the f_j are *continuous* and to impose the following conditions :

(a) every point M of the phase space is the center of a sphere σ such that two different points inside σ have not the same consequent ;

(b) the consequent of every point M of the phase space is the center of a sphere such that every point inside this sphere is the consequent of a point inside σ;

(c) a curve connecting the origin of the phase space to a point at infinite distance cannot have as a consequent a rectifiable curve (having a finite length) (cf |H 3| , p. 363).

When the f_j are differentiable functions, and $p \geqslant 2$, the above mentioned considerations lead to the definition of *two classes of non-invertible maps* (or endomorphisms) having different properties : *the class of endomorphisms with a vanishing Jacobian, and the class of endomorphisms with a non-vanishing Jacobian.* In particular, the properties of the chaotic behaviours are very different. So in the first class, the successive consequents of $J = 0$, called *critical manifolds* play a fundamental role in the definition of a strange attractor and its bifurcations (cf. |G 35|, p. 356-371).

1.7. IMBEDDING OF AN ENDOMORPHISM INTO A DIFFEOMORPHISM WITH A HIGHER DIMENSION

At equal dimensionality, the properties of an endomorphism with a vanishing Jacobian generally are more complex than those of a diffeomorphism. This point will appear from chapter 2. Chaotic behaviours are not possible in the continuous one-dimensional diffeomorphisms, but they appear in the one-dimensional endomorphism T_o, $x_{n+1} = f(x_n, \lambda)$ (λ is a parameter), even in the simplest cases when $f(x, \lambda)$ is a quadratic function of x, for example $f(x, \lambda) \equiv x^2 - \lambda$. It is also known that chaotic behaviours are not possible with two-dimensional autonomous differential equations (1.1). The simplest forms of differential equations which can give such complex dynamical behaviours are (1.11) and (1.12). Consider the associated recurrence T (1.13), which is a diffeomorphism if the data are differentiable, and an arrangement of (1.13) in the form T'_b :

(1.15) $$x_{n+1} = f(x_n,\lambda) + y_n h(x_n,y_n),\ y_{n+1} = bg(x_n,y_n),\ n = 0,1,2,...$$

where λ and b are two particular parameters of (1.11) or (1.12). The simplest analytical form of a two-dimensional diffeomorphism, qualitatively equivalent to (1.15) (from the point of view of the contents of a large part of fundamental properties concerning the chaotic behaviour of (1.11) or (1.12)) is T_b :

(1.16) $\quad x_{n+1} = f(x_n, \lambda) + y_n, \quad y_{n+1} = b\, x_n, \quad n = 0, 1, 2, \ldots$

the Jacobian of which is $J = -b$. In the particular case $f(x,\lambda) \equiv 1-\lambda x^2$ T_b was studied in |H 7| via numerical simulations. Making $b = 0$ into (1.15), or (1.16), leads to the one-dimensional recurrence T_o :

(1.17) $\quad x_{n+1} = f(x_n, \lambda), \quad n = 1, 2, \ldots$, f is continuous,

which appears as imbedded into the two-dimensional diffeomorphism (1.15) or (1.16). A necessary condition of chaotic behaviour for (1.17) is that T_o may be an endomorphism. From these considerations, it results a process of successive simplifications, leading from the differential equations (1.11), (1.12) to the one-dimensional endomorphism (1.17) via (1.15) (1.16). This process can make appear, in a simplest form (isolated from "parasitic effects" due to non-essential analytical complexities), certain fundamental mechanisms of the chaotic dynamics which take place in higher dimensions, and with more elaborate analytical forms. In this book, chapters 2, 3, 4 are devoted to the study of T_o, and chapter 6 is devoted to T_b given by (1.16).

Considering (1.11), in the form :

(1.18) $\quad du_i/dt = f_i(u_1,u_2,u_3), \ \mu du_3/dt = f_3(u_1,u_2,u_3), \ i = 1, 2, 3$

$\mu > 0$ being a parameter, μ is related to b in (1.15) and if $\mu \to 0$, then $b \to 0$ |G 35 |. So the form (1.15) when $b \to 0$ is related to the problem of fast motions in differential equations |P 5||G 23||G 35|.

1.8. THE VALIRON'S RESULTS

The most general form of this problem of embedding is obtained when considering the p-dimensional diffeomorphism (1.10) written |V 1|

(1.19) $X_i = f_i(x_1, x_2, ..., x_p ; b_1,..., b_r), i = 1, 2, ..., p$

$b_1,..., b_r$ representing the components of the parameter vector λ. Let T_b be this diffeomorphism. It is supposed that when $b_j = (b_j)_o$, $j = 1, 2, ..., r$, the Jacobian determinant J of (1.19) is *identically* equal to zero :

(1.20) $J = D(X_1, ..., X_p) / D(x_1, ..., x_p) \equiv 0$

in a whole bounded region Δ of the phase space $(x_1, ..., x_p)$. Let us suppose that one of the minor, $J_1 = D(f_1, ..., f_{p-1}) / D(x_1,...; x_{p-1})$ is different from zero, for a point M_o of Δ. Then, in the neighbourhood of M_o, it is possible to solve the system :

$$X_\ell = f_\ell(x_1, ..., x_p), \quad \ell = 1, 2, ..., p-1$$

for $x_1, x_2, ..., x_{p-1}$, and it is obtained :

$$g(X_1,X_2,...,X_p,x_p) \equiv X_p-f_p(x_1,...,x_{p-1},x_p) \equiv 0$$

$x_1, ..., x_{p-1}$ being functions of $X_1,..., X_{p-1}$ and x_p. It can be shown that g does not depend on x_p, i.e. $\partial g/\partial x_p = 0$. Indeed, $\partial g/\partial x_p$ is (except on the sign) the coefficient of dx_p in $(\partial f_n/\partial x_1)dx_1 + ... + (\partial f_n/\partial x_p)dx_p$ when to the place of the dx_s, $s = 1, ..., p-1$, are taken the solutions of the system :

$$(\partial f_j/\partial x_1)dx_1 + ... + (\partial f_j/\partial x_p)dx_p = 0, \; j = 1, ..., p-1$$

This coefficient is J/J_1, and thus equal to zero. So in the neighbourhood of M_o, the X_j verify a relation and X_p *is a continuous function* of $X_1, X_2, ..., X_{p-1}$. The same property is obtained inside the whole region Δ_p^1 containing M_o, and constituted by the points where $J_1 \neq 0$. This also can occur for other regions $\Delta_j^1 \in \Delta$, in every which X_j is a function of the X_ℓ, $\ell \neq j$.

It results that, for (1.19) with $b_j = (b_j)_o$, an initial condition (rank of iteration $n = 0$) gives rise to an iterated sequence (ranks of iterations $n = 1, 2, \ldots$) which is restricted (from $n = 1$) to a $(p-1)$-dimensional manifold (Γ) given by X_p function of $X_1, \ldots, X_{p-1}$. For the particular forms (1.15), (1.16) of (1.19), $(b_j)_o \equiv b = 0$, and (Γ_1) is $y = 0$ which is attained for $n = 1$. For T given by :

$$(1.21) \qquad x_{n+1} = (\alpha+b\alpha')x_n + (\beta+b\beta')y_n, \ y_{n+1} = (\alpha r+b\gamma)x_n + (\beta r+b\delta)y_n$$

$(b_{jo}) \equiv b = 0$ gives $J \equiv 0$, and (Γ) is the straight line $y = rx$.

If now in a region Δ', which can be confounded with Δ, all the minors of order (number of columns) $p-1$ of J (given by (1.20)) are identically equal to zero, but if one minor of order $p-2$ is different from zero for a point M'_o, it is possible to solve the system of $p-2$ equations :

$$X_j = f_j(x_1, \ldots, x_{p-2}, x_{p-1}, x_p)$$

for $x_1, \ldots, x_{p-2}$. Placing these variables in $X_{p-1} = f_{p-1}$, and $X_p = f_p$, as above it can be shown that X_{p-1} and X_p are functions of the X_j, $j \neq p$, $j \neq p-1$. So there exist regions $\Delta^2_{i,j}$ in every which X_j, X_i are continuous functions of the other X_k.

In the same way, if there exists a region Δ'' in which all the minors of order $p-2$ are identically equal to zero, regions $\Delta^3_{i,j,k}$, where X_i, X_j, X_k are continuous functions of the other X, can exist. More generally, if all the minors of rank $p-1, p-2, \ldots, p-m+1$, of J are identically equal to zero in Δ, with a minor of rank $p-m$ different from zero, in the p-dimensional phase plane the values X_j of f_j constitute a $(p-m)$-dimensional manifold (Γ_m). The following Valiron's theorem |V 1| results :

<u>(P1.2)</u> *If, inside a region Δ of the phase space, the Jacobian determinant J of the map T_b given by (1.19), and its minors of order $p-1, \ldots, p-m+1$, are identically equal to zero, one of the minors of order $p-m$ being not so, the map T_b transforms Δ into a $(p-m)$-dimensional manifold. There exist m relations between the X_j.*

It is worthy of note that, from the preceding proof, the relation between the f_j can be different according to the regions of Δ. So let T be defined for $x_3 > 0$ by : $f_1 \equiv x_3^2(x_1+x_2)$, $f_2 \equiv f_1^2$, $f_3 = x_3^2(x_1-x_2)$; and for $x_3 \leqslant 0$ by : $f_1 \equiv x_3^3(x_1-x_2)$, $f_2 \equiv x_3^2(x_1+x_2)$, $f_3 \equiv f_1^3$. The functions f_1, f_2, f_3 are continuous and continuously differentiable. f_1 and f_3 are independent for $x_3 > 0$, f_1 and f_2 are independent for $x_3 < 0$. Here, J everywhere is identically equal to zero, and the relation between the X_j is $f_2 = f_1^2$ when $x_3 > 0$, and $f_3 = f_1^3$ when $x_3 \leqslant 0$.

1.9. CONSEQUENCES OF THE VALIRON'S RESULTS

Consider the diffeomorphism T_b given by (1.19), and $b_j = (b_j)_o$ such that the Jacobian determinant (1.20) and all its minors of order p-1, ..., p-m+1 are identically equal to zero. Let T_o be the map constrained to the (p-m)-dimensional manifold (Γ_m). T_o is either a (p-m)-dimensional diffeomorphism, or a (p-m)-dimensional endomorphism, and T_o can be considered as embedded into the p-dimensional diffeomorphism T_b.

Let $\phi_1(X_1, \ldots, X_p) = \ldots = \phi_m(X_1, \ldots, X_p) = 0$ be the m relations between the X_j, which defines the (p-m)-dimensional manifold (Γ_m) when $b_j = (b_j)_o$, and $M_1(\alpha_1, \ldots, \alpha_p)$ a point of (Γ_m). T_o is defined by (1.19) and the equations of (Γ_m). Every point M_1 has a continuous set of first rank antecedents M_o (n = 0), which are the points of a m - dimensional manifold Θ_m. This manifold can be considered as the domain of "fast" motions leading to $M(\alpha_1, \ldots, \alpha_p)$, the motion becoming "slow" on (Γ_m) for n = 1, 2, ... In the simple case of (1.16), m = 1, (Γ_1) is $y = 0$, $x_M = \alpha$, $y_M = 0$, and Θ_1 is the curve $y = -f(x,\lambda) + \alpha$. It will appear in chapter 6 that Θ_1, passing through a fixed point, is a degenerate invariant curve of T_b, b = 0, (i.e. the sequence n = 0, 1, 2, ... is considered, while for T_o, n = 1, 2, ...) which is the limit of an invariant curve of T_b, $b \neq 0$, when $b \to 0$. It is conjectured that Θ_m has the same property and is the limit of a manifold, invariant by T_b, when $b_j \to (b_j)_o$.

With the example (1.21), m = 1, (Γ_1) is $y = r\,x$, and Θ_1 is the straight line $y = -ax/b + \alpha/b$. The § 5.2.5. shows the limit properties of an invariant curve of (1.21) when $b \to 0$.

1.10 AUTONOMOUS RECURRENCES AND THE SCHRODER'S EQUATION

1.10.1 Generalities

There exists a strong link between an autonomous recurrence and some functional equations like those of Schröder, or Böttcher, or Abel, or Perron-Frobenius in the special case, or the equation of the automorphic functions. This point is widely shown in the monograph |G 36|. In this paragraph, only the case of the Schröder's functional equation is considered.

The use of the Schröder's equation, for the determination of the solution of a non-linear autonomous recurrence T, appeared from the end of the 19th century in the one, or multi-dimensional problems. The fundamental contributions are those of Grevy |G 11| - |G 13|, Leau |L 5| - |L 7|, Koenigs |K 6| - |K 8|, Lattès |L 1| - |L 4| whose papers concern a "local" study of the solution, i.e. its determination inside a sufficiently small neighbourhood of a fixed point, or a cycle. The first paper about a "global" study is due to Fatou |F 1| in the case of a one-dimensional recurrence with complex variable. Since this publication, it seems that the "global" aspect of this problem has not been considered. The purpose of this paragraph is to show how, from the solution of the Schröder's equation, it is possible to construct a set of non linear autonomous recurrences, the solution of which is obtained using the elementary functions of the mathematical analysis. In general, the solution of a non linear recurrence is a non-classical transcendental function. With the "global" aspect of the problem, two situations are considered. The first one displays the properties of a basin boundary of an attractive fixed point. The second one leads to a method of construction of a class of recurrences giving chaotic behaviours, described from elementary functions, and also leads to the definition of functions having properties similar to those of Chebyshev's polynomials.

The variables, which are considered here, are real. In the one-dimensional case, the recurrence T, and the Schröder equation are respectively :

(1.22) $x_{n+1} = f(x_n)$, $n = 0, 1, 2, \ldots$, x_o = given initial condition

(1.23) $\gamma(x_{n+1}) = S\ \gamma(x_n)$, in general S = multiplier of fixed point of (1.22)

The solution $x_n = x(n, x_o)$ of (1.22) then has the following form :

(1.24) $x_n = \gamma_{-1}\ [S^n \gamma(x_o)]$, γ_{-1} being the inverse function of γ.

In the two-dimensional case, the equivalent relations are :

(1.25) $x_{n+1} = f(x_n, y_n)$, $y_{n+1} = g(x_n, y_n)$, x_o, y_o = initial condition

(1.26) $\gamma_i(x_{n+1}, y_{n+1}) = S_i\ \gamma_i(x_n, y_n)$, $i = 1, 2$

(1.27) $$\begin{cases} x_n = \gamma_{-1}\ [S_1^n\ \gamma_1\ (x_o, y_o), S_2^n\ \gamma_2(x_o, y_o)] \\ y_n = \gamma_{-2}\ [S_1^n\ \gamma_1\ (x_o, y_o), S_2^n\ \gamma_2(x_o, y_o)] \end{cases}$$

where $(\gamma_{-1}, \gamma_{-2})$ is the vectorial function inverse of (γ_1, γ_2).

In (1.24) and (1.27), n can be a continuous variable $n \equiv t$ |G 36|. In the general non linear case, methods for obtaining an expansion of the solution of (1.23) and (1.26) are given in |G 19| |G 20| |M 13|.

1.10.2 Basin boundary and the Schröder equation |M 35|

Let O be an attractive fixed point of (1.22), O(x = 0), or of (1.25), O(x = y = 0), f and g being analytic functions of their arguments. Inside the immediate basin (D_o) (connected part of the basin which contains O), the infinite products, which define γ, γ_1, γ_2 in the general case, converge uniformly and lose this property on the boundary (F) of (D_o) (cf. |F 1| for (1.22) with x = complex variable, and |M 13| for (1.25)). Thus, the boundary (F) constitute a set of singularities for the functions γ, γ_1, γ_2. In the one-dimensional case, (F) is made up of two points. In the two-dimensional case, (F) is an invariant curve of (1.25), H(x, y) = constant, solution of the functional equation of automorphic functions (cf. § 5.1). So the knowledge of (F) gives an information about the convergence domain of γ_1, γ_2.

Now, consider the one-dimensional case. Examples, obtained with the solution (1.24) given by elementary functions (which is exceptional in the non-linear case), illustrates three different situations which characterize the nature of the two points P_1, P_2, boundary (F) of (D_o). So, taking $v = \gamma(x_o) = x_o/(x_o^2-1)$, n = 1 in (1.24) with $0 < S < 1$, and choosing the determination of γ_{-1} which has the sign + before the radical, the equation (1.22) is obtained in the form :

$$(1.28) \qquad x_{n+1} = \left[x_n^2 - 1 + \sqrt{(x_n^2 - 1)^2 + 4\, S^2\, x_n^2}\right] / (2\, S\, x_n)$$

The fixed point x = 0 is attractive because $0 < S < 1$. $f(x_n)$ is a strictly increasing function. (F) *is made up of two repulsive fixed points* $x = \pm 1$. Whatever may be x_o, the solution (1.24) of (1.28) is :

$$x_n = \left[x_o^2 - 1 + \sqrt{(x_o^2 - 1)^2 + 4\, S^{2n} x_o^2}\right] / (2\, S^n x_o)$$

Choosing $v = \gamma(x_o) = x_o / (1 - x_o^2)$, and the determination of γ_{-1} which corresponds to the sign - before the radical, similarly one has :

$$(1.29) \qquad x_{n+1} = \left[1 - x_n^2 - \sqrt{(1 - x_n^2)^2 + 4\, S^2\, x_n^2}\right] / (2\, S\, x_n)$$

Let $-1 < S < 0$, the fixed point x = 0 is attractive. $f(x_n)$ is a strictly decreasing monotonic function. (F) is made up of the two points $x = \pm 1$ of a repulsive cycle of order two. The solution of (1.29) is :

$$x_n = \left[1 - x_o^2 - \sqrt{(1 - x_o^2)^2 + 4\, S^{2n}\, x_o^2}\right] / (2\, S^n\, x_o)$$

Choosing $v = \gamma(x_o) = x_o^2 / (x_o^2-1)$, and the determination of γ_{-1} which corresponds to the sign + before the radical, one has :

$$(1.30) \qquad x_{n+1} = \sqrt{- S\, x_n^2 / \left[(1-S)x_n^2 - 1\right]}$$

Let $0 < S < 1$. The fixed point $x = 0$ is attractive. For $x = 0$, $f(x)$ is non-differentiable, and $f(x)$ is defined for $|x| < (1-S)^{-1/2}$. (F) is made up of the repulsive fixed point $x = 1$, and of its first rank antecedent $x = -1$. In three considered cases, it is evident that a series expansion of $\gamma(x)$ does not converge for the points of (F).

With the two-dimensional case, similarly it is possible to construct examples of (1.25), the solution (1.27) of which can be formulated from the elementary functions of the mathematical analysis (exceptional situation in the non linear case). So, considering $v_1 = \gamma_1(x_o,y_o) = x_o/(x_o^2+y_o^2-1)$, $v_2 = \gamma_2(x_o,y_o) = y_o/(x_o^2+y_o^2-1)$, and choosing the determination of γ_{-1}, γ_{-2}, which corresponds to the sign + before the radical with $0 < S_1 < S_2 < 1$, the recurrence (1.25) is obtained in the form :

$$\begin{cases} x_{n+1} = \left[(x_n^2+y_n^2-1)S_1x_n + S_1Ax_n\right] / B; & A \equiv \sqrt{(x_n^2+y_n^2-1)^2+2B} \\ y_{n+1} = \left[(x_n^2+y_n^2-1)S_2y_n + S_2Ay_n\right] / B; & B \equiv 2(S_1^2x_n^2+S_2^2y_n^2) \end{cases}$$

The fixed point $x = y = 0$ is attractive. (F) is the circle $x^2+y^2-1 = 0$, which is an invariant repulsive curve, where the series expansions of γ_1 and γ_2 do not converge.

1.10.3 One-dimensional recurrences and chaotic behaviour. Generalization of the Chebyshev's polynomials |M 35|

Let $P(v)$, $a \leqslant P(v) \leqslant b$, be a periodic function, the variable $v = \gamma(x)$ of which is a solution of the Schröder equation (1.23), such that S is an integer $K > 1$. Let $g(x)$ and I be respectively a given continuous function, and the set of the x-values such that $a \leqslant g(x) \leqslant b$. In order to simplify the statement, in a first step it is supposed that $P(v)$ and $g(x)$ are analytical function of their argument. From $P(v) = g(x_o)$, let $P_{-1}(x_o)$ be the principal inverse of $P(v)$. Let $x_o = g_{-1}^{(i)}\left[P(v)\right]$ be one of the determinations $g_{-1}^{(i)}$ of the inverse function of $g(x)$, when it is multi-valued, or the inverse function g_{-1} when this inverse exists and is unique in an interval of the variable x.

With $S = K$ = integer > 1, from the properties of the Schröder equation, the recurrence T is constructed :

(1.31) $$x_{n+1} = g_{-1}^{(i)} \{P[K\, P_{-1}(g(x_o))]\}$$

the solution of which is :

(1.32) $$x_n = g_{-1}^{(i)} \{P[K^n\, P_{-1}(g(x_o))]\}$$

I being the interval of x such that $a \leqslant g(x) \leqslant b$. T given by (1.31) is always an endomorphism. The solution (1.32), $x \in I$, is *chaotic* in the sense that the chaos amounts to the selection of ordinates of a periodic function at exponentially (K^n) increasing discrete abscissae. The closeness of a given trajectory $\{x_n\}$, n = 0, 1, 2, ..., with respect to the neighbouring ones is described by the sensitivity coefficient :

$$\partial x_n \,/\, \partial x_o = K^n (dg_{-1}/dP)(dP/dP_{-1})(dP/dg(x_o))(dg(x_o)/dx_o)$$

(dP/dP_{-1}) being periodic function. The average value of this coefficient increases rapidly with n. Two close initial conditions x_o, $x'_o \in I$ yield two bounded sequences $\{x_n\}$, $\{x'_n\}$ via equation (1.31), but the average of the difference $|x_n - x'_n|$ increases rapidly with n. The *sensitive dependence on initial conditions is known as one of the characteristics of chaotic behaviours*. Subject to the practically unavoidable constraint of finite accuracy, the exact solution (1.32) is *not predictive* at all, because the number of significant digits of $|x_n - x'_n|$ diminishes very rapidly with n. Solution (1.32) thus describes a practically chaotic evolution of an intrinsically deterministic origin.

In the particular case $\cos v = g(x_o)$, for $|g(x_o)| \leqslant 1$ $(x \in I)$, with $\cosh v = g(x_o)$, $g(x_o) > 1$, the same recurrence T is obtained :

(1.33) $$x_{n+1} = g_{-1}^{(i)} \{\cos[K \arccos g(x_n)]\} = g_{-1}^{(i)}\{\cosh[K \cosh^{-1} g(x_n)]\}$$

(1.34) $$\begin{cases} x_n = g_{-1}^{(i)}\{\cos[K^n \arccos g(x_o)]\}, & |g(x_o)| \leqslant 1 \\ x_n = g_{-1}^{(i)}\{\cosh[K^n \cosh^{-1} g(x_o)]\}, & g(x_o) > 1 \end{cases}$$

in the interval giving real inverses. Choosing $g(x) = ax + b$, T is defined by a polynomial having the degree K, which for $a = 1$, $b = 0$ is a Chebyshev's polynomial T_K of first species ($T_1 = x$). With $K = 2$, $f(x) = (2a^2x^2 + 4abx + 2b^2 - b - 1) / a$, and the chaotic behaviour occurs for $-(1+b)/a \leqslant x \leqslant (1-b)/a$. Choosing $a = 1/2$, $b = 0$, $f(x) = x^2-2$ and :

$$\begin{cases} x_n = 2\cos\left[2^n \arccos(x_0/2)\right], & -2 \leqslant x_0 \leqslant 2 \\ x_n = 2\cosh\left[2^n \cosh^{-1}(x_0/2)\right], & x_0 > 2 \end{cases}$$

The sensitivity coefficient is $\partial x_n/\partial x_0 = -2^n(1-x_0^2/4)^{-1/2} \sin\left[2^n \arccos(x_0/2)\right]$ for $-2 < x_0 < 2$, interval of a chaotic solution.

When g(x) is non linear function, the right hand side of (1.31) or (1.33) is a function H_K which has properties equivalent to those of the Chebyshev's polynomials, inside an interval $\Delta \in I$. More precisely, the right hand side of (1.33) is a function H'_K, $H'_1 \equiv g(x)$, *with properties equivalent to those of the Chebyshev's polynomial of first species T_K*. With $\sin v = g(x_0)$, $|g(x_0)| \leqslant 1$, $\sinh v = g(x_0)$, $|g(x_0)| > 1$, for K = even integer, the right hand side of (1.31) is a function H''_K which has *properties equivalent to those of a Chebyshev's polynomial of second species*, inside an interval Δ. The case K = odd integer does not give such a polynomial but leads to symmetric properties with respect to those of H'_K. The properties of the right hand sides of (1.31), or (1.33) are the following :

(a) Given a function g(x), the maxima of H_K, $K = 2, 3, \ldots$, have the same ordinate, and it is the same for the minima of H_K. The segment Δ of every H_K is bounded by the same points.

(b) The "chaotic segment" $\Delta \subset I$ does not contain attractor such fixed point, or cycle. Δ is bounded either by a repulsive fixed point A and its first rank antecedent different from A (case of H'_K with K = even integer), or by two repulsive fixed points (case of H'_K with K = odd integer), or by the two points of a repulsive cycle of order two (case of H''_K with K = odd integer), or by two points the first rank consequent of which is the same repulsive fixed point belonging to Δ (case of H''_k, k = even integer). On each side of this fixed point, there are K/2 other repulsive fixed points which belongs to Δ.

(c) If the inverse g_{-1} of g(x) is unique (g_{-1} is single-valued), then the interval I giving the solution (1.32) is such that $I \equiv \Delta$.

(d) If g_{-1} is multi-valued, then Δ is a sub set of I, and the points of I outside Δ are antecedents of Δ.

(e) The functions H'_K satisfy the relation $H'_n[H'_m(x)] = H'_m[H'_n(x)] = H'_{mn}(x)$, m and n = integers, and the recurrence relation $g(H'_{K+1}) - 2g(x).g(H'_K) + g(H'_{K-1}) = 0$.

(f) The Perron-Frobenius equation, associated with every H_K, K = 2, 3,... $x \in \Delta$, has the same particular solution $w(x) = Bdv/dx$, B = arbitrary constant. For H'_K, $w(x) = B\{1 - [g(x)]^2\}^{-1/2}$.

The property (a) results from the right hand side form of (1.33). It is the same for (b), but adding the results of chapter 3. The representation of H_K in the plane (x_n, x_{n+1}) permits to justify the properties (c) and (d). The properties (e) and (f) are conjectures which have been verified from various examples.

<u>Some examples of functions H_K giving rise to chaotic behaviours</u>

-(1) Consider $\cos v = 2x_0/(x_0-1)$, $-1 \leq x_0 \leq 1/3$, $\cosh v = 2x_0/(x_0-1)$ $x_0 < -1$, $x_0 > 1/3$. The function g_{-1} is single-valued and (1.31) with K = 2 is :

(1.35) $$x_{n+1} = (7x_n^2 + 2x_n - 1) / (5x_n^2 + 6x_n - 3)$$

and with K = 3 (1.31) becomes :

(1.36) $$x_{n+1} = [16x_n^3 - 3x_n(x_n-1)^2] / [16x_n^3 - 3(x_n-1)^2(4x_n-1)]$$

For K = integer > 3, functions H'_K, having the above-indicated properties, are obtained. The common Δ is $-1 \leqslant x \leqslant 1/3$ for which the chaotic solution is $x_n = [\cos(K^n v)] / [\cos(K^n v) - 2]$. The point $x = 1/3$ (critical point of rank 1), is the first rank consequent of the maximum $x = 0$ of $f(x_n)$ given by (1.35), $x = -1$ (critical point of rank 2) is a repulsive fixed point first rank consequent of $x = 1/3$.

-(2) Consider $\cos v = x_0 / (x_0-1)$, $x_0 \leqslant 1/2$, K = 2, then :

(1.37) $$x_{n+1} = (x_n^2 + 2x_n - 1) / (4x_n - 2)$$

The solution $x_n = [\cos(2^n v)] / [\cos(2^n v)-1]$, $x \leqslant 1/2$ is a "non-bounded chaotic solution". The critical point of rank 1 is $x = 1/2$, and the critical point of rank 2 is $x = -\infty$.

-(3) Consider $\cos v = 2x_0 - x_0^2$, and I given by $1-\sqrt{2} \leqslant x \leqslant 1+\sqrt{2}$. The real inverse of g(x) does not exist outside I, and for $x \in I$, and K = 2, g(x) has two inverses leading to the two recurrences :

(1.38) $$x_{n+1} = 1 \pm [2 - 2(2x_n - x_n^2)^2]^{1/2}$$

For $x \in I$, the first recurrence (sign + before the radical) has the solution $x_n = 1 + \{1 - \cos[2^n \operatorname{arc}\cos(2x_0 - x_0^2)]\}^{1/2}$, and the chaotic segment Δ is given by $1 \leqslant x \leqslant 1+\sqrt{2}$. The first rank consequent of the interval $1-\sqrt{2} < x < 1$ is Δ. The function f(x) has two smooth maxima ($x = 0$, $x = 2$), and a minimum $x = 1$ which is a cusp. The function f(x) is symmetric with respect to $x = 1$. The second recurrence (sign - before the radical) give symmetric properties.

-(4) Consider $\cos v = x_0 / (x_0^2-1)$, K = 2, inside I defined by $(1-\sqrt{5}) / 2 \leqslant x \leqslant (\sqrt{5}-1) / 2$, $x \leqslant -(1+\sqrt{5}) / 2$, $x \geqslant (1+\sqrt{5}) / 2$, and $\cosh v = x_0 / (x_0^2-1)$, K = 2, inside $(\sqrt{5}-1) / 2 < x < (1+\sqrt{5}) / 2$. Choose the recurrence corresponding to sign - before the radical which is obtained by writing $g_{-1}(x)$:

(1.39) $$x_{n+1} = \{A^2 - [A^4+4(2x_n^2-A^2)^2]^{1/2}\} / [4x_n^2-2A^2], \quad A \equiv x_n^2-1$$

The interval Δ is $(1-\sqrt{5})/2 \leqslant x \leqslant (\sqrt{5}-1)/2$. The solution is $x_n = \{1 - [1+4\cos^2(2^n v)]^{1/2}\} / [2\cos(2^n v)]$ inside Δ and inside $[x \leqslant -(1+\sqrt{5})/2$, $x \geqslant (1+\sqrt{5})/2]$. It has the second form (1.34) inside the complementary intervals.

-(5) Now consider a non analytical case, $P(v)$ being a periodic function of period 2π : $P(v) = -2v/\pi+1$ when $0 \leqslant v \leqslant \pi$; $P(v) = 2v/\pi-3$, $\pi \leqslant v \leqslant 2\pi$. Choose $g(x) \equiv x$, $K = 2$. From $P(2v) = 2P(v)-1$, $0 \leqslant v \leqslant \pi/4$, $P(2v) = -2P(v)-1$, $\pi/4 \leqslant v \leqslant \pi/2$, it is obtained the piecewise linear recurrence :

$$(1.40) \qquad x_{n+1} = -2x_n-1 \quad , x \leqslant 0 \ ; \quad 2x_n - 1, \ x > 0$$

Choosing $K = 3, 4, \ldots$, a *sequence of piecewise linear functions* H_K is obtained.

-(6) Consider the periodic function of period 2π : $P(v) = -4v^2/\pi^2+1$, $-\pi/2 \leqslant v \leqslant \pi/2$; $P(v) = 4v^2/\pi^2-8v/\pi+3$, $\pi/2 \leqslant v \leqslant 3\pi/2$. Choose $g(x) \equiv x$, and $K = 2$. From $P(2v) = -16v^2/\pi^2+1$, $-\pi/2 \leqslant v \leqslant \pi/2$, $P(2v) = 16v^2/\pi^2-16v/\pi+3$, $\pi/2 \leqslant v \leqslant 3\pi/2$, $v = \pi-\pi\sqrt{1+x}/2$, the non-smooth following recurrence T is obtained :

$$\begin{cases} x_{n+1} = 4x_n-3, \ 3/4 \leqslant x \leqslant 1 \ ; \quad x_{n+1} = -4x_n-3, \ -1 \leqslant x \leqslant -3/4 \\ x_{n+1} = 7 - 4x_n-8\sqrt{1-x_n}, \ 0 \leqslant x \leqslant 3/4 \ ; \ x_{n+1} = 7+4x_n-8\sqrt{1-x_n}, \ -3/4 \leqslant x \leqslant 0 \end{cases}$$

T has a minimum for $x = 0$. A *sequence of non-smooth functions* H_K is obtained for $K = 3,4,\ldots$

1.10.4. Generalization to m-dimensional recurrences $m > 1$

The methods of § 1.10.3. have a generalization in the m-dimensional case, with the following process. Let $v_i = \gamma_i(x^{(i)}, \ldots, x^{(m)})$, $i = 1, \ldots m$ be the solution of a m-dimensional Schröder's equation (cf. (1.26) for $m = 2$), and $S_i = K_i$ = integers > 1. Given a vectorial periodic function $P_i(v_i)$, $a_i \leqslant P_i(v_i) \leqslant b_i$, and an analytic vectorial function

$g_i(x^{(i)},\ldots,x^{(m)})$, $i = 1, \ldots, m$, from $P_i = g_i$, it is possible to construct a m-dimensional recurrence (endomorphism), which is a generalization of (1.31), and its solution. The following inequalities are considered : $a_i \leqslant g_i \leqslant b_i$. If they are compatible, they define a domain I of the phase space $(x^{(i)}, \ldots, x^{(m)})$, which contains a "chaotic region" Δ, where the solution can be known in an analytical form which generalizes (1.32). The right hand sides of the so-obtained recurrence T constitute a vectorial function $H_{K_1 \ldots K_m}$, m-dimensional generalization of the function H_K defined in § 1.10.3. They have the following properties :

(a) Given the functions P_i, g_i, the functions $H_{K_1 \ldots K_m}$ obtained from different choices of the set $(K_1, \ldots, K_m)$ have the same region Δ.

(b) The "chaotic region" $\Delta \subset I$ does not contain attractor such fixed point, or cycle. Δ is bounded by parts of critical manifolds which are the rank 1, 2, ... consequents of the manifold defined by $J = 0$, J being the Jacobian determinant of the right hand side of the recurrence (endomorphism).

(c) When the relations $P_i = g_i$ have a unique inverse, one has $I \equiv \Delta$. Else one has $\Delta \subset I$, and the points of I outside Δ are antecedents of points of Δ.

(d) The variables transformation $X^{(j)} = g_j(x^{i)}, \ldots, x^{(m)})$, $j = 1, \ldots, m$, can lead to a recurrence with separated variables.

As example, consider $\cos v_1 = 2x_o / (x_o+y_o-1)$, $\cos v_2 = x_o+y_o$. One has $\Delta \equiv I$, and Δ is bounded by a triangle made up of d_1 $(y = x+1)$, d_2 $(y = -3x+1)$, d_3 $(y = -x-1)$. Choose $K_1 = K_2 = 2$. T is given by :

$$x_{n+1} = [A-8x_n^2][1-(x_n+y_n)^2] / A \ ; \ y_{n+1} = 2(x_n+y_n)^2 - x_{n+1} - 1$$

where $A = (x_n+y_n-1)^2$. The solution is $x_n = \cos(2^n v_1)[\cos(2^n v_2)-1] / 2$, $y_n = \cos(2^n v_2)-x_n$. d_1 is an invariant curve of T, d_2 is an antecedent of d_1. The first rank consequent of d_3 is the repulsive fixed point (0 ; 1). These straight lines are critical curves of T.

With the choice $K_1 = 2$, $K_2 = 3$, T becomes :

$$x_{n+1} = (A-8x_n^2)[1-4B^3+3B] / (2A) \; ; \; y_{n+1} = 4B^3-3B-x_{n+1}$$

where $A = (x_n+y_n-1)^2$, $B = (x_n+y_n)^2$. The solution is given by $x_n = \cos(2^n v_1)[\cos(3^n v_2)-1] / 2$, $y_n = \cos(3^n v_2)-x_n$, and Δ is similarly defined.

1.11 THE PROBLEM OF THE FRACTIONAL ITERATES

The problem of fractional, and continuous iterates can be solved by means of functions satisfying the Abel and the Schröder functional equations |G 36|. So in the one-dimensional case $x_{n+1} = f(x_n)$, these equations are respectively :

(1.41) $\quad w[f(x)] = w(x) + a, \qquad a = 1$

(1.42) $\quad \gamma[f(x)] = S\,\gamma(x)$

After k iterations, there results :

(1.43) $\quad w[f^k(x)] = w(x) + k$

(1.44) $\quad \gamma[f^k(x)] = S^k\gamma(x)$

When the solutions $y = w(x)$, $y = \gamma(x)$ possess at least one inverse $x = w_{-1}(y)$, $x = \gamma_{-1}(y)$, then a particular solution of the functional equation :

(1.45) $\quad f^k(x) = f^{k-1}[f(x)], \; f^1(x) = f(x), \; f^o(x) = x$

is given by :

(1.46) $\quad f^k(x) = w_{-1}[w(x)+k]$

(1.47) $\quad f^k(x) = \gamma_{-1}[S^k\lambda(x)]$

These solutions coincide for k = 0, 1, 2, ... with the solution of the recurrence. If the right-hand sides of (1.46), (1.47) conserve their meaning for fractional and continuous positive values of k, then they represent also fractional and continuous iterates of $f(x)$, except when a specific x is a singular element of the recurrence, or $k \to +\infty$. When $k < 0$, (1.46), (1.47) conserve their validity, provided w_{-1}, γ_{-1}, designate all possible inverses.

Since in general $w(x)$ and $\gamma(x)$ are complicated transcendental functions, as a rule, the study of $w(x)$, $\gamma(x)$ is carried out locally. For example, a preference is given to an attractive fixed point x^*, and after choosing $S = f'(x^*)$, the solution of (1.41), (1.42) is determined in the form of a Mc Laurin series. The equivalence between (1.41), (1.42) implies, however, that the constant S in (1.42) plays a practically useful, but intrinsically an inessential role. In fact, without loss of generality, it can be assumed that $S > 0$. A proper choice of S, different from $f'(x^*)$, may therefore lead directly to a global solution of (1.41), (1.42), as this appears from § 1.10.3. In the same way, fractional iterates may be determined globally, even if the recurrence is an endomorphism (for more details about this point, cf. |M 37||M 53|).

Chapter 2

SOME PROPERTIES OF ONE-DIMENSIONAL RECURRENCES (MAPS)

INTRODUCTION

This chapter gives a general view about the properties of one-dimensional maps T (or first order autonomous reccurences) having the explicit form :

$$x_{n+1} = f(x_n , \lambda) \qquad , \qquad n = 0,1,2,...$$

Where $f(x, \lambda)$ is a single-valued function of both the real variable x and the real parameter λ , which is always a smooth function of λ . The most part of the properties concerns $f(x, \lambda)$ being a smooth function of x. Properties related to a only piecewise differentiable $f(x, \lambda)$ with respect to x, or only piecewise continuous $f(x, \lambda)$, are only summarily indicated. The case, with T defining a map of the real x-axis onto itself, $R \to R$, corresponds to the longest development. However a part of this chapter in devoted to T being a map $S \to S$ of the circle into itself (for example x is an angular variable).

From an initial condition x_0 , successive applications of T generate an iterated sequence (orbit, or discrete trajectory), $x_n = x(x_0,n,\lambda)$. T can be considered as an implicit definition of the function x_n , which plays the role of the recurrence solution for the initial condition x_0 . Such a definition of x_n is theoretically quite satisfactory, but practically it is almost useless, because in general the function $x_n = x(x_0,n,\lambda)$ is unknown, except for the linear case and for very few examples when $f(x,\lambda)$ is a non linear function of x. In all non contrived cases x_n cannot be expressed explicitly in terms of known elementary and transcendental functions. So a meaningful characterization of x_n consists in the identification of its singularities, and in the description of the behaviour of the latter as λ vary, any so obtained change of the singularities, or of their properties being called a *bifurcation.* It is obvious that simple single-valued smooth functions $f(x,\lambda)$ may have complicated multi-valued and not necessarily smooth inverse functions $f^{-1}(x,\lambda)$, or T^{-1} (for example cf. $f(x,\lambda) = x^2 - \lambda$, with $f^{-1}(x,\lambda) = \pm (x+\lambda)^{\frac{1}{2}}$). As a rule the non-uniqueness of T^{-1} triggers a complex branching process. Then T may have infinitely many real singularities with complex composite accumulations. When T is non invertible (T^{-1} is non-unique, or does not exist) T is call an *endomorphism*. When T is invertible, and $f(x,\lambda)$ a smooth function of x, T is called a *diffeomorphism* and has simpler properties.

After the consideration of the linear case, the singularities (fixed points, cycles) are defined with their stability (attractive, or repulsive singularity) and the corresponding bifurcations. The following cases are considered : (i) $f(x,\lambda)$ is a continuous and smooth function with respect to x ; (ii) $f(x,\lambda)$ is continuous function which is not smooth with respect to x ; (iii) $f(x,\lambda)$ is a piecewise continuous function with respect to x. A part of this chapter is also devoted to the domain of convergence (domain of influence, or domain of attraction, or basin) of an iterated sequence toward an attractive singularity. The last part concerns the map (diffeomorphism) of the circle onto itself.

2.1. AUTONOMOUS LINEAR CASE.

Considering the one-dimensional autonomous recurrence (or map) :

$$y_{n+1} = Sy_n + a, \quad S\neq 1 \text{ if } a\neq 0, \quad -\infty \leqslant y \leqslant \infty , \tag{2.1}$$

it has a unique fixed point at finite distance $y^* = a/(S-1) = y_n = y_{n+1}$. The linear change of variable $y = y^* + x$ leads to the form :

(2.2) $$x_{n+1} = S\, x_n$$

From (2.2) the solution for an initial condition x_o can be expressed explicitly in terms of a known elementary transcendental function

(2.3) $$x_n = S^n x_o$$

With the form (2.2) $x^* = 0$ is the unique *fixed point* at finite distance, and its nature is completly defined from the real number S, which is called the *multiplier*, or the eigenvalue of x^*. Indeed, it appears from (2.2) (2.3) that if $|S| < 1$, the iterated sequence generated by (2.2) is such that $\lim_{n\to\infty} x_n = x^* = 0$, whatever x_o may be, and x^* is called an *attractive* (or asymptotically stable) fixed point. If $|S| > 1$ $\lim_{n\to\infty} x_n = \infty$ and x^* is said to be *repulsive* (or unstable).

The nature of the fixed point x^* depends also on the sign of the multiplier S. If $S > 0$, x_n has the sign of x_o whatever n may be, and x^* is called a fixed point of *type 1* . Now if $S < 0$, the sign of x_n is that of x_o if n is even, and that of $-x_o$ if n is odd, and x^* is said of *type 2*.

With the change of variable $x = 1/z$, the point at infinity becomes the fixed point $z^* = 0$ for the recurrence :

(2.4) $$z_{n+1} = (S)^{-1} z_n$$

This point has an opposite stability to that of x^*. When $|S| < 1$, the *influence domain* (or *basin*) of x^* is the set of points x_o giving the convergence of x_n towards $x^* = 0$. This domain is defined by $-\infty < x_o < +\infty$, and it has a boundary point $z^* = 0$ which is the repulsive fixed point at infinity. When $|S| > 1$, the fixed point $z^* = 0$ is attractive and its basin is constituted by the x-axis except the repulsive fixed $x^* = 0$, boundary of this domain. So it appears the the fixed points $x^* = 0$, $z^* = 0$ are the two and only singularities of (2.2).

The case $|S| = 1$, considered as a *critical case* (in the sense of Ljapunov) in non linear situations and so giving rise to typical bifurcations, corresponds here to a simple structurally unstable behaviours of the solutions. If $S = +1$, all the points of the x-axis are fixed points with the same multiplier. If $S = -1$, $x^* = 0$ is a fixed point of type 2, the other points are such that $x_{n+1} = -x_n$, i.e. $x_{n+2} = x_n$, and each pair of these points constitutes a *cycle of order* 2.

2.2. AUTONOMOUS NON LINEAR CASE. THE SINGULARITIES.

Let T be the autonomous recurrence, or map $R \to R$, explicitly defined by :

(2.5) $\quad x_{n+1} = f(x_n, \lambda) \quad$ or $\quad x_{n+1} = Tx_n$

$f(x, \lambda)$ being a single-valued function of both the real variable x and the parameter λ. In the general non linear case, the solution $x_n = x(x_o, n, \lambda)$ cannot be expressed explicitly, as (2,3), in terms of known elementary and transcendental functions. Then, since the recurrence (2.5) contains a simple dependent variable, the singularities, which characterizes the non elementary transcendental function $x(x_o, n, \lambda)$, are zero-dimensional. The simplest point-singularity x^* is a *fixed point*. It is given by an isolated finite root of the equation.

(2.6) $\quad f(x,\lambda) - x = 0 \quad$ or $\quad x = Tx$

Eq. (2.6) may possess more than one root at finite distance. Considering one isolated root x^*, the corresponding stationary state, such that $x_{n+1} = x_n = x^*$ is invariant with respect to T and independant of n. All consequents of a fixed point x^* coincide of course with x^*, but this is not necessarily so for all antecedents. Suppose that x^* admits a non-zero neighbourhood $\Delta : -\varepsilon < x_n - x^* < \varepsilon'$,

$0 < \varepsilon$, $0 < \varepsilon'$, free from other singularities of T. The existence of such a Δ is not always guaranteed a priori, except when the root x^* is known to be isolated. This existence assumption is made provisionally, subject to an a posteriori confirmation. Consider a point $x_o \in \Delta$, and the set of its consequents x_n, $n = 1,2, \ldots$ Three cases are possible as $n \to \infty$:

a) $|x_n - x^*|$ diminishes and $\lim |x_n - x^*| = 0$; x^* is said to be *attractive*

b) $|x_n - x^*|$ increases till one of the x_n reaches the boundary of Δ, or leaves Δ entirely ; x^* is said to be *repulsive*

c) $|x_n - x^*|$ diminishes for $x_n < x^*$ (resp $x_n > x^*$) whith lim $|x_n - x^*| = 0$, and increases for $x_n > x^*$ (resp $x_n < x^*$). In general this situation occurs for particular values of the parameter λ, each of them corresponding to a bifurcation value.

In the terminology of dynamic systems the equivalent statement is : for (a) the constant stationary state (or static equilibrium) x^* is *asymptotically stable*, for (b) it is *unstable*, and for (c) it corresponds to a *critical case in the sense of Ljapunov*. Here stability is understood to be also in the sense of Ljapunov [L 20], extended to discrete dynamic systems. When x^* is an attractive fixed point, contrarily to the linear recurrence (2.1), the basin of x^* is now in general an open part of the x_n -axis, and only exceptionally the domain $-\infty < x_o < +\infty$.

Isolated roots of (2.6) are not the only possible point-singularities of T. Considering the construction of the rank k consequent of x_n (k = integer), and eliminating the "intermediate" variables $x_{n+1}, \ldots, x_{n+k-1}$, from the recurrence (2.5), or map T, the iterated recurrence, or map T^k, is obtained :

$$(2.7) \qquad x_{n+k} = f^k(x_n, \lambda) \qquad \text{or} \qquad x_{n+k} = T^k x_n, \quad f_1 \equiv f$$

Here the function f^k is unambiguously defined, and single valued. The more k increases, the more T^k is difficult to construct, but it remains straightforward. The inverse problem of constructing

T from the sole knowledge of T^k is not straigtforward at all, and its general solution is still unknown. Such a problem is related to the more general one of determining fractional iterates of a given function. The simplest case is that of determining the *half-iterate* f(x) of g(x) = f[f(x)], when g is a given function. As T, T^k may also admit isolated point-singularities, defined by the roots of :

$$f^k(x, \lambda) - x = 0 \quad \text{or} \quad x = T^k x \ , \quad k = 2,3,... \tag{2.8}$$

k = 1 giving (2.6). A set of k roots of (2.8), x_i^*, i = 1,2,...,k, such that

$$x_{i+1}^* = Tx_i^*, \quad x_i^* \neq T^\ell x_i^*, \quad \ell < k, \quad x_{i+k}^* = x_i^* \tag{2.9}$$

is said to be a *cycle* (or a *periodic point*) *of order* k. The relation $x_i^* \neq T^\ell x_i^*$, $\ell < k$, permits to eliminate the roots of (2.8) which are simultaneouly roots of (2.8) when k is replaced by one of its divisors, unity included. Every root defines k - 1 > 0 distinct consequents, obtained by means of T, which are also roots of (2.8), and the k-th consequent of any root coincides with itself.

With regard to neighbouring points the stability situation of a cycle is therefore the same as that of a fixed point : a *cycle of order* k is either *attractive, repulsive,* or corresponds to the above-mentioned case (c).

Unless the form of f(x, λ) is severely restricted (to smooth monotonic functions, for example, T being then a diffeomorphism), the number of different cycles of T is in general not finite. Indeed *accumulations of cycles point-singularities* are possible inside parts of the x-axis, giving rise *to composite singularities.* In general the number of the latter is not finite, which gives rise to new accumulations and thus to *"higher-class" composite singularities* [P 6]. This is so for the quadratic recurrence, or map T, with $f(x,\lambda) \equiv x^2-\lambda$.

With one-dimensional endomorphisms, three different typical situations (but perhaps more) can be defined when such accumulations occur :

- (S_1) T has an attractor which is a fixed point x^*, or a cycle. The boundary of the corresponding basin (B) is constituted by an infinite set ($\bar{B}$) of repulsive cycles of order $k = 1,2,3,\ldots,$ their accumulation points, and the successive antecedents of all these points, which are different from these ones. One of these points q has a minimum abscissa, and one q' has a maximum abscissa. The open interval $]q,q'[$ contains the other points of ($\bar{B}$) (which constitute here a totally disconnected set) and the basin of x^*. Then ($\bar{B}$) plays the role of a *strange repulsor* giving rise for $0 < n < N$ to a *chaotic transient*, when x_0 is in a neighbourhood of ($\bar{B}$). For $n > N$ the convergence toward x^* becomes regular.

- (S_2) T has two attractors, one of them can be located at infinity (it is then a fixed point). The boundary which separates the basins (B_1) (B_2) of the two attractors has the same type as ($\bar{B}$) defined in (S_1), but now each basin is constituted by an infinite set of disjoint open segments of the x-axis. A segment belonging to (B_1) (resp B_2) does not contain points of ($\bar{B}$). It is bounded by two points of ($\bar{B}$), which separates this one from two segments belonging to (B_2) (resp B_1). These points of $\bar{B}$ are antecedents of fixed points or cycles. All these segments jumble together in a very complex way and have for limit points repulsive cycles, and accumulations points of cycles belonging to ($\bar{B}$). Here the *strange repeller ($\bar{B}$) constitutes a chaotic basin boundary or fuzzy boundary ("frontière floue"* french expression introduced for the first time in 1975 [M 23] [M 25] [M 30] [G 35]). Such situations may occur with more than two attractors.

- (S_3) T (or T^k) has a closed segment δ (or k segments) of the x-axis, which containsinfinitely many repulsive cycles of any order and their limit points. Considering the antecedents of any rank of every point of δ, a sub set of these antecedents gives points everywhere dense on δ (in the particular case $f(x,\lambda) \equiv x^2-\lambda$, $\lambda = 2$, the sub set becomes the set of all antecedents, and is the interval [-2;2] cf [J 5] p.186). The structure of the singularities of δ is very complex in the presence of accumulations giving "m-class", m = 1,2,..., composite singularities [P6]. The segment is a *completely invariant set* in the sense of Pulkin [P 6]. It is called also *a chaotic segment* and can be considered as a *strange attractor* with a domain attraction (basin). In the particular case $f(x,\lambda) \equiv x^2-2$, the basin]-2;2[merges into the strange attractor.

2.3. AUTONOMOUS NON LINEAR CASE. STABILITY OF FIXED POINT AND CYCLES.

2.3.1. RECURRENCE DEFINED BY A SMOOTH FUNCTION.

Let x^* be a fixed point of the recurrence (or map) T (2.5), where $f(x,\lambda)$ is a continuous and differentiable function with respect to x ($f \in C_1$). Here the parameter λ plays no role and $f(x,\lambda)$ is written $f(x)$. With the change of variable $x = x^* + X$, a new map is obtained where the linear term can be separated from the non linear part $F(X_n)$ of the map :

(2.9) $$X_{n+1} = SX_n + F(X_n) \; ; \; F(0) = 0 \; ; \; \lim_{x \to 0} F(X)/|x| = 0 \; ; \; S = [df/dx]_{x=x^*}$$

Here S is the *multiplier* (or *eigenvalue*) of the fixed point x^*. Inside a sufficiently small interval containing x^*, the linear approximation of T is $X_{n+1} = SX_n$. So for $|S| < 1$, x^* is *attractive* (locally). It is *repulsive* for $|S| > 1$. The *type* of x^* is *one* if $S > 0$, and the consequents of $x_n \neq x^*$ are all on the same side of x^*. The fixed point x^* is of *type two* if $S < 0$, and the consequents of $x_n \neq x^*$ alternate on both sides of x^*.

Let now x_i^*, i = 1,2,...,k, be the k points of a cycle of order k, solutions of (2.8). The consideration of T^k given by (2.7) permits to write the equivalent of (2.9), with :

$$(2.10)\quad S = [df^k/dx]_{x=x_i^*} = [df/dx]_{x_1^*}[df/dx]_{x_2^*} \dots [df/dx]_{x_k^*}$$

Depending of S, a cycle is attractive or repulsive, and has the type one, or two. It has the same behaviour as a fixed point.

For a fixed point, or a cycle, the values $S = \pm 1$ are critical in two ways :

a) the roots x^* of (2.6) or of (2.8), are not necessarily simple.

b) the stability of the fixed point, or of the cycle cannot be deduced from the linear approximation SX_n. Here the non linear terms of (2.9) settle the behaviour of the iterated sequences generated by an initial condition x_o, whatever small be $|x^*-x_o|$.

For these reasons $|S| = 1$ is said to constitute a *critical stability case in the sense of Ljapunov* equivalent to the situation firstly studied by Ljapunov [L20] for the ordinary differential equations.

A geometrical interpretation of the stability of fixed points can be made in an auxiliary plane (x_n, x_{n+1}), using the properties of the first bisectrix $x_{n+1} = x_n$, via the Königs-Lemeray graphical "staircase" construction (point on x_n-axis $\rightarrow$ point $x_{n+1} = f(x_n)$ $\rightarrow$ point on x_{n+1}-axis $\rightarrow$ transfer to the x_n-axis via the line $x_{n+1} = x_n$, and so forth). This construction is illustrated by the fig.2.1a, b. In fig.2.1a T has an attractive fixed point q_2 $(x_{n+1} = x_n)$ of type one $(0 < S < 1)$, the multiplier S of which being given by the slope of the tangent to the curve $x_{n+1} = f(x_n)$ at the point q_2. The consideration of this slope at the fixed points q_2, q_3 shows that they are repulsive of type one $(S > 1)$. With the fig.2.1b the fixed point q_2 is attractive of type two $(-1 < S < 0)$, q_1 is repulsive of type one $(S > 1)$. The inverse Königs-Lemeray permits to see that for f(x) of the fig.2.1a, the inverse map T^{-1} is single-valued because a point x_n has only a unique real antecedent of rank one. This is not the case for f(x) of the fig.2.1b, where x_n has zero $(x_n < x_c)$ or two real antecedents $(x_n > x_c)$ of rank one, C being the first rank consequent of the extremum of f(x). So the fixed point q_1 is itself its rank one antecedent, the second rank one antecedent being q_1^{-1}. The function f(x) of fig.2.1b defines an endomorphism.

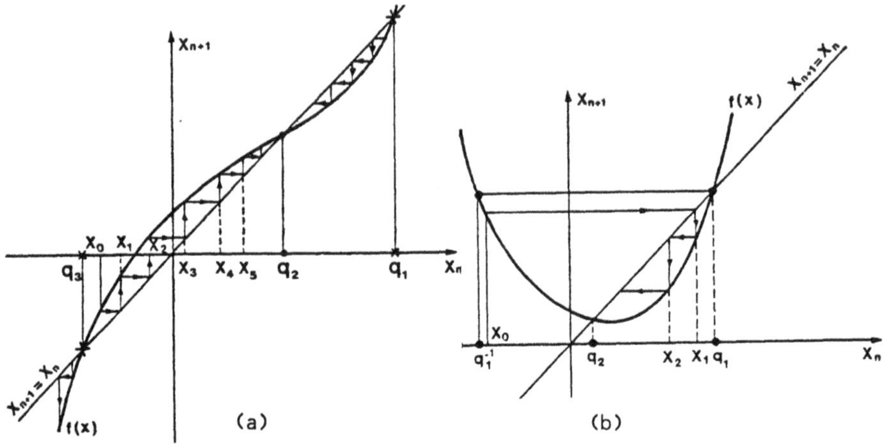

FIG. 2.1

Stability for S = +1

Supposing $f(x, \lambda)$ be a sufficiently differentiable function with respect to x ($f \in C_m$), and $x^* = 0$ be the considered fixed point, (2.9) can be written :

$$(2.11) \quad x_{n+1} = x_n + ax_n^m + F(x_n) \; ; \; m > 1 \; ; \; \lim_{|x|\to 0} F(x)/|x^m| = 0$$

By a geometrical way, using the Königs-Lemeray construction with f(x) tangent to the line $x_{n+1} = x_n$, the conclusion of stability is directly obtained from fig.2.2 (m is even) and from fig.2.3 (m is odd and f(x) has an inflexion point at $x^* = 0$). So the following proposition is obtained [N 2] [G35]:

(P2.1) *When m is even, the fixed point* $x^* = 0$ *of (2.11), is attractive on the left* $(x < 0)$, *and repulsive on the right* $(x > 0)$, *if* $a > 0$ *It is repulsive on the left* $(x < 0)$, *attractive on the right* $(x > 0)$, *if* $a < 0$. *When m is odd,* $x^* = 0$ *is repulsive if* $a > 0$, *and attractive if* $a < 0$. *Here* x^* *is always of type one.*

Stability for S = -1

Now f(x) is tangent to the line $x_{n+1} = -x_n$, and the recurrence T is written with the fixed point at $x^* = 0$:

$$(2.12) \quad x_{n+1} = -x_n + a_o x_n^m + a_1 x_n^{m+1} + F(x_n) ; \; m > 1 \; , \; \lim_{|x|\to 0} F(x)/|x^{m+1}| = 0$$

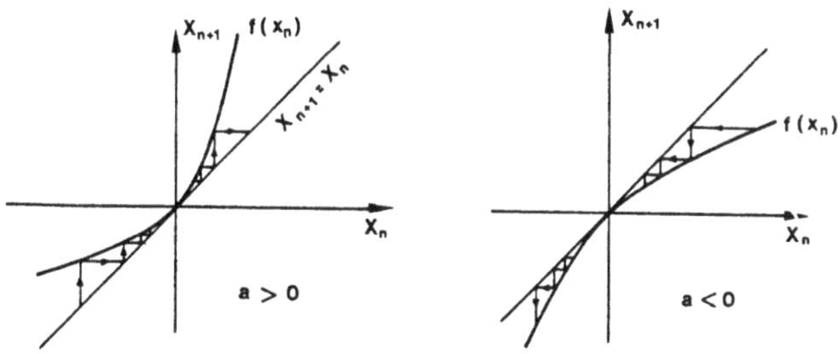

FIG. 2.2 S=+1 , m is even .

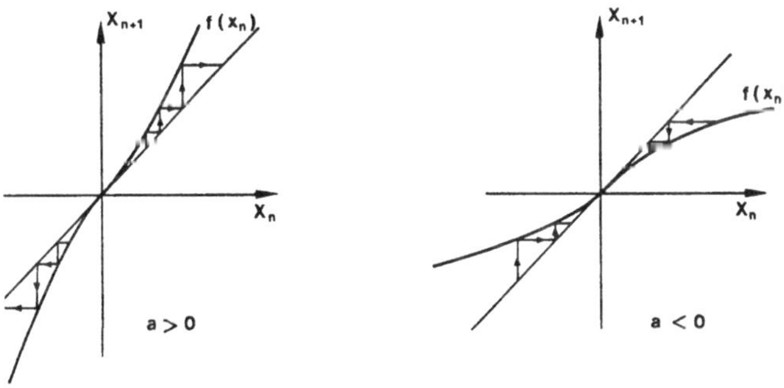

FIG. 2.3 S=+1 , m is odd .

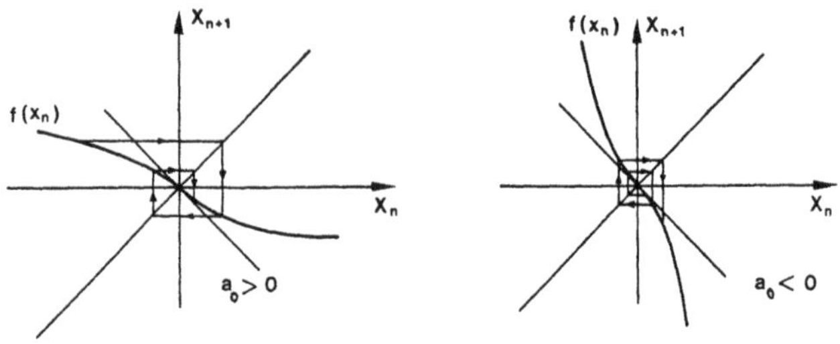

m is odd

FIG. 2.4 S=-1

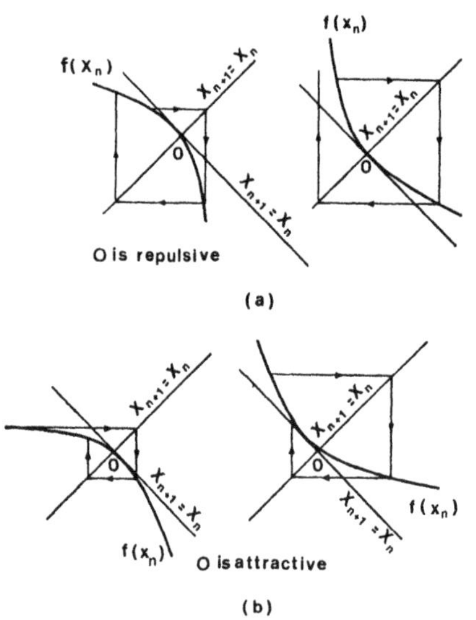

m is even

FIG. 2.5 S=-1

Then T^2 becomes

$$(2.13)\quad \begin{cases} x_{n+2} = x_n - a_o[1-(-1)^m]x_n^m - a_1[1-(-1)^{m+1}]x_n^{m+1} + \bar{F}(x_n) & \text{if } m > 2 \\ x_{n+2} = x_n - (a_1 + 2a_o^2)x_n^3 + \bar{F}'(x_n) & \text{if } m = 2 \end{cases}$$

with $\lim \bar{F}(x)/|x^{m+1}| = 0$, $\lim \bar{F}'(x)/|x^3| = 0$ when $|x| \to 0$. Similarly to the case $S = +1$, and considering that for T^2 the lowest non linear degree is always odd, the conclusion of stability is obtained from the fig.2.4 (m is odd) and 2.5 (m is even) |N 2| |M 9|

(P2.2) *If in (2.12) m is odd, the fixed point* $x^* = 0$ *is attractive for* $a_o > 0$ *, repulsive for* $a_o < 0$*. If* $m = 2$ *,* $x^* = 0$ *is attractive for* $(a_1 + 2a_o^2) > 0$*, and repulsive for* $(a_1 + 2a_o^2) < 0$*. If m is even,* $m > 2$*,* $x^* = 0$ *is attractive for* $a_1 > 0$*, and repulsive for* $a_1 < 0$*. Here* x^* *is always of type 2.*

2.3.2. THE FUNCTION DEFINING THE RECURRENCE IS NOT A SMOOTH ONE.

In the absence of smoothness for $f(x)$ it is impossible to define the multiplier S of a fixed point $x^* = \alpha$. The following theorems (for the proof cf. [U 2] [U 3] [D 4]) give the conclusion of stability.

(P2.3) *If the function* $f(x)$ *which defines the map T is a continuous one, and if a neighbourhood of the fixed point* α *exists such that inside this one* $|f(x)-\alpha| \leqslant a|x-\alpha|$*,* $0 < a < 1$*, then* α *is attractive. If inside this neighbourhood* $|f(x) - \alpha| > a|x - \alpha|$*, then* α *is repulsive.*

For example, with $f(x) = kx \sin(1/x)$ if $x \neq 0$, $f(0) = 0$, $|k| < 1$ one has $|f(x)| \leqslant |k|.|x|$, and so the fixed point $\alpha = 0$ is attractive.

(P2.4) *If* $f(x)$ *is a continuous function, and a neighbourhood of the fixed point exists, such that* $|f(x) - \alpha| < |x - \alpha|$ *for* $x \neq \alpha$*, then* α *is attractive.*

So the function $f(x) = 1 + \sin(x-1)\sin[1/(x-1)]$ if $x \neq 1$, $f(1) = 1$, does not verify the first inequality of (P2.3) with $\alpha = 1$. But this function is such that $|f(x)-1| < |x-1|$ for $x \neq \alpha = 1$, and the fixed point $\alpha = 1$ is attractive.

Let now $\Lambda^+, \lambda^+, \Lambda^-, \lambda^-$, the Dini's numbers defined by

$$\Lambda^+ = \limsup_{x\to\alpha^+} [f(x) - f(\alpha)]/(x-\alpha), \quad \lambda^+ = \liminf_{x\to\alpha^+} [f(x) - f(\alpha)]/(x-\alpha)$$

$$\Lambda^- = \limsup_{x\to\alpha^-} [f(x) - f(\alpha)]/(x-\alpha), \quad \lambda^- = \liminf_{x\to\alpha^-} [f(x) - f(\alpha)]/(x-\alpha)$$

the following theorem gives a conclusion in certain cases for which (P2.3), (P2.4) give no result.

(P2.5) *If $f(x)$ is continuous, and if at the fixed point* $x = \alpha$, $\Lambda^+ < 1$, $\Lambda^- < 1$, $\lambda^+ . \lambda^- < 1$, *then this fixed point is attractive.*

With $f(x) = -2x$ if $x \leqslant 0$, $f(x) = -x/5$ if $x > 0$, (P2.3) and (P2.4) do not give a conclusion of stability for the fixed point $\alpha = 0$. However $f(x)$ is such that $\Lambda^+ = \lambda^+ = -1/5$, $\Lambda^- = \lambda^- = -2$, then α is attractive.

(P2.6) *Let $f(x)$ be the continuous function which defines the map T, having a fixed point* $x^* = \alpha$. *If a neighbourhood U of the fixed point* $x^* = \alpha$ *is such that inside U, $T^2x \neq x$ for $x \neq \alpha$, and :*

$$k \leqslant \frac{f(x)-\alpha}{x-\alpha} < 1 \ \textit{for} \ x < \alpha, \ x \in U, \quad \frac{1}{k} \leqslant \frac{f(x)-\alpha}{x-\alpha} < 1 \ \textit{for} \ x > \alpha, x \in U,$$

k being a negative constant, then α is an attractive point.

As an example, a "complex" function is considered in the interval $-1 \leqslant x \leqslant 1$, where the following sequences are defined :

$$a_n = -7^{-n} \ ; \ b_n = -(1/2)7^{-n} \ ; \quad n = 0,1,2, \ldots$$

$$c_n = 1 \ \text{for} \ n = 0 \ , \ c_n = 2.7^{-n} \ \text{for} \ n > 0 \ ; \ d_n = (2/3)7^{-n}; \quad n = 0,1,2, \ldots$$

Let now $f(x)$ be the following function [U 3] [U4]:

$$\begin{cases} 3x + (5/2)7^{-i} \ \text{for} \ x \in [a_i, b_i[\ ; \ -3x - (1/2)7^{-i} \ \text{for} \ x \in [b_i, a_{i+1}[\\ -x + 7^{-i} \ \text{for} \ x \in]d_i, c_i] \ ; \ (5/4)x - (1/2)7^{-i} \ \text{for} \ x \in]c_{i+1}, d_i] \end{cases}$$

With $f(0) = 0$, $i = 0,1,2, \ldots$ The function $f(x)$ is continuous and the conditions of (P2.6) are satisfied. However it is not the case for the conditions of (P2.3) - (P2.5). The fixed point $x^* = \alpha$ is attractive.

The last theorems give a necessary and sufficient condition of stability for a fixed point.

(P2.7) *The fixed point* α *of a map T, defined by a continuous function f(x), is attractive if and only if the following condition is satisfied in a neighbourhood of* α *:*

- *the function* $f^2(x) = f[f(x)]$ *is such that* $|f^2(x)-\alpha|/|x-\alpha| < 1$

(P2.8) *A necessary and sufficient condition for the attractivity of the fixed point* α *of the map defined by the continuous function f(x), which satisfies* (P2.4) *in a neighbourhood of* α *, is the existence of a continuous function g(x), called a decreasing "involution", solution of the functional equation* $g_2(x) = x$ *, and such that* $|g(x)-\alpha|/|x-\alpha| < |f(x)-\alpha|/|x-\alpha|$ *for* $x \neq \alpha$.

The central idea of (P2.8) is founded on the involutions properties , or product of involutions, having certain non directly visible symmetries. So, for the map defined by $f(x) = (-x)^{\frac{1}{2}}$ if $x \leqslant 0$, $f(x) = -x^3$ if $x > 0$, the theorems (P2.3) - (P2.6) give no result about the stability of the fixed point $\alpha = 0$. But considering the involution $g(x) = x^{\frac{2}{3}}$ if $x \leqslant 0$, $g(x) = -(x)^{\frac{5}{2}}$ if $x > 0$, which satisfies the condition of (P2.8), it appears that this fixed point is attractive.

It is worth of noting that (P2.6) is a particular case of (P2.8) with the linear involution :

$$g(x) = (x-\alpha)/k+\alpha \quad \text{if } x \geqslant \alpha, \quad g(x) = k(x-\alpha)+\alpha \quad \text{if } x < \alpha$$

2.4. AUTONOMOUS NON LINEAR CASE. LOCAL BIFURCATIONS

The considered bifurcations are local, i.e the change of the nature, or of the number of singularities occurs in a sufficiently small neighbourhood of one of these singularities, when the parameter λ varies continuously in the recurrence (or map) T :

$$(2.5) \qquad x_{n+1} = f(x_n,\lambda) \qquad \text{or} \qquad x_{n+1} = Tx_n$$

where the function $f(x,\lambda)$ is continuous with respect to λ . Here, the problem is solved for a reference singularity which is a fixed point. One has the same conclusions for a cycle of order k by considering T^k.

2.4.1. RECURRENCE DEFINED BY A SMOOTH FUNCTION.

$f(x,\lambda)$ is supposed to be a continuous and sufficiently differentiable function with respect to the real variable x in order to permit the different writings of (2.5) in this paragraph. Here the local bifurcations consist in passing through the critical cases $S = \pm 1$ ($\lambda = \lambda_o$), when λ varies continuously in the interval $(\lambda_o-\delta,\lambda_o+\delta)$, $\delta > 0$.

(a) *Bifurcations by passing through the multiplier $S = +1$.*

Let the recurrence T be written :

$$(2.14) \qquad x_{n+1} = A_o(\lambda) + [1 + \varepsilon(\lambda)]x_n + F_1(x_n) + F_2(x_n,\lambda) = f(x_n,\lambda)$$

with $A_o(\lambda_o) = \varepsilon(\lambda_o) = 0$, $F_1(x)$ being a function of x which can be approximated, for sufficiently small $|x|$, by a power of x at least equal to 2, $F_2(x,\lambda_o)$ has the same property but $F_2(x,\lambda_o) = 0$. So :

$$F_1(x) = \alpha x^m + x^m \bar{F}_1(x); \quad F_2(x,\lambda) = \sum_{j=2}^{m-1} b_j(\lambda)x^j + x^m[b_m(\lambda) + \bar{F}_2(x,\lambda)],$$

where $\alpha \neq 0$, $b_j(\lambda) = 0$ if $m = 2$, $\bar{F}_2(x,\lambda_o) = 0$, $b_j(\lambda_o) = 0$ for $2 \leqslant j \leqslant m$, $\bar{F}_1(x)$, $\bar{F}_2(x,\lambda)$ can be approximated, for sufficiently small $|x|$, by a power of x at least equal to 1. Then, inside a sufficiently small neighbourhood of $x = 0$, the fixed points of T are given by the solutions of :

$$(2.15) \qquad \alpha x^m + \sum_{j=2}^{m} b_j(\lambda)x^j + \varepsilon(\lambda)x + A_o(\lambda) = 0$$

The multiplier S of such a fixed point x* is

$$S = (dx_{n+1}/dx_n)_{x=x^*} = 1 + \varepsilon(\lambda) + (dF_1/dx)_{x=x^*} + (dF_2/dx)_{x=x^*}$$

In a neighbourhood of the value $S = +1$, one has $S < 1$ (resp. $S > 1$) if :

(2.16) $\quad (dF_1/dx)_{x=x^*} + (dF_2/dx)_{x=x^*} < -\varepsilon(\lambda) \quad (\text{resp} > -\varepsilon(\lambda))$

The left part of this inequality is the slope of the curve (C), the equation of which is $x_{n+1} = F_1(x_n) + F_2(x_n, \lambda)$. The right part is the slope of the straight line (D) $x_{n+1} = -\varepsilon(\lambda)x_n - A_o(\lambda)$. The intersections of (D) and (C) are the fixed points of T. From (2.15) it appears that attractive and repulsive fixed points alternate. So the following theorem gives the bifurcation [M11] [G35].

(P2.9) *If the power m of (2.15) is even, for* $|\lambda-\lambda_o|$ *sufficiently small, and inside a sufficiently small neighbourhood of* $x = 0$ *, the map (2.14) has for* $\lambda > \lambda_o$ *an even number* p_1 *of fixed points, and for* $\lambda < \lambda_o$ *an even number* p_2 *of fixed points. The numbers* p_1, p_2 *can be such that :* $p_1 \neq p_2$ *(with* $p_1 \neq 0$, $p_2 \neq 0$ *, or* $p_1 \neq 0$, $p_2 = 0$*), or* $p_1 = p_2 \neq 0$. *One of the two extreme fixed points is attractive, the other is repulsive. Attractive and repulsive fixed points alternate. When* $\lambda = \lambda_o$ *, all the fixed points merge into* $x = 0$ *which corresponds to the critical case (P2.1) with m even.*

If the power m of (2.15) is odd, the map (2.14) has for $\lambda > \lambda_o$ *an odd number* q_1 *of fixed points; and for* $\lambda < \lambda_o$ *an odd number* q_2 *of fixed points,* $q_1 \neq q_2$ *or* $q_1 = q_2$. *The two extreme fixed points are attractive if* $\alpha < 0$, *and repulsive if* $\alpha > 0$. *Attractive and repulsive fixed points alternate. When* $\lambda = \lambda_o$ *, all the fixed points merge into* $x = 0$ *which corresponds to the critical case (P2.1) with m odd.*

In the particular case corresponding to $b_j(\lambda) \equiv 0$, $j = 2, \ldots, m-1$, whatever be λ, one has :

m = even, p_1 (resp p_2) = 2 , p_2 (resp p_1) = 0 , or $p_2 = p_1 = 2$

m = odd , q_1 (resp q_2) = 3 , q_2 (resp q_1) = 1 , or $q_2 = q_1 = 3$

Fig.2.6a,b,c illustrate these bifurcations. The first bifurcation (m = even)is often called *fold bifurcation* .

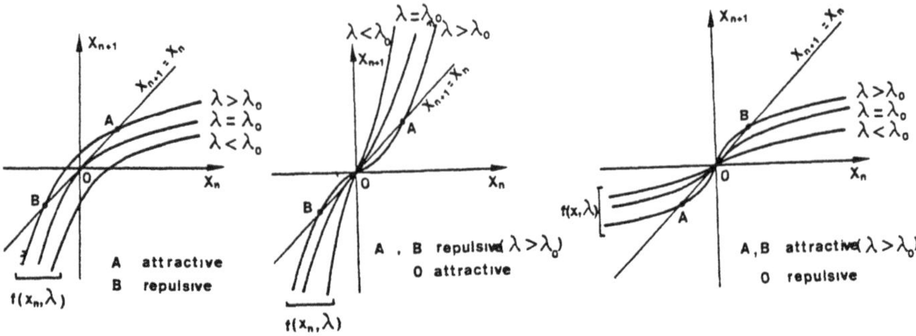

(a) m is even (b) m is odd (c) m is odd

FIG. 2.6

(b) *Bifurcation by passing through the multiplier $S = -1$.*

Let the recurrence T be written :

(2.17) $$x_{n+1} = A_0(\lambda) + [\varepsilon(\lambda) - 1]x_n + F_1(x_n) + F_2(x_n,\lambda)$$

where $A_0(\lambda)$, $\varepsilon(\lambda)$, $F_1(x)$, $F_2(x,\lambda)$ satisfy the hypotheses above mentioned for (2.14). The fixed points are solutions of

(2.18) $$[2 - \varepsilon(\lambda)]x - A_0(\lambda) = F_1(x) + F_2(x,\lambda)$$

or, considering a first approximation of (2.18) :

$$\alpha x^m - [2 - \varepsilon(\lambda) - \sum_{j=2}^{m-1} b_j(\lambda)x^{j-1}]x + A_0(\lambda) = 0 \quad , \quad m \geqslant 2$$

If $|\lambda - \lambda_0|$ is sufficiently small, $|x|$ and $|\varepsilon(\lambda)|$ are also sufficiently small and this equation becomes equivalent to

$$2x - A_0(\lambda) = 0 .$$

Then for $\lambda \neq \lambda_0$ ($\lambda > \lambda_0$ or $\lambda < \lambda_0$) only one fixed point P can exist in a neighbourhood of $x = 0$. *This fixed point merges into* $x = 0$ *for* $\lambda = \lambda_0$.

With this bifurcation no fixed point appears or disappears, when passing through $\lambda = \lambda_o$, but cycles of order two can appear or disappear. These cycles are obtained by considering the fixed points of T^2, written from (2.17) using the expression of x_{n+2}. P is a fixed of T^2. In this case, for T^2, the situation of S = 1 is obtained and the smallest degree of the non linear part independent of λ is here always odd. So using (P2.9) for T^2 one has [M11] [G35].

(P2.10) *Considering the map T (2.17) for $\lambda \neq \lambda_o$, if $\lambda > \lambda_o$ a fixed point P has a central position with respect to the points of ℓ_1 ($\ell_1 = 0$ or $\ell_1 \neq 0$) cycles of order 2. If $\lambda < \lambda_o$ this situation occurs for the points of ℓ_2 ($\ell_2 = 0$, or $\ell_2 \neq 0$; $\ell_1 = \ell_2$, or $\ell_1 \neq \ell_2$) cycles of order two. All these cycles merge into the fixed point P for $\lambda = \lambda_o$. There is alternance of the stability of the fixed points of T^2(points of the cycles of order 2 and of the fixed point.*

It is worth of noting that the points of a cycle of order two are symmetrical with respect to $x_{n+1} = x_n$, in the plane (x_n, x_{n+1}). They are obtained by the intersections (outside of $x_{n+1} = x_n$) of $f(x,\lambda)$ with the symmetric of $f(x,\lambda)$, with respect to $x_{n+1} = x_n$ (cf. fig.2.7).

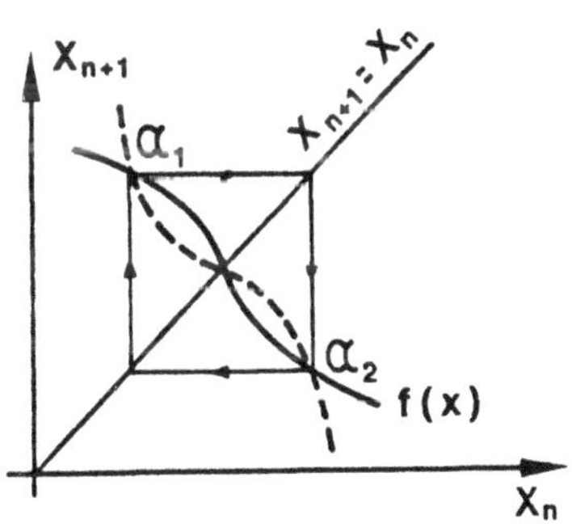

α_1 and α_2 are the two points of the cycle of order two .

FIG. 2.7

In the particular case giving for T^2 all the $b_j(\lambda) \equiv 0$, j = 2,..., m-1 (cf. fig.2.8 obtained for $A_o(\lambda) \equiv 0$), α being the coefficient of x^m, m = odd number smallest degree of the non linear part of T^2 independent of λ, the theorem (P2.10) becomes :

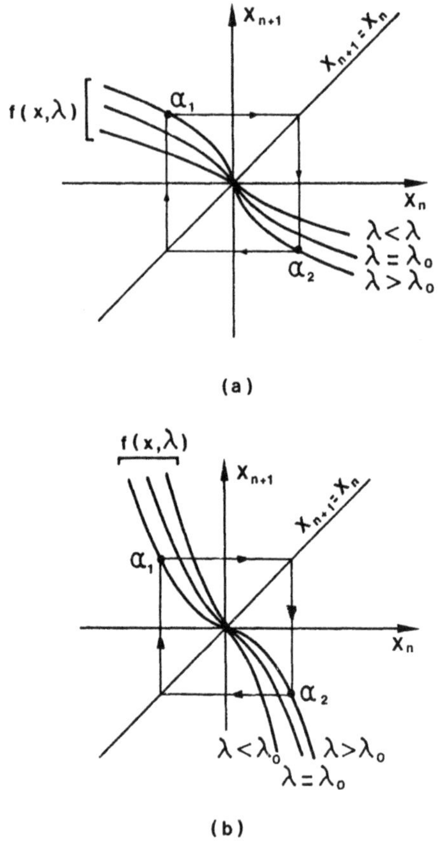

α_1 and α_2 are the two points of the cycle of order two appearing for $\lambda > \lambda_0$.

FIG. 2.8 Bifurcation through S=-1

(P2.11) *Let T be the map (2.17), such that T^2 does not contain non linear terms of degree $j < m$ which depend of λ ($b_j(\lambda) \equiv 0$ in (2.15)). If $\alpha < 0$, the following bifurcation occurs :*

(2.19) $$\text{at}(2)\text{fp} \leftrightarrows \text{rep}(2)\text{fp} + \text{at}(1)C_{k=2}$$

at(i)fp (resp.at(i)$C_{k=2}$) *represents an attractive fixed point (resp. attractive cycle of order* k = 2*) of type i, $i = 1,2$, "rep" is used for "repulsive". In (2.19) when $\lambda = \lambda_o$ the multiplier of the fixed point is $S = -1$, that of the cycle is $S = +1$, and the corresponding critical case gives an attractive fixed point (cf. P2.2). If $\alpha > 0$, the bifurcation becomes :*

(2.20) $$\text{rep}(2)\text{fp} \leftrightarrows \text{at}(2)\text{fp} + \text{rep}(1)C_{k=2}$$

For $\lambda = \lambda_o$, the corresponding critical case leads to a repulsive fixed point (cf. P2.2).

Bifurcations (2.19) and (2.20) are often called *flip bifurcations*, or *period doubling bifurcation*. The first one is illustrated by fig.2.8a, the second one by fig.2.8b, with $A_o(\lambda) \equiv 0$.

(c) *Some remarks*

- (i) If instead of fixed points and cycles of order 2, cycles of order k and cycles of order 2k are considered, the bifurcations (P2.9) - (P2.11) occur. This result is obtained by writing T^k, $x_{n+k} = f^k(x_n,\lambda)$, $f^k(x_n,\lambda)$ being the right part of (2.14) or (2.17).

In particular with $b_j(\lambda) \equiv 0$, $j = 2,\ldots,m-1$, one has :

(2.21) $\phi \leftrightarrows \text{at}(1)C_k + \text{rep}(1)C_k$, m of T^k = even number

(2.22) $\text{at}(1)C_k \leftrightarrows \text{rep}(1)C_k + 2\text{at}(1)C_k$, m of T^k = odd number

(2.23) $\text{rep}(1)C_k \leftrightarrows \text{at}(1)C_k + 2\text{rep}(1)C_k$, m of T^k = odd number

(2.24) $\text{at}(2)C_k \leftrightarrows \text{rep}(2)C_k + \text{at}(1)C_{2k}$, m of T^{2k} = odd number

(2.25) $\text{rep}(2)C_k \leftrightarrows \text{at}(2)C_k + \text{rep}(1)C_{2k}$, m of T^{2k} = odd number

Bifurcations (2.21) - (2.23) correspond to a crossing through $S = +1$, for $\lambda = \lambda_0$, bifurcations (2.24) (2.25) to a crossing through S = -1. ϕ in (2.21) indicates the absence of cycle C_k in the considered neighbourhood.

- (ii) There exist situations such that the crossing through S = +1 , with a continuous and differentiable function $f(x,\lambda)$, leads to the existence of a segment of the x -axis, the points of which are fixed points with S = +1 for $\lambda = \lambda_0$ (cf. for example the fig.2.9).

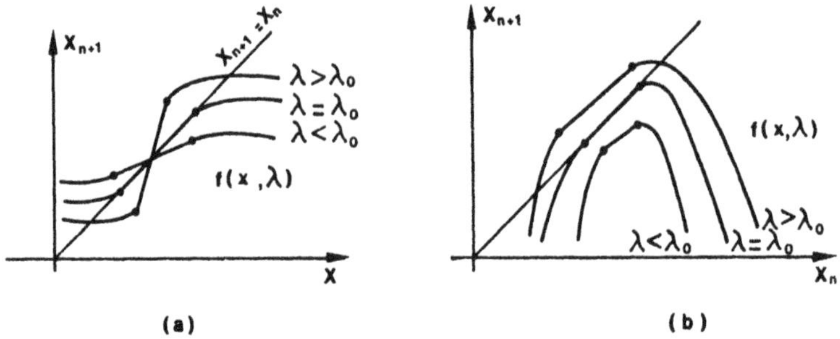

FIG. 2.9

An equivalent situation occurs for S = -1 when $f(x,\lambda)$ has a segment which is symmetrical with respect to $x_{n+1} = x_n$, for $\lambda = \lambda_0$. The points of this segment are points of cycles of order two, and it gives rise to a finite number of cycles of order 2 if $\lambda \neq \lambda_0$.

2.4.2. RECURRENCE DEFINED BY A CONTINUOUS BUT NOT SMOOTH FUNCTION.

It is supposed now that the function $f(x,\lambda)$ of (2.5) is not differentiable at the fixed point for $\lambda = \lambda_0$. So the multiplier of this one cannot be defined. Considering the analytical difficulties of this case, it is better to use a geometrical interpretation, in the auxiliary (x_n, x_{n+1}) plane, to study the corresponding bifurcations.

The bifurcation equivalent to (2.21) with k = 1 is given by fig.2.10a, and to (2.22) (k = 1) by fig.2.10b . In the two cases the stability of the fixed point obtained for $\lambda = \lambda_0$ can be determinated from one of the theorems (P2.3) - (P2.8). With respect to the case when $f(x,\lambda)$ is smooth, a difference concerns the stability : *here the alternance of attractive and repulsive fixed points does not always occur in all the possible situations.*

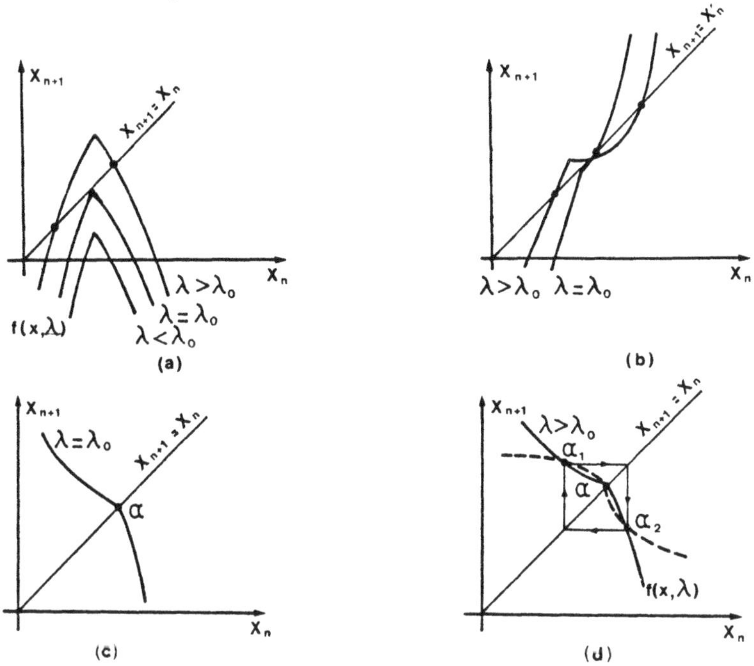

FIG. 2.10

It is the case, in particular, of fig.2.10a for which the passing through $\lambda = \lambda_o$ gives rise to two repulsive fixed points if the absolute values of the slopes on the right and on the left of the point where $f(x,\lambda)$ is not differentiable, are greater than one.

The bifurcation equivalent to (2.24) is illustrated by fig.2.10c,d For $\lambda = \lambda_o$, $[df/dx]_{x=\alpha-} = [df/dx]^{-1}_{x=\alpha+}$, i.e. the two different slopes of $f(x,\lambda)$ at the fixed point α are symmetrical with respect to $x_{n+1} = x_n$. If for $\lambda < \lambda_o$, $|df/dx|_{\alpha-} > |df/dx|^{-1}_{\alpha+}$, there is only one singularity, the fixed point α, in its neighbourhood. If for $\lambda > \lambda_o$, $|df/dx|_{\alpha-} < |df/dx|^{-1}_{\alpha+}$, the fixed point α gives birth to a cycle of order two. This is the qualitative behaviour of (2.24), (2.25), except in certain cases for which the alternance of stability of the singular points does not occur [G35].

For the two considered situations, $\lambda = \lambda_o$ corresponds to an equivalent to a critical case in the sense of Ljapunov.

2.5. BIFURCATIONS OF PIECEWISE CONTINUOUS RECURRENCES.

This paragraph gives a presentation of a part of the Leonov's results [L13] - [L16].

2.5.1. PIECEWISE LINEAR RECURRENCES. BOXES IN FILES BIFURCATIONS STRUCTURE.

Let T be the recurrence, or map

$$(2.26)\qquad x_{n+1} = Tx_n = \begin{cases} T_1 x_n = a + \lambda_1 x_n , & x < 0 \\ T_2 x_n = -b + \lambda_2 x_n , & x > 0 \end{cases}$$

$a > 0$, $b > 0$, $x = 0$ being the discontinuity point of T. The map T depends on three parameters : λ_1, λ_2 and $\Delta_1 = ab^{-1}$. The following cases lead to different situations :

(a) $0<\lambda_i\leqslant 1$, $i = 1,2$; (b) $0<\lambda_1<1$, $\lambda_2>1$; (c) $0<\lambda_1<1$, $\lambda_2<0$
(d) $\lambda_1>1$, $0<\lambda_2<1$; (e) $\lambda_1>1$, $\lambda_2>1$; (f) $\lambda_1>1$, $\lambda_2<0$
(g) $\lambda_1<1$, $0<\lambda_2<1$; (h) $\lambda_1<0$, $\lambda_2>1$; (i) $\lambda_1<0$, $\lambda_2<0$

Here only the results concerning the case (a) are presented. For the other cases, the references [L13] - [L16] can be consulted.

Then let λ_1, λ_2 be such that $0 < \lambda_1 \leqslant 1$, $0 < \lambda_2 \leqslant 1$. Each point of the interval $-b \leqslant x \leqslant a$ has two different first rank antecedents. For the intervals $-\infty < x < -b$, $a < x < \infty$, each point has only one first rank antecedent. With such a property, the two points $x = a$, $x = -b$ can be considered as a generalization of the notion of *critical points in the sense of Julia - Fatou* [J5] [F1] (and no in the sense used at present in many papers : a critical point is an extremum of $f(x,\lambda)$). Moreover an initial condition x_o, $x_o < -b$ or $x_o > a$, generates an iterated sequence which after a certain number of iterations penetrates inside the interval $-b \leqslant x \leqslant a$ and tends towards an attractor belonging to this interval. Leonov showed that this attractor is either a cycle, or for a Cantor set of the parameters space, a limit iterated sequence which is positively stable in the sense of Poisson.

A cycle can only be obtained with its points on $x < 0$ and $x > 0$, and it is attractive of type 1 because $0 < \lambda_1 < 1$, $0 < \lambda_2 < 1$. The simplest cycles are given by the fixed points of the maps $T_2^m T_1$ and $T_1^m T_2$. They are called *cyles of first degree of complexity*. Cycles given by the fixed points of maps such $T_2^m T_1^p$, $m > 1$ and $p > 1$ are not possible. Indeed such maps lead simultaneously to $T(-b) < 0$, $Ta > 0$, which implies $\lambda_1 \lambda_2 > 1$ contrary to the hypothesis (a).

The map $T_2^m T_1$ is considered if $T(-b) > 0$, i.e. $\Delta_1 = a/b > \lambda_1$. Let π_1 the map of the interval $(0,-b)$ onto itself, such that each point of this interval is mapped onto a point x, $-b \leqslant x \leqslant 0$. Two possible cases appear. In the first one π_1 is $T_2^m T_1$, which is a continuous map on $(0,-b)$, mapping this interval onto itself, and having a unique fixed point for each m. In the second case $T_2^m T_1$ maps $(0,-b)$ onto an interval containing $x = 0$, and another application of T_2 is necessary to return inside the interval $(0,-b)$. π_1 is a piecewise continuous map, constituted by $T_2^m T_1$ and $T_2^{m+1} T_1$. The map $T_2^m T_1$ is given by:

$$T_2^m T_1 x = \lambda_2^m a - b (1 - \lambda_2^m)/(1 - \lambda_2) + \lambda_1 \lambda_2^m x$$

which has the fixed point

$$x^* = [\lambda_2^m a - b (1 - \lambda_2^m)/(1 - \lambda_2)]/(1 - \lambda_1 \lambda_2^m).$$

By writing that this point has to satisfy $-b < x^* < 0$, the condition of existence of x^* is :

$$(2.27) \qquad \lambda_2^{1-m} (1 - \lambda_2^{m-1})/(1 - \lambda_2) + \lambda_1 < \Delta_1 < \lambda_2^{-m} (1 - \lambda_2^m)/(1 - \lambda_2)$$

The length of the corresponding closed intervals (or boxes) in Δ_1 is $(1 - \lambda_1 \lambda_2^{-m}) \lambda_2^{-m}$. They are disjointed, the distance between two consecutive intervals (or boxes) being λ_1, and they go to infinity when $m \to \infty$.

The maps $T_1^m T_2$ are possible if $Ta < 0$, i.e. $\Delta_1 < \lambda_1^{-1}$. Using the same approach as for $T_2^m T_1$ the interval in Δ_1, giving the existence of the corresponding fixed point, is given by :

$$(2.28) \qquad \lambda_1^{1-m}(1 - \lambda_1^{m-1})/(1 - \lambda_1) + \lambda_2 < \Delta_1^{-1} < \lambda_1^{-m} (1 - \lambda_1^m)/(1 - \lambda_1)$$

Such intervals (or boxes) are disjointed, their length tends to zero when $m \to \infty$, and $\Delta_1 = 0$ is a point of accumulation when $m \to \infty$. The two files defined from (2.27) (2.28) are called *files of the first degree of complexity.*

Inside the complementary intervals of the set of all the intervals defined by (2.27), a map π_1 constituted from $T_2^m T_1$ and $T_2^{m+1} T_1$ may be defined. Inside the complementary intervals of the set of all the intervals defined by (2.28), a map π_2 constituted from $T_1^{m+1} T_2$ and $T_1^m T_2$ may be defined. Cycles having the first degree of complexity with respect to the maps π_1 or π_2 are called *cycles of second degree of complexity.* Inside these complementary intervals, the sets of intervals, corresponding to the existence of fixed points for the maps $(T_2^{m+1}T_1)^p (T_2^m T_1)$, $(T_2^m T_1)^p (T_2^{m+1} T_1)$, $(T_1^m T_2)^p (T_1^{m+1} T_2)$, $(T_1^{m+1}T_2)^p(T_1^m T_2)$, are defined. Let us consider for example π_1, with a change of variable which leads the origin to the point where π_1 is not continuous. So :

$$(2.29) \qquad y_{n+1} = \pi_1 y_n = \begin{cases} a_1 + \lambda_{12}\, y_n, & y < 0 \\ -b_1 + \lambda_{22}\, y_n, & y > 0 \end{cases}$$

$$\begin{cases} a_1 = \lambda_1^{-1} \lambda_2^{-m}[\lambda_2^m a - b(1 - \lambda_2^m)/(1 - \lambda_2)] , \\ b_1 = b + \lambda_1^{-1}\lambda_2^{-m} [b(1 - \lambda_2^m)/(1 - \lambda_2) - \lambda_2^{-m} a] \\ \lambda_{12} = \lambda_1 \lambda_2^m , \quad \lambda_{22} = \lambda_1 \lambda_2^{m+1}, \ 0 < \lambda_{12} < 1, \ 0 < \lambda_{22} < 1 \end{cases}$$

With respect to the new parameter $\Delta_2 = a_1/b_1$, existence intervals of the fixed point of the maps $(T_2^{m+1} T_1)^p (T_2^m T_1)$, $(T_2^m T_1)^p(T_2^{m+1} T_1)$ are defined as for Δ_1. The relation between Δ_1 and Δ_2 is :

$$(2.30) \qquad \Delta_1 = \lambda_2^{-m} (1 - \lambda_2^m)/(1 - \lambda_2) + \lambda_1 \Delta_2 /(1 + \Delta_2)$$

On a Δ_2 half axis, there is an infinite set of disjointed intervals tending toward $\Delta_2 = 0$ on the left, and toward the infinite on the right. Then inside each of the complementary intervals of the Δ_1 half axis defined by (2.27) for π_1 (which implies $\Delta_1 > 1$), two infinite sets of disjointed existence intervals of cycles of the second degree of complexity are located. These infinite sets, constituting *two files of the second degree of complexity*, have two limit points which are the boundaries of the considered complementary interval. Similarly for the intervals associated with the consideration of π_2, Δ_1 and Δ_2 verify the relation :

$$(2.31) \qquad \Delta_1^{-1} = \lambda_1^{-m} (1 - \lambda_1^m)/(1 - \lambda_1) + \lambda_2/(1 + \Delta_2)$$

and the same kind of cycles is defined, with two corresponding files of the second degree of complexity.

Similarly cycles of the 3rd, 4th,..., N^{th},... degree of complexity are defined, with the corresponding files. The consideration of these cycles leads to the definition of parameters Δ_N, Δ_{N+1}, λ_{1N}, λ_{2N}, such that :

$$(2.32) \qquad \Delta_N = \lambda_{2N}^{-p_N} (1 - \lambda_{2N}^{p_N})/(1 - \lambda_{2N}) + \lambda_{1N} \Delta_{N+1} /(1 + \Delta_{N+1})$$

$$(2.33) \qquad \Delta_N^{-1} = \lambda_{1N}^{-p_N} (1 - \lambda_{1N}^{p_N})/(1 - \lambda_{1N}) + \lambda_{2N}/(1 + \Delta_{N+1}) \text{ , } N - 1,2,\ldots$$

The relation (2.32) corresponds to $\Delta_N > 1$, and (2.33) to $\Delta_N < 1$. Then the parameters space λ_1, λ_2, Δ_1 is constituted by infinitely many files of boxes having the degree N of complexity, N = 1,2,... Let E be the set of all these boxes, which are the domains of existence of an only one attractive cycle. The set E is enumerable. The complementary set $\bar{E}$ of E, inside $0 < \lambda_1 < 1$, $0 < \lambda_2 < 1$, $\Delta_1 > 0$, is a non empty, nowhere dense, perfect, infinite set. It has the structure of a discontinuous Cantor's set, and each of its points is a limit point for the boundaries of the boxes of E. When $(\lambda_1,\lambda_2,\Delta_1) \in \bar{E}$, Leonov [L13] shows that the limit iterated sequence is positively stable in the sense of Poisson. Indeed, after a certain number of iterations, the sequence solution of (2.26) remains located inside the interval (-b,a) which contains no cycle.

The Leonov's result can be formulated by:

(P2.12) *Let T be the recurrence defined by (2.26) with* $a > 0$, $b > 0$, $0 < \lambda_1 \leqslant 1$, $0 < \lambda_2 \leqslant 1$. *The bifurcations structure of T has the "boxes in files" type (which is a devil's staircase structure) characterized by the following properties :*

- a box of the parameters space $(\lambda_1,\lambda_2,\Delta_1)$, $\Delta_1 = a/b$, *is the domain of existence of an only one attractive cycle having a fixed order and a fixed degree of complexity. A repulsive cycle cannot exist.*

- a double sequence of boxes of the first degree of complexity corresponds to the existence of a fixed point of $T_1^m\ T_2$, *or* $T_2^m\ T_1$. *When* $m \to \infty$, *the limit of boxes related to* $T_1^m\ T_2$ *is* $\Delta_1 = 0$, *and for* $T_2^m\ T_1$ *is* $\Delta_1 = \infty$.

- between two consecutive boxes having the same degree N *of complexity, a double infinite sequence of boxes (double file), having the degree* N+1 *of complexity, exists. A cycle of degree* N+1 *has the first degree with respect to a map defining cycles of degree* N.

- the boxes are disjointed. The two boundaries of a box of the degree N *of complexity are limits for the two sequences (files) of boxes of the degree* N+1 *of complexity.*

- in the domain $0 < \lambda_1 \leqslant 1$, $0 < \lambda_2 \leqslant 1$, $\Delta_1 > 0$, *if the set E of all the boxes associated with the degrees* N = 1,2,3,.. *of complexity are taken off, the remaining set* $\bar{E}$ *is non empty, perfect, nowhere dense, and has the structure of a Cantor's discontinuous set. When* $(\lambda_1,\lambda_2,\Delta_1) \in \bar{E}$, *the iterated sequences (orbits) are positively stable in the sense of Poisson, inside the "absorbing" segment* $-b \leqslant x \leqslant a$.

The fig.2.11 gives the aspect of the boxes for the degrees N = 1,2 in the parameters space $(\lambda_1,\lambda_2,\Delta_1)$.

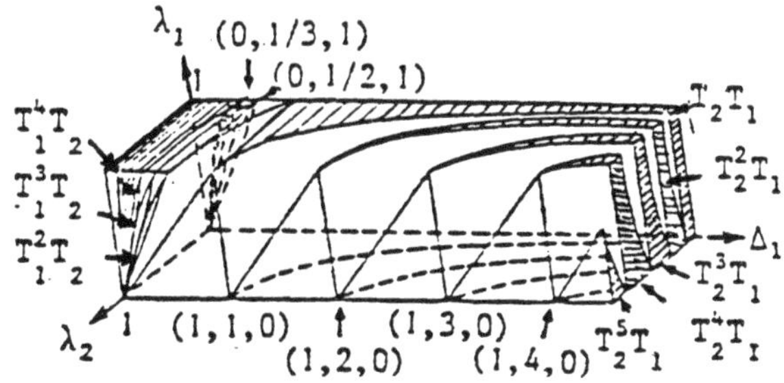

Boxes of the first degree of complexity

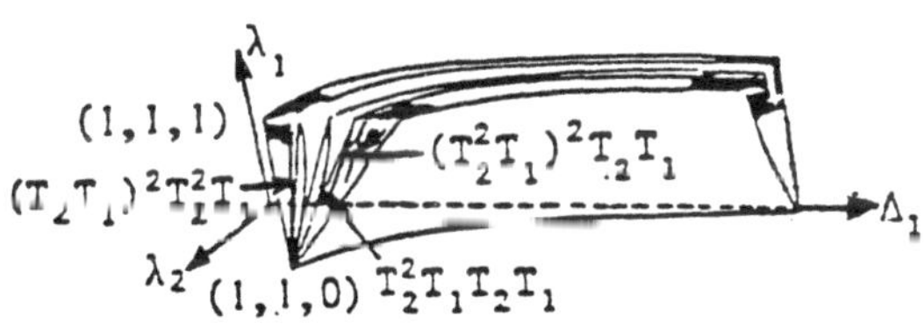

Boxes of the second degree of complexity

FIG. 2.11 - The boxes in files bifurcations structure.
(According to |L 13|)

2.5.2. PIECEWISE CONTINUOUS RECURRENCES.

Let T be the recurrence (or map)

$$(2.34) \qquad x_{n+1} = Tx_n = \begin{cases} T_1(x_n) = a(\lambda) + f_1(x_n,\lambda) & x < 0 \\ T_2(x_n) = b(\lambda) + f_2(x_n,\lambda) & x > 0 \end{cases}$$

depending on the parameter λ, with $f(\lambda,0) = f(\lambda, 0) = 0$, $a(\lambda) > 0$, $b(\lambda) < 0$. If T_1 and T_2 are defined by two continuous monotonically increasing functions, the first one with $x < 0$, the second one with $x > 0$, such as $0 < \partial f_1/\partial x < 1$, $0 < \partial f_2/\partial x < 1$, a result identical to (P2.12) is obtained.

With more general hypotheses about (2.34), the behaviour of the iterated sequences are given in [L16] or in [G35](cf.p.421-423).

2.6. PROPERTIES OF INVERTIBLE ONE-DIMENSIONAL RECURRENCES.

Let T be the recurrence (or map) (2.5). It is now supposed the existence and the uniqueness of the inverse recurrence T^{-1} given by $x_n = f^{-1}(x_{n+1},\lambda)$, $f^{-1}(x,\lambda)$ being a single-valued function.

The following theorems are proved in [L13].

(P2.13) *Let* T *be the map* $x_{n+1} = f(x_n,\lambda)$, $R \to R$, $x \in (a,b)$, $f(x,\lambda)$ *being a continuous, increasing function with respect to* x *in the interval* (a,b) *Then the singularities of* T *are only fixed points of type one, the stability of which alternates, if* T *corresponds to a non-critical case in the sense of Liapunov. If a fixed point corresponds to such a critical case, this fixed point must be considered as the merging of several ordinary fixed points having alternating stability with all other fixed points.*

(P2.14) *Let* T *be the map* $x_{n+1} = f(x_n,\lambda)$, $R \to R$, $x \in (a,b)$, $f(x,\lambda)$ *being a continuous, decreasing function with respect to* x *, in the interval* (a,b). *Then the singularities of* T *are only : one fixed point of type two, and cycles of order two of type one. The stability of all these points alternates.*

For (P2.13), (P2.14) if $f(x,\lambda)$ is a differentiable function with respect to x, T is a diffeomorphism.

2.7. PROPERTIES OF NON-INVERTIBLE ONE-DIMENSIONAL RECURRENCES.

2.7.1. THE CRITICAL POINTS.

Let T be the recurrence (or map) $x_{n+1} = f(x_n,\lambda)$, $f(x,\lambda)$, being a continuous, single-valued function of the real variable x, and the real parameter λ. The single-valuedness of f for n = 0,1,2,... does not imply anything about the existence and uniqueness of x_n for n = -1,-2,..., because the inverse map T^{-1} :

$$x_n = f^{-1}(x_{n+1},\lambda) \quad , \quad n = -1,-2,... \tag{2.35}$$

where $f[f^{-1}(x)] = f^{-1}[f(x)] = x$, may not exist, or it may be multi-valued. For example, with $f(x,\lambda) = x^2 - \lambda$, T^{-1} is given by $f^{-1}(x_{n+1},\lambda) = \pm(x_{n+1}+\lambda)^{\frac{1}{2}}$ and there are no real x for $x_{n+1} < -\lambda$, while for $x_{n+1} > -\lambda$ the x_n are double-valued. The boundary between these two situations is given by the point C ($x_c = -\lambda$), which is such that it has two merging first rank antecedents x = 0, or the equation $f(x,\lambda) = x_c$ has a double root. The point C is call the *critical point* of T^{-1}(defined by $f(x,\lambda) = x^2 - \lambda$) in the sense of Julia-Fatou [J 5] [F 1]. It is the first rank consequent of the extremum $x = x_e$ of $f(x,\lambda)$.

More generally, $f(x,\lambda)$ being a continuous function of x, the *critical points* of T^{-1}, or *first rank critical points*, are given by the first rank consequent of every extremum of $f(x,\lambda)$. If $f(x,\lambda)$ is a continuous and continuously differentiable function with respect to x, the critical points belong to the first rank consequent of the real roots of $(\partial f/\partial x) = 0$. If T is an invertible map, T^{-1} has no real critical point. Critical points plays an obvious role in the branching of inverse iterates, and a less obvious, but an equally important one, in the maximum number of attractive fixed points and cycles, and also in the appearance of certain bifurcations leading to a qualitative change of the iterated sequence occuring inside an interval of the x_n-axis which is not small (*global bifurcation*).

Remark. Now many publications call "critical point" an extremum of $f(x, \lambda)$. In this book this expression will be always used as above indicated, i.e. in the original sense, that of Julia-Fatou.

2.7.2. FATOU'S THEOREMS ABOUT CRITICAL POINTS.

These theorems are given in [F 1] for a map T defined by a *complex rational function* f(x) *of the complex variable* x. The under-mentioned formulations are an adaptation of the original ones, in order to introduce simultaneously the real case.

(P2.15) *(a) If* f(x) *is a rational function of the degree* d, *then the number* N *of critical points at finite, or infinite, distance is* $N = 2(d-1)$, *when* x *is a complex variable, and* $N \leqslant 2(d-1)$, *when* x *is a real variable.*

(b) If f(x) *is a polynomial of the degree* d, *the point at infinity is critical with the multiplicity* (d-1), *and the number* N' *of critical points at finite distance is then* $N' \leqslant d-1$, *when* x *is a real variable.*

Let f(x) be a continuous, single-value function. Let d be the maximum number of intersections of the curve defined by f(x) and a straight line. The rule $N' \leqslant d-1$ is evident by considering a critical point as the first rank consequent of an extremum of f(x).

(P2.16) *(a) If* f(x) *is a rational function of the complex variable the number of attractive cycles is at most the number of critical points.*

(b) If f(x) *is a polynomial of the degree* d *of the real variable* x, *the number of attractors (fixed points, cycles, or more complex attractors) at finite distance is at most* d-1.

The part (a) is a Fatou's theorem (cf. [F 1] p.63). The part (b) is a conjecture, derived from (a), which is its equivalent in terms of real variable, for the case of more complex attractors, which may be a Cantor, or a chaotic segment as indicated in § 4.4.2 and 4.4.3.

Considering now the map T^r, $r > 1$ being an integer value, and T being defined by the continuous single-valued function $f(x,\lambda)$ of the real variable x and the real parameter λ, the following property is founded on Fatou's results |G35|.

<u>(P2.17)</u> *Let* T *be the continuous map defined by* $f(x,\lambda)$, x *and* λ *being real. When the parameter value* λ *is such that the set of the real critical points of* T^r*, on the finite part of the x -axis, is the "highest" (i.e. complex critical points do not exist), these ones are the critical points of* T^{-1}*, and all their consequents up to the rank* $r-1$*. For other values the set of the real critical points of* T^{-r} *is made up of either the critical points of* T^{-1} *("smallest" set), or these last points and the consequents of some of them up to a rank* $i \leqslant r-1$.

(P2.17) is related to the number of extrema of the curve T^r in the plane (x_n,x_{n+1}), having different ordinates, number which depends on the λ value. This property is illustrated by $f(x,\lambda) \equiv x^2-\lambda$, x being real. For $\lambda < 0$, the curve which represents T^2 in the (x_n,x_{n+1}) plane has only one extremum, the minimum $x = 0$, and so only one real critical point C at finite distance, $x_c = -\lambda$. When $\lambda > 0$, the curve T^2 has two minima $x = \pm\sqrt{\lambda}$ with the same ordinate $-\lambda$, which is the ordinate of the minimum of T, and a maximum $x = 0$ corresponding now to a second real critical point C of T^{-2}, $x_{c_1} = \lambda^2 - \lambda$, first the rank consequent of C. The ordinate of this maximum is x_{c_1}. The value $\lambda = 0$, for which C merges into the fixed point $x = 0$ having the multiplier $S = 0$, separates two different shapes of the curve T^2. So for $\lambda < 0$ the first rank consequent C_1 of the critical point C of T^{-1}, is not a critical point of T^{-2}, x being a real variable.

From now the first rank consequent of an extremum will indiferently be called a critical point of T or T^{-1} and a m rank consequent, a *critical point of the rank m of* T.

2.7.3. <u>CONSTRUCTION OF T^r FROM THE MAP T.</u>

This construction is founded on [G35].

<u>(P2.18)</u> *The abscissae of the extrema of* T^r *are the abscissae of the extrema of* T*, and the abscissae of their antecedents up to and including the rank* (r-1). *The ordinates of these extrema in the* (x_n,x_{n+1}) *plane are the abscissae of the critical points of* T^{-r}.

So with $f(x,\lambda) \equiv x^2-\lambda$, and $\lambda < 0$ the antecedents of the minimum A of T are always complex. The curves T^r have only one minimum A, the more "flattened" as r is large, giving only one real critical point of T^{-r} (or T^r). From $\lambda = 0$, $\lambda = \varepsilon > 0$ having a sufficiently small value, it appears, on the both sides of A, 2(r-1) new extrema of T^r, associated with r critical points of T^{-r}, C, C_1, ... , C_{r-1}. When $r \to \infty$ the ordinates of these extrema tend toward the ordinate of the attractive fixed point $x = 1/2 - (1/4 + \lambda)^{\frac{1}{2}}$, with a negative multiplier S = 2x. When λ increases from $\lambda = 0$, other new extrema appear. Indeed the degree of T^r being 2^r, the maximum number of extrema for T^r is $m = 2^r - 1$, and new extrema are created up to m is obtained. These new extrema appear when certain antecedents of A, having a rank $i \leqslant r-1$, become real, which occurs in particular when critical points of T^{-r} merge into an attractive cycle of order r, with the multiplier S = 0. So for $\lambda < 0$, T^3 has only one extremum. For $0 < \lambda < 1$, T^3 has five extrema which are symmetrical with respect to $x_n = 0$. When $\lambda = 1$, two new rank-two antecedents of A become real, and the critical points C, C_1 are the points of a cycle of order two with S = 0. For $\lambda > 1$, T^3 has $m = 2^3-1 = 7$ extrema with ordinates equal to the abscissae of C, C_1, C_2. Considering now T^4, the corresponding curve in the (x_n, x_{n+1}) plane has only one extremum if $\lambda < 0$. For $\lambda = \varepsilon > 0$ having a sufficiently small value, 6 new extrema of T^4 appear. They are associated with the critical points C, C_1, C_2, C_3, and an extremum has an ordinate equal to the abscissa of one of these points. From $\lambda = 1$, 6 new extrema appear with an ordinate equal to the abscissa of one of the points C, C_1, C_2, C_3. From $\lambda = \lambda^0 \simeq -1.754877666$, for which C, C_1, C_2 constitute a cycle of order 3 with S = 0, two new extrema appear. Then T^4 has the maximum number of possible extrema, i.e. $2^4-1 = 15$.

For $f(x,\lambda) \equiv x^2-\lambda$, the following rule related to T^r is obtained from the above consideration. Passing through a λ-value, such as the critical points C, C_1,...,C_{r-2} constitute a cycle of order r-1 with S = 0, gives to 2 new extrema of T^r. Passing through a λ-value such as C, C_1,..., C_{r-3} constitute a cycle of order r-2, S = 0, gives rise to $2 + 2^2 = 6$ new extrema, if r > 3. Passing through a λ-value, such as C, C_1,..., C_{r-4} constitute a cycle S = 0 of order r-3, gives rise to $2 + 2^2 + 2^3 - 2 = 12$ new extrema if r > 4.

Example of construction of T^r *from* T.

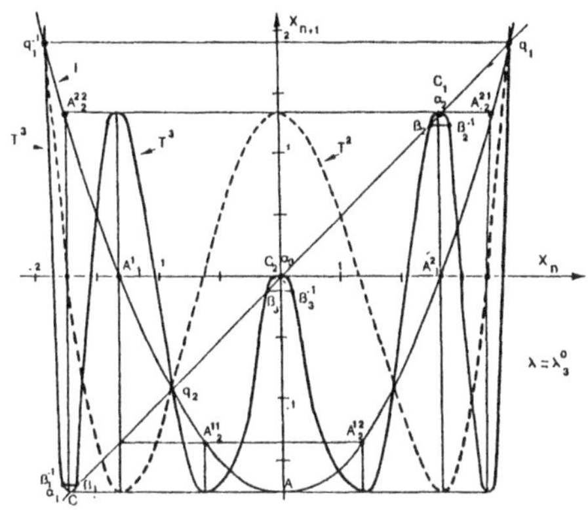

FIG. 2.12 - Construction of T^r from T, r=2, r=3 .

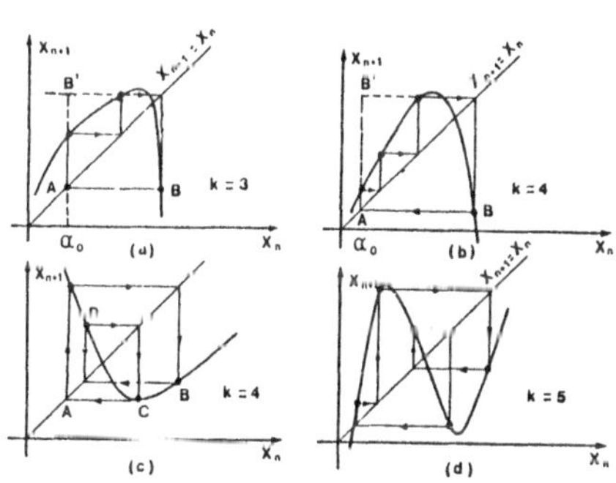

FIG. 2.13 - Cycles of order k represented in the (x_n, x_{n+1}) plane

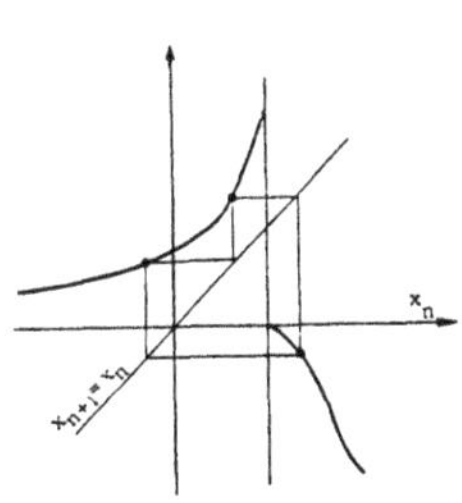

FIG. 2.14 - Cycle of order k> 2 obtained with a map having a unique inverse .

This construction is obtained from (P2.17) (P2.18). Consider $f(x,\lambda) = x^2-\lambda$, $\lambda = \lambda_3^0 \simeq 1.754877666$ value of the parameter for which C, C_1, C_2 are the points of the cycle of order 3 with the multiplier $S = 0$. Fig.2.12 shows the construction of T^2 and T^3 from T. The minimum A of T has two first rank antecedents A^i_{-1}, $i = 1,2$ and four rank two antecedents A^{ij}_{-2}, $i,j = 1,2$ the abscissae of which are equal to the abscissae of the extrema. The ordinate of each of the extrema is the abscissa of one of the points C, C_1, C_2. Moreover, the curves T^r pass through the fixed points q_1, q_2 of T^r on $x_{n+1} = x_n$, and their first rank antecedent q_1^{-1}, q_2^{-1} located on $x_{n+1} = f(x,\lambda)$.

2.7.4. CYCLES OF ONE-DIMENSIONAL ENDOMORPHISMS.

For one-dimensional endomorphisms T the properties (P2.13) (P2.14) do not occur in general. In particular, *the non-invertibility of* T *is a necessary condition of existence of cycles with an order* $k > 2$ *when* T *is defined by a continuous function.* In the (x_n,x_{n+1}) plane, such a cycle is represented by a closed curve made up of straight lines segments parallel to x_n and x_{n+1}-axes, which is obtained by the Koënigs-Lemeray graphical construction (fig.2.13). *Some maps* T *defined by a piecewise continuous function* $f(x)$ *are such that* T *has a unique inverse* T^{-1}, *and nevertheless can have cycles of order* $k > 2$. This point is illustrated by fig. 2.14 where : $\lim f(x) = 0^+$ if $x \to -\infty$.

The existence discussion of cycles of different orders, and also the construction of the curves T^r is made easier from the following theorem due to Koenigs [K 6].

(P2.19) *(a) If an integer* p *is a divisor of an integer* k, *the equation* $x_{n+k} = x_n$ *(or* $x = T^k x$*), has all the roots of the equation* $x_{n+p} = x_n$ *(or* $x = T^p x$*).*

(b) If c *is the remainder of the division of the integer* a *by the integer* b, *every common root to the equations* $x_{n+a} = x_n$, $x_{n+b} = x_n$, *is also a root of* $x_{n+c} = x_n$.

(c) If d *is the highest common factor of the two integers* a,b, *the common roots to* $x_{n+a} = x_n, x_{n+b} = x_n$, *are roots of* $x_{n+d} = x_n$.

From (P2.13) (P2.14) and the impossibility to obtain a cycle of order $k > 2$, if the continuous curve $f(x)$ defining T has not at least an extremum, it results that in the (x_n, x_{n+1}) plane the curve $f(x)$ has at least one intersection point with $x_{n+1} = x_n$. This point is located between the two points of the cycle having the least and the largest abscissae. Several important results obtained by Sharkovski [S 1] concerns the existence of cycles in one-dimensional endomorphisms. They are given by the proposition.

(P2.20) *Let* T *be a recurrence (or map) defined by a single-valued continuous function* $f(x)$ *of the real variable* x.

(a) If T *has a cycle of order* $k > 2$, *then* T *has at least one cycle of order* 2, *and at least one fixed point.*

(b) If T *has a cycle of order* 2^m, $m > 1$, *then* T *has the cycles of all orders* 2^i, $i < m$.

(c) If T *has a cycle of odd order* $N > 1$, *then* T *has the cycles of all orders* 2^i, $i = 1,2,3,\ldots$, *cycles of all even orders* , *and cycles of all odd orders* $M > N$.

2.7.5. ABSORBING SEGMENT.

When an endomorphism has one or several attractors (fixed points, cycles, chaotic segments, or Cantor set occuring as the limit of a sequence of order doubling (flip) bifurcations described in(2.24)), the critical points of T up to a certain rank bound one or several segments E_i of the x_n-axis such that every initial condition x_0 outside E_i, but in a sufficiently small neighbourhood of the E_i, gives rise to an iterated sequence x_n which enters into the E_i for $n > N$, the E_i becoming no-escape intervals. Such an E_i,which is a non invariant "attracting set" except in the case of chaotic segments which are invariant,is called an *absorbing segment* [G35]. It contains one, or several, attractors.

Different situations corresponding to absorbing segments are given by fig.2.15. For the fig.2.15a, $\overline{CC_1}$ is an absorbing segment limited by the critical points of rank one and two, and containing only one attractor if the Schwarzian derivative of the function $f(x)$ defining T is negative for every point of $\overline{CC_1}$. The Koenigs-Lemeray graphical construction shows that for every point x_n, belonging to the open segment $]q_1, q_1^{-1}[$, T gene-

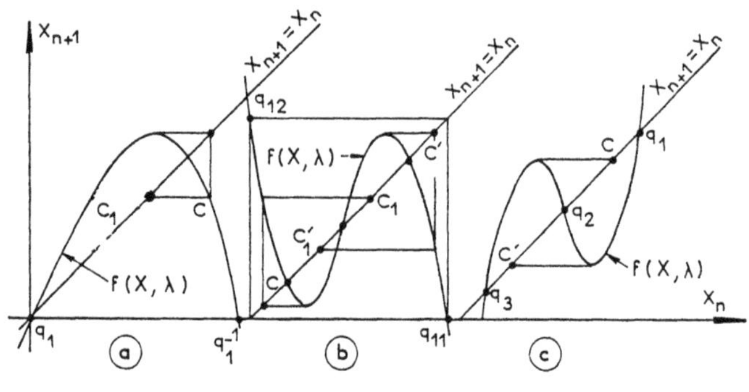

FIG. 2.15

rates an iterated sequence x_n which for $n > N$ enters into $\overline{CC}_I$ and remains inside this segment. With the fig.2.15b,c two attractors may exist inside the absorbing segment $\overline{CC'}$, limitated by two critical points of the rank one. It will be shown in chapter 3 that cyclic absorbing segments exist with a given order (periodicity).

2.8. DOMAIN OF ATTRACTION (BASIN) OF AN ATTRACTIVE FIXED POINT. SOME GLOBAL BIFURCATIONS.

Let x^* be an attractive fixed point of T, $R \to R$, defined by $f(x,\lambda)$ which is a continuous function with respect to x and λ. The domain of attraction (or domain of influence, or basin) of x^* is the open set of initial points x_n, the iterated sequences of which converge towards x^* when $n \to \infty$. Such a basin may be either all in one block, or made up of several disjointed parts. It will be shown that the merging of a critical point and a repulsive fixed point gives rise to a *global bifurcation*, i.e. the behaviour of the iterated sequence (orbit) undergoes a qualitative change inside an interval of the x_n-axis which is not small.

2.8.1. ALL IN ONE BLOCK BASIN WITH A BOUNDARY MADE UP OF TWO POINTS.

The boundary of such a basin is made up of two points corresponding to three different situations, when $f(x,\lambda)$ is a continuous function of x.

a) The boundary is made up of two repulsive fixed points of type one (multiplier $S > 1$*), located on both sides with respect to the considered attractive fixed point. The two repulsive fixed points are situated in a part of the x -axis where a point has a unique first rank antecedent. The open interval bounded by the two repulsive fixed points is free from other singularity.* In particular it is the case of the fixed points of (P2.13) at finite or infinite distance, which separate the basins of the attractive fixed points. This occurs when T is invertible and also for certain endomorphisms.

b) The boundary is made up of repulsive fixed point q_1 *of type one and the first rank antecedent* q_1^{-1} *of* q_1, $q_1^{-1} \neq q_1$. *This first rank antecedent* q_1^{-1} *is unique and* q_1 *is located in a region where a point has two real first rank antecedents. The m-rank antecedents,* $m > 1$, *of* q_1 *are not real, and the open interval* $]q_1, q_1^{-1}[$ *is free from other singularities than the considered attractive fixed point.* It is the case of q_1, q_1^{-1} in the fig. 2.1b. Here T is always an endomorphism.

c) The boundary is made up of the two points of a repulsive cycle of order 2 having the type 1 ($S > 1$*). Each of these two points is the first rank antecedent of the other and has no other real first rank antecedent. The basin is free from other singularities than the considered attractive fixed point.* It is the case of the cycle of order two α_1, α_2 in fig. 2.8b.

The situations a),b) and c) result directly from the Köenigs-Lemeray graphical construction which gives the conditions of convergence toward the considered attractive fixed point.

2.8.2. BASIN MADE UP OF A FINITE, OR AN INFINITE NUMBER OF DISJOINTED PARTS WITH ONLY ONE ACCUMULATION POINT. FIRST EXAMPLE OF GLOBAL BIFURCATION

This occurs only if T is an endomorphism. The simplest example illustrating the case of a basin made up of infinitely many disjointed parts corresponds to $f(x,\lambda)$ with a "cubic" aspect given by fig.2.16a. Here the map T has three fixed points : q_1, q_3 which are repulsive of type one ($S > 1$), q_2 which is attractive ($|S| < 1$). With such a $f(x,\lambda)$ a point

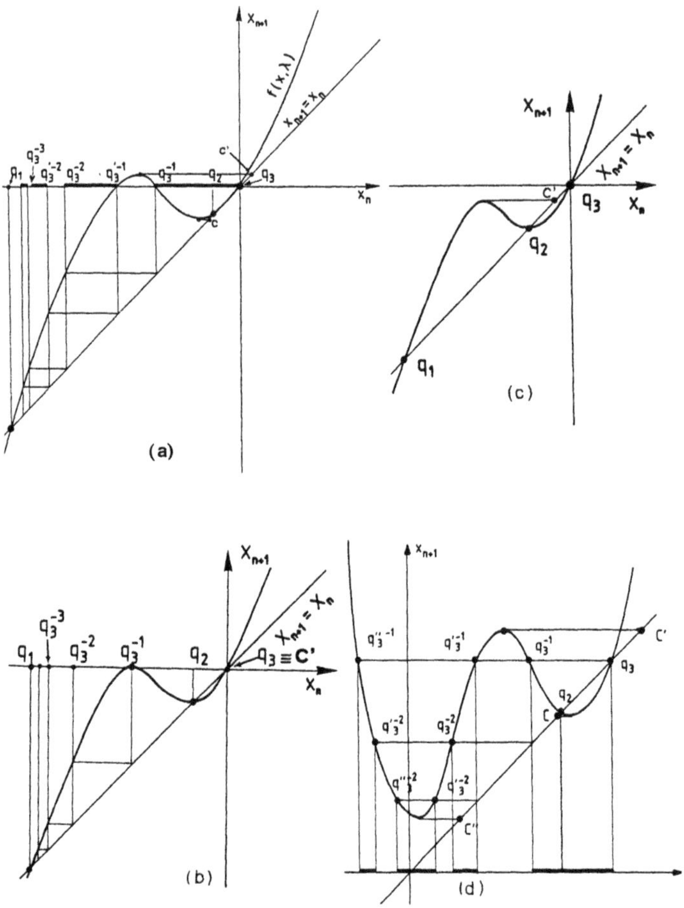

FIG. 2.16 - In fig. (a) the basin of q_2 is made up of an infinite number of disjoint parts .Fig. (b) corresponds to the bifurcation . In fig. (c) the basin of q_2 is all in one block (]q_1,q_3[) .In fig. (d) the basin is made up of a finite number of disjoint parts .

x_n has either three real antecedents of the first rank (x_n is located between the two critical points C, C' first rank consequents of the two extrema of $f(x,\lambda)$), or only one antecedent of the first rank (x_n is located outside the segment $\overline{CC'}$). So the fixed point q_1 is its own and unique antecedent, and the fixed point q_3, $q_3 \in \overline{CC'}$, is its own antecedent, but has two other first rank antecedents q_3^{-1}, $q_3'^{-1}$. The open segment $(D_o) \equiv]q_3, q_3^{-1}[$, containing q_2, corresponds locally to the situation of §2.8.1b (cf. fig.2.1b). Every initial condition $x_o \in (D_o)$ generates from T an iterated sequence (orbit) converging toward q_2. But here each boundary point of (D_o) having a first rank antecedent different from itself, (D_o), *called immediate basin* of q_2, is only a part of the *total basin* (D) of this fixed point ("immediate" and "total" are used in the sense of Julia-Fatou [J 5] [F 1]). Considering the successsive rank i, i = 1,2,..., antecedents of q_3, noted $q_3^{-i}, q_3'^{-i}$, it appears that the successive antecedents $(D_{j-1}) \equiv]q_3^{-j}, q_3'^{1-j}[$, j = 2,3,..., of (D) belong to the total basin (D) of q_2. Because $\lim q_3^{-i} = \lim q_3'^{-i} = q_1$ for $i \to \infty$, the fixed point q_1 is an accumulation point for the partial domains (D_i) when $i \to \infty$. So the total basin (D) of q_2 is a *multiply connected domain* $(D) = \bigcup_{i=0}^{\infty} (D_i)$. Outside (D) the iterated sequences converge toward the point at infinity on the x_n-axis.

The basin (D) is a multiply connected domain as for the parameter λ in $f(x,\lambda)$ is such that $x_{c'} > x_{q_3}$. When $x_{c'} < x_{q_3}$ (fig.2.16c), the repulsive fixed point q_3 is its own and unique first rank antecedent, and it is obtained the situation of §2.8.1a. Here the basin is all in one block $(D) \equiv]q_1, q_3[$. *So the parameter value* $\lambda = \lambda_b$, *such that* $C' = q_3$, *corresponds to a global bifurcation limiting two qualitatively different situations : the basin* (D) *is a multiply connected domain, and* (D) *is all one block. For* $\lambda = \lambda_b$ *(fig. 2.16b) the boundary of* (D) *is made up of the two repulsive fixed points* q_1, q_3 *and the infinitely many m-rank antecedents* q_3^{-m} *of* q_3, $q_3^{-m} \neq q_3$, *with* $\lim_{m\to\infty} q_3^{-m} = q_1$. *The points* q_3^{-m} *are the limit of the holes* $[q_3^{-m}, q_3'^{-m}]$, m = 1,2,..., *obtained for* $x_{c'} > x_{q_3}$, *and which do not belong to* (D) *inside* $]q_1, q_3[$.

Changing the shape of $f(x,\lambda)$, in such a way that $f(x,\lambda)$ does not cut $x_{n+1} = x_n$ in q_1 (fig.2.16d), leads to the basin of q_2 made up of a finite number of disjointed parts. $]q_3^{-1}, q_3[$, $]q_3^{-2}, q_3'^{-1}[$, $]q_3''^{-1}, \bar{q}_3^{-2}[$, $]q_3''^{-2}, q_3'^{-2}[$. This is obtained by creating a new critical point C" such that a point

x_n has no real antecedent for $x_n < x_{c''}$, $f(x,\lambda)$ having a "four degree" aspect. All the results of this paragraph have been published in [M8][M23] [M25] [M30] [G35] (cf. p.48) and [G36] (cf. p. 31).

2.8.3. COMPLEX BASIN OF THE FIRST TYPE. SECOND EXAMPLE OF GLOBAL BIFURCATION. CHAOTIC TRANSIENT.

Let α_m (resp.α_M) be the point of the basin boundary having the smallest (resp.largest) abscissa. *Such a complex basin (of the first type) is such that the open interval* $]\alpha_m,\alpha_M[$ *contains a strange repeller* $(\bar{F})$ *made up of infinitely many repulsive cycles of any order having the type one or two, of their successive antecedents of rank 1,2,3,..., of the limit points of all these points, and* $]\alpha_m,\alpha_M[$ *does not contain open intervals belonging to a different basin.* This occurs only when T is an endomorphism.

The simplest example illustrating this basin boundary corresponds to $f(x,\lambda)$ having a "four degree" aspect given by fig.2.17b, A being the only attractive fixed point at finite distance, q_1, q_2 being two repulsive fixed points of type one ($S > 1$), C_1 being the critical point of rank two (first rank consequent of C critical point of T) such that $x_{q_2} < x_{c_1} < x_{q_1}$ (fig.2.17b). Then $\alpha_m \equiv q_2$, $\alpha_M \equiv q_2^{-1}$ first rank antecedent of q_2 different of q_2 which is its own antecedent, and the strange repeller $(\bar{F})$ is contained inside the segment $\overline{q_1q_1^{-1}}$, q_1^{-1} being the first rank antecedent different from q_1.

The existence of the set $(\bar{F})$ will be established from the results of the § 3.2.5., via a global bifurcation corresponding to a $\lambda = \lambda_1^*$ value such that $q_1 \equiv C_1$. Indeed considering $f(x,\lambda)$ which corresponds to a λ value ($\lambda < \lambda_1^*$) such that $x_{q_1} < x_{c_1}$ (fig.2.17a) locally in the segment $\overline{q_1q_1^{-1}}$ one has qualitatively the situation of a quadratic $f(x,\lambda)$ described in the § 3.2. . Then $\overline{CC_1}$ is an absorbing segment containing only one attractor, and a strange repeller. For $\lambda \gg \lambda_1^*$, all the fixed points of T^k, $k = 1,2,3,\ldots$, are now real, new cycles cannot be created by bifurcation, and all the existing cycles are repulsive and are located in $]q_1,q_1^{-1}[$. For $\lambda > \lambda_1^*$ ($x_{c_1} < x_{q_1}$) these cycles, of order $k = 1,2,3,\ldots$, their succesive antecedents, and the limit points of all these points constitute

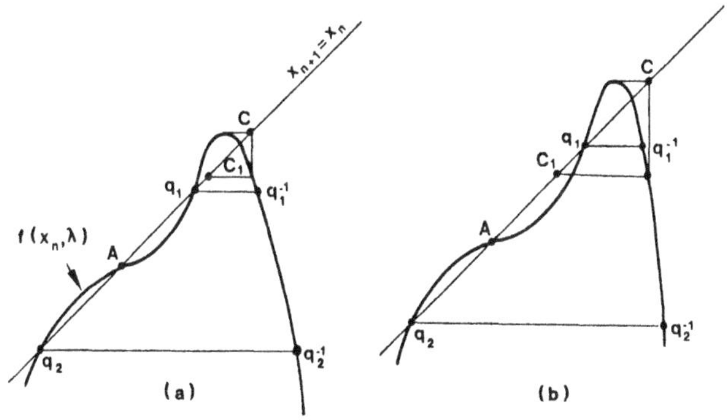

FIG. 2.17 - In fig. (a) T has two attractors, the fixed point A and an attractor on $\overline{CC_1}$, the basin of which is bounded by q_1, q_1^{-1}. In fig. (b) this attractor has disappeared, giving rise to a strange repeller located on $\overline{q_1 q_1^{-1}}$. On this interval a chaotic transient occurs before a regular convergence towards A.

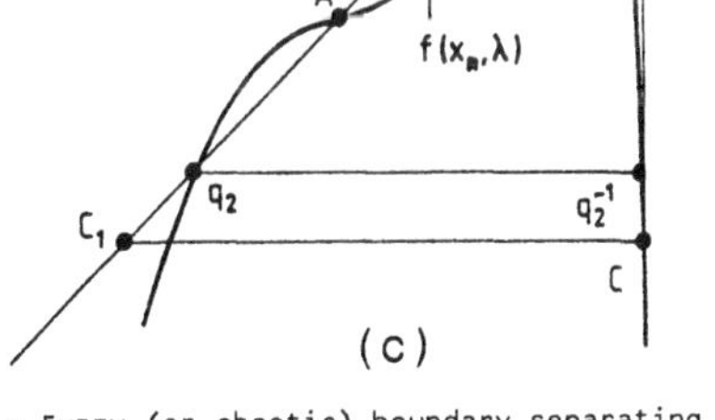

FIG. 2.18 - Fuzzy (or chaotic) boundary separating the basin of A from the basin of the point at infinity. These basins jumble together in avery complex manner on the interval $\overline{q_1 q_1^{-1}}$

$(\bar{F})$. It is asked the reader to admit provisionally this fact.

When $x_{c_1} < x_{q_1}$ the basin boundary of A is the set $(F) = q_2 \cup q_2^{-1} \cup (\bar{F})$, and the immediate basin of A is the open interval $]q_2, q_1[$, the first rank antecedent of which is $]q_1^{-1}, q_2^{-1}[$. If $x_{q_2} < x_{c_1} < x_{q_1}$, only the part $[C, q_1^{-1}[$ $(x < x_c)$ of $]q_1^{-1}, q_2^{-1}[$ has two first rank antecedents $]q_1'^{-2}, C^{-1}]$ and $[C^{-1}, q_1^{-2}[$ the union of which gives the open interval $]q_1'^{-2}, q_1^{-2}[$. The successive antecedents of $]q_1'^{-2}, q_1^{-2}[$ give rise to an "arborescent" infinite sequence of open intervals, the m-rank antecedent corresponding to 2^m open intervals. For $m \to \infty$ all these antecedents cover completly $]q_1, q_1^{-1}[$ except the points of the set $(\bar{F})$. Indeed if it is not the case, either at least one attractor exists inside $]q_1, q_1^{-1}[$, which is contrary to the above indicated result of § 3.2.5. , or $]q_1, q_1^{-1}[$ contains a part of the basin of an attractor located outside $]q_1, q_1^{-1}[$, which is impossible because C_1 belongs to the immediate basin of A.

Any initial condition x_o located inside $]q_1, q_1^{-1}[$, in a sufficiently small neighbourhood of the strange repeller $(\bar{F})$, gives rise to an iterated sequence (orbit) x_n with an apparently erratic evolution for $n < N$ as far as $x_n \in]q_1, q_1^{-1}[$. This evolution is called a *chaotic transient*. As soon as $x_n \in]q_2, q_1[$, for $n > N$, the convergence of x_n toward A becomes regular.

Returning now to $\lambda < \lambda_1^*$, $x_{q_1} < x_{c_1}$, (fig.2.17a), $\overline{CC_1}$ is an absorbing segment containing only one attractor. Then the repulsive fixed point q_1 is one of the points separating the basin of this attractor from the basin of A. Here the open interval $]q_1, q_2[$ gives a situation locally equivalent to the one of fig.2.1a, but this interval, which is the *immediate basin* of A, is only a part of the *total basin* of A. *The origin of this situation is the existence of first rank antecedents* q_1^{-1}, q_2^{-1} *of the boundary points* q_1, q_2 *different from these points.* So the open interval $]q_1^{-1}, q_2^{-1}[$ is the first rank antecedent of the immediate basin of A. The total basin of A is here constituted only by the two disjointed open intervals $]q_1, q_2[$ and $]q_1^{-1}, q_2^{-1}[$, because this last interval, with abscissae greater than the abscissa of the critical point C, has no real antecedent.

The two situations $\lambda < \lambda_1^*$ $(x_{q_1} < x_{c_1})$ *and* $\lambda > \lambda_1^*$ $(x_{q_1} > x_{c_1})$ *are qualitatively different. The parameter value* λ_1^* *is a boundary between these two situations, and it is characterized by the merging of two singulari-*

ties having different nature : the critical point C_1, *and the repulsive fixed point* q_1. *Then* λ_1^* *is related to a global bifurcation corresponding to a sudden increase of the basin of* A, *when crossing through* λ_1^* *from* $\lambda < \lambda_1^*$ *to* $\lambda > \lambda_1^*$, *the basin boundary becoming made up of infinitely many repulsive singularities after the bifurcation.*

All the elements of this chapter can be found in [M31] [G35](cf.p.396), [M46] [M47].

2.8.4. COMPLEX BASIN OF SECOND TYPE. CHAOTIC (OR FUZZY) BASIN BOUNDARY. THIRD EXAMPLE OF A GLOBAL BIFURCATION.

Consider α_m *and* α_M *of §2.8.3. The complex basin* (D_A) *of second type is such that the open interval* $]\alpha_m,\alpha_M[$ *contains a strange repeller* $(\bar{F})$ *made up of inifinitely many repulsive cycles of any order, of their successive antecedents, of the limit points of all these points. The interval* $]\alpha_m,\alpha_M[$ *contains also infinitely many disjointed open interval belonging to basins different from* (D_A). *All these disjointed open intervals jumble togethor in a very complex manner with a complementary infinite set of disjointed open intervals belonging to the basin* (D_A). *The accumulation points of the disjointed open intervals belonging, or not, to* (D_A) *are points of* $(\bar{F})$. Here T is always an endomorphism, and, as for the complex basin of first type, the accumulations of repulsive cycles, when their orders tend toward infinite values, are complex.

The simplest example of basin of second type is given by the function $f(x,\lambda)$ with a "four degree" aspect, the parameter λ being now such that $x_{c_1} < x_{q_2}$ (fig.2.17c), the half-open interval $[C_1,q_2[$, its first rank antecedent $]q_2^{-1},C]$, and the antecedents of rank m, m = 1,2,3,..., of this latter belong to basin (D_∞) of the attractive fixed point at infinity on the x_n-axis. $]q_2^{-1},C]$ has two first rank antecedents $[C^{-1},q_2^{-2}[$ and $]q_2'^{-2}C^{-1}]$, the union of which gives the open interval $]q_2'^{-2},q_2[$ belonging to (D_∞). Here C^{-1} is the maximum of $f(x,\lambda)$. Having abscissae smaller than that of the critical point C, all the points considered here have two different first rank antecedents (except C). Each of these antecedents in its turn gives also two different first rank antecedents. It results that the successive m-rank antecedents of $]q_2'^{-2},q_2^{-2}[$ give rise to an "arborescent" infinite sequence of disjointed open intervals, the accumulation points of which are elements of the set $(\bar{F})$. This infinite arborescent sequence of disjointed open intervals jumbles together in a very complex manner with a complementary infinite arborescent sequence made up of the successive antecedents of the open interval $]q_2,q_1[$ (which is the immediate basin of the attractive fixed point A). So the total basin of A is obtained. This second arborescent sequence has also elements of $(\bar{F})$ as accumulation points.

Theoritically the boundary separating the basin of A from D_∞ is well defined from the two infinite arborescent sequences. Practically with the constraint of finite accuracy (the significant digits of x_n, $n = 0,1,2,...$ are unavoidably limited), it is impossible to predict the convergence of x_n toward A, or its divergence, when x_0 is located in a sufficiently small neighbourhood of the set $(\bar{F})$. It is the reason why this set (strange repeller) is said to be a *chaotic*, or *fuzzy*, *basin boundary* ("frontière floue" in French) [G31]. [M23] [M25] [M46].[M31] [G10]

Consider now the *two qualitatively different situations :* $x_{q_2} < x_{c_1} < x_{q_1}$ (§ 2.8.3: the open interval $]q_2,q_2^{-1}[$ contains a strange repeller but no part of the basin (D_∞)) and $x_{c_1} < x_{q_2}$ (§ 2.8.4: $]q_2,q_2^{-1}[$ contains a strange repeller and parts of (D_∞)). *Then the parameter value* $\lambda = \lambda_b$, *such that* $q_2 \equiv C_1$, *corresponds to a global bifurcation which is the limit between these two situations. Crossing through* λ_b *gives rise to a chaotic (or fuzzy) basin boundary, without giving rise to new local singularities.*

2.8.5. SOME REMARKS.

- (a) Consider that the immediate basin is the total one when the basin is all in one block. It is possible to sum up the most important results of the § 2.8 with (P2.21). The proposition (P2.22) is an obvious adaptation of a Fatou's result [F 1].

(P2.21) *Let* T *be the map defined by* $f(x,\lambda)$, *a continuous real function of the real variable* x, *and the real parameter* λ. *Every attractive fixed point* A *has an immediate basin the two boundary points of which are : either the nearest repulsive fixed points of type one located on both sides of* A , *or the nearest repulsive fixed point of type one and its first rank antecedent different from this one, or the nearest repulsive cycle of order two and type one. If these two boundary points have no other first rank antecedent than themselves, the considered basin is all in one block. If it is not the case, the basin has a more complex nature, the simplest situations corresponding to the § 2.8.2 - 2.8.4, and the merging of a critical point into a repulsive fixed point, or a repulsive cycle of order two may lead to a global bifurcation.*

(P2.22) *Let* T *be the map defined by a continuous function of the real variable, and* $\alpha_1, \alpha_2, \ldots, \alpha_k$ *the points of an attractive cycle of order* k. *Then there exists at least a critical point* C *of* T *belonging to the immediate basin of this cycle.*

- (b) Results of §2.8 become general for the basin of an attractive cycle of order k by considering the map T^k.

- (c) Results of §2.8 are given with the hypothesis of $f(x,\lambda)$ being a continuous function of x. If $f(x,\lambda)$ is a piecewise continuous function of x, T may be a non invertible map. So let p_i be a point of discontinuity of $f(x,\lambda)$, $C_i^- = f(p_i-0,\lambda)$, $C_i^+ = f(p_i+0,\lambda)$ the values of the function on the right and on the left of p_i. *The two points* C_i^- *and* C_i^+ *play the role of critical points* in the sense of Julia-Fatou. Let now b_i a point p_i such that $f(b_i-0,\lambda) < b_i < f(b_i+0,\lambda)$. Then one has :

(P2.23) *Let* T *be a map defined by* $f(x,\lambda)$ *a piecewise continuous function of* x. *Every attractive fixed point* A *has an immediate basin the two boundary points of which are : either the ones given by (P2.21), or the two nearest points of discontinuity of the type* b_i *located on both sides of* A, *or the nearest repulsive fixed point of type one and the nearest point* b_i *on both sides of* A, *or the nearest point* b_i *and one of its first rank antecedent. If these two boundary points have no other first rank antecedent, the considered basin is all in one block. If it is not the case the basin has a more complex nature, the simplest cases being equivalent to the situations of § 2.8.2 - 2.8.4. The merging of a critical point, or a consequent of* (b_i-0), (b_i+0), *into a repulsive fixed point, or cycle of order two, may lead to a global bifurcation.*

When the basin is all in one block, the results of (P2.23) are established via the Kœnigs-Lemeray graphical construction.

2.9. INVERTIBLE MAP OF THE CIRCLE ONTO ITSELF.

2.9.1. DIFFEOMORPHISM OF THE CIRCLE.

Let τ be the diffeomorphism of the circle :

$$\theta_{n+1} = \theta_n + \lambda + g(\theta_n) = f(\theta_n,\lambda) \quad , \tag{2.36}$$

θ being a real angle, λ a parameter such that $0 \leqslant \lambda \leqslant 2\pi$, $g(\theta)$ a smooth periodic function of period 2π having only one extremum for $0 < \theta < 2\pi$, such that $g(0) = g(2\pi) = 0$, $g'(0) = g'(2\pi) = 0$, $g' = (dg/d\theta)$, $|g'(\theta)| < 1$. The two equalities with g, g' ensure the smoothness of τ for $\theta = 0$ (modulo 2π), the inequality with g' ensures the invertibility of the map. A good many publications are devoted to this problem (cf. for example [P 3] [D 1] [A 4] [J 1] [S10] [H 8] and also a general presentation in [G18] p.295-306. This paragraph gives a different presentation with the object of obtaining a description of the λ-intervals corresponding to a rational rotation number μ, $\mu = \lim (\theta_n - \theta_o)/(2\pi n)$ for $n \to \infty$. Then it appears that the organisation of the bifurcations on the λ-axis is of the "boxes in files" type [M27] [G35] [M42].

In the case of a simple rotation, i.e. $g(\theta) \equiv 0$, for $-2\pi < \theta < 2\pi$ (2.36) is identical with the piecewise linear map T (2.26), considering $\lambda_1 = \lambda_2 = 1$, $a+b = 2\pi$, $\theta_n \equiv x_n$, $a = \lambda$, $0 \leqslant b \leqslant 2\pi$. The map τ may be considered as a smooth distortion of (2.26) giving qualitatively equivalent results for the bifurcations structure, when $g(\theta) \not\equiv 0$, with $\lambda_1 = \lambda_2 = \gamma$, $0 < \gamma < 1$, a being equivalent to λ, $a+b = 2\pi$, $0 \leqslant b \leqslant 2\pi$. τ may be also considered as a particular case of (2.34). This qualitative equivalence appears by considering T^k in the (x_n, x_{n+1}) plane, and τ^k in the (θ_n, θ_{n+1}) plane, k = integer. Indeed inside an interval $\ell = 2\pi$-long, T^k is piecewise linear, with k points of discontinuity. The curve T^k is made up of k segments d_i of straight lines, $i = 1,2,\ldots,k$. In general there exist several intervals of the a-values such that T^k intersects $x_{n+1} = x_n$ in k points, belonging to a cycle of order k which is always attractive. A boundary point of one of these intervals, associated with a determined degree of complexity, corresponds to one of the two ends for every d_i , located on $x_{n+1} = x_n$.

The map τ^k obtained from (2.36) has particular properties due to the special form (2.34) of τ, with $a+b = 2\pi$, $a = \lambda$, $g(0^-) = g(2\pi^-)$, $g(0^+) = g(-2\pi^+)$, the part $-2\pi < \theta_n < 0$ of $f(\theta_n,\lambda)$ being obtained by a translation of $\theta = -2\pi$ parallel to the θ_n-axis, with respect to the part $0 < \theta_n < 2\pi$. In this particular situation of (2.34), τ^k is continuous all along an interval having the length 2π,belonging to $-2\pi < \theta < 2\pi$, and has only two points of discontinuity for $\theta = \theta^{(j)}$, $j = 1,2$, such that $|\tau^k(\theta^{(i)+})-\tau^k(\theta^{(i)-})| = 2\pi$. The points $\theta^{(j)}$ are the two rank-k antecedents of $\theta = 0$, obtained inside $-2\pi < \theta < 2\pi$. In general, there are p_k intervals $[\lambda]_k^j$, $j=1,2,\ldots,p_k$, of the λ-values such that τ^k has 2k intersections with $\theta_{n+1} = \theta_n$, whereas the k points of discontinuity of T^k lead to only k intersection with $x_n = x_{n+1}$. The so obtained 2k intersections are the points of an attractive cycle of order k, and of a repulsive one. The two cycles of order k exist inside an interval, or box, $[\lambda]_k^j$ of λ-values, the two boundary points of which are defined when one cycle merges into the other. Then τ^k is tangent to $\theta_{n+1} = \theta_n$ with k points of contact of order $m \geqslant 1$. Fig.2.19a and c show the two situations of tangency for $k = 1$, and fig.2.19b gives the two cycles of order $k = 1$ (fixed point), the attractive one q_1, and the repulsive

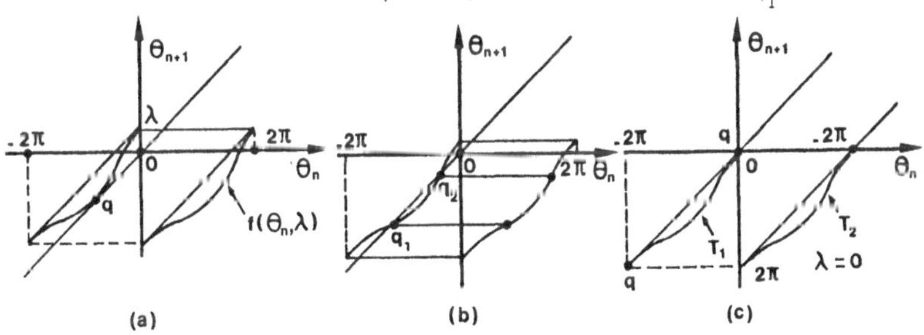

FIG. 2.19

one q_2. The indicated equivalent occurs for $k > 1$. For a given j, the corresponding interval $[\lambda]_k^j$ is associated with a given rational rotation number μ,the denominator of which is the order k of the two cycles, and the numerator of which depends on the degree of complexity of the box $[\lambda]_k^j$. The numerator is then directly related to the index j. Outside $[\lambda]_k^j$ other cycles having the same rotation number do not exist.

Let T_1 be the part of τ corresponding to $\theta_n < 0$, T_2 the part associated with $\theta_n > 0$. Considering the results of § 2.5, it appears that the boxes associated with the maps $T_1^m T_2$ correspond to the rotation numbers $\mu = 1/(m+1)$, and the ones associated with $T_2^m T_1$ correspond to $\mu = -1/(m+1)$. The boxes having one boundary point which is 0 (or $\pm 2\pi$) are such that $\mu = 1$. *All the rotation numbers* $\mu = \pm 1/k$ *correspond to the set of the cycles having the degree* $N = 1$ *of complexity.* Between two consecutive boxes with $N = 1$, associated with $T_1^{m-1}T_2 (\mu = 1/m)$ and $T_1^m T_2$ ($\mu = 1/(m+1)$), there is a double infinite sequence of boxes (double file) having the degree $N = 2$ of complexity. One sequence is associated with the maps $(T_1^{m-1}T_2)^{\alpha_1} T_1^m T_2$ (order of the two cycles $k = m\alpha_1 + m + 1$), the other with the maps $(T_1^m T_2)^{\alpha_1} T_1^{m-1} T_2$, ($k = (m+1)\alpha_1 + m$), $\alpha_1 = 1,2,\ldots$ For τ the numerator of μ is then the number of products $(T_1^{m-1}T_2)$ and $(T_1^m T_2)$ appearing in the expressions of these maps. So the rotation numbers for $N = 2$, respectively for the for the first and second sequence, are :

$$
(2.37) \qquad \begin{cases} \mu_2' = (1+\alpha_1)/(m\alpha_1+m+1), & \lim\limits_{\alpha_1\to\infty} \mu_2' = 1/m \\ \mu_2'' = (1+\alpha_1)/[(m+1)\alpha_1+m)], & \lim\limits_{\alpha_1\to\infty} \mu_2'' = 1/(m+1) \end{cases}
$$

For $\alpha_1 = 1$, $\mu_2' = \mu_2''$ is called the *central* μ of the files $N = 2$ located between $\mu = 1/m$, $\mu = 1/(m+1)$.

Between two consecutive boxes of the considered double file $N = 2$, there is a double file of degree $N = 3$. So between the two consecutive boxes corresponding to μ_2' and the parameters $(\alpha_1 - 1)$, α_1, there is the double file $N = 3$ associated with the maps $\{(T_1^{m-1}T_2)^{\alpha_1} T_1^m T_2\}^{\alpha_2} \{(T_1^{m-1}T_2)^{\alpha_1 - 1} T_1^m T_2\}$ and $\{(T_1^{m-1}T_2)^{\alpha_1 - 1} T_1^m T_2\}^{\alpha_2} \{(T_1^{m-1}T_2)^{\alpha_1} T_1^m T_2\}$ $\alpha_2 = 1,2,\ldots$, giving respectively the rotation numbers:

$$
(2.38) \qquad \begin{cases} \mu_3' = [\alpha_2(\alpha_1+1)+\alpha_1]/[\alpha_2(m\alpha_1+m+1)+m\alpha_1+1] \\ \mu_3'' = [\alpha_2\alpha_1+\alpha_1+1]/[\alpha_2(m\alpha_1+1)+m\alpha_1+m+1] \end{cases}
$$

with $\lim \mu_3' = \mu_2'(\alpha_1)$, and $\lim \mu_3'' = \mu_2'(\alpha_1 - 1)$, if $\alpha_2 \to \infty$. Similarly the rotation numbers of the double file $N = 3$ located between two consesecutive boxes defined by μ_2'' are obtained. Carrying on the process of division of the remaining λ intervals, the rotation numbers with $N = 4,5,\ldots$, are obtained from the following rule [M27] [G35] :

(P2.24) *Given two consecutive boxes of* τ *associated with cycles of the degree of complexity N, and the rotation numbers :*

$$\mu_N^{(1)} = \{A(\alpha_{N-1}-1) + B\}/\{C(\alpha_{N-1}-1) + D\},\quad \mu_N^{(2)} = \{A\alpha_{N-1}+B\}/\{C\alpha_{N-1}+D\}.$$

The double infinite sequence (double file) corresponding to the degree of complexity N+1, *and located between* $\mu_N^{(1)}$ *and* $\mu_N^{(2)}$, *is associated with the rotation numbers*

$$\mu'_{N+1} = \frac{\alpha_N(A\alpha_{N-1}+B)+A(\alpha_{N-1}-1)+B}{\alpha_N(C\alpha_{N-1}+D)+C(\alpha_{N-1}-1)+D},\quad \mu''_{N+1} = \frac{\alpha_N\{A(\alpha_{N-1}-1)+B\}+A\alpha_{N-1}+B}{\alpha_N\{C(\alpha_{N-1}-1)+D\}+C\alpha_{N-1}+D}$$

with $\lim \mu'_{N+1} = \mu_N^{(2)}$, $\lim \mu''_{N+1} = \mu_N^{(1)}$ *when* $\alpha_N \to \infty$. *For* N=1, A=D=0, α_o = m+1, B=C=1.

From (P2.24), it may be obtained a recurrence relation which gives the rotation numbers μ'_{N+1}, μ''_{N+1} of a double file with the degree N+1, from the rotation $\mu_N^{(1)}$, $\mu_N^{(2)}$ of two consecutive boxes belonging to a file of degree N. Then a boundary point of each boxes is an accumulation point of one of the sequence μ'_{N+1}, μ''_{N+1}. This reccurence relation, uses the rotation number μ_{N-1}, limit value (degree N-1) of the sequence $\mu_N^{(i)}$, i = 1,2,.., when $i\to\infty$. It is given by [M58] :

$$(2.39)\quad \begin{cases} \mu'_{N+1} = [\mu_N^{(1)}\mu_N^{(2)}(1+\alpha)-\mu_{N-1}(\alpha\mu_N^{(1)}+\mu_N^{(2)})]/[\alpha\mu_N^{(2)}+\mu_N^{(1)}-\mu_{N-1}(1+\alpha)] \\ \mu''_{N+1} = [\mu_N^{(1)}\mu_N^{(2)}(1+\alpha)-\mu_{N-1}(\alpha\mu_N^{(2)}+\mu_N^{(1)})]/[\alpha\mu_N^{(1)}+\mu_N^{(2)}-\mu_{N-1}(1+\alpha)] \end{cases},$$

where $\alpha \equiv \alpha_N$, $\lim \mu'_{N+1} = \mu_N^{(1)}$, $\lim \mu''_{N+1} = \mu_N^{(2)}$ if $\alpha\to\infty$, $\alpha=1$ giving a central μ_{N+1}, N = 1,2,3,..., and μ_{N-1} = 0 for N=1. As for the piecewise linear map (2.26), the bifurcation structure of τ is of the type "boxes in files" and *the graphical representation of the rotation number* μ *as a function of the parameter* λ *is the devil's staircase*. From the above results and the results concerning (2.26) the following proposition can be formulated [M27] [G35].

(P2.25) *Let τ be the diffeomorphism of the circle defined by (2.36). As the piecewise linear map (2.26), the bifurcation structure is of the type "boxes in files" on the parameter λ-axis. Every box is the domain of existence of an attractive cycle of order k and a repulsive one having the same order.*

One double infinite sequence (file) of boxes has the first (N=1) degree of complexity. The corresponding boxes are made up of the existence domains of the cycles with the rotation numbers $\mu = \pm 1/k$, $k = 1,2,3,\ldots$ Between two consecutive boxes $\mu = \pm 1/k$, $\mu = \pm 1/(k+1)$ of one of these two sequences, there is a double file of boxes of the degree N=2, existence domains of the cycles with the rotation numbers given by (2.37). More generally, between two consecutive boxes of a sequence associated with the degree N of complexity, there exists a double file of boxes of the degree N+1, the rotation numbers of which are given by (P2.24) and (2.39), N = 1,2,3,...

If on the λ-axis, the set E of all the files of degree N = 1,2,3,... is removed, it remains a non empty, perfect, nowhere dense set $\bar{E}$ which has the structure of a Cantor's discontinuous set. For $\lambda \in \bar{E}$, the corresponding rotation numbers are irrationals. The graphical representation of the rotation number as a function of the parameter λ is the devil's staircase.

The so obtained sequences of rotation numbers verify the Poincaré's theorem which states the continuity of the rotation number with respect to the parameter λ .

2.9.2. GENERALIZATIONS.

- (a) Using the results of [A 4], and in particular the theorems 5, 6 and 7, it is possible to widen the above results to the bifurcations structure in the functional space of the diffeomorphism of the circle with a real variable.

- (b) If $g(\theta)$ has 2r-1 extrema in the interval $0 < \theta < 2\pi$, $2r' \leqslant 2r$ intersections of τ^k with $\theta_{n+1} = \theta_n$ are now possible, giving 2r' fixed points of τ^k alternately attractive and repulsive. They corresponds to $r' \leqslant r$ pairs of cycles of order k, with a given rotation number, the existence domain of every pair being a box on the λ-axis. A pair of cycles is related to a determined extremum of $g(\theta)$ which leads to a given "boxes in files" structure. So r "boxes in files" structures, with overlapping of boxes corresponding to different extrema of the same nature, take place on the λ-axis.

- (c) If the condition $g'(0) = g'(2\pi)$ associated with (2.36) does not occur, or if $g(\theta)$ is only a continuous, piecewise differentiable function such that $g(0) = g(2\pi)$, $|g'(\theta)| < 1$, but verifying or not $g'(0) = g'(2\pi)$ τ is no longer a diffeomorphism, but only an invertible map. Then the same qualitative behaviour is obtained with r "boxes in files" structures. The only difference between this case and the above considered case is due to the definition of the boundary points of a box. These points are now defined by a contact of order zero between τ^k and $\theta_{n+1} = \theta_n$, giving the merging of an attractive cycle of order k into a repulsive one, and not by a tangency condition.

2.9.3. BOXES IN FILES BIFURCATIONS STRUCTURE AND FAREY' SEQUENCES.

The sequences of rotation numbers, appearing in the "boxes in files" bifurcations structure of τ, correspond to a particular ordering of the Farey's sequences of fractions in their lowest terms. So two consecutive rotation numbers $\mu^{(1)} = a/b, \mu^{(2)} = c/d$ of a given file of the degree N are neighbouring fractions in the sense of Farey's sequences ($bc-ad = \pm 1$). Then (P2.24) and (2.39) may be considered as some less known properties of the Farey's sequences with respect to those given firstly by Farey and after by Cauchy. These properties are obtained by introducing the notion of the degree of complexity associated to a given file. With use, the "boxes in files" ordering appears to be more fitted to the solution of dynamics problems.

2.9.4. BOXES IN FILES BIFURCATIONS STRUCTURE AND THE PHENOMENON OF PHASE INTERMITTENCY.

From (P2.25) it appears that each boundary of a box with the complexity degree N is an accumulation point of files of boxes with a degree > N, when the parameter λ is taken outside the considered box. Let λ_b be one of the boundary of the box of degree N, $\lambda = \lambda_b - \varepsilon$, $\varepsilon \to 0$ being sufficiently small, a λ-value outside this box which is the existence domain of a pair of cycles with $\mu = p/q$, one being attractive, the other being repulsive. For this λ-value, the curve τ^k in the (θ_n, θ_{n+1}) plane presents q points located at a minimum distance from $\theta_{n+1} = \theta_n$. A situation equivalent to that of fig.4.10 (cf. also [M22]) is obtained. So the "density" of the points of the iterated sequence is much larger in a neighbourhood (δ) of each of the q points, and this density increases when $\varepsilon \to \infty$. The maximum of "density" of the iterated points results from the concentration of the most part of the q' points of an attractive cycle such that $\mu = p'/q'$ (degree > N), in the neighbourhood (δ), with $q' \to \infty$ and $p'/q' \to p/q$ when $\varepsilon \to 0$. On the whole, for $\lambda = \lambda_b - \varepsilon$ the iterated sequence presents oscillations which are interrupted by bursts becoming less frequent as $\varepsilon \to 0$. The observed behaviour is the result of a transient toward the attractive cycle with $\mu = p'/q'$. This phenomenon was firstly described in 1976 (cf. [M22]) for the recurrence $x_{n+1} = x_n^2 - \lambda$. In 1978 (cf. [M27]) the discovery of the boxes in files bifurcations structure in files bifurcations structure in the diffeomorphism of the circle, permitted to make appear the role of the accumulation of boxes in files in its formation. The described phenomenon was rediscovered in 1980 [P 4], and it is now called *intermittency*.

Chapter 3

MYRBERG' S RESULTS ON THE ONE-DIMENSIONAL QUADRATIC RECURRENCES. THEIR CONSEQUENCES

3.1. INTRODUCTION.

This chapter is devoted to the properties of the solutions (orbits, iterated sequences, discrete trajectories) generated by the unimodal one-dimensional endomorphism (recurrence, map, or iteration) defined by a quadratic function $f(x,\lambda) = x^2-\lambda$.

(3.1) $$x_{n+1} = x_n^2-\lambda \quad , \quad n = 0,1,2,\ldots$$

Here x is a real variable, λ a real parameter.

As previously indicated, this map can be considered as the last step resulting from a process of successive analytical simplifications. These simplifications consist of a decrease in the dimension of an initial equation, and a reduction of the "non linearities degree". They lead to the simplest mathematical model having a set of characteristical properties which are common to a whole class of equations including the initial one. In this sense (which is not exactly the one of the differential topology), equation (3.1) is said a *generic equation*. Then the common characteristical properties are :

- the fractal organisation of the bifurcations for the equations belonging to the considered class (called *box-within-a-box bifurcations structure*) in the parameters space (cf. chap.4).
- the existence of an infinite number of repulsive cycles on the x_n-axis, when the model parameters belong to a certain region of the parameters space.
- the existence of a "strict" chaotic attractive solution on the x_n-

axis, for characteristic points of the parameters space.

So the role of the quadratic map (3.1) is fundamental. All other quadratic maps :

$$(3.2) \qquad y_{n+1} = ay_n^2 + 2by_n + c$$

have the same behaviour as (3.1). Indeed a linear change of variable, $y = \alpha x + \beta$, gives (3.1) with $\lambda = b^2 - ac - b$, $a\alpha = 1$, $a\beta = -b$ [M55] [M25] [G35]. Starting now from (3.1), and inversing the described simplifications process, a first degree of complication is obtained with the one-dimensional endomorphism T' :

$$(3.3) \qquad x_{n+1} = f(x_n, \lambda) \qquad n = 0,1,2,\ldots,$$

$f(x,\lambda)$ being a continuous and differentiable function in x, with only one extremum. Given one of the more restrictive conditions :

- (H1) between a maximum and a minimum of T'^k (k being an integer), which are consecutive, only a point of inflexion exists [M21] [M25] [G35].
- (H2) $f(x,\lambda)$ is a continuously three times differentiable function with respect to x, and has a negative Schwarzian derivative :

$$f'''(x)/f'(x) - 3/2\,[f''(x)/(f'(x)]^2 < 0$$

- (H3) $x_1 < x_2 < x_3 < x_4$ being four points in a segment on which f is monotone, and denoting by $R(x_1,x_2,x_3,x_4)$ the crossratio $(x_4-x_1)(x_3-x_2)/[(x_4-x_3)(x_2-x_1)]$, $R(x_1,x_2,x_3,x_4) < R[f(x_1,\lambda), f(x_2,\lambda), f(x_3,\lambda)\ f(x_4,\lambda)]$ [C 2]. In [C 2], the x_i, $i = 1,2,3,4$, are considered as points of the segments $\overline{CC_1}$, C being the critical point in the Julia-Fatou's sense (C is the first rank consequent of the extremum of $f(x,\lambda)$), C_1 the first rank consequent of C.

Then *an extension of a Fatou's theorem* (cf § 2.72 and [M21] [G35] [C 2]) *permits to assert the existence of at most one attractive cycle of order* k = 1,2,... (cycle k=1 is a fixed point) *on* $\overline{CC_1}$. If one of the conditions (H1), (H2), (H3) is not satisfied, then the existence of more than one attractor on $\overline{CC_1}$ becomes possible (one attractive cycle of order k=2,3,..., and one attractive fixed point, for example) for certain ranges of the parameter.

Remark : Most of the contemporary publications about (3.3) present this endomorphism as a map of an interval [0,1], or [-1,1] onto itself, where its limits are C, C_1. In this book we would rather use a larger interval than $\overline{CC_1}$. This interval contains the repulsive fixed point q_1 having its multiplier (eigenvalue) S > 1 and its first rank antecedent $q_1^{-1} \neq q_1$. Indeed such a point q_1 has a fundamental function for the dynamical properties of (3.3), because it belongs, (with its first rank antecedent q_1^{-1}) to the boundary of the influence domain (basin) of the *absorptive segment* $\overline{CC_1}$ (cf. § 2.8).

A second degree of complication in the above described inverse process of simplifications takes place when f(x,λ) in (3.3) has more than one extremum (cf. [G35] p.140-144, 401-418 and [M24] [F 4]). This case, and the situation of f(x, λ) having a unique extremum but with $\overline{CC_1}$ containing more than one attractor, are not considered here.

Considering now the quadratic generic equation (3.1), it is worthy of note the importance of the Myrberg's contribution [M54]-[M56]. Indeed this author was the first to give a complete set of results concerning the properties of (3.1) on the x_n-axis, and a description of the bifurcations on the parameter λ-axis. In particular, Myrberg was the first to give the following important results :

- the phenomenon of cascade of bifurcations by order (period) doubling on a definite interval of λ, with an accumulation value of the corresponding bifurcations when the order tends to an infinite value. This interval is called a *spectrum* in [M55] (1963). Here the term *Myrberg's chain* (or cascade) is also used (cf. the reference to Myrberg given for the study of (3.3) [G31] [M21]). About this phenomenon, some authors speak of Feigenbaum's cascade refering to its rediscovery at a latter date.

- the characterization of a cycle, by using a binary code (with k-2 symbols +,- in [M55]) which gives the transfer order of its points by successive applications of (3.1).This characterization is called here a *binary rotation sequence* [r]. The same representation was used later (1973) in [M 5] with k-1 symbols R,L, and the equivalence R → -, L → +. The first symbol R of the Metropolis representation, being always the same, is omitted in [r]. Implicitly, or explicitly, the Myrberg's papers [M54] [M55] give an extension of the rotation sequence notion to the limit cases when the order k of the cycle is such that $k \to \infty$, which correspond to an accumulation of λ-values called by Myrberg a *singular value*.

- the representation of each point $q^{(i)}$, $i = 1,\dots,k$, belonging to a given cycle of order k, by a sequence $[q^{(i)}]$ of k signs + and -. Here, the sequence $[q^{(i)}]$ of the point $q^{(i)}$, first rank consequent of $q^{(i-1)}$, $q^{(i)} = Tq^{(i-1)}$, is obtained by a cyclic permutation from $[q^{(i-1)}]$ [M54] (1958).

- the law of classification (ordering law) of all the binary rotation sequences for finite, or infinite, order k [M55] (1963). It is shown in this chapter that the now very popular notions of *invariant coordinate* and *kneading invariant* [M 6] are only variants of the Myrberg's ordering law.

From 1977, there is an explosive growth of the papers devoted to the problems (3.1) (3.3). These papers give generally a classical and a wide bibliography about this subject, which passes over in silence the Myrberg's contribution. In this book, such a complete bibliography is voluntarily omitted. The only references considered here are those which are fundamental for the exposed matter.

The purpose of this chapter is the presentation of the Myrberg's results and of some developments of these results.

3.2. SOME GENERAL PROPERTIES.

3.2.1. FIRST SET OF MYRBERG'S RESULTS.

Consider the quadratic recurrence (3.1) $f(x,\lambda) \equiv x^2-\lambda$. Writing that a point of a cycle of order k is located at the extremum $x_e = 0$, Myrberg [M55] [G35] [M25] gives the proposition:

(P3.1) *Let* T *be the quadratic recurrence (3.1). The cycles of order* k *with multiplier* $S = 0$ *correspond to a set of the parameter λ-values, which are the real zeros of* $G_k(\lambda)$, *with* $G_{j+1}(\lambda) = G_j^2(\lambda)-\lambda$, $G_1(\lambda) = -\lambda$.

The complete set of the real zeros of $G_k(\lambda)$ corresponds to the λ-values of all the cycles of order k(S=0) and also to the λ-values of all the cycles having an order which is a divisor of k, including one ($\lambda=0$ is always a zero). The roots corresponding only to the cycles of order k are called by Myrberg *primitive zeros*. An obvious generalization of (P3.1) is (P3.2).

(P3.2) *Let T be recurrence (3.3) defined by a smooth function* $f(x,\lambda)$ *having a single extremum* x_e *and associated with conditions ensuring the existence of one attractive cycle at most. The cycles of order* k *with multiplier* $S = 0$ *correspond to the set of* λ*-values, which are the real roots of* $G_k(\lambda) = x_e$ *with* $G_k(\lambda) = f^k(x_e,\lambda)$, f^k *being given by* $x_{n+k} = f^k(x_n,\lambda)$. *Here* $G_{k+1}(\lambda) = f[G_k(\lambda)]$, $G_1(\lambda) = f(x_e,\lambda)$.

The equation $G_k(\lambda) = x_e$ means that the rank k consequent $T^k x_e$ of x_e is x_e itself when $S = 0$, it will be call the *Myrberg's equation of the consequents.* For (3.1) the following Myrberg's theorems [M55] give more informations about the cycles of order k with $S = 0$.

(P3.3) *Let T be the quadratic recurrence (3.1).*

(a) All the real zeros of $G_k(\lambda)$ *are located inside the interval* $0 \leqslant \lambda < 2$.

(b) Between two (real) zeros λ_{k-2} *of* $G_{k-2}(\lambda)$ *and* λ_{k-1} *of* $G_{k-1}(\lambda)$ *there exists at least one zero of* $G_k(\lambda)$, *for* $\lambda > 1$.

(c) Between two zeros λ_{k-2} *of* $G_{k-2}(\lambda)$ *and* λ_k *of* $G_k(\lambda)$, *there exists at least one zero of* $G_{k+1}(\lambda)$, *for* $\lambda > 5/4$.

(d) Let m *and* p *be two incommensurable integers,* $m > p$. *Then between the zeros* λ_m *of* $G_m(\lambda)$ *and* λ_p *of* $G_p(\lambda)$, *there is at least one zero of* $G_\ell(\lambda)$, $\ell = m\mu+1$.

A direct generalization of (P3.3) for (3.3) is not possible. For (P3.3) (b)-(d), a generalization implies conditions about the shape evolution of $f(x,\lambda)$ with respect to the variation of λ. From (P3.3) it results that all the bifurcation values of (3.1) occur inside the interval $-1/4 \leqslant \lambda \leqslant 2$. For $\lambda = \lambda_{(1)_0} = -1/4$, the parabola $x_n^2-\lambda$ is tangent to $x_{n+1} = x_n$ and the bifurcation (2.21) with $k = 1$ occurs giving rise to the repulsive fixed point q_1 (fig.3.1a), $x_{q_1} = 1/2 + \sqrt{1/4+\lambda}$, $S_{q_1} = 2x_{q_1} = 1 + \sqrt{1+4\lambda}$, and the attractive (for $-1/4 < \lambda < 3/4$) fixed point q_2, $x_{q_2} = 1/2 - \sqrt{1/4+\lambda}$, $S_{q_2} = 2x_{q_2} = 1 - \sqrt{1+4\lambda}$. One has $S_{q_1} = S_{q_2} = 1$, $x_{q_2} = x_{q_1}$ for $\lambda = -1/4$, $S_{q_2} = 0$ for $\lambda = 0$, which is the unique real zero of $G_1(\lambda)$. For $\lambda < \lambda_{(1)_0}$ (3.1) has neither fixed point, nor cycle at finite distance. For (3.3), q_1 is the repulsive fixed point with $S > 1$ which merges into q_2, fixed point

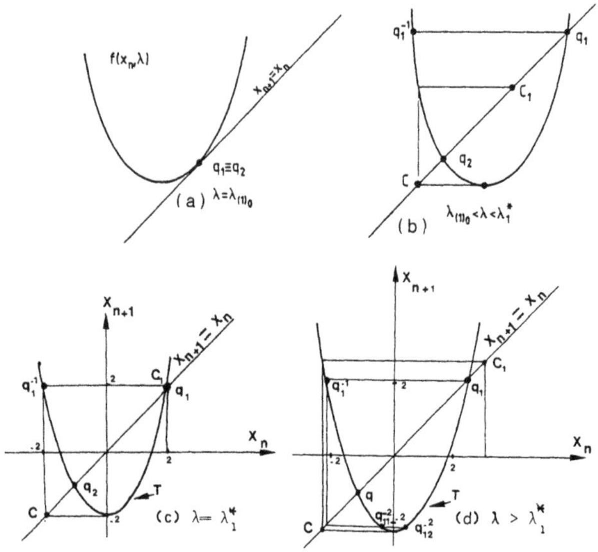

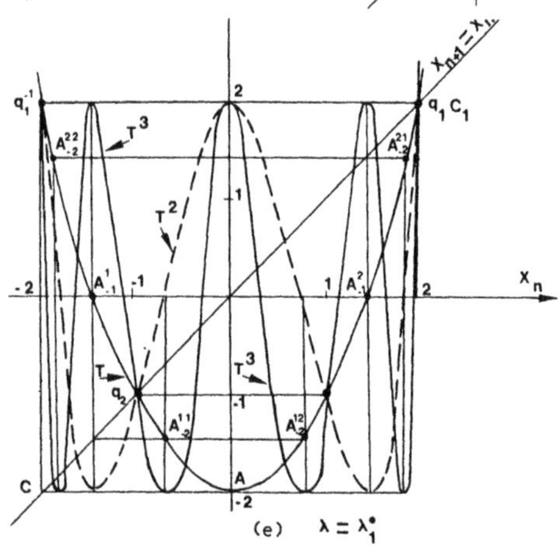

$\lambda = \lambda_1^*$, all the roots of $T^k x = x$ become real.

FIG. 3.1 a,b,c,d,e.

with $S < 1$, when $\lambda > \lambda_{(1)_0}$.

3.2.2. NUMBER OF ALL THE POSSIBLE CYCLES OF ORDER k.

For (3.1) the parameter value $\lambda = \lambda_1^* = 2$ is defined when the critical point of second rank C_1, $x_{C_1} = \lambda^2 - \lambda$, merges into the repulsive fixed point q_1 with $S > 1$. Then $[q_1^{-1}, q_1] \equiv [C, C_1]$, $x_{q_1^{-1}} = -1/2 - \sqrt{1/4+\lambda}$, $x_C = -\lambda$. For the recurrence (3.3) associated with conditions ensuring the existence of no more than one attractive cycle, λ_1^* is defined for $q_1 \equiv C_1$, $x_{C_1} = f^2(x_e,\lambda) = T^2 x_e$ (cf. fig.3.1c). Using the § 2.7.3 results it is possible to construct in the (x_n, x_{n+1}) plane the curves T^k, $k = 2,3,\ldots$, from T for $\lambda = \lambda_1^*$. Indeed with $\lambda = \lambda_1^*$ this construction is the simplest, because the extrema of T^k are located on the two straight lines $x_{n+1} = x_c$ (=-2 for (3.1)), $x_{n+1} = x_{c_1}$ (=2 for (3.1)). The abscissae of all the extrema of T^k are the abscissae of all the antecedents of the extremum $A(x = x_e)$ of T up to the rank (k-1), x_e being also the abscissa of an extremum of T^k (cf. fig.3.1e for k = 2,3). In the considered case all these antecedents are real. So T^k has $2^k - 1$ extrema marking the boundaries of 2^k segments of T^k, which are by turns strictly increasing and strictly decreasing. With such an arrangement of these extrema, whatever may be k the curve T^k has 2^k intersections with $x_{n+1} = x_n$. Then the equation $x = T^k x$ has 2^k real solutions, i.e. all the roots of this equation are real. For (3.1), except the point q_1, $(df^k/dx) = \pm 2^k$ at each of the 2^k intersections. Consider (3.1), or (3.3) with additional conditions about the shape evolution of $f(x,\lambda)$ with respect to the variation of λ, and about the existence of one attractor at most in the finite part of the x_n-axis. *For $\lambda \geqslant \lambda_1^*$ all the (real) fixed points and cycles have been created and no bifurcation occurs*. Then it is said also that *hyperbolicity* holds, for $\lambda \geqslant \lambda_1^*$, i.e. all the multipliers are $S \neq 1$.

From this last result, and considering that all the fixed points of T^r are fixed points of T^p, if the integer r is a factor of the integer p, it is possible to determine the number N_k of cycles of a given order k for $\lambda \geqslant \lambda_1^*$, and the number $N_\lambda(k)$ of λ-bifurcation values giving rise to these cycles. Then two different cases are to be considered [G35].

- (a) k *is a prime number*.

The number of the points of all the cycles of order k is $(2^k - 2)$, i.e.

2^k intersections of T^k with $x_{n+1} = x_n$, the two fixed points q_1, q_2 being removed. Then the number N_k of cycles of order k is :

$$(3.4) \qquad N_k = (2^k-2)/k \quad , \quad \text{if } k = \text{prime number}$$

- (b) k *is not a prime number*.

$k = k_1^{\alpha_1} k_2^{\alpha_2} \ldots k_n^{\alpha_n}$, $k_1 > k_2 > \ldots > k_n$ being prime numbers, the α_1, $\alpha_2, \ldots, \alpha_n$ being integers. Given $\ell_a = k/k_a$, $a = 1,\ldots,n$, and $\ell_b,\ldots,\ell_\mu,\ldots$, being determined similarly for the indexes $b,c,\ldots,\mu$. Given $(\ell_a\ell_b)$ the highest common factor of ℓ_a and ℓ_b ; $(\ell_a\ell_b\ldots\ell_\mu)$ the highest common factor of $\ell_a,\ell_b,\ldots,\ell_\mu$. From (P2.19), the number N_k of cycles of order k is [Z 1] :

$$(3.5) \qquad N_k = [2^k - \sum_{a=1}^{n} 2^{\ell_a} + \sum_{b=2}^{n}\sum_{a=1}^{b-1} 2^{(\ell_a\ell_b)} - \sum_{c=3}^{n}\sum_{b=2}^{c-1}\sum_{a=1}^{b-1} 2^{(\ell_a\ell_b\ell_c)} + \ldots$$
$$(-1)^\beta \sum_{\mu=\beta}^{n} \ldots \sum_{b=2}^{c-1}\sum_{a=1}^{b-1} 2^{(\ell_a\ell_b\ldots\ell_\mu)} + \ldots]/k \quad ,$$

β being the running number of the rank of the expressions related to the symbols Σ between the square brackets, the last expression being obtained for $\beta = n$. In the particular case $k = k_1^\alpha$, N_k has a simpler expression :

$$(3.6) \qquad N_k = [2^k - 2^\ell]/k \quad , \quad \ell = k_1^{\alpha-1}$$

3.2.3. NUMBER OF THE BIFURCATIONS GIVING RISE TO THE CYCLES OF ORDER k.

Consider the quadratic recurrence (or map) (3.1), $f(x,\lambda) \equiv x^2-\lambda$. The following conjecture is now made : in the interval $\lambda_{(1)_0} < \lambda < \lambda_1^*$, there exist only the bifurcations giving rise to the N_k cycles given by (3.4)-(3.6). This conjecture may be also considered as a consequence of the *Myrberg's "hypothesis H"* [M55] which is under-mentioned. For now, the elements which permit to accept this conjecture are of geometrical nature by considering the construction of T^k in the (x_n,x_{n+1}) plane when λ varies in the indicated interval (cf. § 2.7.3). Indeed such a construction makes appear :

- (i) The abscissae of the extrema of T^k are $x_e = 0$ of T, and all the antecedents of x_e up to the rank k-1. Their ordinates are the abscissae of

the critical points C_i, $i \leqslant k-1$, $C_o \equiv C$, i.e. the values $G_{i+1}(\lambda)$ of (P3.1).
- (ii) When λ increases from $\lambda_{(1)_0}$ every antecedent of x_e which becomes real, remains real. So such an increase may not lead to the vanishing of extrema of T^k. It leads only to the birth of new extrema when $x_{c_i} = x_e$ (r consecutive critical points constitutes a cycle of order $r \leqslant k-1$ with a multiplier S=0), for $\lambda = \lambda_r^{j_o}$. Crossing through this "bifurcation" value for the shape of T^k, an extremum changes its nature (max,min) and gives rise to two new extrema on the both sides of the first one.
- (iii) Consider, with an increasing λ, two consecutive "bifurcations" related to the T^k shape, i.e. two consecutive primitive zeros of the set $G_j(\lambda)$, $j = 1,2,\ldots,k$. The difference, between the ordinates of an extremum of T^k and of the related two new extrema born from the first shape bifurcation, is a monotic function between the two consecutive bifurcations. The simplest case to show this point corresponds to the last shape bifurcation of T^k, $\lambda = \lambda_b$, when λ increases. As indicated in § 2.7.3, this bifurcation is such that the critical points $C, C_1, \ldots, C_{k-2}$ constitute a cycle of order k-1, and gives rise to 2 new extrema on the both sides of the extremum $x_e=0$ of T^k, $T^k x_e$ = ordinate $G_k(\lambda)$ of C_{k-1}. The ordinate of the 2 new extrema is $-\lambda$. So the difference of ordinates is $F(\lambda) = G_k(\lambda) + \lambda = G_{k-1}^2(\lambda)$, and $dF/d\lambda = 2G_{k-1}G'_{k-1}$. For $\lambda > \lambda_b, G'_{k-1} \neq 0$, $G_{k-1} > 0$, $G_{k-1}(\lambda_b) = 0$, and then $F(\lambda)$ is a strictly increasing function. This last property is a little harder to establish for the other cases. The proof can be lead with similar way.

With the above conjecture it is possible to determine the number $N_\lambda(k)$ of the λ bifurcation values, leading to the N_k cycles of order k defined for $\lambda = \lambda_1^*$. These bifurcations have only either the type (fold) (2.21) with multiplier S = +1, giving rise from "nothing" to two cycles of order k, one being attractive, the other repulsive, or the type (flip) (2.24) with S = -1, an attractive cycle of order k becoming repulsive and giving rise to an attractive cycle of order 2k. *The number* $N_\lambda(k)$ *is equal to the number of the real primitive zeros of* $G_k(\lambda)$. So if k is odd, the only possible bifurcations are (2.21) with S = +1, from which it may be concluded :

$$N_\lambda(k) = N_k/2 \quad , \quad k = \text{odd integer.} \tag{3.7}$$

If k is even, the corresponding cycles may have appeared either from a bifurcation (2.21) with S = +1, or from a bifurcation (2.24) with S = -1 giving a doubling order (period). In this last case the total number of the bifurcation values is $N_\lambda(k/2)$, and then :

$$N_\lambda(k) = [N_k - N_\lambda(k/2)]/2 + N_\lambda(k/2)$$

$$(3.8) \qquad N_\lambda(k) = [N_k + N_\lambda(k/2)]/2 \qquad , \quad k = \text{even integer.}$$

In the particular case $k = 2^a$,(3.8) becomes :

$$(3.9) \qquad N_\lambda(k) = 2^{k-1}/k \qquad , \quad k = 2^a$$

Equivalent formula is given in [M55] with :

$$(3.10) \qquad \begin{cases} N_\lambda(k) = [2^{k-1} - \varepsilon_k]/k & , \ \varepsilon_k = 0 \text{ if } k = 2^a,\ \varepsilon_k = 1 \text{ if } k = \text{prime integer} \\ \varepsilon_k = k \sum_m [N_\lambda(k/k_m)]/k_m & , \text{ if } k \text{ is even ,} \\ \varepsilon_k = 1 + k \sum_m [N_\lambda(k/k_m)]/k_m & , \text{ if } k \text{ is odd ,} \end{cases}$$

where k_m belongs to the complete set of the odd divisors of k. The table given here shows the important increase of N_k and $N_\lambda(k)$ with k. So with k = 30, N_k = 35790267 and $N_\lambda(k)$ = 17895679.

k.....	2	3	4	5	6	7	8	9	10	11	12	13	14	15	16	...	20
N_k....	1	2	3	6	9	18	30	56	99	186	335	630	1161	2182	4080	...	52377
$N_\lambda(k)$.	1	1	2	3	5	9	16	28	51	93	170	315	585	1091	2048	...	26214

The generalization of these results to the recurrence (3.3) implies some additional conditions about the function $f(x,\lambda)$ having only one extremum : the condition the non-existence of more than one attractive cycle, and a condition about the shape variation of $f(x,\lambda)$ with respect to the parameter λ ruling out the possibility of other bifurcations than the ones giving the numbers $N_\lambda(k)$ of (3.7)-(3.10).

From (P3.3) and (3.7) - (3.10) it appears that *the set of the bifurcation values has infinitely many accumulation points inside* $\lambda_{(1)_0} < \lambda < \lambda_1^*$

when $k \to \infty$. These accumulation points are called by Myrberg *singular values* of λ. Let $\lambda_{(1)_s}$ be the first singular value from $\lambda_{(1)_0}$ with an increasing λ. *Given a parameter* λ*-value such that* $\lambda > \lambda_{(1)_s}$. Consider the relations (3.4)-(3.6), and the existence of infinitely many accumulation of λ-values inside $\lambda_{(1)_0} < \lambda < \lambda_1^*$. *It results that inside the open interval* $]q_1^{-1}, q_1[$ (q_1 *repulsive fixed point with* $S > 1$, q_1^{-1} *its first rank antecedent*), *there are infinitely many accumulation points for the points set of the cycles of order* k *with* $k \to \infty$. This property has been used in § 2.8.3 and 2.8.4 before this justification. Whatever $\lambda > \lambda_{(1)_s}$ may be (in particular $\lambda > \lambda_1^*$), the set of all the cycles of order k, and their antecedents of any rank are inside $]q_1^{-1}, q_1[$, because with a continuous variation of λ no cycle point can pass through one of these two boundary points.

3.2.4. PROPERTIES RELATED TO $\lambda = \lambda_1^*$.

The parameter value $\lambda = \lambda_1^*$ corresponds to a non-classical bifurcation resulting from the merging into one another of two singular points of different nature : the repulsive fixed point q_1 with $S > 1$, and the critical point of rank two C_1. For the quadratic recurrence (3.1), $\lambda_1^* = 2$, $C_1 \equiv q_1 (x=2)$, $C \equiv q_1^{-1}$ (x=-2). The first paper devoted to this problem is due to Herschel |H 9| (1814), and after there are the ones of Julia [J 5] Fatou [F 1], Pincherle [P 2] (1917-1920), Ulam and Von Neuman [U 1] (1947), Pulkin [P 6] (1950). The equivalent case for a piecewise linear recurrence was studied by Ulam, Von Neuman [U 1] (1947) and by Baayen, Kuyk, Maurice in [B 1] (1962). Many papers, written after 1975 and which, for the most part, are ignorant of a part of these basic contributions, are not mentioned.

- a) *Herschel - Pincherle's results.*

The value $\lambda = \lambda_1^* = 2$ in (3.1) is very special in the sense that a general solution in terms of elementary functions becomes possible. Let $x_n = 2\cos y_n$, where y_n is a new dependant variable. After insertion into (3.1), there results the recurrence $\cos y_{n+1} = \cos 2 y_n$, which is modulo-$\pi$ equivalent to $y_{n+1} = 2y_n$. Since the solution of this last recurrence is $y_n = C.2^n$, C being an arbitrary constant, the general solution of $x_{n+1} = x_n^2 - 2$ is (cf. for example [H 9] [P 2] [G36]) :

(3.11) $$x_n = 2\cos\,[2^n \text{ arc}\cos(x_o/2)] \quad , \quad -2 < x_o < 2 \quad ,$$

where an appropriate branch of (arc cos) is chosen for each initial value x_o. The chaos generated here amounts to the selection of ordinates of a periodic function at exponentially increasing discrete abscissae. The closeness of a given iterated sequence $\{x_n\}, n = 0,1,2,\ldots$, with respect to the neighbouring ones is described by the sensitivity coefficient [G36] :

(3.12) $$\partial x_n/\partial x_o = [-2^{n+1}/\sqrt{1-(x_o/2)^2}]\,\sin\,[2^n \text{ arc}\cos\,(x_o/2)]$$

This coefficient is a rapidly increasing oscillating function of n. Hence two initially close iterated sequences diverge rapidly from each other as n increases (for this property cf. also [G18]). The same situation prevails for other recurrences involving a single extremum.

- b) *Fatou - Julia's results.*

For $x_{n+1} = x_n^2-2$, it was established (cf. for example [J35] p.186) that the points of all the cycles of order k = 1,2,3,..., and their antecedents of rank i = 1,2,... constitute a real set (E), (E) = [-2,2], the derived set (set of the limit points) (E') of which is perfect, and the antecedents of (E) are everywhere dense on (E'). In the particular case $\lambda = 2$, the set (E) belongs entirely to the real axis. In the complex plane, (E) constitutes the boundary of the atraction domain (basin) of the fixed point at infinity.

From these results it appears that the segment $\overline{CC_1} \equiv [-2,2]$ is invariant with respect to T. Every $x_o \in \overline{CC_1}$, generates an iterated sequence which remains inside $\overline{CC_1}$ without converging toward an attractive cycle (such a cycle does not exist for $\lambda = \lambda_1^*$). Inside $\overline{CC_1}$ the points x_n are repeled by the repulsive cycles in a very irregular way. Such a segment, and its equivalents when a critical point of rank i > 2 merges into a repulsive fixed point, or cycle,were called in 1975 *chaotic segment* ("segment stochastique" in French [M21] [M25]).

It appears also that all the sequences of antecedents of rank i=1,2,... of a point x_n belonging to $\overline{CC_1}$ remain inside this segment. The number of these antecedents is 2^i.

-c) *Pulkin's theorem.*

This result correponds to the general situations $\lambda = \hat{\lambda}$ for which a critical point of rank $i \geqslant 2$ merges into a point of a repulsive cycle of order $k = 1,2,\ldots$, or a limit point of the set of these cycles [M34]. The considered map is also more general. *Here* $f(x,\lambda)$ *may be a piecewise continuous function of the real variable* x (then a critical point is considered in the extented sense of § 2.8.5.c), or *a continuous function which is differentiable, or piecewise differentiable.* Several definitions are necessary before the formulation of the theorem.

A *limit point of class 1* is an isolated accumulation point for the set of the iterated points generated by $x_{n+1} = f(x_n \lambda)$. A *limit point of class p* is an accumulation point for the set of the limit points of class $q < p$. An *iterated sequence belongs to the class p* if the number of its limit points of class p is finite. An *iterated sequence is said to oscillate with respect to a set of repulsive cycles*, if all the points of these cycles, and only these ones, are limit points of the highest class for the considered iterated sequence. The *neighbourhood of a point ξ of a cycle* is the one-sided neighbourhood which is defined by a half open interval B_ξ, the extremity of which is $\xi \in B_\xi$.

Let $\{C_u\}$, $u = 1,2,\ldots,m$, be a set of m repulsive cycles. Let $\xi_1^u,\ldots,\xi_k^u$ be the points of a cycle C_u of $\{C_u\}$. A *regular neighbourhood* Δ_u of the cycle C_u is the union of the neighbourhoods of all the cycle points which satisfies the following conditions :

- (i) the neighbourhoods $\Delta(\xi_i^u)$ of the points of the cycle C_u do not overlap.
- (ii) Δ_u does not contain points belonging to the other cycles of $\{C_u\}$, and does not overlap with the neighbourhoods of the other cycles of $\{C_u\}$.
- (iii) $f(x, \lambda)$ is monotonic and continuous inside every $\Delta(\xi_i^u)$.
- (4i) among the $\Delta(\xi_i^u)$, there exists one neighbourhood designated by Δ_i^u (let the one which corresponds to i=1) such that $f[\Delta(\xi_i^u)] = \Delta(\xi_{i+1}^u)$, $i = 1,2,\ldots,k-1$. It is called *principal neighbourhood* of the points of the cycle C_u.
- (5i) whatever may be $\bar{x} \in \Delta_i^u$, the point $T^k\bar{x}$ belongs to the interval $[\xi_1,\bar{x}]$.

Let $\{\Delta_u\}$ be the set of the regular neighbourhood of $\{C_k\}$. Pulkin calls *completely invariant set* F_Δ, the set of the real numbers $\Delta = \bigcup \Delta_u$ such that :

- (i') all the Δ_u are regular.
- (ii') whatever may be $x_o \in F_\Delta$, the corresponding iterated sequence belongs to F_Δ, and every Δ_u contains a point of this sequence.
- (iii') whatever may be $x_o \in F_\Delta$, there exists a sequence of antecedents belonging to F_Δ, and at least one of its points belongs to a given Δ_u.

<u>(P3.4)</u> *Let* $\{C_u\}$ *be a set of repulsive cycles and a set of regular neighbourhoods* Δ_u, $u = 1,\ldots,m$, *such that a completely invariant set* F_Δ *exists, and such that no one cycle is contained in the closure of the set* $F_\Delta - \sum_1^m \Delta_u$. *Whatever be* $q \geq 2$, F_Δ *contains a non enumerable set of points which generate oscillating iterated sequences of class* q *with respect to* $\{C_u\}$. F_Δ *may be decomposed as follows :* $F_\Delta = E_o + \sum_{k=2}^{\infty} E_k + E_\infty$, *where* E_o *is the set of cycles points of finite order, and their antecedents of any order,* E_k *the set of points generating iterated sequences of class* k, E_∞ *the set of points generating sequences of class* ∞. *The set* E_o *is enumerable, the sets* E_k *and* E_∞ *have the power of the continuum. Every interval belonging to* F_Δ *contains points of* E_o, E_k , E_∞, $k = 2,3,\ldots$, *and points generating oscillating iterated sequences of class* q *with respect to the set of cycles whatever be* $q \geq 2$.

The initial proof [P6] of (P3.4) is not quite satisfactory. A complete proof is given in [C3]. This not very known theorem is particularly important, because it gives a classification of the iterated sequences in a chaotic behaviour, an a better understanding of the structure of the corresponding chaotic attractor. This occurs whenever a critical point of rank $i \geq 2$ merges into a repulsive cycle,or a limit point of the set of these cycles [G32] [M25] [G35] [M34]. It will appear that (P3.4) is a consequence of the fractal organization of the bifurcations set of an endomorphism. Moreover the proof given in [C3] permits to obtain the existence of an invariant measure for the maps which satisfy the conditions of the theorem.

- (d) *Baayen, Kuyk and Maurice's results.*

On the interval $I(-2;2)$ the quadratic recurrence (3.1), $f = x^2 - 2$, with $\lambda = \lambda_1^* = 2$ is conjugate to the piecewise linear map $x_{n+1} = g(x_n)$:

(3.13) $x_{n+1} = 2x$ if $0 \leq x \leq 1/2$, $x_{n+1} = 2-2x$ if $1/2 \leq x \leq 1$, considered on the interval J(0;1), i.e. there exists a single-valued uniquely inversible transformation u, I onto J such that $g(x) = u[f(u^{-1}(x))]$ [U 1], [G18]. So the non-wandering set of $f = x^2-2$ is in relation with the non-wandering set of g(x) via this transformation.

In [B1] a point $a \in J$ is written in a dyadic expansion : $a = 0.a_1 a_2 a_3 a_4 \ldots$, where each a_k is either 0 or 1, and where $a = \sum_1^\infty a_k / 2^k$. Let $\theta(x) = \{g^k(x) : k = 0,1,2,\ldots\}$ be the iterated sequence (orbit) of x, when $x \in J$, [B1] gives the following results.

(P3.5) *Let* $a \in J$, *and let* $0.a_1 a_2 a_3 \ldots$ *be a dyadic expansion of a.*

- *(a) Then for* $k = 1,2,3,\ldots$, $f^k(a) = 0.a_{k+1} a_{k+2} \ldots$ *if* $a_k = 0$, $f^k(a) = 0.\bar{a}_{k+1} \bar{a}_{k+2} \ldots$ *if* $a_k = 1$, $\bar{a}_i = 1-a_i$.
- *(b)* $\theta(a)$ *is finite if and only if a is rational,* $a = 0.a_1 a_2 \ldots a_m (b_1 b_2 \ldots b_p)^\infty$, *this last parenthesis with the index* ∞ *indicating the periodical repetition of* $b_1 b_2 \ldots b_p$. *The set of such numbers a is dense in* J. *The orbit is here made up of the points of a cycle of order* p, *and of* m *successive antecedents of one of these points.*
- *(c) There are irrational numbers* $a \in J$ *such that* $\theta(a)$ *is everywhere dense.*
- *(d) Let* E *be the set of all* $a \in J$ *such that* $\theta(a)$ *is everywhere dense. The set* E *is itself everywhere dense.*
- *(e) There is a set* N *of irrational numbers* $a \in J$ *such that* $\theta(a)$ *is nowhere dense. This set* N *is everywhere dense.*

For (e) an example of the corresponding irrational number is $a = \sum_1^\infty 2^{-(n+1)n/2}$, or $a = 0.1010010001000010\ldots$ It is easily checked that the interval $0.101001 < x < 0.1010001$ contains no point $\theta(a)$, hence $\theta(a)$ is not everywhere dense. The set N contains all numbers having a dyadic expansion of the form $0.a_1 a_2 \ldots a_p\, 1010010001000010\ldots$ Here $\theta(a)$ is a Cantor set.

The dyadic expansion of the fixed point q_1 (x = 0) is 0.000..., that of x = 1 is 0.1111..., that of the fixed point q_2 (x = 2/3) is 0.1010101..., that of the point α_1 (x = 2/5) of the cycle of order two (with $\alpha_2 = 4/5$) is 0.01100110011...

- (e) *Remarks.*

- The results of § 3.2.4.a - d are more and less complete variants of the Pulkin's theorem (P3.4).

- *The sequence* $a_1 a_2 a_3 \ldots$ *of the dyadic expansion of the point* $a \in J$ *in* (P3.5) *is the now very popular invariant coordinate introduced later in* [M]. Indeed considering the definition of a point a by the sequence $a_1a_2a_3\ldots$, a point b by the sequence $b_1b_2b_3\ldots$, the ordering $1 > 0$, a common part $b_1 = a_1,\ldots,b_m = a_m$, one has $a < b$ if $a_{m+1} < b_{m+1}$. This result remains the same if $\lambda \neq \lambda_1^*$, i.e. by considering the map : $f(x,\lambda)=(1-\lambda)2x+\lambda$, if $x \leqslant 1/2$, $f(x,\lambda)=2-2x$ if $x \geqslant 1/2$, $\lambda = \lambda_1^* = 0$ giving (3.13). In this case a point a must be defined by the sequence $a_1a_2a_3\ldots$, and not by a dyadic expansion. It will be shown in § 3.3.4. that the notion of invariant coordinate is itself an equivalent formulation of the *Myrberg's ordering law* [M55]. This law was published en 1963, the dyadic expansion for (3.13) in dec.1962, but these fundamental results remain unknown in so many contemporary publications which quote only [M 6] as an unpublished work made in 1977, i.e. 14 years after.

- The whole interval $[q_1,q_1^{-1}]$ contains transversal homoclinic orbits (degenerated case here) made up of the set of the rank i antecedents, i = 1, 2,..., of the points of the cycles of order k = 1,2,3,... (cf.|G35| p.397-400, and chapter 6).

3.2.5. PROPERTIES RELATED TO $\lambda > \lambda_1^*$.

When $\lambda > \lambda_1^*$, the Julia-Fatou's set (E) constitutes the non-wandering set at finite distance, and belongs to the interval $[q_1^{-1},q_1]$. All the cycles of T, given by (3.1), or (3.3) with additional conditions, are real and repulsive, and the number of those of order k is given by (3.4)-(3.6). The unique attractor is the fixed point at infinity and every point of $[q_1^{-1}q_1]$ different from (E) (it is the case of the critical points) has its successive consequents which tend toward this fixed point. In [F 1] [J 5] it is shown that the set (E) is *completely disjointed*, and is a Cantor set. Indeed considering the fig.3.1d, all the $2^{i-1}+1$ antecedents of rank i of the repulsive fixed point ($S > 1$) q_1, i = 1,2,..., are real. Let q_1^{-2}, $q_1'^{-2}$ be the antecedents i = 2, and $I_0 = [q_1^{-2},q_1]$, $I_1 = [q_1^{-1},q_1'^{-2}]$, $J =]q_1^{-2},q_1'^{-2}[$. J and its successive antecedents i = 1,2,... define an arborescent sequence of holes (open intervals) inside $I = [q_1^{-1},q_1]$ belonging to the basin of the fixed point at infinity. The closure of this sequence when $i\to\infty$ is (E).

Let T|(E) the recurrence (3.1) or (3.3) constrained to the set (E). It is conjugate to a full one-sided shift on two symbols (cf.[H10]). Denoting by Q^+ the set of semi-infinite sequences $q = \{q_n\}_{n=0}^{\infty}$ of symbols 0(or+), 1(or-), according to the rule $q_n(x) = 0$ if $T^n x \in I_0$, $q_n(x) = 1$ if $T^n x \in I_1$, with $x \in (E)$, consider the shift map $\sigma : Q^+ \to Q^+$. Then:

$$\sigma q = \{q_n\}_{n=1}^{\infty} \quad , \quad \sigma^{\alpha} q = \{q_n\}_{n=\alpha}^{\infty} \quad . \tag{3.14}$$

This notion of shift map seems due to Hadamard |H 2 |. It was used by Birkhoff [B 9], Morse and Hedlung [M52], and after by many authors.

As indicated in [H10] the dynamics of $T|(E)$ are easily read from the model provided by the shift. There are countably many cycles (periodic sequences) in Q^+, and the number of those of order k is given by (3.4)-(3.6). There are countably many sequences which are obtained from the antecedents of rank 1,2,... of the points of cyclic sequences. There are an uncountable number of aperiodic sequences corresponding to non-periodic non-wandering points of T. The cycles points of $\sigma|Q^+$ are dense in Q^+, so the cycles of T are dense in the non-wandering set (E).

3.3. THE MYRBERG'S RESULTS.

3.3.1. THE MYRBERG'S CHARACTERIZATION OF A CYCLE POINT.

Let $q^{(i)}$ be a point of a given cycle of order k, $i = 1,2,\dots,k$, with a multiplier $S > 0$, or $S < 0$. Denote by the same representation $q^{(i)}$ the abscissa of the point $q^{(i)}$. In [M54] Myrberg proves that $q^{(i)}$ satisfies the relation :

$$q^{(i)} = \underset{1}{\pm}\sqrt{\lambda \underset{2}{\pm}\sqrt{\lambda \pm \dots \underset{k}{\pm}\sqrt{\lambda + q^{(i)}}}} \quad , \quad \text{or } q^{(i)} = f_i^{-k}(q^{(i)}, \lambda) \tag{3.15}$$

from the equation $q^{(i)} = T^k q^{(i)}$. The first relation is related to the quadratic recurrence (3.1), and concerns the problem studied by Myrberg. The second relation (3.15) is a natural enlargement, related to the more general recurrence (3.3) with the above indicated additional conditions, and based on the antecedents of $q^{(i)}$ up to the rank k. So in (3.15) the choice of the sequence of the signs (+,-) before the radicals is given by T_i^{-k}, where the index i indicates that the determination of the inverse map T^{-k} is the one which follows the points of the considered cycle with $S \neq 0$, from the point $q^{(i)}$. In [M54] Myrberg gives the convergence conditions for (3.15) which is a "periodical radicals (or antecedents) series" by replacing successively $q^{(i)}$ by (3.15). Such a series is convergent if $\lambda > \lambda_a$, where the λ_a are different for two cycles born of the same (fold) bifurcation (2.21). The λ_a-value is such that the considered cycle, with multiplier $S > 1$ or $S < 1$, is always repulsive for $\lambda = \lambda_a$.

Let $[q^{(i)}]$ be the sequence of the k signs (+,-) in (3.15). *The first sign in* $[q^{(i)}]$ *is the one of the slope of* $f(x,\lambda)$ *at the point* $q^{(i)}$. *The following signs are given by the slopes of* $f(x,\lambda)$ *considered at the successive consequents of* $q^{(i)}$ *up to the rank* k-1. Independently of the definition of the abscissa $q^{(i)}$, and the convergence of (3.15), the sequence of signs $[q^{(i)}]$ can be considered as a binary characterization of the cycle point $q^{(i)}$. For the two cycles $S > 1$ and $S < 1$, born of the same (fold) bifurcation (2.21), an ambiguity occurs for distinguishing two equivalent points $q^{(i)}$, one belonging to the cycle $S > 1$, the other belonging to the cycle $S < 1$, as far as the multiplier of the latter is $0 < S < 1$. Indeed when $0 < S < 1$ the two corresponding sequences are identical, and the antecedents series (3.15) is divergent. When $S < 0$ the two corresponding sequences are different. Considering the form (3.15), it appears that $q^{(i)}$ is also given by an infinite sequence of (+,-) constituted by all the elements of $[q^{(i)}]$. This infinite sequence will be represented by $[q^{(i)}[\, \equiv [q^{(i)}]^{\infty}$ which is the periodic repetition "on the right" of all the elements of $[q^{(i)}]$. In order to distinguish the sequence $[q^{(i)}]$ from the definition of the abscissa $q^{(i)}$ the notation 0,1, with the correspondence $0 \to +$, $1 \to -$, will be now used for $[q^{(i)}]$ which then presents itself as a *binary word* associated with the point $q^{(i)}$, and constituted by a set of k digits 0,1.

So let $[q^{(i)}] = [a_1 a_2 \ldots a_k]$, $a_j = 0$ or 1, $j = 1,\ldots k$, be the binary word of $q^{(i)}$, point of the considered cycle of order k ($S \neq 0$). Then the binary word associated with the rank-one consequent $q^{(i+1)}$ of $q^{(i)}$, $q^{(i+1)} = Tq^{(i)}$, is $[q^{(i+1)}] = [a_2 a_3 \ldots a_k a_1]$, or :

(3.16) $\quad [q^{(i+1)}] = \sigma [q^{(i)}] \qquad \sigma$ = rotational shift operation on $[q^{(i)}]$

and $[q^{(i+j)}] = [a_{1+j}a_{2+j}\ldots a_k \ldots a_j]$, $[q^{(i+j)}] = \sigma^j[q^{(i)}]$, $j = 1,\ldots,k-1$, the index of a being at most equal to k.

3.3.2. THE MYRBERG'S ROTATION SEQUENCE.

Let $q^{(1)}$ be a point of a cycle of order k with S = 0, which has the minimum (resp. maximum) abscissa when the extremum of $f(x,\lambda)$ is a minimum (resp. maximum). One has $q^{(1)} = f(x_e,\lambda)$, and $q^{(1)} = -\lambda$, for (3.1). By inverse iteration, $f_j^{-(k-1)}(x_e,\lambda)$, of the extremum x_e, where j indicates the determination of $T^{-(k-1)}$ which follows the points of the considered cycle, one has :

$$f_j^{-(k-1)}(x_e,\lambda) = f(x_e,\lambda) = q^{(1)} \tag{3.17}$$

For the quadratic map (3.1), (3.17) is the *Myrberg's radical equation,* or equation of the antecedents :

$$\lambda = \underset{0}{\sqrt{}}\lambda \pm \underset{1}{\sqrt{}}\lambda \pm \ldots \pm \underset{k-2}{\sqrt{\lambda}} \qquad \text{or} \qquad \lambda = r(\lambda) \tag{3.18}$$

In (3.17), $f_j^{-(k-1)}$ defines a given sequence of (k-2) signs (+,-). Among of all the possible permutations [p] of these (k-2) signs, only some of them correspond to the real roots of the Myrberg's equation of consequents $G_k(\lambda) - x_e$. The corresponding [p] are associated with all the possible cycles of order k with S = 0. These particular permutations constitute sequences [r] of (k-2) signs (+,-) :

$$[r] = \overset{+}{m}_1\overset{-}{n}_1\overset{+}{m}_2\overset{-}{n}_2\ldots\overset{+}{m}_q\overset{-}{n}_q \tag{3.19}$$

m_i(resp.n_i), $i = 1,2,\ldots,q$, indicating the number of consecutive signs + (resp.-) in [r]. They are named the *Myrberg's rotation sequences* (binary sequence). In the more general case (3.3), [r] is *constituted by the sequence of the slopes signs of* $f(x,\lambda)$ *at the points* $q^{(2)}, q^{(3)}\ldots q^{(k-1)}$, $q^{(i+1)} = Tq^{(i)}$, $i = 1,\ldots,k-1$. *The first sign of* [r] *before the radical 1 in* (3.18), *is always* +. [r] characterizes completely the cyclic transfer of the points of a given cycle by successive iterations using T. The values of λ, which are limit points for the real roots of $G_k(\lambda) = x_e$, when $k \to \infty$, are called by Myrberg *singular values of* λ. For these singular values, sequences [r] are also defined |M55|.

The above defined point $q^{(1)}$ of a cycle of order $k > 2$ is represented by the sequence $[q^{(1)}] = [1[r]z]$, with $z = 0,1$ according to the sign of its multiplier S. For a cycle with $S > 1$, [r] represents the rotation sequence of the "parallel cycle" with $S < 1$ born of the same fold bifurcation (2.21). Conventionally $[q^{(1)}]$ will be called the *Myrberg's sequence of this cycle*. Besides its characterization of the cyclic tranfer of the cycle points by successive iterations, $[q^{(1)}]$ gives the sign of the multiplier (> 0 if the number of 1 is even, < 0 if it is odd).

3.3.3. THE MYRBERG'S ORDERING LAW. CONSEQUENCES.

This law was obtained in [M55] directly from the expression of two radical equations (3.18) corresponding to two Myrberg's rotation sequences [r] [r'] with the hypothesis that each radical is real. By arguing about antecedents, its generalization to (3.3) with the above indicated additional conditions becomes evident.

(P3.6) *Myrberg's ordering law. Let* $[r] = m_1^+ n_1^+ \ldots m_q^+ n_q^-$, $[r'] = m_1'^+ n_1'^- \ldots m_q'^+ n_q'^-$ *be two sequences of rotation, either associated to two cycles of order* k, k' (k = k',*or* k ≠ k'), *or having an infinite number of signs (association to singular values). Let* $[r_i] = m_i^+ n_i^- \ldots m_q^+ n_q^-$, $i = 2,3,\ldots,q$. *One has* [r'] > [r] *in the following cases :*

a) if $m_1' > m_1$; *b) if* $m_1' = m_1$, *with* $n_1' < n_1$ (n_1' *is even*),*or* $n_1 < n_1'$ (n_1 *is odd*) ; *c) if* $m_1' = m_1$, $n_1' = n_1$, *with* $[r_2'] < [r_2]$ (n_1 *is odd*), *or* $[r_2'] > [r_2]$ (n_1 *is even*) ; *d) if* $m_j' = m_j$, $n_j' = n_j$, $j = 1,2,\ldots,q$, *it is proceeded as in c) with* $[r_i]$, $[r_i']$, $i = 3,4,\ldots,q$. *In this process,* $n_q' = 0$,*or* $n_q = 0$, *are considered as even number.*

Let (A) (B) be the concatenation of two sequences of signs, i.e. the sequence (AB) made up of the writing of A followed by B. An equivalent formulation to (P3.6) having a more "compact" form, and so being easier to use is (cf. [C 2] p.65-66) :

(P3.6') *Let* [r] = (A)(B), [r'] = (A)(B') *be two Myrberg's rotation sequences, such* (A) *in* [r], [r'] *represents the same sequence of signs (+,-).*

Let b_1 *the first sign of* (B), b_1' *the first sign of* (B'), ϕ *taking the place of this sign if* [r], *or* [r'], *is only constituted by* (A). *The following ordering relation between* b_1, b_1' *is set* : $(-) < \phi < (+)$. *Then one has* $[r'] > [r]$:

a) if (A) *contains no sign* −, *or an even number of signs* −, *with* $b_1' > b_1$.

b) if (A) *contains an odd number of signs* −, *with* $b_1' < b_1$.

In [M55] Myrberg gave also the following important result :

<u>(P3.7)</u> *Equation* (3.17), *or* (3.18), *has a real root associated with a cycle defined by* [r], *if and only if* $[r] > [r_i]$, $i = 2,3,\ldots q$.

<u>(P3.7')</u> *Or equivalently :*

The necessary and sufficient condition for a permutation [p] *of* (k-2) *signs (begining with* +*) to be a rotation sequence* [r] *of a cycle of order* k *is* $[p] > [p_i]$, $i = 2,3,\ldots,q$. *Then* [p] *is said a r-admissible sequence (or permutation).*

So $[p] = m^+ n^-$, $n^- = 0,1,2,3,\ldots$, is always a r-admissible permutation, because $[r_i] = \phi$, $i > 1$.

The unicity of the real root of (3.17) (3.18), for a given [r], constitutes the *Myrberg's hypothesis H*. Indirectly, this hypothesis is justified because , if it is not the case, it results the existence of attractive cycles, which cannot be derived from cycles with S = 0, by a continuous variation of λ (cf. [M55]63, p.16). Such a situation is excluded by the properties of the curve T^k in the (x_n, x_{n+1}) plane, as indicated in § 3.2.3 (i) - (iii) (here (iii) necessitates a proof for the general case). A direct justification of unicity using the radical equation (3.18) is given in [R 1] for the particular case of $[r] = m^+ n^-$, $n^- = 0,1,2,\ldots$. For the quadratic map (3.1) the results of [D 3] obtained in the complex map constitutes a proof of the unicity of every real root of (3.18) for a given [r] (cf. also [H10]). This ensures the validity of the Myrberg's hypothesis H for the quadratic maps and justify the following Myrberg's result [M55]63:

<u>(P3.8)</u> $[r'] > [r]$ *is equivalent to* $\lambda' > \lambda$ *when* $f(x,\lambda) \equiv x^2-\lambda$.

<u>Consequence of the Myrberg's ordering law : ordering on the x-axis.</u>

Being obtained from the radical equation (3.18), this law has an evident application when the radical equation is the one (3.15) which defines a cycle point $q^{(i)}$. Then it becomes possible to compare the relative positions (on the x-axis) of the points of a given cycle, or the points belonging to different cycles, using the corresponding binary words $[q^{(i)}]$ with $0 \to (+)$, $1 \to (-)$. So implicitely, the Myrberg's papers [M54] [M55] contain the following result:

<u>(P3.9)</u> *On the x-axis, the relative positions of two points of a given cycle, or of two points belonging to two different cycles, of order* $k = 1, 2, \ldots$ *are given by using the Myrberg's law* (P3.6) (P3.6') *applied to the corresponding binary words* ($0 \to +$, $1 \to -$). *Then in* (P3.6') *the absence of a common part* (A) *is noted* Φ *considered as even, the ordering of* b_1, b'_1 ($b_1 = \Phi$ *if* b_1 *is absent*) *is given by* $1 < \Phi < 0$. *The so obtained ordering of the binary words is that of the corresponding abscissae if the extremum of* $f(x,\lambda)$ *is a minimum. It is the inverse of the ordering of the corresponding abscissae if the extremum is a maximum.*

So let the cycle of order k=5 (S < 1) related to $[r] = + - -$, when S=0 for $\lambda = \lambda°$. For $\lambda > \lambda°$, S becomes negative, and the points $q^{(i)}$, $i = 1,\ldots,5$, $q^{(i+1)} = Tq^{(i)}$, verify :

$$[q^{(1)}] = [10110] < [q^{(4)}] = [10101] < [q^{(3)}] = [11010] < [q^{(5)}] = [01011] < [q^{(2)}] = [011101].$$

If the extremum of $f(x,\lambda)$ is a minimum, the abscissae of the points $q^{(i)}$ verify $q^{(1)} < q^{(4)} < q^{(3)} < q^{(5)} < q^{(2)}$. If it is a maximum : $q^{(1)} > q^{(4)} > q^{(3)} > q^{(5)} > q^{(2)}$. The map (3.1), or (3.3), has two fixed points : q_1 with $[q_1]=0$, q_2 with $[q_2]=1$ as soon as its multiplier is < 0. So one has $[q_1] > [q^{(i)}]$ where $q^{(i)}$ is a point of whatever cycle. For $[r] = + - -$, k=5, one has $[q^{(4)}] < [q_2] < [q^{(3)}]$.

If the infinite sequence $[q^{(i)}[$ is used, instead of the binary word $[q^{(i)}]$, (P3.9) has a similar formulation. It appears that (P3.9) is a variant of the result given in § 3.2.4 e. So (P3.9) may be applied for points which do not belong to cycles, by considering the binary iterated sequences generated by (3.1) with these points as initial conditions.

3.3.4. THE MYRBERG'S ORDERING LAW AND ITS RELATION TO THE NOTION OF INVARIANT COORDINATE AND KNEADING INVARIANT.

The law (P3.9) has the evident following equivalent formulation. Let $[q^{(i)}[$, $[q'^{(j)}[$ be two semi-infinite binary sequences which are not necessarily related to cycles . Let (A) be the common initial part of these semi-infinite sequences, (A) being made up of α digits 0, 1, and $\theta_{\alpha+1}(q^{(i)})$, resp. $\theta'_{\alpha+1}(q'^{(i)})$, the products of the $(\alpha+1)$ first digits of $[q^{(i)}[$, resp. $[q'^{(j)}[$, according to the rule $0 \times 0 = 0$, $1 \times 1 = 0$, $0\times1 = 1\times0 = 1$, $\theta_{\alpha+1} = \phi$ if $f^{\beta}(q) = x_e$, $0 \leqslant \beta \leqslant \alpha$, $1 < \phi < o$. If the abscissae are such that $q^{(i)} < q'^{(j)}$, then $\theta_{\alpha+1}(q^{(i)}) < \theta_{\alpha+1}(q'^{(j)})$ when the extremum of $f(x,\lambda)$ is a minimum, and $\theta_{\alpha+1}(q^{(i)}) > \theta_{\alpha+1}(q'^{(j)})$ when it is a maximum.

In [M 6] (cf. also [G18]) the *invariant coordinate* of a point x is defined by $\Theta(x) = \{\theta_n(x)\}_{n=0}^{\infty}$, $\theta_n(x)$ being the product of the n first digits of x, in order to obtain a lexicographic ordering of the sequences, with the property that the mapping $x \to \Theta(x)$ is monotone decreasing when the extremum of $f(x,\lambda)$ in a maximum. So it appears that this result is a simple variant of the Myrberg and Baayen-Kuyk-Maurice's results of [M55] [B 1]

In [M 6] (the extremum x_e of $f(x,\lambda)$ being a maximum, $f(x,\lambda)$ verifying conditions ensuring the existence of one attractor at most) the *kneading invariant* of $f(x,\lambda)$, is defined by $\mathfrak{I}(\lambda) = \lim_{x \uparrow x_e} \Theta(x)$. This notion is very popular since [M 6], and is considered as a discovery of outstanding importance. Nevertheless this notion, and its consequences about the ordering of the bifurcation set of (3.1), or (3.3), is contained into the Myrberg's ordering law (P3.6) (P3.6') published long before [M6]. Indeed consider the sequence $[q^{(k)}] = [+ -[r]] \equiv [0\ 1[r]]$, $Tq^{(k)} = q^{(1)}$, $q^{(k)}$ being the first point of the sequence, located at $x \to x_e$, such that the first sign of the sequence is + (0). Due to the presence of the sign - before [r] in $[q^{(k)}]$, $I(\lambda) = \Theta(q^{(k)})$ gives the inverse order of that of the rotation sequences [r] obtained by applying the Myrberg's ordering law (P3.6), or (P3.6'), which long before [M 6] gave the ordering of the bifurcation set of (3.1) via (P3.8), from the hypothesis H. Consequently this ordering occurs for (3.3), $f(x,\lambda)$ verifying additional conditions about its shape evolution when λ increases (cf. [M21] and the general result of [G15] defining completely the topological equivalence class of $f(x,\lambda) \equiv x^2-\lambda$).

The notion of admissible kneading invariant, defined in [M6], is also contained in the Myrberg's results because it is a variant of (P3.7'), which is a more "compact" form, and so easier to use. Indeed if a sequence [p] is r-admissible then the invariant coordinate of [0 1 [p]] is an admissible kneading invariant, and applying (3.7') leads to less operations than those implied by the definition the admissible kneading invariant [M6].

Let $[q^{(m)}[$ be a two-symbols semi-infinite sequence. Suppose that a finite number of shift operations permits to obtain a sequence $[q^{(1)}[$, which is the smallest in the sense of (P3.9). Writing $[q^{(1)}[= [1\ [p]\ z[$, $[q^{(m)}[$, $m = 1,2,3,\ldots$, *is said to be* $\bar{r}$*-admissible if* [p] *is r-admissible.* It was proved in [M 6] (cf. also [G16]) that if $\mathcal{I}$ is an admissible kneading invariant then there is a map $f(x,\lambda)$ for which $\mathcal{I} = \mathcal{I}(\lambda)$. Consequently if a sequence $[q^{(m)}[$ is a $\bar{r}$-admissible there is a map $f(x,\lambda) = x^2-\lambda$ for which an iterated sequence corresponds to a signs sequence identical with $[q^{(m)}[$.

The notion of kneading invariant being a variant of the Myrberg's laws, its use in order to obtain the bifurcation set of (3.1), *or* (3.3), *implies also the above mentioned hypothesis* H.

3.3.5. FUNDAMENTAL BIFURCATIONS AND MYRBERG'S SINGULAR PARAMETER VALUES OF THE 1st TYPE.

The quadratic recurrence (3.1) is considered. As already indicated, two fundmental bifurcations occur in (3.1), or (3.3). The first one (cf. (2.21)) is:

$$\phi \to \text{at}(1)\ C_k + \text{rep}(1)\ C_k \qquad (3.20)$$

in which an attractive cycle (type 1) of order k, with multiplier $S < 1$, merges into a repulsive (type 1) one, with $S > 1$, for $\lambda = \lambda^{j'}_{(k)_0}$ when the multiplier is $S = +1$. Here the index j' is related to the type of cyclic transfer of the cycle points by applications of T, and ϕ indicates the absence of cycle with the same cyclic transfer for $\lambda < \lambda^{j'}_{(k)_o}$ when (3.1) is considered. The second type of bifurcation is the Myrberg's period doubling cascade [M55] of bifurcations (2.24).

$$\text{(3.21)} \qquad at(2)\ C_{k2^i} \to rep(2)\ C_{k2^i} + at(1)\ C_{k2^{i+1}}$$

which occurs for $\lambda = \lambda^{\ell'_{i+1}}_{(k2^{i+1})_o}$, when the multiplier of the cycle of order $k2^i$ crosses through the value S = -1, after a monotone decrease in the interval 1 > S > -1, when λ increases in (3.1). Here i = 0,1,2,..., and the bifurcation values for (3.21) constitute an increasing sequence with a limit value $\lambda^{j'}_{(k)_s} < 2$ [M55]. *A cycle of* (3.21) *is such that its multiplier* S *decreases monotonically from* +1 *to* $-\infty$, *when* λ *increases from* $\lambda^{j'}_{(k.2^i)_o}$ *to* ∞. *In the same conditons, the multiplier* S *of the repulsive cycle in* (3.20) *increases from* +1 *to* ∞. These properties are obtained explicitly for k = 1,2,3, i = 0, because the analytic expression of S can be obtained as indicated in § 4.4.1. For the general case they are justified from the properties of the curve T^k in the (x_n, x_{n+1}) plane, and particularly from § 3.2.3 (iii), considering the slope of T^k at its intersection with $x_{n+1} = x_n$. The bifurcation set of (3.21), i = 0,1,2,... was called by Myrberg a *spectrum*. This result is now wrongly attributed to a paper writen in 1977 [F 3].

The cycles generated by (3.21) with S = 0 are such that (cf. [M55] :

$$\text{(3.22)} \qquad [r^{\ell'_{i+1}}_{(k.2^{i+1})}] = [r^{\ell'_i}_{(k.2^i)}]\ [\pm\ -]\ [r^{\ell'_i}_{(k.2^i)}]$$

where $[r^{\ell'_i}_{(k.2^i)}]$ is the rotation sequence of the cycle of order $k.2^i$, with a cyclic transfer defined by ℓ'_i, and where the double sign $\pm$ is + (resp.-) if the number of signs - into $[r^{\ell'_i}_{(k.2^i)}]$ is even (resp. odd). Here it is considered that $[r^{\ell'_i}_{(k.2^0)}] \equiv [r^{j'}_{(k)}]$ cannot be put in the form (3.22). Using (P3.6), (P3.6'), the $[r^{\ell'_i}_{(k.2^i)}]$, i = 0,1,2,..., constitute an increasing sequence of rotation sequences. For i = ∞, a *singular rotation sequence* $[r^{j'}_{(k)_s}]$ is defined for the singular value $\lambda^{j'}_{(k)_s}$ [M55].

The interval $\omega^{j'}_k = [\lambda^{j'}_{(k)_o}, \lambda^{j'}_{(k)_s}]$ also is called a *spectrum* by Myrberg who gives the following theorem [M55]:

(P3.10) *For every* $\lambda \in \omega_k^{j'}$ *, only one attractive cycle exists. Its order is* $k2^i$, *or* $k2^{i+1}$, *into the interval bounded by two zeros of* G_{k2^i}, $G_{k2^{i+1}}$

The relation (3.22) between two consecutive rotation sequences of the Myrberg's period doubling cascade can be written:

$$[r^{\ell'_{i+1}}_{(k.2^{i+1})}] = \{[r^{j'}_k](\pm\ -)\}^a[r^{j'}_k]\ ,\ a = 2^i-1,\ k = 1,3,4,\ldots, \tag{3.23}$$

Here $\{---\}^a$ indicates the repetition a-times of the sequences $[r^{j'}_k]$ (± -). For every one of these sequences, the double sign ± inside (± -) is + (resp. -) if the number of signs - on the left of the considered ± is zero, or even (resp odd). The sequence $[r^{j'}_k]$ is that of the cycle given by the fold bifurcation (3.20). Considering now the variant of the Myrberg ordering law using the kneading invariant, if $[\beta^{\ell'_i}_{k.2^i}]$ is the kneading invariant corresponding to $[r^{\ell'_i}_{k.2^i}]$ one has :

$$[\beta^{\ell'_{i+1}}_{k2^{i+1}}] = [\beta^{\ell'_i}_{k2^i}]\ [\bar{\beta}^{\ell'_i}_{k2^i}]\ ,\qquad k = 1,3,4,\ldots,\quad i = 0,1,2,\ldots, \tag{3.24}$$

which is obtained by the concatenation of two sequences of $k2^i$ signs, $\bar{\beta}$ being the opposite sequence with respect to β, i.e. the sequence obtained from β by changing + into -, and - into + [H10].

Remarks

- (i) If the more general recurrence (3.3) is now considered, the properties of this paragraph occur when the function $f(x,\lambda)$ is such that its shape evolution with respect to λ leads to the same sequence of bifurcations as the one of (3.1).

- (ii) If the rule, concerning the choice of the first sign in (± -) for (3.23), is reversed (sign - (resp +) with an even (resp odd) number of - on the left of ±), the equation (3.17) of the antecedents has still a real root. However this real root is related to a λ-value corresponding to a non-primitive root of $G_m(\lambda) = x_e$, $m = k2^{i+1}$, i.e. it corresponds to a cycle having an order which is a divisor of 2^{i+1}. So one has the following correspondences for sequences of k-2 signs :

+--+ → [r] = + (k=3); $(+-)^3$ → [r] = +- (k=4)

++--++ → [r] = ++ (k=4); +----+-- → [r] = +-- (k=5)

++-+-++- → [r] = ++- (k=5)

3.3.6. VARIANT OF THE SINGULAR PARAMETER VALUES OF THE (1st) TYPE.

Instead of the rotation sequences $[r_{k2}^{\ell'_i} i]$ resulting from the period doubling cascade, now let us consider the rotation sequences $[r_{k_1(k_2)}^{m'_i} i]$, associated with a sequence of cycles of order $k = k_1(k_2)^i$, $i = 1,2,\ldots$, and given by the following rule :

$$(3.25) \qquad [r_{k_1(k_2)}^{m'_{i+1}} i+1] = \{[r_{k_1(k_2)}^{m'_i} i](\pm -)\}^{k_2-1} [r_{k_1(k_2)}^{m'_i} i] ;$$

where $k_1 = 1,3,4,\ldots$, $k_2 = 3,4,5,\ldots$, $i = 0,1,2,\ldots$ These sequences are r-admissible. Here (3.25) differs from (3.22) in that it does not correspond to a bifurcation. The rule of choice of the first sign into ($\pm$ -) is is the same as the above indicated for (3.22). For an increasing i, an increasing set of $[r_{k_1(k_2)}^{m'_i} i]$ is obtained (cf.(P3.6')), and so for the quadratic map (3.1) it corresponds an increasing set of λ-values giving cycles of order $k_1(k_2)^i$ with multiplier $S = 0$. When $i \to \infty$, these λ-values have an accumulation point $\lambda_{k_1}(s_{k_2})$.

The sequence (3.25) are not due to Myrberg. The first descriptions are posterior to Myrberg (cf.[D2] [H10]) and use the kneading invariant notion .

If the rule of choice of the first sign into ($\pm$ -) is reversed, as in § 3.3.5 the equation (3.25) of the antecedents has still a real root, corresponding to a non-primitive root of $G_m(\lambda) = x_c$, $m = k_1(k_2)^{i+1}$. Here the sequence given by (3.25) corresponds to a cycle of order $k_1(k_2)^i$. So $(+--)^2+ \to [r] = +(k-3)$.

3.3.7. CONSEQUENCE OF THE MYRBERG ORDERING LAW : FIRST NUMBERING OF THE CYCLES OF THE SAME ORDER. NON-EMBEDDED REPRESENTATION.

The Myrberg's ordering law permits to arrange the cycles with the same order k in an increasing set. From this point of view, it is noted that two cycles, generated from (3.20), have the same cyclic transfer characterized by the rotation sequence $[r_k^{j'}]$, defined for the attractive cycle when its multiplier is $S = 0$. It is noted also that this cyclic transfer does not change, whatever be the multiplier of the considered cycle. It is the same for a cycle generated from (3.21), the cyclic transfer of which remains the same (defined by $[r^{\ell'_i}_{(k.2^i)}]$ whatever be the multiplier. So the cyclic transfer related to cycles can be characterized by $[r]$ for multipliers $S \neq 0$.

Then, *whatever be the multiplier* S, *a cycle will be represented by the double symbol* (k,j'), k *being the order of the cycle,* j' *being the running number of the corresponding* [r] *into the increasing set of the* [r] *associated with all the cycles having the same order* k. Bifurcation (3.20) gives two cycles (k,j'). Bifurcation (3.21) gives only one cycle $(k.2^{i+1}, \ell'_{i+1})$, ℓ'_{i+1} taking the place of j'. So the running number j' (or ℓ'_i) defines the type of cyclic transfer related to the considered cycle. The representation (k,j') does not take into account the fractal "box-within-a-box" bifurcation structure of T defined by (3.1),or (3.3). It will be called *non embedded representation*. Here $j' = 1,2,\ldots,p'_k = N_\lambda(k)$ (cf. § 3.2.3).

Let $[q^{(1)j'}_k]$ be the binary word of the point of a cycle (k,j'), which has the minimum (resp.max.) abscissa if the extremum of $f(x,\lambda)$ into (3.3) is a minimum (resp.max.). As above indicated this binary word can be used to characterize the considered cycle, because $[q^{(1)j'}_k] \equiv [1[r^{j'}_k]z]$, z=0,1 according to the sign of the multiplier S. This sign is positive (resp. negative) if the number of characters 1 into the binary word is even (resp.odd).

With the non-embedded representation, the λ-bifurcation values related to a cycle (k,j') are indicated by an equivalent symbolism. So the λ-value, giving the fold bifurcation (3.20) for a pair of cycles (k ,j'), is $\lambda^{j'}_{(k)_0}$.

3.3.8. MYRBERG' SINGULAR PARAMETER VALUES OF THE 2nd TYPE.

Let $\breve{\lambda}$ be a λ-value such that the critical point $C_{\alpha-1}$ of rank α (consequent of rank α of the extremum x_e) is the first critical point of the least rank,which merges into a point q belonging to a repulsive cycle (k,j'), i.e. $f^\alpha(x_e,\breve{\lambda}) = q = f^{-k}_{j'}(q,\breve{\lambda})$ by considering (3.15). $\breve{\lambda}$ is said to be a *singular value of the* 2nd *type*. For the quadratic recurrence (or map) (3.1), $x_e = 0$, $f(0,\breve{\lambda}) = -\breve{\lambda}$, and $\breve{\lambda}$ satisfies a "periodical radical series" equation given by [M42]:

(3.26) $\check{\lambda} = -f_c^{1-\alpha}[f_{j'}^{-k}(q,\check{\lambda})]$, $q = f_{j'}^{-k}(q,\check{\lambda})$, $\alpha > 1$, or $\check{\lambda} = r(\check{\lambda})$

where the index c indicates the determination (or the branch) of the map $T^{1-\alpha}$ corresponding to the critical points different from the cycle (k,j'). These $\check{\lambda}$-values are always such that $\check{\lambda} > \lambda_a$, λ_a ensuring the convergence of the "radical series" (3.15). So (3.26) is similar to the radical equation (3.18), the argument q of $f_{j'}^{-k}$ being a Myrberg's periodical infinite series of $\check{\lambda}$. If the more general recurrence (3.3) (with the above indicated conditions) is now considered, $\check{\lambda}$ verifies a "periodical antecedents series" obtained from $q = f^{\alpha}(x_e,\check{\lambda})$. Then $\check{\lambda}$ is associated with a sequence $[\check{r}]$, made up of an infinite numbers of sign ±, and which is periodical from a certain rank :

(3.27) $$[\check{r}] = (m_1^+ n_1^- \ldots m_s^+ n_s^-)(m_{s+1}^+ \; n_{s+1}^- \ldots\ldots m_t^+ n_t^-)^{\infty} \equiv (A)(B)^{\infty}$$

where $m_{s+1}^+ + n_{s+1}^- + \ldots + m_t^+ + n_t^- = k$ number of the components of (B). In (3.27) m_s, or n_s, or m_{s+1}, or m_t, or n_t can be equal to zero. *The signs of the first set* (A) *are given by the determinations of* $f_c^{1-\alpha}$, *i.e. by the slopes signs of* $f(x,\lambda)$ *at the critical points* $C_1, C_2, \ldots, C_{\alpha-2}$, consequents of rank $2,\ldots,\alpha-1$, of the extremum x_e. Moreover (B) ≡ [q] (cf.§ 3.3.1), and the power index "∞" of (B) indicates the infinite periodical repetition of the set (B), i.e. $(B)^{\infty}$ ≡ [q[. With $\check{\lambda} > \lambda_a$, the $[\check{r}]$ may be ordered as the [r], and also compared with the [r], via the Myrberg's ordering law.

Some examples of Myrberg's infinite periodical sequences.

- a) Let q_1 be the fixed point of (3.3), $q_1 = 1/2 + (1/4 + \lambda)^{\frac{1}{2}}$. The radical equation of q_1 is $q_1 = \sqrt{\lambda + q_1}$, $\lambda_a = 0$ [M 54], and $[q_1[\equiv (+)^{\infty}$. Let $\check{\lambda} \equiv \lambda_1^* = 2$ be the λ-value such that $C_1 (x = \lambda^2 - \lambda)$ merges into q_1. Here in $f_c^{1-\alpha}$, $\alpha = 2$, $[\check{r}] \equiv (+)^{\infty}$.
- b) Let q_2 be the fixed point of (3.3), $q_2 = 1/2 - (1/4 + \lambda)^{\frac{1}{2}}$. Here $q_2 = -\sqrt{\lambda + q_2}$, $\lambda_a = 1$ [M54], and $[q_2[\equiv (-)^{\infty}$. Let $\check{\lambda} \equiv \lambda_{2^1}^* = 1.543689013$ [G32], the λ-value such that C_2 $(x = (\lambda^2-\lambda)^2-\lambda)$, critical point of rank 3, merges into q_2. One has $\alpha = 3$, (A) ≡ (+), and $[\check{r}] \equiv (+)(-)^{\infty}$.

Let now $\check{\lambda} = 1.892910988$ [M34] [C 6] be the λ-value such that the critical point of rank 4, C_3, merges into q_2. One has $\alpha = 4$, $(A) = (++)$ and $[\tilde{r}] \equiv (++)(-)^{\infty}$ for $\lambda = \tilde{\lambda}(C_3 \equiv q_2)$.

- c) Let $\alpha_1, \alpha_2, \alpha_3$, $\alpha_1 < \alpha_3 < \alpha_2$, be the abscissae of the cycle points of order 3, with the multiplier $S < 1$. When λ increases this cycle becomes repulsive with $S < -1$, S being a decreasing function of λ. Here $[\alpha_1] \equiv (-++)$, $[\alpha_3] \equiv (+-+)$, $[\alpha_2] \equiv (++-)$ and $\lambda_a \simeq 1.92$ [M54].

Let $\beta_1, \beta_2, \beta_3$, $\beta_1 < \beta_3 < \beta_2$, be the abscissae of the cycle points of order 3 with $S > 1$, S being an increasing function of λ. Here $[\beta_1] \equiv (-+-)$, $[\beta_3] \equiv (--+)$, $[\beta_2] \equiv (+--)$, and $\lambda_a \simeq 1.77$ [M54].

Let $\tilde{\lambda} \equiv {}^*\lambda_3^1 = 1.924661063$ [M33] [M42] be the λ-value such that C_2 is the least rank critical point belonging to the cycle $\beta_1, \beta_2, \beta_3$. For this λ-value, C_2 merges into β_2. Here $(A) \equiv (+)$, and $[\tilde{r}] \equiv (+)(+--)^{\infty} \equiv ++(--+)^{\infty} \equiv ++-(-+-)^{\infty}$.

Let $\tilde{\lambda} = \lambda_3^{*1} = 1.790327493$ [G32] [M21] be the λ-value such that C_3 is the least rank critical point belonging to the cycle $\beta_1, \beta_2, \beta_3$. For this λ-value C_3 merges into β_1. Here $(A) \equiv (++)$ and $[\tilde{r}] \equiv ++(-+-)^{\infty} \equiv ++-(+--)^{\infty} \equiv ++-+(--+)^{\infty}$.

Let $\tilde{\lambda} = 1.952133665$ [C 6] be the λ-value such that C_2 is the least rank critical point belonging to the cycle $\alpha_1, \alpha_2, \alpha_3$. For this λ-value, C_2 merges into α_2. Here $(A) \equiv (+)$ and $[\tilde{r}] \equiv +(++-)^{\infty} \equiv ++(+-+)^{\infty} \equiv +++(-++)^{\infty}$.

- d) Let $x_1, \ldots, x_5$ be the points of the cycle (5;1) with the multiplier $S > 1$, born of the fold bifurcation (3.20) for $\lambda = \lambda^1_{(5)_0} \simeq 1.6229$. Here $[x_1] \equiv (-+---)$, $[x_2] \equiv (--+--)$, $[x_3] \equiv (---+-)$, $[x_4] \equiv (----+)$, $[x_5] \equiv (+----)$ and $\lambda_a \simeq 1.63$.

Let $\tilde{\lambda} \equiv {}^*\lambda_5^1 = 1.707364499$ [M33] be the λ-value such that C_2 is the least rank critical point belonging to this cycle (5;1). For this λ-value C_2 merges into x_2. Here $(A) \equiv (+)$ and $[\tilde{r}] \equiv +(--+--)^{\infty}$.

3.3.9. MYRBERG'S SINGULAR PARAMETER VALUES OF THE 3rd TYPE.

Let $\lambda = \lambda^i_{(p)_0}$ be the bifurcation value giving rise to two cycles (p;i) according to (3.20). For this λ-value the two cycles (p;i) merge into themselves (the multiplier $S = +1$), and every point $q_0^{(\gamma)}$, $\gamma = 1, \ldots, p$ of this structurally unstable "double" cycle is attractive on one side of q_γ, and repulsive on the other side, by application of T^p. The attractive side is the one where are located the successive consequents of the extremum x_e of T. So:

(3.28) $$\lim f^{\alpha}(x_e, \lambda^i_{(p)_0}) = q_0^{(\gamma)} \quad \text{when} \quad \alpha \to \infty .$$

Consider the quadratic recurrence (3.1), $f(x_e,\lambda) = -\lambda$. In |M55 |63 (cf.p.15) it is shown :

(P3.11) *As the end point* $\lambda^{j'}_{(k)s}$, *the initial point* $\lambda^{j'}_{(k)o}$ *of every spectrum* $\omega^{j'}_{k}$ *is singular, except the point* $\lambda_{(1)_o} = -1/4$ *of the first spectrum obtained for* $k = 1$. *The point* $\lambda^{j'}_{(k)o}$ *is only singular on its left side, i.e. it is a limit point for an infinite set of spectra for* $\lambda < \lambda^{j'}_{(k)o}$. *The point* $\lambda^{j'}_{(k)s}$ *is singular on the both side,* $\lambda >$ *or* $< \lambda^{j'}_{(k)s}$.

Now one intends to show that $\lambda^{i}_{(p)_o}$ satisfies in the case of (3.1) the following "periodical radical series" [M38] [M42] :

$$(3.29) \qquad \lambda^{i}_{(p)_o} = r_o(\lambda^{i}_{(p)_o}),\ [r_o] \equiv [r^{i}_{op}] \equiv (B_o)^{\infty} \equiv (m_1^+ n_1^- \ldots m_u^+ n_u^-)^{\infty} \quad ,$$

In (3.29) $m_1^+ + n_1^+ + \ldots + m_u^+ + n_u^- = p$, the first sign of (B_o) is that of the second rank critical point C_1, the following signs are those of C_2, $C_3,\ldots,C_p$. If [r] is the rotation sequence of the cycle (p;i), then it appears that : $(B_o) = ([r]+-)$ *if the number of signs - of* [r] *is odd, and* $(B_o) = ([r]--)$ *if this number is even or zero*. (B_o) satisfies (P3.7) and so is r-admissible.

Formaly (3.29) can be obtained considering the similitude of form with the relation $f^{\alpha}(0,\hat{\lambda}) = q$ of § 3.3.8, which defines the signs sequence (A) of (3.27). Here $\alpha \to \infty$, but by applications of $T^{\ell p}$, $\ell = 1,2,\ldots$, a point $f^{\gamma}(0,\lambda^{i}_{(p)_o})$, $\gamma = 1,2,\ldots,p$, is always located on the segment $C_{\gamma-1}q_o^{(\gamma)}$, with $C_o \equiv C$, and $C_{\gamma-1}$ being the critical point of rank γ, $q_o^{(\gamma)}$ being the point of the considered double cycle which is the nearest with respect to $C_{\gamma-1}$ [M42].

Now for the proof of (3.29) the set of the rotation sequences $(B_o)^m$, $m = 1,2,\ldots$ ((B_o) being r-admissible, the $(B_o)^m$, $m > 1$, satisfy (P3.7) and so are r-admissible) is considered. $(B_o)^m$ is the rotation sequence of a cycle of order $k = 2+mp$, $m = 1,2,\ldots$, which may be represented in the (x_n,x_{n+1}) plane, from a *"reduced" Koenigs-Lemeray graphical construction*, by a closed curve using only 2+m points. This construction simultaneously uses the curve T and the segment of the curve T^p giving a tangency with $x_n = x_{n+1}$, for $\lambda = \lambda^{i}_{(p)_o}$, and corresponding to the minimum of T^p which has the least abscissa, when the extremum of T is a minimum. This abscissa is equal to $r(\lambda)$. Here the construction is made for the multi-

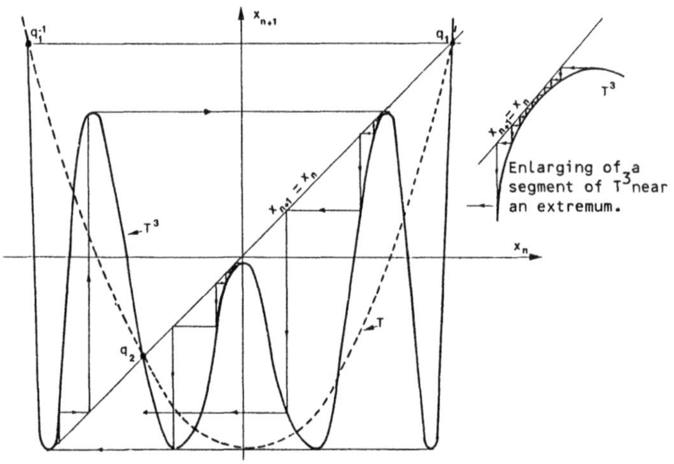

Behaviour of the iterated sequence for $\lambda = \lambda_{(3)_0} - \varepsilon$, $\varepsilon > 0$ is small

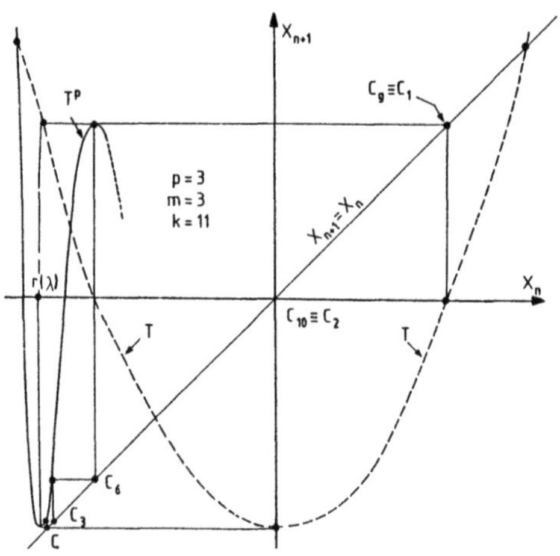

FIG. 3.2- Reduced Koenigs-Lemeray graphical construction . Cycle of order k=11 with a multiplier S=0 .

$$x_c = -\lambda_m^0 = -(B_0)^m \quad , \quad \lim_{m\to\infty} \lambda_m^0 = \lambda_{(p)_0}^i$$

plier S = 0. The cycle corresponds to a closed curve, having two points on T, and m points on the T^p segment; and the first rank consequent of the extremum $x_e = 0$ of T is $-\lambda_m^o$, λ-value of the considered cycle of order 2+mp giving S = 0 (cf.fig.3.2 for p = m = 3). When $m \to \infty$, the curve T^p tends to the tangency to $x_{n+1} = x_n$, which defines the bifurcation value $\lambda^i_{(p)_o}$, and $\lim \lambda_o^m = \lambda^i_{(p)_o}$ when $m \to \infty$, with $\lambda_m^o = (B_o)^\infty$. Indeed as long as the considered segment of T^p does not cut $x_n = x_{n+1}$, the number of the angles of the closed curve is finite. The abscissae of these angles are those of the critical points $C, C_p, C_{2p}, \ldots, C_{mp}, C_{mp+1} \equiv x_e = 0$. When $m \to \infty$, $\lim C_{mp} = q_o^{(1)}$ which is the smallest point abscissa of the double cycle obtained for $\lambda = \lambda^i_{(p)_o}$. So (3.29) is justified.

Examples of "periodical radical series" related to singular parameter values of the 3rd type are : $[r_{03}^{\ 1}] = (+--)^\infty$; $[r_{04}^{\ 1}] = (++--)^\infty$; $[r_{05}^{\ 1}] = (+----)^\infty$; $[r_{05}^{\ 2}] = (++-+-)^\infty$; $[r_{05}^{\ 3}] = (+++--)^\infty$.

3.4. CONVERGENCE OF ROTATION SEQUENCES TOWARD A SINGULAR VALUE.

3.4.1. CONVERGENCE TOWARD A SINGULAR VALUE OF THE 1st TYPE.

The rotation sequences (3.22) with i = 0,1,2,... constitute for (3.1) an *increasing sequence* having a limit. The corresponding values $\lambda^{o\ell'_i}_{(k2^i)}$ of the parameter λ (which gives cycles $(k2^i; \ell'_i)$ with multiplier S = 0) obtained via (3.18) constitute an increasing sequence with the limit $\lambda^{j'}_{(k)_s}$. It is the same for the bifurcation values $\lambda^{\ell'_i}_{k2^i}$ corresponding to the multiplier S =-1. The convergence speed of these parameter values may be defined by the ratio :

$$(3.30) \qquad \rho = \lim_{i\to\infty} [(\lambda^{j'}_{(k)_s} - \lambda^{o\ell'_{i+1}}_{(k2^{i+1})})/(\lambda^{j'}_{(k)_s} - \lambda^{o\ell'_i}_{(k2^i)})] = 1/\delta$$

The number δ = 4.669201609... was determined by Feigenbaum [F3]. This number remains the same whatever be k = 1,3,4,..., and f(x,λ) in (3.3), if at its extremum x_e, the series expansion of f(x,λ) begins with x^2. The ratio ρ plays the role of a multiplier for the sequence of parameters having the fixed point $\lambda^{j'}_{(k)_s}$.

For the variant of the singular parameter value (cf. § 3.3.6) the convergence speed ρ parameter values $\lambda^{om'_i}_{k_1(k_2)^i}$, i = 1,2,..., giving a multiplier S = 0 is similarly defined. The ρ-value was determined in [D 2], $\rho \sim 0.0181$ for $k_1 = 1$, $k_2 = 3$ (cf. also [C 7]).

3.4.2. CONVERGENCE TOWARD A SINGULAR VALUE OF THE 2nd TYPE.

Consider the rotation sequence of a cycle of order $\alpha + k\ell + 2$:

$$(3.31) \qquad [r_\ell] = (m_1^+ n_1^- \dots m_s^+ n_s^-)(m_{s+1}^+ n_{s+1}^- \dots m_t^+ n_t^-)^\ell \equiv (A)(B)^\ell \quad , \quad \ell = 1,2,\dots$$

extracted from $[\tilde{r}]$, and constituted by the signs of (A) followed by the $k\ell$ first signs of $(B)^\infty$ in (3.27), k being the order of the repulsive cycle defining $[\tilde{r}]$. By construction $[\tilde{r}]$ is r-admissible. Then the $[r_\ell]$ are also r-admissible. The sequence $[\tilde{r}]$ may be given with k different writings $[\tilde{r}] = (A_h)(B_h)^\infty$, h = 0,1,...,k-1, $A_o \equiv A$, $B_o \equiv B$, where A_h is made up of the signs of A followed by the h first signs of B, and B_h is made up of the k-h remaining last signs of B, followed by the h first signs of B. From these k different writings it is possible to extract k sets of rotation sequences [M42] :

$$(3.32) \qquad [r_\ell]_h = (A_h)(B_h)^\ell \quad ; \quad h = 0,1,\dots, k-1 \quad ; \quad \ell = 1,2,\dots$$

Though $[\tilde{r}]$ is r-admissible, all the sequences (3.32) are not rotation sequences. So $[r_\ell]_3 = (+--+)\ (----+)^\ell$ extracted from $[\tilde{r}] = +(--+--)^\infty$ is not r-admissible. However *if* $[r_1]_h$ *is a rotation sequence, the* $[r_\ell]_h$, $\ell > 1$, *are rotation sequences*. So $\delta \leqslant k$ rotation sequences of cycles may be extracted from $[\tilde{r}]$ for a given index ℓ, and $\lim [r_\ell]_h = [\tilde{r}]$ when $\ell \to \infty$. Applying the ordering law (P3.6), or (P3.6'), the rotation sequences $[r_\ell]_h$, $\ell = 1,2,\dots$ (h being given) constitutes either an *increasing sequence*, or a *decreasing sequence*, or an *alternating sequence* (the position of $[r_\ell]_h$ with respect to $[\tilde{r}]$ depending of the evenness, or oddness, of ℓ) converging toward $[\tilde{r}]$. One has [M38] [M42]:

<u>(P3.12)</u> *If in (3.31)* B *has an even (resp. odd) number of signs –, the*

different B_h *have also an even (resp. odd) number of signs* -, *and the convergence of the* $[r_\ell]_h$, $\ell = 1,2,\ldots$, *(h given) toward* $[\hat{r}]$ *gives an increasing, or decreasing sequence (resp. alternating sequence). An increasing, or a decreasing sequence, occurs when a* m+1-*rank critical point* C_m *becomes point of a cycle with a multiplier* $S > 1$. *An alternating sequence occurs when* C_m *becomes point of a cycle with a multiplier* $S < -1$.

In chapter 4 some $\hat{\lambda}$-values, designated by $^*\lambda_k^{j'}$, are such that the 3rd rank critical point C_2 is the first critical point belonging to the cycle (k,j') with S > 1. The B_h obtained from such a value $^*\lambda_k^{j'}$, and also the $\lambda_k^{*j'}$ values playing an equivalent role to that of λ_1^*, always have an even number of signs -. Then the convergence is not of the alternating kind, but a given h can lead to an increasing sequence of rotation sequences, and other h to a decreasing one. All the sequences here considered are also constituted by the corresponding parameter values giving cycles with S = 0, or by the bifurcation values giving rise to these cycles via (3.20). From $[\hat{r}]$ other rotation sequences may be determined with the following form :

$$(3.33) \qquad [r_\ell]'_h \equiv (A_h)(B_h)^\ell(C) \quad , \quad \ell = 1,2,\ldots$$

A_h and B_h being defined as for (3.32), (C) being a set of α' signs such that $[r_\ell]'_h$ is r-admissible. When $\ell \to \infty$ such rotation sequences tend also toward $[\hat{r}]$.

For a given value of ℓ, the rotation sequences (3.32), or (3.33), may be ordered from (P3.6) (P3.6'). So for cycles having an order equal to the signs number of [r] plus two, it is obtained :

(a) $(+)^\infty \to [\lambda_1^*] > (+)^\ell > (+)^\ell - > (+)^{\ell-1}-- > (+)^{\ell-1}- > (+)^{\ell-1}-+ > (+)^{\ell-2}$
$(+)^{\ell-2} > (+)^{\ell-2}--+ > (+)^{\ell-2}-- > (+)^{\ell-2}--- > (+)^{\ell-2}- > (+)^{\ell-2}-+- >$
$(+)^{\ell-2}-+ > (+)^{\ell-2}-++ > (+)^{\ell-3}$ etc...

(b) $\ldots, ++(--+)^\ell+- > ++(--+)^\ell > +(+--)^\infty \to [^*\lambda_3^1] > ++-(-+-)^\ell > ++-(-+-)^\ell+ >$
$+(+--)^\ell > +(+--)^\ell--+ > +(+--)^\ell--$ etc. (cf. § 3.3.8 for the definition of $^*\lambda_3^1$).

(c) $\ldots, ++(-+-)^{\ell} > ++-(+--)^{\ell}- > ++-(+--)^{\ell} > ++(-+-)^{\infty} \rightarrow [\lambda_3^{*1}] > ++-+$ $(--+)^{\ell} \ldots$ (cf. § 3.3.8 for λ_3^{*1}).

For these three examples the convergence toward the singular value of the 2nd type is regular. But for the following examples the convergence is alternating (here λ_{21}^{*} is such that $C_2 \equiv q_2$ with $[\hat{r}] = +(-)^{\infty}$, and $\hat{\lambda}$ ($C_3 \equiv q_3$) corresponds to $[\hat{r}] = (++)(-)^{\infty}$

(d) $+(-)^{\ell-4}+-+- > +(-)^{\ell-4} > +(-)^{\ell-2}+- > +(-)^{\ell-2} > +(-)^{\ell} > +(-)^{\infty} \rightarrow [\lambda_{21}^{*}]$ $> +(-)^{\ell'} > +(-)^{\ell'-2} > +(-)^{\ell'-2}+- > +(-)^{\ell'-4} > +(-)^{\ell'-4}+-+- > +(-)^{\ell'-4}+-$ etc..., ℓ being even, ℓ' being odd.

(e) $++(-)^{\ell-4}+-+ > ++(-)^{\ell} > ++(-)^{\ell-2}+ > ++(-)^{\ell-2}+- > ++(-)^{\ell-2} > ++(-)^{\ell}$ $++(-)^{\infty} \rightarrow [\hat{\lambda}(C_3 \equiv q_2)] > ++(-)^{\ell'} > ++(-)^{\ell'}+ > ++(-)^{\ell'-2} > ++(-)^{\ell'-2}+-+ >$ $++(-)^{\ell'-2}+-$ etc... ℓ being even, ℓ' being odd.

The $[r_\ell]_h$ of (3.32), or (3.33), are associated with the λ-values giving a cycle with multiplier $S = 0$. Let $\lambda_{\ell,h}^{o}$ be such λ-values. Then $\hat{\lambda} = \lim \lambda_{\ell,h}^{o}$ when $\ell \rightarrow \infty$. Every $\lambda_{\ell,h}^{o}$ is a zero of the Myrberg's polymonial $G_{\alpha+\beta+\ell k}(\lambda)$, where α is given by (3.26), k is the order of the repulsive cycle (k;j') of (3.26), and $\alpha+\beta+2$ is the sum of the components of (A_h) for (3.32), or the sum of the components of (A_h) and (C) for (3.33). So β is determined from h. Every h gives an infinite sequence of $\lambda_{\ell,h}^{o}$, $\ell = 1,2,\ldots$, and of $G_{\alpha+\beta+\ell k}(\lambda)$ which is the abscissa of the critical point of rank $\alpha+\beta+\ell k-1$. These critical points being cyclic with the order k when $\lambda = \hat{\lambda}$:

$$G_{\alpha+\beta+\ell k}(\hat{\lambda}) = G_{\alpha+\beta+(\ell+1)k}(\hat{\lambda}) = \bar{G}(\hat{\lambda}) \tag{3.34}$$

If the multiplier of the cycle (k;j') is such that $S > 1$, for a given h, the $\lambda_{\ell,h}^{o}$ give rise to a stictly increasing, or decreasing, sequence according to the considered h. If $S < -1$, the $\lambda_{\ell,h}^{o}$ give rise to an alternating sequence with respect to $\hat{\lambda}$.

The relation (3.34) indicates that all the curves $G_{\alpha+\beta+\ell k}(\lambda)$ in the $(\lambda, G_v(\lambda))$ plane ($v = \alpha+\beta+\ell k$) pass through the same point $(\hat{\lambda}, \bar{G}(\hat{\lambda}))$. The

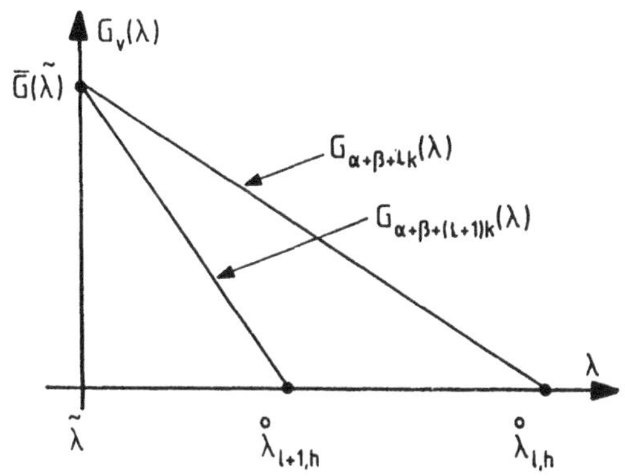

FIG. 3.3 - Convergence toward a singular value of the 2nd type .

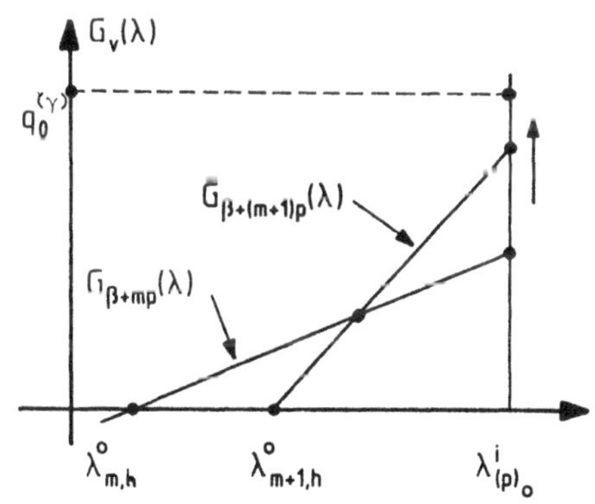

FIG. 3.4 - Convergence toward a singular value of the 3rd type .

curve $G_{\alpha+\beta+\ell k}(\lambda)$ intersects the λ-axis at $\lambda = \lambda^o_{\ell,h}$, which is the nearest intersection point of this curve with respect to $\check{\lambda}$. Inside the interval $(\lambda^o_{\ell,h},\check{\lambda})$, $G_{\alpha+\beta+\ell k}(\lambda)$ is strictly monotone. The convergence of the $\lambda^o_{\ell,h}$ toward $\check{\lambda}$, $\ell \to \infty$, is characterized by $\rho = \lim [(\check{\lambda}-\lambda^o_{\ell+1,h})/(\check{\lambda}-\lambda^o_{\ell,h})]$ when $\ell \to \infty$. In the $(\lambda,G_v(\lambda))$ plane, for $\ell \to \infty$ it is possible to write (cf. fig.3.3) |M38| |M42|.

$$\check{\lambda}-\lambda^o_{\ell,h} \simeq G(\check{\lambda})/[(dG_{\alpha+\beta+\ell k}/d\lambda)_{\lambda=\lambda^o_{\ell,h}}] \quad ,$$

with $G_{\alpha+\beta+\ell k}(\lambda) = f^{\alpha+\beta+\ell k}(x_e,\lambda) = f(f(f(\ldots)\ldots))$. Then :

$$[df^{\alpha+\beta+\ell k}(x_e,\lambda)/d\lambda] = \left[\prod_{j=1}^{\alpha+\beta+\ell k-1} (\partial f/\partial x)_{x=f^j(x_e,\lambda)}\right] (\partial f/\partial \lambda)$$

(3.35) $$\rho = \left[\prod_{j=\alpha+\beta}^{\alpha+\beta+k-1} (\partial f/\partial x)_{x=f^j(x_e,\check{\lambda})}\right]^{-1} = 1/S$$

(P3.13) *The convergence ratio* ρ *of the* $\lambda^o_{\ell,h}$ *toward a singular value of the* 2nd *type is the inverse of the multiplier* S *which defines this singular value* $\check{\lambda}$. [M 38] [M42].

Such a result was proved in [G 1] for a particular value of h and particular values of $\check{\lambda}$. It also was obtained numerically by C.Gillot (1982).

3.4.3. CONVERGENCE TOWARD A SINGULAR VALUE OF THE 3rd TYPE.

The set of the rotation sequences $(B_o)^m$, or of the λ^o_m (cf. § 3.3.9), converge toward $(B_o)^\infty$, or $\lambda^i_{(p)o}$, when $m \to \infty$. The sequence $[r^i_{op}] = (B_o)^\infty$ may be given with p different writings : $[r^i_{op}] = (A_h)(B_h)^\infty$, $h = 0,1,\ldots, p-1$, $A_o \equiv \phi$ = empty set, where A_h is made up of the h first signs of B_o, and B_h is made up of the p-h remaining last signs of B_o followed by the h first signs of B_o. From these p different writings it is possible to extract p sets of rotation sequences [M 38] [M 42] :

(3.36) $$[r_m]_h = (A_h)(B_h)^m \quad ; \quad h = 0,1,\ldots, p-1 \quad ; \ell = 1,2,\ldots$$

Though $[r^i_{op}]$ is r-admissible, all the p sequences corresponding to different h are not rotation sequences of cycles. So $[r_m]_1 = +(--+)^m$ extracted from $[r_{o3}^1] = (+--)^\infty$ is such that $+(--+)^m \to [r] = +$ (cf. § 3.3.5ii), which is the rotation sequence of the cycle (3;1). It is the same for $[r_m]_2 = ++(--++)^m$ extracted from $[r_{o4}^1] = (++--)^\infty$, because $++(--++)^m \to [r] = ++$ rotation sequence of the cycle (4;1). Then $\delta_o \leqslant p$ rotation sequences of cycles may be extracted from $[r_{op}^i]$ for a given index m, with $\lim [r_m]_h = [r_{op}^i]$, and $\lim \lambda_{m,h}^o = \lambda_{(p)_o}^i$ when $m \to \infty$. Considering the definition of $[r_{op}^i]$, *the* $[r_m]_h$ *always constitute increasing sequences* for m=1,2,..., and it is the same for the $\lambda_{m,h}^o$. As for (3.33), other rotation sequences may be determined with the following form :

$$[r_m]_h' = (A_h)(B_h)^m(C) \quad ; \quad m = 1,2,... \tag{3.37}$$

So from $[r_{o4}^1]$ the following ordering is obtained:

$$(++--)^\infty \to [\lambda_{(4)_o}^1] > ++-(-++-)^m > +(+--+)^m > +(+--+)^m- > (++--)^m >$$

$$(++--)^m-- > (++--)^m- > (++--)^m-+ \text{ etc...}$$

Every $\lambda_{m,h}^o$ associated with $[r_m]_h$ is a zero of $G_{\beta+mp}(\lambda)$, β-2 being the sum of the signs of (A_h) in (3.36), and the sum of the signs of (A_h) and (C) in (3.37). So β depends on h. Every h gives an infinite sequence of $\lambda_{m,h}^o$, m = 0,1,..., and of $G_{\beta+mp}(\lambda)$ abscissa of the critical point of rank $\beta+mp-1$. For $\lambda = \lambda_{(p)_o}^i$ these critical points tend toward a point of the "double" cycle with S = +1, when $m \to \infty$ (fig.3.4), giving a decreasing (resp.increasing) sequence of critical points if the "double" cycle point is on a segment of T^p with a minimum (resp.maximum). So one has (cf.(3.28)) :

$$\lim G_{\beta+mp}(\lambda_{(p)_o}^i) = \bar{G}(\lambda_{(p)_o}^i) = q_o^{(\gamma)}, \gamma = 1,...,p, \text{ when } m \to \infty \tag{3.38}$$

In the $(\lambda, G_v(\lambda))$ plane, $v = \beta+mp$, the curve $G_{\beta+mp}(\lambda)$ intersects the λ-axis at $\lambda = \lambda_{m,h}^o$, which is the nearest intersection point of this curve from the point $\lambda_{(p)_o}^i$. Then inside the interval $(\lambda_{(\mu)_o}^i, \lambda_{m,h}^o)$, $G_{\beta+mp}(\lambda)$ is strictly monotone. The convergence of the $\lambda_{m,h}^o$ toward $\lambda_{(p)_o}^i$, when $m \to \infty$, is characterized by $\rho = \lim [(\lambda_{(p)_o}^i - \lambda_{m+1,h}^o)/(\lambda_{(p)_o}^i - \lambda_{m,h}^o)]$. In the $(\lambda, G_v(\lambda))$ plane, for $m \to \infty$ it is possible to write (cf.fig.3.4) :

$$\lambda^{i}_{(p)o} - \lambda^{o}_{m,h} \simeq G_{\beta+mp}(\lambda^{i}_{(p)o})/[(dG_{\beta+mp}/d\lambda)_{\lambda=\lambda^{o}_{m,h}}]$$

with $G_{\beta+mp}(\lambda) = f^{\beta+mp}(x_e,\lambda)$. Let $D_{mp} = dG_{\beta+mp}(\lambda^{o}_{m,h})/d\lambda$, $\rho = \lim[G_{\beta+(m+1)p}(\lambda^{i}_{(p)o})/G_{\beta+mp}(\lambda^{i}_{(p)o})][D_{mp}/D_{(m+1)p}]$. The limit of the first square bracket is 1^-. The limit of the second square bracket is $\prod_{j=\beta}^{\beta+p-1}(\partial f(q_o^{(j)},\lambda^{i}_{(p)o})/\partial x)$, multiplier of the double cycle (p;i) equal to 1. Then [M38] [M42]:

<u>(P3.14)</u> *The convergence ratio* ρ *of the* $\lambda^{o}_{m,h}$ *toward a singular value of the third type is equal to* 1^-. *This corresponds to the least convergence speed.*

3.4.4. <u>REMARK</u>.

Considering the convergence of the λ-values associated with cycles having a multiplier S = 0, it is evident from the Myrberg's results that the same rules occurs for the whole Myrberg's spectrum in the case of the variant of singular values of 1st type, and of singular values of 2nd and 3rd type.

3.5. <u>THE ALGORITHMS GIVING THE SET OF ALL THE ORDERED CYCLES OF ORDER K.</u>

3.5.1. <u>NOTATIONS AND DEFINITIONS</u>.

In this part, various operations on the binary word $[q_{1k}^{j}]$ are considered. This word is composed with the elements (or bits) (0,1). The opposite element to z is noted $\bar{z}$, and one has $\bar{0} = 1$, $\bar{1} = 0$. A word of n bits is written $[a] = [a_1 \ldots a_n]$. With this word, the following operations are considered :

* rotational shift on the left $\sigma[a] = [a_2, a_3 \ldots a_n, a_1]$
* rotational shift on the right $\sigma^{-1}[a] = [a_n, a_1, \ldots, a_{n-1}]$
* inversion $[a]^{-1} = [a_n, a_{n-1}, \ldots, a_1]$
* concatenation of two words [a] [b] represented by the word [a b].

So one has : $\sigma \circ \sigma^{-1} = \sigma^{-1} \circ \sigma = \text{id.}$, $\sigma^{-n} = \sigma^{n} = \text{id.}$, $\sigma^{-1} = \sigma^{n-1}$ and $\sigma = \sigma^{1-n}$. The repetition of m successive identical characters will be written $[a\ b^m] = [a\ b\ \ldots\ b]$ (m times repetition of b after a).

So the Myrberg's rotation sequence [r] of a cycle of order k is constituted by a word of (k-2) bits, and as above mentionned $[q^{(1)j'}_{k}]$ is the concatenation $[1[r]z]$, z = 0 or 1, binary word of the cycle (k;j') with a positive (resp.negative) multiplier if the number of 1 into the word is even (resp.odd), which is also the binary word of the point of (k;j') having least (resp.largest) abscissa when the extremum of $f(x,\lambda)$ is a minimum (resp.maximum).

3.5.2. FIRST ALGORITHM.

(a) It is considered an initial ordered set A_1. A_1 *is constituted by two words of one bits* $A_1 = \{[0], [1]\}$, *with the order* $[0] > [1]$.

(b) A set $A_m = \{a_1, a_2, \ldots, a_{2^m}\}$ *is considered, each* a_i, $i = 1,2,\ldots,2^m$, *being a word of m bits.* A_m *is supposed to be ordered,* $a_1 > a_2 > \ldots > a_{2^m}$, *from the law (P3.9), which is here applied independently of the consideration of the x-axis.*

(c) The set A_{m+1}, *constituted by all the words having* (m+1) *bits, is constructed from* A_m :

$$A_{m+1} = \{[0\ a_1],\ \ldots\ [0\ a_{2^m}],\ [1\ a_{2^m}],\ \ldots\ [1\ a_1]\}$$

(d) So from A_1 , *by recurrence, the ordered sets* $A_2 \ldots A_n$ *are obtained,* A_n *being constituted by the words of n bits* a_i, $i = 1,2,\ldots,2^n$, *with* $a_1 > a_2 > \ldots > a_{2^n}$. (cf. [K3]).

For example, with n = 3, the ordered set A_3 is $A_3 = \{[000], [001], [011], [010], [110], [111], [101], [100]\}$.

3.5.3. SECOND ALGORITHM.

This algorithm gives all the words related to the Myrberg's rotation sequences [r] of cycles of order k > 4

(a) With the first algorithm, the ordered set A_{k-3}, $A_{k-3} = \{a_1, \ldots, a_{2^{k-3}}\}$ *is constructed.*

(b) For each word a_i *of* A_{k-3}, *it is made the concatenation* $a'_i = [0\ a_i]$,

so the set $A'_{k-3} = \{a'_1, a'_2, \ldots, a'_{2^{k-3}}\}$, *constituted by words of* (k-2) *bits is ordered.*

(c) Into A'_{k-3}, *all the words, which do not verify* $[r] \geqslant [r_\ell]$, $\ell = 2,\ldots q$ *(cf.P3.7), are cut out. So it is obtained an ordered set* $A'' = \{a''_1, a''_2, \ldots, a''_{p'_k}\}$ *where* p'_k *is the number of rotation sequences* [r], *obtained for the set of the cycles of order* k (p'_k *is given by (3.7)-(3.10),* $p'_k = N_\lambda(k)$).

(d) For each word of A'', *it is made the concatenation* $b_i = [1\ a''_i]$, $i = 1,2,\ldots,p'_k$, *and the set* $B_k = \{b_1, b_2, \ldots, b_{p'_k}\}$ *is considered as ordered,* $b_1 > b_2 > \ldots > b_{p'_k}$

(e) The running number j' *of the rotation sequence* [r] *of a given cycle* (k;j'), *into the increasing set of all the* [r] *with the order* k, *is* $j' = p'_k+1-i$. (cf.[K3] [K4]).

So is obtained the binary word $[q^{(1)}{}^{j'}_{k}] = [b_{p'_k+1-j'}, z]$, $z = 0,1$.

An illustration of the application of the two algorithms is given by the table1($x \equiv z$). In this table the columns [u] and "decomposition" will be defined in the § 4.2. From (P3.7), one has the following results :

(P3.15) $[10^m1\ldots0^\ell\ldots]$ *with* $\ell > m$, *cannot be a binary word* $[q^{(1)}{}^{j'}_{k}]$ *of a cycle generated by* (3.3). *It is the same for a word begining with* 1010. *The binary word* $[1,a'_i] = [1011\ldots0z]$, $z = 0,1$, *cannot be the binary word of a cycle generated by* (3.3).

Indeed, for $[10^m1\ldots0^\ell\ldots]$, one has $[r] = [0^m1\ldots0^\ell\ldots]$ and a $[r_i] = [0^\ell\ldots]$ exists such that $[r] < [r_i]$ if $\ell > m$. For a binary word $[1011\ldots0z]$ one has the last $[r_i]$, $[r_q]$, in (P3.7) which is [0] ($m_q = 1, n_q = 0$) and $[r] = [011\ldots0]$. Then $[011\ldots0] < [0]$, $[r] < [r_q]$, and the permutation $[011\ldots0]$ cannot be the Myrberg's rotation sequence of a cycle generated by (3.3). For a binary word [1010...], with [r] = [010...], $[r_1] = [0\ldots]$, ..., applying the same rule in all the possible permutations of a set of 0,1 after [010], it appears that it exists always a $[r_i]$, $i = 1,2,\ldots,q$, such that $[r] < [r_i]$.

Considering that [1010...] cannot be a binary word of a cycle generated by (3.3), it is sufficient to apply the second algorithm for a set of $3 \times 2^{k-5}$ words, obtained from the table 1 and resulting from the elimination of the "way" giving the words which begin with [1010]. Then the rule $[r] \geqslant [r_\ell]$, $\ell = 1,2,\ldots q$, is applied to the set D of the binary

Table I ($k=8$)

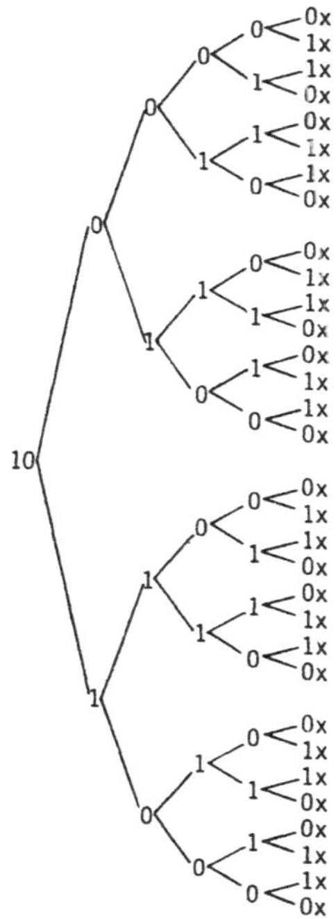

[c]	j'
1000000x	16
1000001x	15
1000011x	14
1000010x	13
1000110x	12
1000111x	11
1000101x	10
1000100x	9
~~1001100~~	
1001101x	8
1001111x	7
1001110x	6
1001010x	5
1001011x	4
~~1001001x~~	
~~1001000x~~	
~~1011000x~~	
~~1011001x~~	
1011011x	3
~~1011010x~~	
~~1011110x~~	
1011111x	2
1011101x	1
~~1011100x~~	
~~1010100x~~	
~~1010101x~~	
~~1010111x~~	
~~1010110x~~	
~~1010010x~~	
~~1010011x~~	
~~1010001x~~	
~~1010000x~~	

j' is the number of the cycle with the increasing Myrberg's order.

Table II

x ≡ z of the text

j'	j	λ	[u]	[r]	[b]	Adj. $x=0$	Adj. $x=1$	Décomposition
1	*	1,3815	18452736	+ − − − + −	1011101	Auto	Auto	[12] * [12] * [12]
2	*	1,5218	18453627	+ − − − − −	1011111	Auto	Auto	[12] * [1432]
3	1	1,7110	18527436	+ − − + − −	1011011	Auto	Auto	
4	2	1,8100	18627435	+ + − + − −	1001011	Auto	7(0)	
5	3	1,8517	18635274	+ + − + − +	1001010	Auto	8(0)	
6	4	1,8700	18634275	+ + − − − +	1001110	Auto	8(1)	
7	5	1,8969	18743526	+ + − − − −	1001111	4(1)	Auto	
8	6	1,9170	18742635	+ + − − + −	1001101	5(1)	6(1)	
9	*	1,9417	18642753	+ + + − + +	1000100	Auto	12(0)	[1432] * [12]
10	7	1,9607	18752634	+ + + − + −	1000101	Auto	11(0)	
11	8	1,9722	18763425	+ + + − − −	1000111	10(1)	Auto	
12	9	1,9816	18753264	+ + + − − +	1000110	9(1)	Auto	
13	10	1,9887	18764253	+ + + + − +	1000010	Auto	14(0)	
14	11	1,9943	18765324	+ + + + − −	1000011	13(1)	Auto	
15	12	1,9979	18765423	+ + + + + −	1000001	Auto	Auto	
16	13	1,9997	18765432	+ + + + + +	1000000	Auto	Auto	

*j': number of the non embedded cycle with the increasing Myrberg's order. j: number of the embedded representation. * indicates that the sequence is decomposable. λ: $S=0$ value of λ. [u]: decimal representation of the sequence. [r]: Myrberg's representation of the sequence. [b]: binary representation. adj.: adjoint cycle. auto: self adjoint cycle. $j'(x)$: adjoint cycle of the cycle of the line.*

Columns [u] and "Decomposition" are defined in § 4.2

TABLES 1 and 2 (according [K3])

words having k-2 bits.

$D = \{[000\ a_1], \ldots [000\ a_{2^{k-5}}], [001\ a_{2^{k-5}}], \ldots, [001\ a_1], [011\ a_1],$

$\ldots [011\ a_{2^{k-5}}]\}$

where $A_{k-5} = \{a_1, \ldots a_{2^{k-5}}\}$ is given by the first algorithm. This remark decreases considerably the number of words to be considered in order to define B_k. So with k = 12, 10 ! combinaisons are possible, and the algorithm proposed in [G] works with 66324 words. Here only 384 words are to be considered, which gives 170 possible Myrberg's rotation sequences [K3] [K4].

3.6. ADJOINT, AND SELF ADJOINT, BINARY WORDS OF THE CYCLES OF UNIMODAL ONE DIMENSIONAL ENDOMORPHISMS. SYMMETRY.

Let C_k be the set of all the binary words $[q^{(1)j'}_k]$ related to the cycles of order k generated (3.3), and $c \in C_k$. One has the following proposition [K3] [F 9]

(P3.16) *If* $c \in C_k$, *then there exists an unique integer* ℓ *such that* $\sigma^{\ell}([c]^{-1}) = c^* \in C_k$, $\ell = 0,1,2,\ldots,k-1$.

Let $c = [10^m1\ldots10^m1\ldots10^m1\ldots]$, m = 1,2,..., be the general form of $c \in C_k$, which contains $[0^m]$ α times. Then $\sigma^{\ell_i}([c^{-1}]) = [10^m1\ldots]$, i = 1,2,...,$\alpha$. Among the α binary words so obtained, one of them is the *smallest* in the sense of (P3.9), with $\ell_i = \ell$. So, $\sigma^{\ell}([c]^{-1}) = c^*$ is unique, it verifies $[r] > [r_j]$, j = 1,2,...q, and it is the binary word of a cycle generated by (3.3). The *binary word* c^* is called *adjoint* of c if $c^* \neq c$. *A binary word* c such that $c = c^*$ is called *self adjoint*. In all the cases, one has $(c^*)^* = c$ for all $c \in C_k$. Each binary word of a cycle generated from (3.3), $c \in C_k$, either has an adjoint, or is self adjoint. The corresponding *cycles* are also called *adjoint* , or *self adjoint* (cf. table II for k = 8).

Another property of a cycle binary word is that of symmetry [K3] [F 8] [K4] [K5]. Let the elements (bits) of a binary word $[q^{(1)j'}_k]$ be considered as k points, equally spaced and put with this order on a circle. If the set of these elements has an axis of symmetry, the *binary word* is called *symmetrical*. If k is odd, only one element is on the axis. If k is even, two of these elements are on the axis of symmetry, or none. If such an axis does not exist, the *binary word* is called *dissymmetrical*. So $[q^{(1)}{}_6^1] = [101110]$ is symmetrical and $[q^{(1)}{}_6^2] = [100101]$ is dissymmetrical.

(P3.17) *A binary word* c *is self adjoint if and only if it is symmetrical.*

Indeed, if an axis of symmetry exists, on the considered circle, the points of c* are the symmetrical of the points of c with respect the axis.

(P3.18) *Let* $c = [10^m 1\ a]$ *be a binary word such that* [a] *does not contain* 0^m *; then :*

1) $c^* = [10^m 1\ [a]^{-1}]$.

2) the binary word c *is self adjoint if and only if* $[a] = [a]^{-1}$.

3) For the last bit z = 0, *or* 1, *at least one the two binary words so obtained is such that* $c \neq c^*$. [K3].

When the one-dimensional endomorphism (3.3) is embedded into a two-dimensional diffeomorphism as indicated in chapter 1, it will appeared that the *notions of adjoint, or self adjoint, cycles of (3.3) play a fundamental role in the understanding of the bifurcations of the two-dimensional diffeomorphism.*

3.7. THE COMPLETE NON-EMBEDDED REPRESENTATION OF A CYCLL.

More information about the nature of a cycle can be given with a more complete representation than that corresponding to (k;j'). This new representation adds an information about the symmetry of the binary word $[q^{(1)}{}_k^{j'}]$, and about the sign of the multiplier of the considered cycle (k;j').

The complete representation is constituted by four symbols. The first and the second one are respectively k,j'. The third is "s" if $[q^{(1)}{}_k^{j'}]$ is symmetrical, and then self adjoint, and "d" if it is dissymmetrical. The last one is "p" (resp."n") if the number of 1 into $[q^{(1)}{}_k^{j'}]$ is even, or multiplier S > 0 (resp.odd, or S < 0).

So a pair of cycles (k;j') born of the same bifurcation (3.20) can be : either (kj'sn), (kj'sp); or (kj'sp), (kj'dn); or (kj'sn), (kj'dp); or (kj'dn), (kj'dp), as soon as the multiplier $S < -1$.

Such a representation is fundamental in the study of the bifurcations plane of the two-dimensional diffeomorphism.

Chapter 4

THE BOX-WITHIN-A-BOX BIFURCATIONS STRUCTURE AND ITS CONSEQUENCES

4.1. INTRODUCTION.

This chapter is devoted to the structure description of the set of all the bifurcations occuring in the quadratic recurrence (3.1), $x_{n+1} = x_n^2 - \lambda$, or in the endomorphism T (3.3), $x_{n+1} = f(x_n,\lambda)$, with conditions ensuring the existence of no more than one attractor at finite distance. The *box-within-a-box bifurcations structure*, or *embedded boxes structure* (cf. [G35] and [G14]), characterizes the organization of this set, not only for one-dimensional recurrences such as (3.3), $f(x,\lambda)$ having only one extremum, but also for a whole class of dynamical systems such as : (3.3) with $f(x,\lambda)$ having more than one extremum [M24] [G35] [F 4], multi-dimensional recurrences [M28] [M32] [M36], m-dimensional differential equations, $m \geqslant 3$ in the autonomous case, $m \geqslant 2$ in the non autonomous case with periodical coefficients (internal excitations giving parametric resonances), or with an external periodical excitation. At least one "strong" non-linearity is at the origin of this organization, which can be encountered as well in engineering systems (cf. for example [M44]...), as in other dynamical systems [G35].

This bifurcations set has a fractal structure in the parameters space of the considered dynamical system, i.e. the organization of this set is similar to that of its parts, even if these parts are infinitesimal. It is related to the presence of a chaotic behaviour in the sense of § 1.1, i.e. either a non-constant and non-periodic stable stationary state (strange attractor), or a stable periodic stationary state, the period of which is larger than the duration of the observations, or a *fuzzy (or chaotic) boundary* (related to the existence of a strange repeller (cf. § 2.8.4)) separating the basins (or influence domains) of two(or more) attractors, or a chaotic transient (cf. § 2.8.3).

The first description of the box-within-a-box bifurcations structure was given in 1975 [G32] [M21], after the publications [M25]-[M48] have completed these first results. It was shown in § 2.9.3 that *the boxes in files bifurcations structure corresponds to an ordering of the Farey's sequences of fractions in their lowest terms*. Now *the box-within-a-box bifurcations structure corresponds to ordering of the Myrberg's spectra* (or Myrberg's cascades by order (or period) doubling) introduced in § 3.3.5. Such an ordering is obtained from the definition of a non-classical bifurcation corresponding to the merging of two sets of singular points having an essentially different nature : the critical points of T from a certain rank, and the points of a repulsive cycle. The box-within-a-box structure is the direct consequence of decomposition properties of a sub-set of the whole set of rotation sequences (set of the decomposable rotation sequences). The decomposition operations by using the decimal representation [u] of the rotation sequences appear more convenient to manipulate. This way was used for the numbering of the embedded boxes as early as the first papers [G32] [M21], and was published after in [M25] (1978), where it is shown that for the considered sub-set it is possible to break down [u] into partial rotation sequences. Then this fractal bifurcations structure leads to a non-commutative *symbolic code of representation of the embedded boxes* presented in this chapter.

With the binary representation using the kneading invariant υ, [J3] (1981) shows that υ may be factorized in a certain way, so that each factor corresponds to an invariant basic subset on the non-wandering set of T, leading to a spectral decomposition of this set. Taking into account the relations between [u], [r], υ, and the data of chapter 3, this result appears as a different formulation of the possibility to break down some [u] into partial rotation sequences [M25]. Besides this remark is made in [G17], where it is said that "the decomposition of non-wandering sets of maps, studied by Jonker, Rand, can be explained in different terms making use of Mira's concept of the embedded box".

A first part deals with the characterization of a cycle by a decimal rotation sequence. It is followed by the description of operations of decompositon, and composition, of such rotation sequences. It is then possible to present the box-within-a-box bifurcations structure, and to show that it coexists with a boxes in files bifurcations structure. The resulting properties on the x-axis are given, and it is shown that the now popular phenomena of *chaotic attractor in crisis*, and *intermittency*, are a consequence of the mentioned bifurcations structure, and are rediscovery of results published in 1976 [M22]. A paragraph is devoted to "non-classical" invariant measures, another to the topological entropy for $f(x,\lambda) \equiv x^2-\lambda$ in (3.3), in relation with the coexisting properties of boxes in files bifurcations. The final part concerns generalizations such as : maps defined by a function with only one extremum, having more than one attractor at finite distance, or by a function with two extrema, or by a piecewise linear function (hat function).

4.2. CHARACTERIZATION OF A CYCLE BY A DECIMAL ROTATION SEQUENCE.

4.2.1. DEFINITION OF THE DECIMAL ROTATION SEQUENCE.

Let (k;j') be a cycle of (3.1), or (3.3), without any restrictive condition about its multiplier S, which may be positive or negative. If $f(x,\lambda)$ has a minimum (resp.maximum), the k points $q^{(i)}$ of this cycle are numbered from 1 to k with increasing (resp.decreasing) abscissae. The *decimal rotation sequence* [M25] [M33] [G35] [M42] of this cycle is the set [u] made up of the k first integers:

$$[u] = [1,k,\alpha_3,\alpha_4,\dots,\alpha_k] \quad , \quad \alpha_1 \equiv 1, \alpha_2 \equiv k \ ,$$

the α_i, i = 1,2,...,k, being the running numbers of the points obtained by successive applications of T, beginning with the point $q^{(1)}$ defined in § 3.3.2. This decimal sequence is invariant when λ, or S, varies. The two cycles generated by the bifurcation (3.20) have the same [u]. Other definition of a decimal rotation sequence is given in [G2] [G3].

4.2.2. DEDUCTION OF THE BINARY SEQUENCE FROM THE DECIMAL ONE.

Taking into account the definition of [u], one has the following rule of deduction [M38] [M41] [M42]:

(P4.1) *The Myrberg's binary rotation sequence* [r] *is obtained from the decimal rotation sequence* [u] *by omitting* 1 *and* α_k *into* [u], *by replacing the* $\alpha_i < \alpha_k$ *by the sign* –, *and the* $\alpha_i > \alpha_k$ *by the sign* +.

This is the direct consequence of the definition of [u] and [r]. So, for the three pairs of cycles (5;j'), j' = 1,2,3, generated from the bifurcation (3.20), one has : j' = 1 → [u] = [15324], [r] = [+--] ; j' = 2 → [u] = [15423], [r] = [++-] ; j' = 3 → [u] = [15432], [r] = [+++].

4.2.3. DEDUCTION OF THE DECIMAL ROTATION SEQUENCE FROM THE BINARY ONE.

Taking into account the definition of [u], one has also :

(P4.2) *From* [r], *one writes* $[q^{(1)}] = [1[r]z]$, z = 0, *or* 1, *binary word of the cycle, and of the point having the running number "one" for this cycle. The words* $[q^{(i)}]$, *with* $q^{(i+1)} = Tq^{(i)}$, $i = 1,2,\dots,k$, *are written and ordered by (P3.9) in an increasing set of* $[q^{(i)}]$. *This order is independant of* z. *So each* $[q^{(i)}]$ *is numbered by a running number* α_i *into this increasing set. Starting from* $[q^{(1)}]$ *which is always such that* $\alpha_1 = 1$, $[q^{(2)}]$ *which always corresponds to* $\alpha_2 = k$, *the sequence of* $[q^{(i)}]$ *give the sequence of* α_i, *with* $(\alpha_k - 2)$ = *(number of* 1 *into* [r]*). Then* $[u] = [\alpha_1,\alpha_2,\alpha_3,\dots,\alpha_k]$ *is obtained.*

For example, considering [r] = [+--], one has $[q_1]$ = [10110], if z = 0, and $[q_1] < [q_4] < [q_3] < [q_5] < [q_2]$ (from the end of § 3.3.3). It results that $\alpha_1 = 1$, $\alpha_2 = 5$, $\alpha_3 = 3$, $\alpha_4 = 2$, $\alpha_5 = 4$, and [u] = [15324].

4.2.4. NECESSARY AND SUFFICIENT CONDITION FOR A PERMUTATION OF THE k FIRST INTEGERS TO BE A DECIMAL ROTATION SEQUENCE OF A MAP DEFINED BY A FUNCTION HAVING A SINGLE EXTREMUM.

Let [p] be a permutation of the k first integers. A necessary condition for [p] to be a decimal rotation sequence is that the two first numbers of [p] are $\alpha_1 = 1$, $\alpha_2 = k$. Another necessary condition is obtained by transforming [p], with $\alpha_1 = 1$, $\alpha_2 = k$, and by verifying $[r] > [r_i]$, $i = 2,3,\dots,q$ (cf.(P3.7). A necessary and sufficient condition for [p]

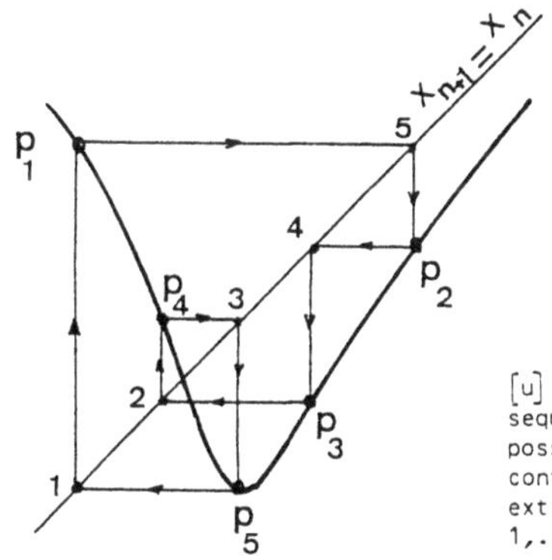

(a)

[u] = [1 5 4 2 3] is a rotation sequence for a cycle of T'. It is possible to pass an infinity of continuous curves with only one extremum by the points p_i, i = 1,...,5.

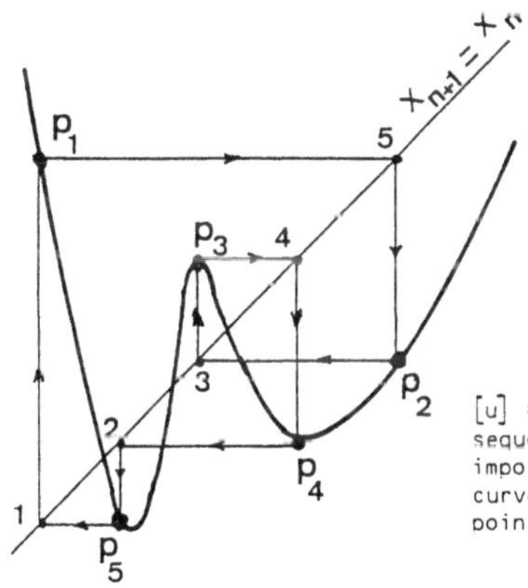

(b)

[u] = [1 5 3 4 2] is not a rotation sequence for a cycle of T'. It is impossible to pass a continuous curve with only one extremum by the points p_i.

FIG. 4.1

being a decimal rotation sequence is given from a very simple geometrical construction [M38] [M42], which itself is a prove. This condition is :

(P4.3) *Let* $[p] = [\alpha_1,\ldots,\alpha_k]$ *be a permutation of the* k *first integers with* $\alpha_1 = 1$, $\alpha_2 = k$. *On the line* $x_{n+1} = x_n$ *of the* (x_n, x_{n+1}) *plane,* k *points, numbered from* 1 *to* k,*with increasing abscissae,are considered. The points* $\alpha_1 = 1$ *and* $\alpha_2 = k$ *are joined by two segments, one from the point* 1, *parallel to the* x_{n+1}*-axis, the other from the point* k *parallel to the* x_n*-axis, which intersect at the point* p_1.
Following the order given by [p] *the same process is used for joining the other points of* [p], *and the last point* α_k *is also joined to the first one* $\alpha_1 = 1$ *similarly. So is obtained a closed curve, constituted by segments of straight lines with* k *angular points* p_i, $i = 1,2,\ldots k$, *outside* $x_{n+1} = x_n$. *The necessary and sufficient condition for* [p] *being a decimal rotation sequence* [u],*associated with a cycle of (3.3), is that it should be possible to pass through the* k *points* p_i *a continuous curve with only one extremum.*

The figures 4.1 give an example of application of (P4.3). So [p] = [15423] (fig.4.1a) is a decimal rotation sequence for an endomorphism defined by a continuous function with a only one extremum. It is not the case for [p] = [15342] (fig. 4.1b). The example of fig.4.1b shows that the condition $[r] > [r_i]$, $i = 2,3,\ldots,q$, ([p] after the use of (P4.1) is r-admissible) obtained when passing from [p] to [r], is only a necessary condition. Indeed [15342] gives [r] = [+++].

4.2.5. THE SUBSET OF THE ROTATION SEQUENCES DIRECTLY RELATED TO THE POINCARE'S ROTATION NUMBERS.

Into the set U of the rotation sequences, there exists a subset U_p, the elements of which are directly related to the Poincare's rotation-numbers [M32] [M42]. The corresponding relation is given by :

(P4.4) *Let* $\mu = p/k$, (p, k = *integers*, $p < k/2$) *be a rotation number. On a circle* k *points* P_i, $i = 1,2,\ldots k$, *limiting* k *equal segments, are considered. From the center, these segments define* k *angles* $2\pi/k$. *From a reference point* P_i, *numbered* 1 *(polar angle* $\theta = 0$*), it is traced a chord with its second extremity, numbered* k, *having the polar angle* $\theta = -2p\pi/k$. *This*

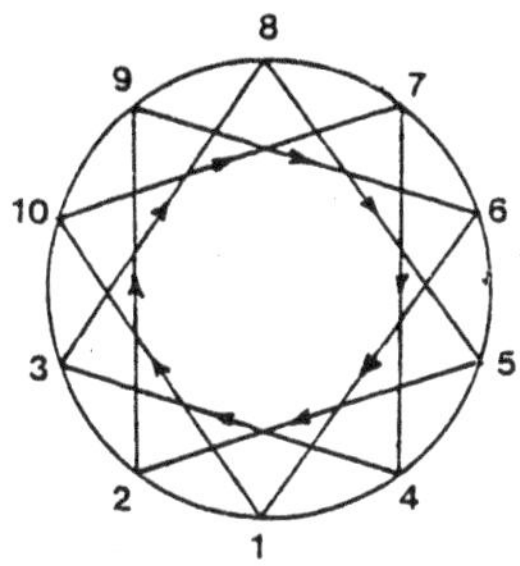

FIG. 4.2

The Poincaré's rotation number is defined by p=3 k=10 . The rotation sequence is given in the decimal form by : (1,10,7,4,3,8,5,2,9,6) .

chord subtends a circle segment consisting of p *equal segments, defined by points* P_i *which are numbered* 1,2,...,p-1,p,k. *Starting from the point numbered* k, *the* (k-p-1) *other points* P_i *are numbered* k-1,k-2,...,p+1. *From the point* 1, *the points separating the angle* $2p\pi/k$ *are joined. Then the sequence of those* k *points* [u] = [1,k,k-p,...,2p], *given by closing the so obtained curve, is a decimal rotation sequence related to the rotation number* $\mu = p/k$. *The obtained curve is a polygon if* p = 1, *and is star-shaped if* p > 1.

The so obtained sequence is a rotation sequence because it satisfies (P4.3). An example is given by the figure 4.2, where $\mu = 3/10$ is related to [u] = [1, 10, 7, 4, 3, 8, 5, 2, 9, 6], or [r] = [++--+--+] , or $[q^{(1)}]_{10}^{20'}$ = [100110110z], with z = 0 or 1.

The last number of such a decimal rotation sequence is always α_k = 2p. The considered circle is divided into two segments. The first one contains the points numbered $\alpha_i > 2p$, associated with the bit 0 in the binary word $[q^{(1)}]_k^{j'}$. The second one contains the points numbered $\alpha_i < 2p$, associated with the bit 1. The number $\alpha_k = 2p$, common point of the two segments is associated with the bit z = 0 or 1, according to the sign of the multiplier S of the cycle (k;j'). On the considered circle, the points α_i, i = 1,2,...,k, associated with the bit 0, or 1, have an axis of symmetry, whatever be z = 0 or 1. Placing these points, equally spaced on an other circle with the order given by [u], one has also an axis of symmetry on account of the construction given by (P4.5). It results:

(P4.5) *A binary word* $[q^{(1)j'}_{k}]$ *related to a rotation number* $\mu = p/k$ *is always symmetrical, and thus self adjoint, whatever be its last bit* z = 0 *or* 1.

Consider now the results of § 2.9.1, and (P4.4). Let $[u_\ell]$, $\ell = 1,2,\ldots,$ be the decimal rotation sequences belonging to U_p, and associated with the rotation numbers $\mu = p_\ell/k_\ell$ belonging to a same file, having the degree of complexity N, $\ell = 1$ corresponding to the central μ. Let $\mu = q/m$ the rotation number associated with the degree of complexity N-1, which is the accumulation point of the considered file. Let us arrange the set of the $[u_\ell]$ on lines having the running numbers $\ell = 1,2,\ldots,$ the last integer of every $[u_\ell]$ (here equal to $2p_\ell$) constituting a column which is the first from the right, the other integers constituting columns from this one. Then it exists m integers $\beta_1,\ldots,\beta_m$, defined from $\mu = q/m$, such that the elements of the last column on the right constitute an arithmetic progression with the common difference β_1, and the elements of the m-1 following columns constitute arithmetic progressions with common differences $\beta_2,\ldots,$ β_m. After the mth column, continuing from the right side to the left one, the following columns have the same properties as those of the m first columns, with a periodical repetition of common differences $\beta_1,\ldots,\beta_m$, the set of which is called the *rhythm* of the considered file of $[u_\ell]$, m being the *period* of this rhythm [M33]. The set of these $[u_\ell]$, $\ell = 1,2,\ldots,$ with the dregree N of complexity, is designated by $U_N(\beta_1,\ldots,\beta_m)$ and is called a *rhythmic set*.

Let $\mu_\ell = 1/k$, $\ell = k = 1,2,\ldots,$ be the rotation numbers corresponding to the unique file with the degree one of complexity. The corresponding decimal rotation sequences are such that $[u_\ell] = [1\ k\ k\text{-}1\ \ldots\ 3\ 2]$. Here U_1 is such that the rhythm is degenerated and is made up of a set of zeros. The μ of the degree two of complexity, $1/k < \mu < 1/(k+1)$, corresponds to two files given from (2.37), $\alpha_1 = 1$ giving the central μ. Then from (P4.4) it appears that : *the* $[u_\ell]$, $\ell = \alpha_1$, *associated with the* μ''_2 *of (2.37) having the complexity degree two, constitute a set* $U_2(2\ 3\ \ldots\ k\ 1\ 1)$ *of period* k+1, corresponding to cycles of order $\ell(k+1)+k$. Let now $[r_\ell]$ be the binary rotation sequences corresponding to these $[u_\ell]$.

The $[r_\ell]$ constitute always an increasing sequence (cf.P3.6) when ℓ increases, and they have the form (3.36) with a given h, their limit being $[r^{p_{k+1}}_{o,k+1}]$ when $\ell \to \infty$ (cf.§3.3.7 for p_{k+1}). Let λ^o_ℓ the λ-values associated with these $[r_\ell]$, one has $\lim \lambda^o_\ell = \lambda^{j'}_{(k+1)_o}$, $j' = p_{k+1}$, when $\ell \to \infty$. Here the λ^o_ℓ are the λ-values giving cycles of order $k' = \ell(k+1)+k$ with multiplier $S = 0$. *The* $[u_\ell]$, $\ell = \alpha_1$, *associated with the* μ'_2 *of* (2.37) constitute a *set* $U_2(2\ 3\ \ldots\ k\text{-}1\ k\ 0)$ *of period* k, corresponding to cycles of order $\ell k+k+1$. The associated $[r_\ell]$, $\ell = 1,2,\ldots$, constitute always a decreasing sequence (cf.(P3.6)), and have the form (3.32) with a given h, their limit being a $[\check{r}]$ when $\ell \to \infty$, and the limit of the corresponding λ^o_ℓ is a $\check{\lambda}$. This $\check{\lambda}$-value is always such that the critical point of rank 3 $C_2(x = (\lambda^2-\lambda)^2-\lambda)$ is the first which merges into a point of the cycle $(k;j')$ (for the complexity degree 2) with $S > 1$. Such $\check{\lambda}$-values plays a particular role, and in the general case $[u] \notin U_p$, or $[u] \in U_p$, they are designated by $*\lambda^{j'}_k$ (cf. § 3.3.8d).

More generaly it is possible to said : *let be two consecutive rotation numbers of a file of the complexity degree* N, *and two files of degree* N+1 *between these rotation numbers. The rotation numbers of each of these files correspond to a sequence of* $[r_\ell]$ *and* λ^o_ℓ. *One is increasing, the other decreasing according to (P3.6). When* $\ell \to \infty$ *the increasing sequence tends toward a fold bifurcation value* $\lambda^i_{(p)_o}$. *The decreasing sequence tends toward a* $\check{\lambda}$*-value,* $\check{\lambda} = *\lambda^j_k$, *such that the critical point* C_2 *is the first which merges into a cycle* $(k;j')$ *with* $S > 1$. The under-mentioned table 1 gives an example with the set $U_3(4\ 1\ 3\ 5\ 0)$ (complexity degree=3), the two consecutive rotation numbers of complexity degree 2 being $\mu = 2/5$, $\mu = 3/8$, and the rotation numbers associated with the set U_3 being $\mu = (2\alpha+3)/(5\alpha+8)$, $\alpha = 0,1,2,\ldots$, $\ell = \alpha+1$.

TABLE 1

													...	β_4	β_3	β_2	β_1	β_5	β_4	β_3	β_2	β_1	l	μ
...	...	...	0	5	3	1	4	0	5	3	1	4	0	5	3	1	4	0	5	3	1	4		
															1	8	5	2	7	4	3	6	1	3/8
										1	13	8	3	11	6	5	9	2	12	7	4	10	2	5/13
					1	18	11	4	15	8	7	12	3	16	9	6	13	2	17	10	5	14	3	7/18
1	23	14	5	19	10	9	15	4	20	11	8	16	3	21	12	7	17	2	22	13	6	18	4	9/23
...	...	...	...	...	...	...	...	...	...	...	...	...	...	...	...	...	...	...	...	...	...	...	...	...

$D=3,\ \mu=(2\alpha+3)/(5\alpha+8)$

For $\ell = 5$, $\lambda^{o}_{\ell} = 1.707364739399$, $\mu = 11/28$; $\ell = 6$, $\lambda^{o}_{\ell} = 1.7073645205$, $\mu = 13/33$; $\ell = 7$, $\lambda^{o}_{\ell} = 1.70736450093$, $\mu = 15/38$; $\ell = 8$, $\lambda_{\ell}=1.7073644992$ $\mu = 17/43$; $\tilde{\lambda} = {}^{*}\lambda^{1}_{5} = 1.7073644991$ (cf. § 3.3.8d).

The table 2 gives some examples of rhythmic set with N = 1,2,3.

N	μ	$[r_{\ell}]$	$\ell \to \infty$	$U_N(\beta_1 \ldots \beta_m)$
1	$1/\ell$	$(+)^{\ell}$	λ^{*}_{1}	0 (degenerate)
2	μ'' of (2.37) $m = 3$, $\alpha_1 = \ell$	$+(+--+)^{\ell}$	$\lambda^{1}_{(4)_{o}}$	1132
2	μ' of (2.37) $m = 3$, $\alpha_1 = \ell$	$++(--+)^{\ell}$	${}^{*}\lambda^{1}_{3}$	032
2	μ' of (2.37) $m = 4$, $\alpha_1 = \ell$	$+++(--++)^{\ell}$	${}^{*}\lambda^{1}_{4}$	0432
2	μ'' of (2.37) $m = 4$, $\alpha_1 = \ell$	$++(+--++)^{\ell}$	$\lambda^{3}_{(5)_{o}}$	11432
2	μ' of (2.37) $m = 2$, $\alpha_1 = \ell$	$(+--)^{\ell}$	$\lambda^{1}_{(3)_{o}}$	112
2	μ'' of (2.37) $m = 2$, $\alpha_1 = \ell$	$+(--)^{\ell}$	${}^{*}\lambda_{2^1} \equiv \lambda^{*}_{2^1}$	02
3	μ' of (2.39) $\mu^{(1)}_{2} = 2/5$ $\mu^{(2)}_{2} = 3/8$, $\mu_1 = 1/3$	$+(--+--)^{\ell}$	${}^{*}\lambda^{1}_{5}$	05314
3	μ'' of (2.39) $\mu^{(1)}_{2} = 3/7$ $\mu^{(2)}_{2} = 2/5$, $\mu_1 = 1/2$	$(+----)^{\ell}$	$\lambda^{1}_{(5)_{o}}$	22314
3	μ'' of (2.39) $\mu^{(1)}_{2} = 2/5$ $\mu^{(2)}_{2} = 3/8$, $\mu_1 = 1/3$	$(+--)(+4^{-}+--)^{\ell}$	$\lambda^{1}_{(8)_{o}}$	33517426

Table II : N = degree of complexity of the considered file ;

μ = rotation number of the file components ;

$[r_{\ell}]$ = binary rotation sequence ;

$(\ell \to \infty)$ = singular $\tilde{\lambda}$-values given with the embedded representation (cf. § 4.2.2) ;

$U_N(\beta_1 \ldots \beta_m)$ = rhythmic set.

4.3. DECOMPOSABLE AND INDECOMPOSABLE ROTATION SEQUENCES [u].

4.3.1. DEFINITION AND EXAMPLES.

The disposition of the components of a decimal rotation sequence [u] is directly related with the box-within-a-box bifurcations structure of

T [M21] [G32] [M25] [M32]. Fundamental property of the [u] about decomposition was published for the first time in [M25] (1978), but was used before to number the boxes of the structure [M21] [G32] (1975). It is shown in [M25] (1978) that, for a subset of the rotation sequences, it is possible to break down [u] into partial sequences (decomposition of [u] cf. [G35] p.128 and [M42]). For the operations of decomposition and composition, it seems that the form [u] is more convenient to manipulate that the other above mentioned decimal form [g], which uses the "evolution matrix" tool [G 2] [G 3] (1981). However, the form [g] seems to be better adapted for the problem of topological entropy. The results related to the decimal rotation sequences [u] are immediatly transposed for the binary one [r] via (P4.1). With the binary representation such results were described in [J 3] (1981), and where presented as a conseqence of the box-within-a-box structure (or embedded boxes) in [G17].

Definition.

Let $[u] = [1k\alpha_3 \ldots \alpha_k]$ *be the rotation sequence associated with the cycle* (k;j). *Let* β *be one of the divisor of* k *(if it exists) such that the partial sequence* $[u_{k/\beta}] = [1\alpha_{\beta+1}\alpha_{2\beta+1}\ldots\alpha_{p\beta+1}]$, *of order* $m = k/\beta$, $p = m-1$, *extracted from* [u] *is a rotation sequence (cf. (P4.3)). Let* $[u_\beta]$ *(remainder) be the partial sequence of order* β, *which is obtained by forming* β *sets of* k/β *consecutive points, the* k *points having the numbering used to define* [u]. *Similarly, the* β *sets are numbered from* 1 *to* β. *The sequence* $[u_\beta]$ *is written from* [u] *by following the exchange order of the sets with successive applications of* T. *If, as* $[u_{k/\beta}]$, $[u_\beta]$ *is a rotation sequence,* [u] *is said decomposable ; then* $[u_{k/\beta}]$ *and* $[u_\beta]$ *are said the components factors, or the components of* [u]. *In the opposite case,* [u] *is said indecomposable.*

Example.

Let [u] = [1,15,9,4,12,3,13,7,6,10,2,14,8,5,11]. Among the divisors of k = 15, β = 5 gives $[u_{15/5}]$ = [1,3,2] which verifies (P4.3) (corresponding cycle : k = 3, j' = 1). Let $g_1 \doteq (1,2,3)$, $g_2 = (4,5,6)$, $g_3 = (7,8,9)$, $g_4 = (10,11,12)$, $g_5 = (13,14,15)$ be the β = 5 sets of points. Taking into account [u], the exchange order of these sets is $[u_5] = [g_1 g_5 g_3 g_2 g_4] =$

[1 5 3 2 4], rotation sequence of the cycle (k = 5 ; j' = 1). From this step, $[u_{15/3}]$ is indecomposable. Afterwards, β will be said the *decomposing factor* and $[u_{k/\beta}]$ the *quotient*.

Remarks.

a) If k is prime number, [u] is always indecomposable. Nevertheless, [u] can be indecomposable with k different from a prime number. So [1 6 5 3 2 4], (k = 6 ; j' = 3) is indecomposable, but [1 6 3 4 2 5] associated with the cycle (k = 6 ; j' = 1) is decomposable with β = 2 giving $[u_{6/2}]$ = [1 3 2].

b) Let us consider the Myrberg's bifurcations cascade given by (3.21). With k = 1, a cycle of order 2^i born from (3.21) is such that :

$$(4.1) \qquad [u] = [u]_{2^i} = [1,\gamma-1,\alpha_1,\gamma-\alpha_1,\alpha_2,\gamma-\alpha_2,\ldots,\alpha_{m/2-1},\gamma-\alpha_{m/2-1}]$$

$\gamma = 2^i+1$, $m = 2^i$, $\alpha_1 = 2^{i-1}$, and $[u]_{2^{i-1}} = [1,\alpha_1,\alpha_2,\ldots,\alpha_{m/2-1}]$. Applying i times the decomposing factor β = 2, or once the decomposing factor $\beta = 2^i$, we obtain the indecomposable quotient [u] = 1.

4.3.2. SECOND NUMBERING OF THE CYCLES. EMBEDDED REPRESENTATION.

Let U_I be the set of the indecomposable rotation sequences [u]. Using (P4.1), from (P3.6), or (P3.6'), it is possible to arrange every set of the indecomposable [u], having the same order k, in an increasing set of elements. Let j, j = 1,2,...,$p_k \leqslant N_\lambda(k)$ (cf § 3.2.3), be the running number of such a [u] into this set. If k is a prime number $p_k = N_\lambda(k)$. A cycle corresponding to $[u] \in U_I$ will be represented by (k;j) ; for example, [u] = [1 6 5 3 2 4] corresponds to (k = 6 ; j' = 3) with the first numbering, and to (k = 6 ; j = 1) with the second numbering, [u] being the first indecomposable rotation sequence in the increasing set of the indecomposable [u] of order k = 6. The table of § 3.5 shows what occurs for k = 8, the numbering with j', the one with j, and the decomposition for j' = 1,2,9.

The embedded representation appears in the following situations when [u] is decomposable with $[u] \neq [u]_{2^i}$.

- <u>(a) $[u_{k/\beta}]$ and $[u_\beta]$ are indecomposable.</u> Let $\beta = k_1$, $k/\beta = k_2$. The rotation sequence is only once decomposable with the decomposing factor k_1. As above indicated the indices j_1, j_2 are respectively associated with $[u_\beta]$ and $[u_{k/\beta}]$ which are indecomposable. Then $[u]$ is the rotation sequence of a cycle, which will be represented by $(k_1.k_2 \; ; \; j_1, \; j_2)$. For the example of § 4.2.1 with $k = 15$, $k_1 = 5$, $k_2 = 3$, $j_1 = j_2 = 1$. It is possible to write $[u] = [u_{k_1}^{j_1}] \circ [u_{k_2}^{j_2}]$, or more compactly $[u] = [u_{k_1.k_2}^{j_1,j_2}]$.
- <u>(b) $[u_{k/\beta}]$ is indecomposable and $[u_\beta] \equiv [u]_{2^i}$.</u> Let $k/\beta = k_2$. Then $[u]$ is the rotation sequence of a cycle, which will be represented by $(2^i k_2 \; ; \; 1, \; j_2)$. Such a cycle appears into a "box", of the fractal bifurcations structure, called Ω_{2^i} in § 4.3.3., at the boundary of the box $\Omega_{2^i.k_2}^{1, j_2}$.
- <u>(c) $[u_\beta]$ is indecomposable, and $[u_{k/\beta}] \equiv [u]_{2^i}$ $(\gamma = k/\beta + 1)$.</u> Let $\beta = k_1$. The corresponding cycle is born from (3.21) with $k = k_1$. This cycle will be represented by $(k_1 2^i \; ; \; j_1)$. So $[u] = [1,12,8,4,9,5,2,11,7,3,10,6]$ is the rotation sequence of the cycle $(3.2^2;1)$, because $[u_{12/3}] = [1\ 4\ 2\ 3]$ is the rotation sequence of the cycle having the order 2^2 $(m = 2)$.
- <u>(d) General case, $[u_{k/\beta}]$ and (or) $[u_\beta]$ are decomposable.</u> Among the set of all the divisors of k, all the decomposing factors of $[u]$ are extracted. They appear in the form of products : k_1, $k_1.k_2, \ldots, k_1.k_2 \ldots k_{a-1}$, and we have $k = k_1.k_2 \ldots k_{a-1}.k_a$. The smallest, k_1, gives the first indecomposable component of $[u]$, represented by $[u_{k_1}^{j_1}]$, the index j_1 being the corresponding running number in the increasing set of the indecomposable $[u]$ of order k_1. After the quotient $[u_{k/k_1}]$ is considered, with the corresponding factors k_2, $k_2.k_3, \ldots, k_2.k_3 \ldots k_{a-1}$. The smallest is k_2 giving the 2nd indecomposable component $[u_{k_2}^{j_2}]$ with the index j_2. Continuing till the rank (a-1), the last decomposing factor k_{a-1}, the index j_{a-1}, and the component $[u_{k_{a-1}}^{j_{a-1}}]$ are defined. The quotient of the decomposition is $[u_{k/k_1/\ldots/k_{a-1}}]$, which is now either indecomposable with the order k_a and the index j_a, or having the form $[u]_{2^i}$.

In the first situation $[u]$ is the rotation sequence of a cycle represented by $(k_1.k_2 \ldots k_a \; ; \; j_1, j_2, \ldots j_a)$, and $[u]$ is written in the form $[u] = [u_{k_1}^{j_1}] \circ [u_{k_2}^{j_2}] \circ \ldots \circ [u_{k_a}^{j_a}]$.

In the second situation, $[u]$ is the rotation sequence of a cycle born from (3.21), and represented by $(k_1.k_2 \ldots k_{a-1} 2^i \; ; \; j_1, j_2, \ldots, j_{a-1})$.

In the described decomposition, if one of the factors is $k_n = 2^\ell$,

$n = 1,2,\ldots,a-1$, ℓ = integer, and such that $[u_{k_n}] \equiv [u]_{2\ell}$, it is not decomposed for obtaining the smallest decomposing factor. With $[u_{k_n}]$ the index $j_n = 1$ is associated, and it is proceeded as for the case (b). Then the corresponding cycle is represented by $(k_1\ldots k_{n-1}2^{\ell}k_{n+1}\ldots k_a \;;\; j_a,\ldots, j_{n-1},1,j_{n+1},\ldots j_a)$, or by $(k_1\ldots k_{n-1}2^{\ell}k_{n+1}\ldots k_{a-1}2^{i} \;;\; j_a,\ldots,j_{n-1},1,j_{n+1},\ldots,j_{a-1})$ in the second above mentioned situation.

So $[u]$ = [1,24,16,8,17,9,4,21,13,5,20,12,3,22,14,6,19,11,2,23,15,7, 18,10] is the rotation sequence of the cycle $(3.2^1.4 \;;\; 1,1,1)$, and $[u]$ = [1,24,16,8,17,9,6,19,11,4,21,13,2,23,15,7,18,10,5,20,12,3,22,14] is the rotation sequence of the cycle. $(3.4.2^1 \;;\; 1,1)$.

Remarks: the operations of § 4.2.2 constitute a decomposition of $[u]$ into a set of indecomposable rotation sequences. So it is obtained a representation of all the cycles of (3.1), or (3.3). This decomposition is unique. These operations indicate that the embedded representation $(k_1\ldots k_a;j_1\ldots j_a)$ is not commutative. In the representation symbolism of cycles: $2^2 \neq 4$, $2^3 \neq 8$,..

4.3.3. THE COMPOSITION OF INDECOMPOSABLE ROTATION SEQUENCES.

It is the inverse problem of the problem treated in § 4.2.2. Composing two rotation sequences $[u_{k_1}^{j_1}]$, $[u_{k_2}^{j_2}] \in U_I$ amounts to define the arrangements of the rotation sequence components associated with the cycle $(k_1k_2; j_1j_2)$, with $[u_{k_1.k_2}^{j_1,j_2}] = [u_{k_1}^{j_1}].[u_{k_2}^{j_2}]$.

The solution of this problem must take into account the representation of the curve T^{k_1} in the (x_n,x_{n+1}) plane, for the λ-values just after the birth of the cycle $(k_1k_2 \;;\; j_1j_2)$. For these λ-values, k_1 segments (C_ℓ) of T^{k_1}, $\ell = 1,\ldots k_1$, with a parabolic aspect, have a projection on the x_n-axis, everyone containing a set of k_2 points of the cycle $(k_1k_2 \;;\; j_1j_2)$. These k_2 points have a cyclic transfer defined by the index j_2, when applying successively T^{k_1}. The k_2 points of a set are numbered with increasing abscissae, if the extremum of the corresponding (C_ℓ) is a minimum, and with decreasing abscissae if it is a maximum. The rotation sequence $[u_{k_1.k_2}^{j_1,j_2}]$ is then directly obtained from $[u_{k_1}^{j_1}]$ and $[u_{k_2}^{j_2}]$, as it is shown from the example of this paragraph.

The final solution of this problem is given when the nature (maximum or minimum) of each extremum of (C_ℓ) is determinate In order to obtain this solution, it is supposed that the extremum of $f(x,\lambda)$ is a minimum (the case of a maximum is solved similarly by a symmetric argument). Then the k_1 segments (C_ℓ) of T^{k_1} are numbered with increasing abscissae, $\ell = 1$ corresponding to the smallest abscissa of the extrema of the considered (C_ℓ) (cf. fig.4.3 for $k_1 = 3$).

With this hypothesis, the first extremum E_1 of (C_1) is always a minimum. Then, for the multiplier $S = 0$ of the cycle $(k_1 ; j_1)$, the successive consequents of E_1 by T, till the rank k_1-1, define the nature of each extremum E_ℓ of (C_ℓ), via the binary word $[q^{(1)}{}_{k_1}^{j_1}] = [10\ldots z] \equiv [\delta_1\delta_2\ldots\delta_{k_1}]$. Here $q^{(1)} \equiv E_1$, δ_i, $i = 1,2,\ldots,k_1-1$, is a bit 0 or 1. To this binary word corresponds the decimal rotation sequence $[u_{k_1}^{j_1}] = [\alpha_1\alpha_2\ldots\alpha_{k_1}]$, $\alpha_1 = 1$, $\alpha_2 = k$. Starting from the minimum E_1, the evident rule for obtaining the nature of the E_ℓ, $\ell > 1$, is the following [K9] :

<u>(P4.6)</u> *Every extremum* E_ℓ, $\ell = \alpha_i$, *is asssociated with a bit* δ_i. *If* $\delta_i = 0$ (E_ℓ *is located on the segment of* $f(x,\lambda)$ *with a positive slope*), *the consequent* $E_{\ell'} = T\,E_\ell$ *of* E_ℓ, $\ell' = \alpha_{i+1}$, *has the same nature (minimum or maximum) as that of* E_ℓ. *If* $\delta_i = 1$ (E_ℓ *is located on the segment of* $f(x,\lambda)$ *with a negative slope*), $E_{\ell'}$ *has the opposite nature to that of* E_ℓ.

Practically, let $\delta_i' = +1$ *if* $\delta_i = 0$, $\delta_i' = -1$ *if* $\delta_i = 1$, *and* $p(\alpha_s) = \prod_{i=0}^{s-1} \delta_i'$, $\delta_o' = +1$, $p(\alpha_s)$ *being associated with the extremum* E_ℓ, $\ell = \alpha_s$. *If* $p(\alpha_s) = +1$, E_ℓ *is a minimum and the* k_2 *points of the cycle* $(k_2 ; j_2)$, *located on the segment* (C_ℓ) *are numbered with increasing abscissae. If* $p(\alpha_s) = -1$, E_ℓ *is a maximum and the considered set of* k_2 *points is numbered with decreasing abscissae.*

Applying (P4.6) gives the following set of operations :

* creating of the sets $g(\ell)$, $\ell = 1,2,\ldots k_1$, $g(\ell) = [(\ell-1)k_2+1 ; (\ell-1)k_2+2, \ldots, \ell k_2]$, every set containing k_2 points ;
* a matrix M, $k_1 \times k_2$, is constructed with the lines $g(\alpha_1), g(\alpha_2),\ldots,g(\alpha_k)$ written with this order ;
* a new matrix $\bar{M}$, $k_1 \times k_2$ is considered by conserving the numbered order of a line $g(\alpha_s)$ if $p(\alpha_s) = +1$, and inversing this order if $p(\alpha_s) = -1$, i.e. the new line is $\bar{g}(\alpha_s) = [\alpha_s k_2, \ldots, (\alpha_s-1)k_2+2, (\alpha_s-1)k_2+1]$.
* Let $[u_{k_2}^{j_2}] = [1\ k_2\ \alpha_3' \ldots \alpha_{k_2}']$, $[u_{k_1.k_2}^{j_1,j_2}]$ is obtained by writing the columns of $\bar{M}$ on a line with the order given by $[u_{k_2}^{j_2}]$.

Example : Let $[u^1_5] = [1\ 5\ 3\ 2\ 4]$ and $[u^1_3] = [1\ 3\ 2]$. The composed rotation sequence $[u^{1,1}_{5.3}] = [u^1_5].[u^1_3]$ must be constructed. One has :

$$M = \begin{bmatrix} g_1 \\ g_5 \\ g_3 \\ g_2 \\ g_4 \end{bmatrix} = \begin{bmatrix} 1 & 2 & 3 \\ 13 & 14 & 15 \\ 7 & 8 & 9 \\ 4 & 5 & 6 \\ 10 & 11 & 12 \end{bmatrix} \quad \begin{matrix} p(\alpha_1) = 1 \\ p(\alpha_2) = -1 \\ p(\alpha_3) = -1 \\ p(\alpha_4) = +1 \\ p(\alpha_5) = -1 \end{matrix} \qquad \bar{M} = \begin{bmatrix} g_1 \\ \bar{g}_5 \\ \bar{g}_3 \\ g_2 \\ \bar{g}_4 \end{bmatrix} = \begin{bmatrix} 1 & 2 & 3 \\ 15 & 14 & 13 \\ 9 & 8 & 7 \\ 4 & 5 & 6 \\ 12 & 11 & 10 \end{bmatrix}$$

The writing order of the 3 columns is [1 3 2]. It is given by the arrows of $\bar{M}$.

Finally : $[u^{1,1}_{5.3}] = [1,15,9,4,12,3,13,7,6,10,2,14,8,5,11]$.

The general case, with more than two indecomposable rotation sequences, is treated similarly by successive compositions of pairs of [u].

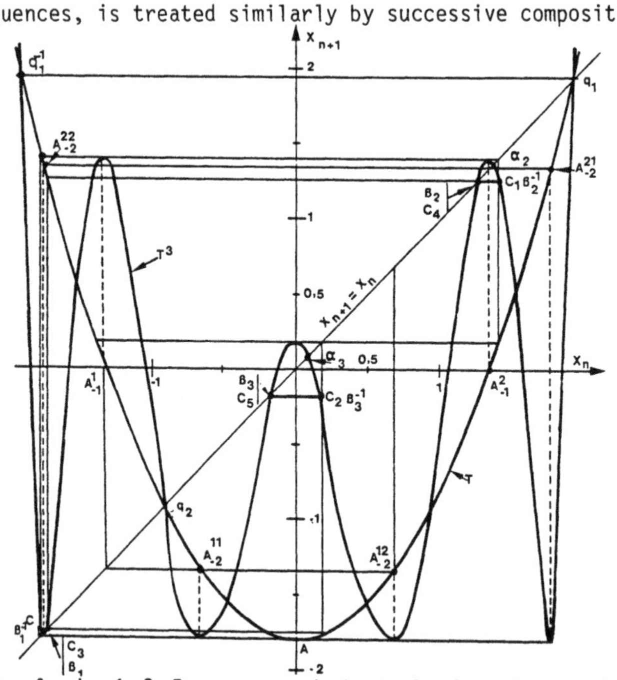

FIG. 4.3

$\lambda = \lambda_3^{1*}$

α_i, β_i, i = 1, 2, 3, are respectively the 3 points of the cycle k = 3 with S < 1, and of the cycle k = 3 with S > 1. C_j, j = 0, 1, ..., 5, $C_0 \equiv C$ is the consequent of rank j of the minimum A. The projections of the parabolic segments (C_l), l = 1, 2, 3, of T on x_n-axis are respectively $\overline{CC_3}$, $\overline{C_2C_5}$, $\overline{C_1C_4}$. (C_1) has a minimum, (C_2), (C_3) have each one a maximum.

4.4. THE BOX-WITHIN-A-BOX-BIFURCATIONS STRUCTURE (FRACTAL STRUCTURE).

4.4.1. SIMPLE (OR NON-EMBEDDED) BOXES OF THE STRUCTURE.

Consider the quadratic recurrence (3.1), $f(x,\lambda) \equiv x^2-\lambda$. Let $C(x = -\lambda)$ be the critical point of rank 1, $C_1(x = \lambda^2-\lambda)$ that of rank 2. As indicated in § 3.2.4, C_1 merges into the repulsive fixed point q_1 with multiplier $S > 1$, for $\lambda = \lambda_1^* = 2$. With $k = 1$ the fold bifurcation (3.20) occurs for $\lambda = \lambda_{(1)_0} = -1/4$. Inside the interval $\lambda_{(1)_0} \leq \lambda \leq \lambda_1^*$ all the possible bifurcations of (3.1) occur (cf. § 3.2). The interval $[\lambda_{(1)_0}, \lambda_1^*]$ is called *box* Ω_1.

Consider now *the set* U_I *of indecomposable rotation sequences* $[u]$ (or $[r]$ obtained via (P4.1)). Every $[u]$ (or $[r]$) $\in U_I$ characterizes a pair of cycles $(k;j)$, $k = 1,3,4,\ldots$, $j = 1,2,\ldots$, $p_k \leq N_\lambda(k)$, born from the fold bifurcation (3.20). They are called *basic cycles of the simple box* Ω_k^j, corresponding to the interval $[\lambda_{(k)_0}^j, \lambda^{*j}_k]$. Here $\lambda_{(k)_0}^j$ is the λ-value giving the bifurcation (3.20) ; it is such that k segments of T^k are tangent to $x_{n+1} = x_n$ in the (x_n, x_{n+1}) plane. λ^{*j}_k is the least λ-value such that the critical points $C_k, C_{k+1}, \ldots, C_{2k-1}$ of rank $k+1, k+2, \ldots, 2k$, merge into the k points of the basic cycle $(k;j)$ with multiplier $S > 1$. This λ-value is the equivalent of λ_1^*. By consideration of T^k in the (x_n, x_{n+1}) plane fig.4.4 et 4.5) λ_k^{*j} gives k segments of T^k bounded by critical points $C, C_1, \ldots, C_{2k-1}$ reproducing the situation of fig.3.16, and having some equivalent properties to those of § 3.2.5. Comparatively with the case $\lambda = \lambda_1^*$, now *only all the cycles, having rotation sequences* $[r] < [\hat{r}]$ *associated with* λ^{*j}_k*, have been created for* $\lambda = \lambda^{*j}_k + \varepsilon$, $\varepsilon > 0$, $\varepsilon \to 0$. The k segments $\widehat{CC_k}, \widehat{C_1C_{k+1}}, \ldots, \widehat{C_{k-1}C_{2k-1}}$ of T^k constitute a *chaotic cyclic segment* when $\lambda = \lambda_k^{*j}$, with a cyclic transfer characterized by the index j from successive applications of T.

When λ varies from $\lambda_{(k)_0}^j$ to λ^{*j}_k, in the (x_n, x_{n+1}) plane, k segments of T^k, with a parabolic shape, are in equivalent situations of those of T, when λ varies from $\lambda_{(1)_0}$ to λ_1^*. So in the box Ω_k^j, T^k undergoes with the same order the bifurcations of T in the box Ω_1. For T^k, Ω_k^j reproduces the structure of Ω_1 for T. The Myrberg's cascade (3.21), $i = 0,1,2,\ldots$, occurs in the interval $\omega_k^j \equiv [\lambda_{(k)_0}^j, \lambda_{(k)_s}^j]$ which is the *spectrum* $(k;j)$. Let $\omega_1 \equiv$

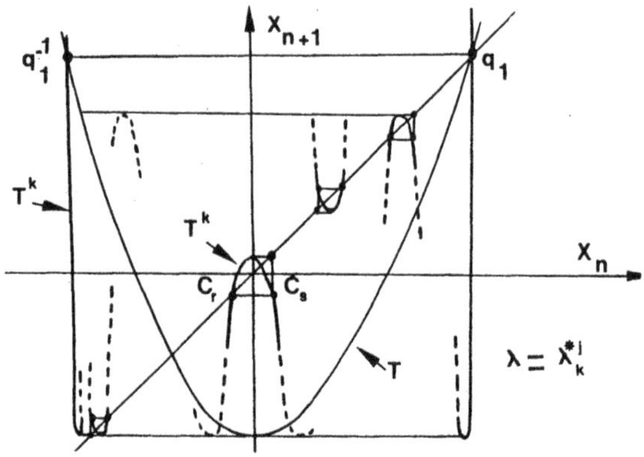

FIG. 4.4

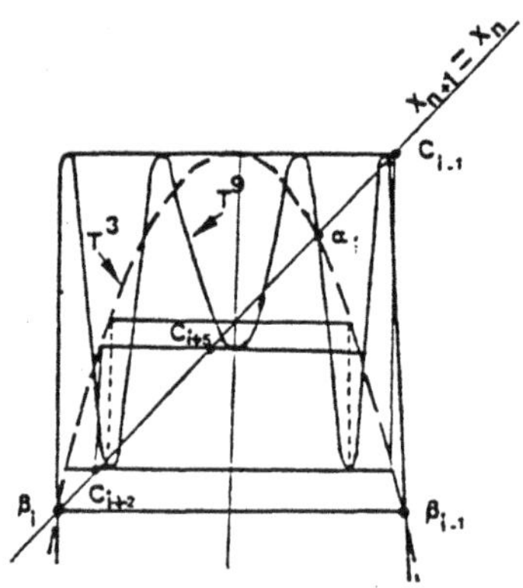

FIG. 4.5 - Representation of T^9 from that of T^3 ,on the segment $\overline{\beta_i\beta_i^{-1}}$ i=2,3 .For i=1 ,the figure has the same aspect, but the segment $\overline{\beta_1\beta_1^{-1}}$ has a minimum . T^9 on $\overline{\beta_i\beta_i^{-1}}$ has the same shape as T^3 on the segment $\overline{q_1q_1^{-1}}$.

$[\lambda_{(1)_0},\lambda_{(1)_s}]$, $\Delta_1 \equiv]\lambda_{(1)_s},\lambda_1^*]$, $\Delta_k^j \equiv]\lambda_{(k)_s}^j,\lambda_k^{*j}]$. Due to the existence of at most one attractive cycle at finite distance, one has $\Omega_k^j \subset \Delta_1$.

Let $\lambda_{2^1}^*$ be the λ-value such that C_2 merges into the fixed point q_2 with multiplier $S < -1$. It will appear in § 4.4.3 that simple boxes ($[u] \in U_I$) do not exist outside of the interval $\Gamma_{2^0} \equiv \Gamma_1 \equiv]\lambda_{2^1}^*,\lambda_1^*[$, so $\Omega_k^j \subset]\lambda_{2^1}^*,\lambda_1^*[\equiv \Gamma_1 \subset \Delta_1$. The values $\lambda_{2^1}^*$ and λ_1^* are accumulation values for two subsets of all the simple boxes when $k \to \infty$.

4.4.2. EMBEDDED BOXES OF FIRST KIND.

Let $(k_1;j_1)$ be the basic cycles associated with the indecomposable rotation sequence $[u_{k_1}^{j_1}]$, and with the simple box $\Omega_{k_1}^{j_1}$, $k_1 \neq 2^i$, $i = 1,2,\ldots$ (it is recalled that the notation $k_1 = 2^i$ implies that the corresponding cycle is born from the flip bifurcation (3.21) ($i-1 \to i$). With this symbolism $2^2 \neq 4$, $2^3 \neq 8,\ldots$). Let now $[u_{k_1.k_2}^{j_1,j_2}]$ be the composed rotation sequence obtained from $[u_{k_1}^{j_1}]$,and from the indecomposable rotation sequence $[u_{k_2}^{j_2}]$, $k_2 \neq 2^i$, $i = 1,2,\ldots$, $j_1 = 1,2,\ldots,p_{k_1}$, $j_2 = 1,2,\ldots,p_{k_2}$. This composed rotation sequence is related to a pair of basic cycles $(k_1.k_2$; $j_1,j_2)$ of order $k = k_1.k_2$,and gives rise to the *embedded box* $\Omega_{k_1.k_2}^{j_1,j_2} \equiv [\lambda_{(k_1.k_2)_0}^{j_1,j_2},\lambda_{k_1.k_2}^{*j_1,j_2}]$. Here this symbolism has the same sense as that defined for Ω_k^j. Considering $T^{k_1.k_2}$ in the (x_n,x_{n+1}) plane, when λ varies from $\lambda_{(k_1.k_2)_0}^{j_1,j_2}$ to $\lambda_{k_1.k_2}^{*j_1,j_2}$, inside everyone of the k_1 segments $\overline{C_iC_{i+k_1}}$, $i = 0, 1,\ldots,k_1-1$, $C_0 \equiv C$, (corresponding to k_1 segments of T^{k_1} with a parabolic shape) $T^{k_1.k_2}$ has k_2 parabolic segments at the origin of the bifurcations. These have a cyclic transfer characterized by the index j_2 from k_2 successive applications of T^{k_1}. One has $\Omega_{k_1.k_2}^{j_1,j_2} \subset \Delta_{k_1}^{j_1}$. The box $\Omega_{k_1.k_2}^{j_1,j_2}$ is the box $\Omega_{k_2}^{j_2}$ of T^{k_1},considered for λ-values giving rise to the box $\Omega_{k_1}^{j_1}$.

Let $\lambda_{k_1 2^i}^{*j_1}$, $i = 0,1,2,\ldots$, be the λ-value such that the critical point C_{2^i} of rank 2^i+1 merges into the point of the cycle of order $k_1 2^i$ $(S < -1)$,appearing from the flip bifurcation (3.21) with $k = k_1 2^{i-1}$, and which has the least abscissa. As for § 4.4.1 $\Omega_{k_1.k_2}^{j_1,j_2} \subset]\lambda_{k_1.2^1}^{*j_1},\lambda_{k_1}^{*j_1}[\subset \Delta_{k_1}^{j_1}$. For $T^{k_1k_2}$, $\Omega_{k_1.k_2}^{j_1,j_2}$ reproduces the bifurcations structure of $\Omega_{k_1}^{j_1}$ considered for T^{k_1}, and then that of Ω_1 for T.

Repeating compositions of indecomposable rotation sequences $[u_{k_m}^{j_m}]$, $m = 1,2,\ldots,a$, or considering $T^{k_1k_2\ldots k_a}$, it appears that every pair of

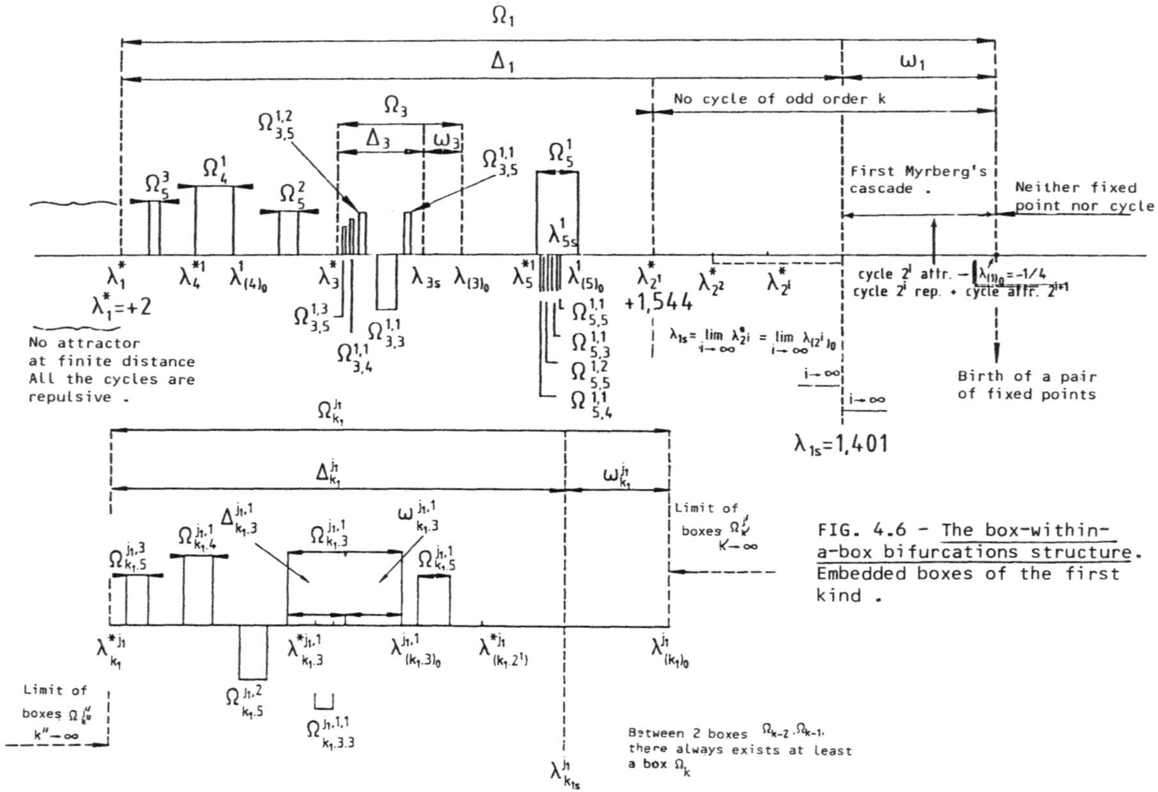

FIG. 4.6 - The box-within-a-box bifurcations structure. Embedded boxes of the first kind .

basic cycles $(k_1.k_2...k_a;\ j_1,j_2,...,j_a)$ defines an embedded box such that :

$$(4.2)\qquad \Omega^{j_1,\dots,j_a}_{k_1\dots k_a} \equiv \left[\lambda^{j_1,\dots,j_a}_{(k_1\dots k_a)_0}, \lambda^{*j_1,\dots,j_a}_{k_1\dots k_a}\right]$$

$$\Omega^{j_1,\dots,j_a}_{k_1\dots k_a} \subset \left]\lambda^{*j_1,\dots,j_{a-1},1}_{k_1\dots k_{a-1}.2^i}, \lambda^{*j_1,\dots,j_{a-1}}_{k_1\dots k_{a-1}}\right[\subset \Delta^{j_1,\dots,j_{a-1}}_{k_1\dots k_{a-1}}$$

with $j_\alpha = 1,2,\dots,p_{k_\alpha}$, $\alpha = 1,2,\dots,a$, $k_\alpha = 1,3,4,\dots$, and

$$(4.3)\qquad \Omega^{j_1,\dots,j_a}_{k_1\dots k_a} = \omega^{j_1,\dots,j_a}_{k_1\dots k_a} \cup \Delta^{j_1,\dots,j_a}_{k_1\dots k_a},$$

Here the interval ω is the Myrberg's spectrum generated from the basic cycle $(k_1\dots k_a\ ;\ j_1,\dots,j_a)$ with the multiplier $S < 1$, and the interval Δ is the remaining part of Ω. Such a process may be carried on till $a\to\infty$. The box given by (4.3) is called *embedded box of the first kind at the degree* a-1 *of nesting*. The two parts of fig.4.6 gives the general aspect of the bifurcations structure for the first kind.

4.4.3. EMBEDDED BOXES OF SECOND KIND.

Such boxes are not generated from basic cycles, but from cycles obtained with the flip bifurcation (3.21). Let $\lambda_{(2^i)_0}$ be the λ-value giving rise to the cycle of order 2^i via (3.21) ($k = 1$, $i \to i-1$, $S = -1$ for the cycle of order 2^{i-1}). Let $\lambda^*_{2^i}$ be the λ-value such that the critical point C_{2^i} of rank 2^i+1 merges into the point of the cycle of order 2^{i-1} which has the smallest abscissa. The critical points $C_{2^i+1},\dots,C_{2^i+2^i-1}$ merge into the other points of this cycle. The simplest embedded boxes of second kind correspond to intervals $\Omega^1_{2^i} \equiv [\lambda_{(2^i)_0}, \lambda^*_{2^i}]$, $i = 1,2,\dots$ In the general case they correspond to :

$$(4.4)\qquad \Omega^{j_1,\dots,j_a,1}_{k_1\dots k_a.2^i} \equiv \left[\lambda^{j_1,\dots j_a,1}_{(k_1\dots k_a 2^i)_0}, \lambda^{*j_1,\dots,j_a,1}_{k_1\dots k_a.2^i}\right] \subset \Omega^{j_1,\dots,j_a}_{k_1\dots k_a},$$

$k_\alpha = 1,3,4,\dots,$; $\alpha = 1,2,\dots,a$; $j_\alpha = 1,2,\dots,p_{k_\alpha}$; $i = 1,2,\dots$ Let now $\Gamma_{2^i} \equiv \left]\lambda^*_{2^{i+1}}, \lambda^*_{2^i}\right]$, $i = 0,1,2,\dots$, $\Gamma_{2^0} \equiv \Gamma_1$, $\Gamma_{2^i} \subset \Omega^1_{2^i}$.
The interval Γ_{2^i}, $i = 1,2,\dots$, contains boxes $\Omega^{1;j_2}_{2^i.k_2}$, which are the boxes $\Omega^{j_2}_{k_2}$ of T^r, $r = 2^i$. The boxes $\Omega^{1;j_2}_{2^i.k_2}$ are generated from a pair of basic

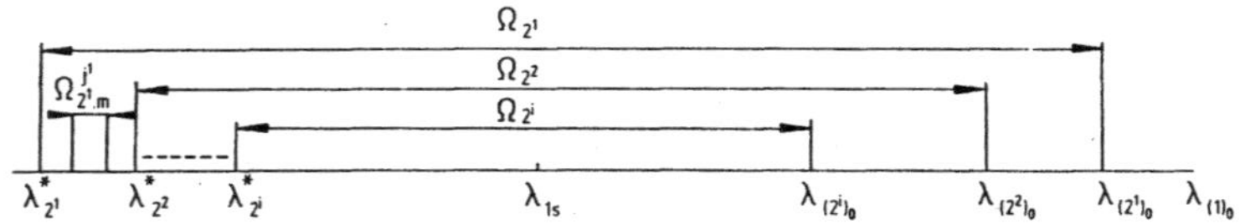

FIG. 4.7 - The box-within-a-box bifurcations structure. Embedded boxes of the second kind.

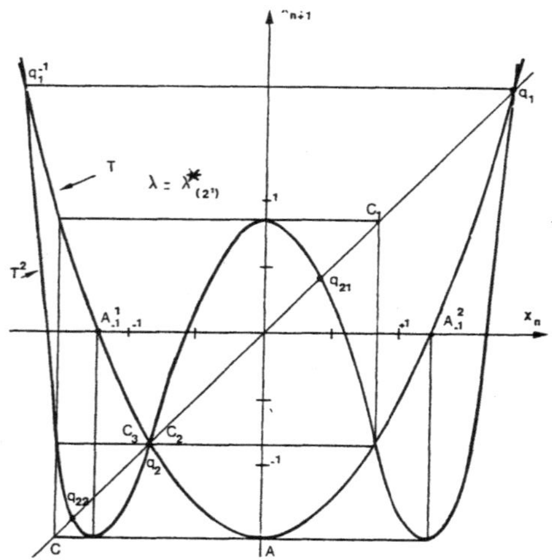

FIG. 4.8.a

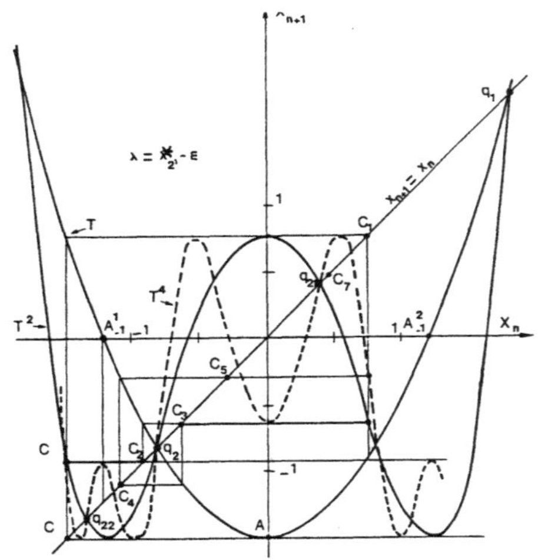

FIG. 4.8.b

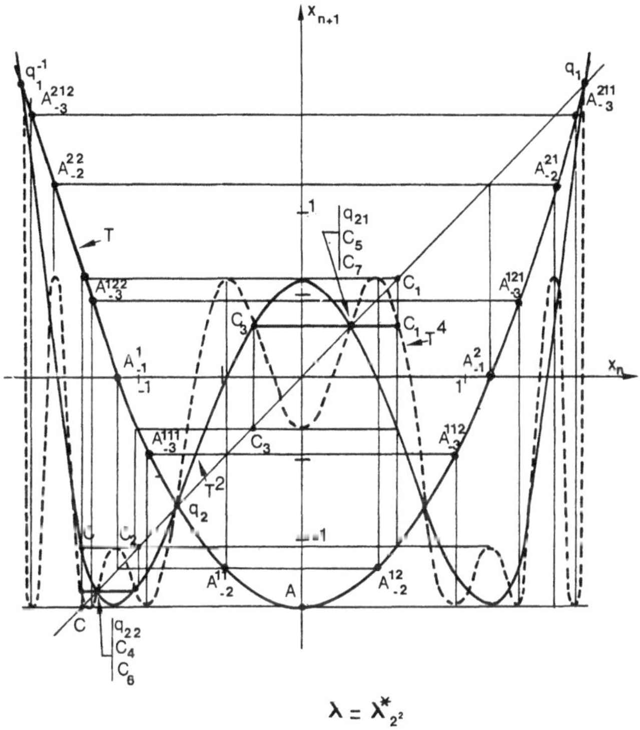

x_{n+1}
x_n
q_1^{-1}
q_1
A_{-3}^{212}
A_{-3}^{211}
A_{-2}^{22}
A_{-2}^{21}
1
q_{21}
C_5
C_7
T
C_1
A_{-3}^{122}
A_{-3}^{121}
C_3
C_1
T^4
A_{-1}^1
A_{-1}^2
-1
1
A_{-3}^{111}
C_3
A_{-3}^{112}
T^2
q_2
C_2
-1
A_{-2}^{11}
A
A_{-2}^{12}
C
q_{22}
C_4
C_6
$\lambda = \lambda^*_{2^2}$

FIG. 4.8.c

cycles $(2^i.k_2\ ;\ 1,j_2)$, the rotation sequence $[u]'$ of which results from the composition of § 4.2.2.b. So one has :

$$(4.5)\qquad \Omega^{1,j_2}_{2^i.k_2} \subset \Gamma_{2^i} \equiv \left]\lambda^*_{2^{i+1}}, \lambda^*_{2^i}\right] \quad , \qquad i = 1,2,\ldots$$

Equivalent relations may be obtained for embedded boxes at higher degree, such that :

$$(4.6)\qquad \Omega^{j_1,\ldots,j_{n-1},1,j_{n+1},\ldots,j_a}_{k_1\ldots k_{n-1}2^i k_{n+1}\ldots k_a} \subset \Gamma^{j_1,\ldots,j_{n-1},1}_{k_1\ldots k_{n-1}2^i} \quad ,$$

the box Ω corresponding to one of the composition of §4.2.2.d .

Fig.4.8 give an illustration of the simplest case. With fig.4.8 a ($\lambda = \lambda^*_{2^1}$), two segments of T^r, $r = 2^1$, having a parabolic shape are in the situation of $\lambda = \lambda^*_1$ for T. With fig.4.8 c ($\lambda = \lambda^*_{2^2}$), four segments of T^r, $r = 2^2$ are in the situation of $\lambda = \lambda^*_1$ for T |G32| |M21|. The structure of second kind is given by the fig. 4.7 .

4.4.4. ACCUMULATION POINTS OF THE BOXES ($f(x,\lambda) \equiv x^2-\lambda$).

From § 3.3.8, 3.4.2, it appears that the singular parameters values $\hat{\lambda}$ of 2nd type are accumulation points on either side of $\hat{\lambda}$ ($\lambda > \hat{\lambda}$, and $\lambda < \hat{\lambda}$) for a subset of boxes, since they are accumulation points for λ-values giving rise to basic cycles. From § 3.3.9, 3.4.3 it appears that the singular values $\lambda^j_{(k)_0}$ of 3rd type, $k \neq 1$, are accumulation points on the left side ($\lambda < \lambda^j_{(k)_0}$) for a subset of boxes. Let E_c be the set of the consequents of the critical points with the ranks 2,3,4,... Let E'_c be the set of the limit points of E_c. Let E be the set of the repulsive cycles of T, E' the set of the limit points of E. From the Fatou's results (cf.§ 30 of |F 1|) it results |M34| |M42| :

(P4.7) *When the recurrence* T *defined by* $f(x,\lambda) \equiv x^2-\lambda$ *has an attractive cycle, then a point of* E_c, *or of* E'_c , *cannot be a point of* $E \cup E'$. *When* T *is such that a point of* E_c, *or of* E'_c, *belongs to* $E \cup E'$, *for* $\lambda = \hat{\lambda}$, *then* T *has no attractive cycle at finite distance, but has a strange attractor. This attractor is a cyclic chaotic segment for* E_c *belonging to* E ($\lambda = \hat{\lambda}$).

The values $\tilde{\lambda}$ are limit points for simple boxes Ω_k^j, $k \to \infty$. or for embedded boxes. Some $\tilde{\lambda}$-values are limit points for other $\tilde{\lambda}$-values. As in § 3.2.4 c, but here on the parameter axis, these limit points must be associated with the notion of class of accumulation. *A $\tilde{\lambda}$ limit point of class* p *is an accumulation point for a set of limit points of class* q < p. The λ-values $\hat{\lambda} \neq \tilde{\lambda}$ appear as limit points of class p = ∞. Among the $\tilde{\lambda}$, *the λ_k^{*j}-value defined in* § 4.4.1 *plays a particular role, as boundary of the box Ω_k^j, this boundary being limit point on either side for boxes with order* k' $\to \infty$. *The $\tilde{\lambda}$-values, $\tilde{\lambda} \neq \lambda_k^{*j}$, are limit points for a set of boxes, without being boundaries of boxes.*

The limit point $\lambda_{(k)_s}^j$, k = 1,3,4,..., of a Myrberg's cascade corresponding to the spectrum ω_k^j, is on the right side ($\lambda > \lambda_{(k)_s}^j$) an accumulation point for $\lambda_{(k2^i)}^{*j}$-values i = 1,2,... defined in § 4.4.2. This result was proved the first time in 1975 [G32] (cf. also [M21] [M25]) by arguing for k = 1 about the cyclic chaotic segment of order 2^i obtained when $\lambda = \lambda_{2^i}^*$. For i $\to \infty$ this cyclic chaotic segment tends toward the Cantor set generated by (3.21) with k = 1, when $\lambda = \lambda_{(1)_s}$. Another way to prove this result consists in this that $[r_{(k)_s}^j]$ (corresponding to $\lambda_{(k)_s}^j$ and given by (3.23) with i = ∞) is the limit of $[\hat{r}_{(k2^i)}^j]$ associated with $\lambda_{k2^i}^{*j}$ when i $\to \infty$, with $[r_{(k)_s}^j] < [\hat{r}_{(k.2^i)}^j]$. Indeed, when i $\to \infty$ from (3.27) : $[\hat{r}_{(k2^i)}^j] \equiv (A)(B)^{\infty} \simeq (A)$, (A) being given by the slopes signs of f(x,λ) at the critical points $C_1, C_2, \ldots, C_{\alpha-2}$, $\alpha = 2^{i+1}$, and then lim (A) = $[r_{(k)_s}^j]$ when i $\to \infty$.

<u>(P4.8)</u> *The boundary $\lambda_{(k)_s}^j$ of the Myrberg's spectrum ω_k^j is for $\lambda > \lambda_{(k)_s}^j$ the limit point of the singular values $\lambda_{k2^i}^{*j}$ of 2nd type, when* i $\to \infty$, *with* k = 1,3,4,..., i.e. $\lambda_{(k)_s}^j$ *is a limit point for the boundaries of the boxes* $\Omega_{k2^i}^j$, lim $\Omega_{k2^i}^j = \lambda_{(k)_s}^j$ *if* i $\to \infty$. *This property recurs in a fractal way inside boxes having higher degree of nesting, and given by (4.2), (4.4), (4.6).*

All the cycles generated inside a box $\Omega_{k2^i}^j$, i = 1,2,..., k = 1,3,4,.., have an even order. *So $\lambda_{2^i}^*$, boundary of such a box with* k = i = 1, *is the limit of existence of cycles of odd order for $\lambda > \lambda_{2^1}^*$.* This property was given firstly in 1975 [G32] [M21], as also (P4.9) which results from § 3.4.2., 4.4.2., 4.4.3.

<u>(P4.9)</u> *For the quadratic recurrence (3.1) one has :*

$\lim \lambda^{*1,j_2}_{2^i.k_2} = \lim \Omega^{1,j_2}_{2^i.k_2} = \lambda^*_{2^i}$, $i = 1,2,\ldots,j_2 = p_{k_2}$, *when* $k_2 \to \infty$, *the limit being defined for* $\lambda < \lambda^*_{2^i}$, *and* $\lim \lambda^{*1,j'_2}_{2^i-1.k'_2} = \lim \lambda^{1,j'_2}_{(2^{i-1},k'_2)_0} =$ $\lim \Omega^{1,j'_2}_{2^i-1.k'_2} = \lambda^*_{2^i}$, k'_2 = *odd integer*, $j'_2 = 1$, *when* $k'_2 \to \infty$, $(\lambda > \lambda^*_{2^i})$. *Here* $j_2 = 1$ (*resp.*p_{k_2}) *is the first* (*resp.last*) *running number of the embedded representation,* $j'_2 = 1$ *being associated with the Myrberg's rotation sequence* $[r] = +(-)^{\ell}$, $k'_2 = \ell+3$; $j = p_{k_2}$ *with* $[r] = (+)^{\ell}$, $k_2 = \ell+2$. *The index* $i = 0$ *corresponds to* $\lambda^*_{2^0} \equiv \lambda^*_1$, $\Omega_{2^0} \equiv \Omega_1$, $\lim \lambda^{*j_1}_{k_1} = \lambda^*_1$, $j_1 = p_{k_1}$, $k_1 \to \infty$, *and* $i = 1$ *to* $\lim \lambda^{*j_1}_{k_1} = \lambda^*_{2^1}$, $j_1 = 1$, $k_1 \to \infty$. *All these propertiesrecur equivalently in a fractal way inside embedded boxes having higher degrees of nesting and given by* (4.2), (4.4), (4.6).

Among the $\hat{\lambda}$-values, a ${}^*\lambda^j_k$ -value , $k = 3,4,\ldots$, is the smallest λ-value such that the critical point C_2 of rank 3 is the point of the cycle $(k;j)$ with $S > 1$. In the particular case when j is associated with a rotation sequence $[u]$ directly related to a rotation number, $[u] \in U_p$ (§ 4.1.5), the value ${}^*\lambda^j_k$ is the limit point of a decreasing sequence of boxes belonging to one of the two files having a given degree of complexity (cf. table II of § 4.1.5). *On the* λ*-axis the simple boxes* Ω^j_k, $[u^j_k] \in U_p$, with the degree $N = 1$ of *complexity are such that,* $j = p_k$ $([r]=(+)^{k-2})$, $k = 3,4,5,\ldots$; Ω_{2^1}, having $\lambda^*_{2^1}$ as one of the boundaries,can be considered as belonging to $N = 1$.Since $\lim \lambda^{*j}_k = \lambda^*_{2^1}$, $j = 1$ $([r] = +(-)^{k-3} \to [u] \in U_p)$, $\lambda^*_{2^1} \equiv {}^*\lambda_{2^1}$ in this particular case. Between two consecutive boxes with the degree $N = 1$ of complexity, the simple boxes Ω^j_k, $[u^j_k] \in U_p$ with $N = 2$, are located inside the intervals [M38] [M42]:

$$(4.7)\qquad \Phi^{i_1}_2 \equiv [{}^*\lambda^{j_1}_{i_1+1}, \lambda^{j_2}_{(i_1+2)_0}[, \quad i = 1,2,3,\ldots, \quad {}^*\lambda_2 \equiv {}^*\lambda_{2^1} \equiv \lambda^*_{2^1}.$$

Here the index i_1, correspond to boundaries of boxes with $N = 1$, $j_m = p_\alpha$, $\alpha = i_1+m$, $m = 1,2$. Inside the interval $\Gamma_{2^0} \equiv \Gamma_1 \equiv]\lambda^*_{2^1},\lambda^*_1]$, and between two consecutive boxes related to U_p with $N = 1$, the intervals (4.7) leave holes defined by :

$$(4.8)\qquad \Pi^{i_1}_2 \equiv]\lambda^{*j_1}_{i_1+1}, {}^*\lambda^{j_1}_{i_1+1}[, \quad i_1 = 2,3,\ldots, \quad i_1 = 1 \to \Pi^1_2 \equiv 0$$

and which do not contain any simple boxes Ω^j_k such that the rotation sequence of the cycle $(k;j)$ belongs to U_p. More generally by considering

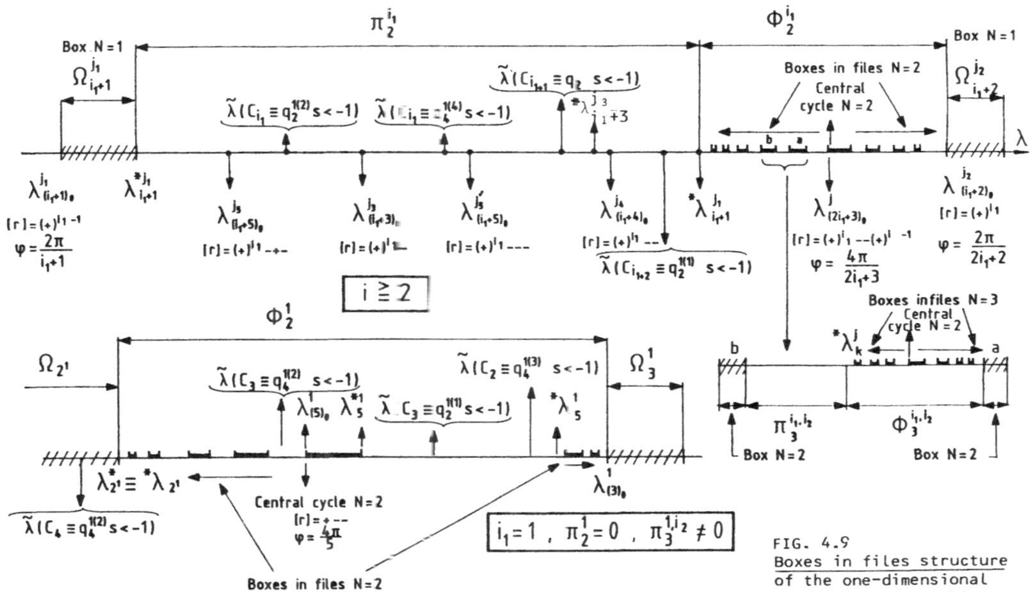

FIG. 4.9 Boxes in files structure of the one-dimensional quadratic endomorphism

$q_k^{j(\alpha)}$ s<-1 or s>-1 : point numbered α with increasing abscissae belonging to the cycle (k;j) having the multiplier s
c_β : critical point of rank β+1 (rank β=2 consequent of the extremum of f(x, λ))
[r] : Myrberg's rotation sequence. φ=2pπ/q ; p/q : Poincaré's rotation number

the simple boxes belonging to $\Phi_2^{i_1}$, with basic cycles such that $[u] \in U_p$ and the degree of complexity $N = 3,4,\ldots,$ intervals $\Phi_N^{i_1,\ldots,i_{N-1}}$ and $\Pi_N^{i_1,\ldots,i_{N-1}}$ are then defined with :

$$(4.9) \qquad \bigcup_{i_{N-1}=1}^{\infty} \Phi_N^{i_1,\ldots i_{N-1}} \quad \bigcup_{i_{N-1}=1}^{\infty} \Pi_N^{i_1,\ldots,i_{N-1}} \subset \Phi_N^{i_1,\ldots,i_{N-2}}.$$

Here (4.9) indicates that two files of degree N and the corresponding holes Π are located between two consecutive boxes having the degree N-1 of complexity.

Inside the intervals (4.7), (4.8), $i_1 \geqslant 2$, the limit points of boxes are ordered and recur with laws obtained from (P3.6) or (P3.6'). For the lowest orders fig.4.9 gives a general idea of such laws. *The diagram of fig.4.9 recur equivalently in a fractal way inside embedded boxes having higher degrees of nesting, and given by* (4.2), (4.4), (4.6), *and so limit points* $\tilde{\lambda}$ *with an increasing class of accumulation* (cf. § 3.2.4 c) *are obtained*. For the basic cycles with $[u] \in U_p$, consider those corresponding to the rotation number of the complexity degree $N, \mu_N^{(1)}$ given by (P2.24) (cf. § 2.9.1). Then $\alpha_{N-1} = 2,3,4,\ldots$ gives a set $[r_\ell]_h = (A_h)(B_h)^\ell$, having the form (3.32), or (3.36), with h = a fixed value and $\ell = 1,2,3\ldots$ Let now μ_{N+1} be one of the rotation number belonging to the files of degree N+1 given by (P2.24), α_N having a fixed value. Then $\alpha_{N-1} = 2,3,4,\ldots$ gives a set $[r_\ell]'_h = (A_h)(B_h)^\ell(C)$ having the form (3.33), or (3.37), with h = a fixed value, and $\ell = 1,2,3,\ldots$ When $\alpha_{N-1} \to \infty$, i.e. $\ell \to \infty$, $\lim [r_\ell]_h = \lim [r_\ell]'_h = (A_h)(B_h)^\infty$.

For basic cycles with $[u] \notin U_p$, i.e. for cycles (k;j) such that $\lambda^j_{(k)_0} \in \Pi_N^{i_1,\ldots,i_{N-1}}$, equivalent boxes in files structure can be defined by considering that (3.32), or (3.36), with a fixed h, generates a file of a certain degree N of complexity, from $\ell = 1,2,3,\ldots,$ and that (3.33), or (3.37), gives elements of different files having the degree N+1 of complexity, a file corresponding to a fixed ℓ and to $[r_p]''_{h'} = (A_{h'})[(B_{h'})^\ell (C_{h'})]^m$, $m = 1,2,3,\ldots$ So *it appears that the box-within-a-box bifurcations structure implies the coexistence of boxes in files structures on the* λ*-axis*.

From a theorem of [J 2] it results that *the set of the parameters-*

values $\hat{\lambda}$ *in* Ω_1*, defined in* (P4.7), *has a positive Lebesgue measure,* and not a zero measure.

4.5. RESULTING PROPERTIES ON THE X-AXIS.

4.5.1. THE PARAMETER VALUE λ IS NOT SINGULAR IN THE SENSE OF MYRBERG.

Consider $\lambda \in \Delta_1 \equiv]\lambda_{(1)_s}, \lambda_1^*]$. For these non-singular values the segment $\overline{CC}_1$ contains an attractive cycle $(k2^i;j)$, $k = 3,4,\ldots$, $i = 0,1,2,\ldots$ Considering T^k, this cycle gives rise to k attractive cycles of order 2^i. The immediate domains of attraction (basins) of every of these k cycles consist of k intervals, bounded by the k points of the basic repulsive cycle (k;j) with $S > 1$, and the k (rank k) antecedents which are nearest. For T, the basin of $(k2^i;j)$ has the type described in § 2.8.3 : $]q_1, q_1^{-1}[$ (q_1 = fixed point $S > 1$, q_1^{-1} = its 1st rank antecedent) contains a strange repeller.

The most particular case corresponds to $k = 3$, $j = 1$ $(p_3 - 1)$. The abscissae of the two cycles $k = 3$ are the real roots of :

$$\lambda x^3 + (t_j+\lambda-1)x^2 + (t_j-\lambda^2-1)x - \lambda t_j = 0 \quad , \quad j = 1,2.$$

Here $t_j = 1-(\lambda/2)[1\pm(4\lambda-7)^{1/2}]$, and thus these cycles are real for $\lambda \geqslant \lambda^1_{(3)_0} = 7/4$. The multiplier of one of the cycles is $S_j = 8t_j$. Let $\alpha_1, \alpha_2, \alpha_3$ (j=1) the points of the cycle with $S < 1$, $\beta_1, \beta_2, \beta_3$ (j=2) the points of the cycles with $S > 1$. In this particular case, $S_j = 8t_j$ permits to verify that S_1 is a strictly increasing function of λ, and S_2 a strictly decreasing one. Let λ be such that $\lambda \in \omega_3^1$, or $\omega \in \omega_k^j$ in the general case.

Considering T^3, the immediate basins of the 3 attractive cycles of order 2^i are given by the open intervals $]\beta_\ell, \beta_\ell^{-1}[$, $\ell = 1,2,3$ (cf. fig. 2.12 for i = 0), β_ℓ^{-1} being the rank 3 antecedent of β_ℓ which is the nearest from β_ℓ. This basin is free of other singularities than the attractive cycle. All the repulsive singularities resulting from the set of the bifurcations occuring in the parameter interval $[\lambda_{(1)_0}, \lambda^1_{(3)_0}[$ are located outside these immediate basins but inside the interval $[q_1, q_1^{-1}]$.

Let d_3 be the domain of the x-axis which contains these singularities. The total basin of every attractive cycle of order 2^i consists of the immediate basin and its successive antecedents with increasing ranks, giving disjointed open intervals. All the so-obtained 3 sets of open intervals jumble together in a very complex manner in the situation of § 2.8.4, the accumulation points being the cycles born from the bifurcations of the interval $[\lambda_{(1)_0}, \lambda^1_{(3)_0}[$, their antecedents, the limit points of all these points (the set of these limit points is the set S defined in [L17] as indicated in [G35]). Then *the boundary separating the basins (complex of the 2nd type) of the 3 considered cycles of* T^3 *is of chaotic (or fuzzy) type*. Considering now T, the basin of the cycle $(3.2^i;j)$ is in the situation of § 2.8.3 : $]q_1, q_1^{-1}[$ contains the strange repeller located inside d_3 and corresponding to all the repulsive singularities which have been just described. The basin is complex of the first type (§ 2.8.3), which gives rise to chaotic transients. The same results occur for $k \neq 3$ [M21] [M25] [G35] in the non embedded, or the embedded situation. Inside $]q_1, q_1^{-1}[$, but outside the immediate basins of the k cycles of order 2^i, *chaotic transients* are obtained.

The simplest cases of chaotic transient and fuzzy boundary are described § 2.8.3, 2.8.4, $f(x,\lambda)$ being given by fig.2.17b,c.

4.5.2. THE PARAMETER VALUE λ IS SINGULAR OF THE 1st TYPE.

One has : $\lambda = \lambda^j_{(k)s}$, $k = 1,3,4,\ldots$, or $\lambda = \lambda_{k_1}(s_{k_2})$ (cf. § 3.3.6). The attractor is then a Cantor set defined from the Myrberg's spectrum ω^j_k in the first case. Let $C(\omega^j_k)_s$ be this Cantor set, it is called in [G18] *critical attractor*. The extremum of $f(x,\lambda)$ lies in $C(\omega^j_k)_s$, which is the limit set of the points of the cycles $(k2^i;j)$ when $i \to \infty$. The description of the non-wandering set for $\lambda = \lambda^j_{(k)s}$ and $\lambda_{(k_1)}(s_{k_2})$ is given in [H10] p.56 and in [J 3].

4.5.3. THE PARAMETER VALUE λ IS SINGULAR OF THE 2nd TYPE.

Let be $\lambda = \lambda^{*j}_k$, $k = 1,3,4,\ldots$, or $\lambda^{*j}_{k2^i}$, $i = 1,2,3,\ldots$ Considering T^r, $r = k$ or $k2^i$, the open interval $]q_1, q_1^{-1}[$ contains r *chaotic segments*

$\overline{C_m C_{r+m}}$, $m = 0,1,\ldots,r-1$, $C_o \equiv C$, (a part of these critical points merge into a repulsive cycle) having equivalent properties to those of § 3.2.4 obtained for $\lambda = \lambda_1^*$. With the last remark of § 3.2.4 e (transversal homoclinic orbits inside these segments), and the property equivalent to that of § 3.2.4 a (sensitivity to initial conditions), *every of these r chaotic segments is for* T^r *a strange attractor* (cf. the definition of § 1.1). Concerning the basins of these attractors, one has the same situation as that of § 4.5.1 inside $[q_1, q_1^{-1}]$, i.e. the basins are separated by chaotic (or fuzzy) boundaries. Considering now T, the $\overline{C_m C_{r+m}}$, $m = 0,1,\ldots,r-1$, constitute an *cyclic chaotic segment* (strange attractor) (k;j) of order r, with a transfer order (by successive applications of T) which is characterized by the index j. With regard to the structure of the basin boundary, it is that related to an attractive cycle (cf. § 4.5.1), i.e. the basin is complex of the first type (§ 2.8.3) giving rise to chaotic transients.

Let now $\beta_1,\ldots,\beta_k$ be the points of the repulsive cycle (k;j), $S > 1$, which is used for the definition of λ^{*j}_k, $\beta_1^{-1},\ldots,\beta_k^{-1}$ the rank k antecedents which give also the boundaries of the cyclic chaotic segment (k;j): $\overline{\beta_1\beta_1^{-1}},\ldots,\overline{\beta_k\beta_k^{-1}}$ ($\equiv \overline{C_m C_{k+m}}$, $m = 0,1,\ldots,k-1$, when $\lambda = \lambda^{*j}_k$). For $\lambda^{j_0}_k < \lambda < \lambda^{*j}_k$ ($\lambda^{j_0}_k$ corresponding to $S = 0$) each segment $\overline{C_m C_{k+m}}$, $m = 0,1,\ldots k-1$, is inside one $\overline{\beta_1\beta_1^{-1}},\ldots,\overline{\beta_k\beta_k^{-1}}$, and they constitute a *cyclic absorbing segment* (§ 2.7.5) (k;j), merging into the cyclic chaotic segment (k;j) when $\lambda = \lambda^{*j}_k$. Considering $f(x,\lambda) \equiv x^2-\lambda$, for $\lambda > \lambda^{*j}_k$ the $\overline{C_m C_{k+m}}$ are outside $\overline{\beta_1\beta_1^{-1}},\ldots,\overline{\beta_k\beta_k^{-1}}$. When $\lambda = \lambda^{*j}_k + \varepsilon$, $\varepsilon > 0$ being sufficiently small, λ belongs to a box $\Omega^{j'}_{k'}$, $k' \to \infty$ if $\varepsilon \to 0$, j' being defined from § 3.4.2. Then if $x_o \in \overline{\beta_\ell\beta_\ell^{-1}}$, $\ell = 1,\ldots,k$, the iterated sequence remains in general inside the disjoint union of these segments for $n < N$, N being a large integer. After, for $n > N'$, $N' > N$, $x_n \in \overline{\beta_{\ell'}\beta_{\ell'}^{-1}}$, $\ell' = 1,\ldots,k'$. With regard to the limit properties of § 3.4.2, the k' segments, so obtained, reproduce the "shape" of the cyclic chaotic segment (k;j) when $\varepsilon \to 0$, i.e. $k' \to \infty$ For $\lambda = \lambda^{*j}_{k2^i}$ one has equivalent properties.

All these properties were firstly described from 1975 in [G32] [M21] [M22] [G35], with an extent to the two-dimensional endomorphism [G38] in 1977. More particularly [M22] (1976) gives an extended description of the phenomenon. Since 1982, it was rediscovered and many authors now speak of *chaotic attractor in crisis*, or *boundary crisis*, when $\lambda = \lambda^{*j}_k$, or $\lambda =$

$\lambda^{*j}_{k2^i}$

Results concerning the cases $\breve{\lambda} \neq \lambda^{*j}_k$ are given in § 4.6 .

4.5.4. THE PARAMETER VALUE λ IS SINGULAR OF THE 3rd TYPE. PHENOMENON OF INTERMITTENCY.

One has $\lambda = \lambda^j_{(k)_o}$, $k = 3,4,\ldots$, $k \neq k_1 2^i$, which gives the critical case in the Liapunov sense (P2.1). Here a double cycle (k;j) exists and results from the merging of an attractive cycle and a repulsive one, via the fold bifurcation (3.20). Let T be the quadratic recurrence (3.1), $f(x,\lambda) \equiv x^2-\lambda$. As in § 2.9.4, for $\lambda = \lambda^j_{(k)_o}-\varepsilon$, $\varepsilon > 0$ being a sufficiently small value, the attractive cycle (k;j) is "announced" by an iterated sequence behaviour, which makes appear statistically a periodicity k with a rotation sequence characterized by the index j. Indeed with $\lambda = \lambda^j_{(k)_o}-\varepsilon$, k segments of T^k, having a parabolic shape in the (x_n,x_{n+1}) plane, are located close to $x_{n+1} = x_n$, before the tangency occuring when $\varepsilon = 0$, and k points P_i are located at a minimum distance from $x_{n+1} = x_n$. The Koenigs-Lemeray graphical construction (cf.fig.4.10 for k = 3) shows that the "density" of the iterated points is the most important in a neighbourhood (δ_i) of each of the k points P_i, as $\varepsilon \to 0$. Then the k neighbourhoods (δ_i) tend toward the k points of the double cycle (k;j) obtained when $\varepsilon = 0$.

As in § 2.9.4, the k maxima of "density" of the iterated points result from the concentration of the most part of the points of an attractive cycle $(2^i k';j')$, $i = 0,1,2,\ldots$, belonging to a box $\Omega^{j'}_{k'}$ which may be an embedded box. As indicated in § 3.4.3, the two basic cycles (k';j') verify (3.36), or (3.37), and $\lim \Omega^{j'}_{k'} = \lambda^j_{(k)_o}$, $k' \to \infty$, $\lim [r^{j'}_{k'}] = [r^j_{ok}]$ when $\varepsilon \to 0$, $[r^{j'}_{k'}]$ being the Myrberg's rotation sequence of the cycle (k',j'), $[r^j_{ok}]$, defined in § 3.3.9 and 3.4.3, is the rotation sequence associated with $\lambda^j_{(k)_o}$. These "limit properties" explain the k maxima of points "density". On the whole, for $\lambda = \lambda^j_{(k)_o}$, the solution seems made up of episodes of almost periodic behaviour *intermittent*, because separated by apparently chaotic portions, and it is so described in the classical litterature. In reality the observed behaviour is a very long transient toward an attractive cycle (k';j'), or more exceptionally toward a chaotic segment (k';j'). For the first time this phenomenon was described in 1976 [M22] (cf. also [G35] p.122-124), where the intermittent behaviour was called

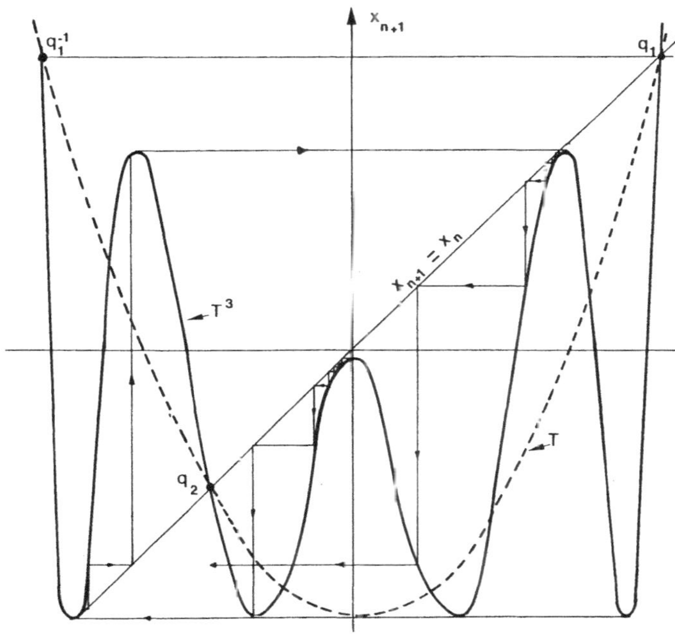

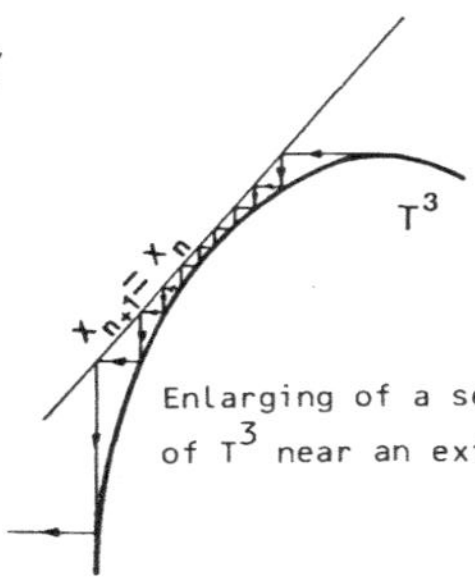

Enlarging of a segment of T^3 near an extremum .

$\lambda = \lambda_{(3)0} - \varepsilon$, $\varepsilon > 0$ is small .

FIG. 4.10 -(according to |M22| 1976) <u>Phenomenon of intermittency</u> .

The concentration of the iterated points is the most important in a neighbourhood of k=3 points located at a minimum distance from $x_{n+1}=x_n$.

"average value cycle" ("cycle en valeur moyenne" in French). It was re-discovered in 1980 [P 4], and the classical litterature refers now to this latter publication, calling *intermittency* the corresponding behaviour. A direct analysis for establishing the number of iterates, spent near the k points P_i, is very difficult on account of the complex basin of first type associated with the attractive cycle (k';j') (cf. § 2.8.3 and 4.5.1).

4.6. INVARIANT MEASURES ASSOCIATED WITH SINGULAR PARAMETERS VALUES OF THE 2nd TYPE

Let T be the recurrence $x_{n+1} = f(x_n,\lambda)$ with only one extremum. When a recurrence possesses a chaotic stationary state, such as a chaotic segment $\overline{X}$, then some implicit regularity may become discernable by going over to a suitably chosen "collective" representation of the chaotically evolving "microscopic" x_n. In principle the transition from microscopic to collective, or "macroscopic" variables can be carried out in various ways, but up to now only two practical approaches were used, one inspired by ergodic theory and the other by statistical physics. In the first the objective is to determine some physically meaningful invariant quantities, the simplest of which appear to be *invariant densities*, related to some *invariant measures*. The second is related to the notion of *topological entropy*. The invariant density $\rho(x) \in L^1[C,C_1]$, C,C_1 being the critical points of rank 1 and 2, is usually sought in the form of a solution of the Perron-Frobenius functional equation:

$$(4.10) \qquad \rho = \frac{d}{dx}\int_{s_o}^{s} \rho(u)du, \qquad s_o \leqslant s \leqslant s_1, \qquad x \in \overline{X}$$

where s_o, s_1 are constants, $s = s(x,\lambda) = f^{-1}(x,\lambda)$. If $\rho(x)$ is a solution, $A\rho(x)$ (A = arbitrary constant) is also a solution. In [M49] [J2] the existence of an absolutely continuous T-invariant measure was proved for $\lambda = \hat{\lambda}$ (singular value of the 2nd type).

Let now $f(x,\lambda)$ be a quadratic polynomial in the real variable x. When $\hat{\lambda} = \lambda^{*j}_{k}$ (resp. $\lambda^{*j}_{k2^i}$), i.e. $\hat{\lambda}$ is one of the boundaries of the box Ω^j_k (resp. $\Omega^j_{k2^i}$), the properties of $\rho(x)$ are known [R 7] [C 4] [C 5]. The

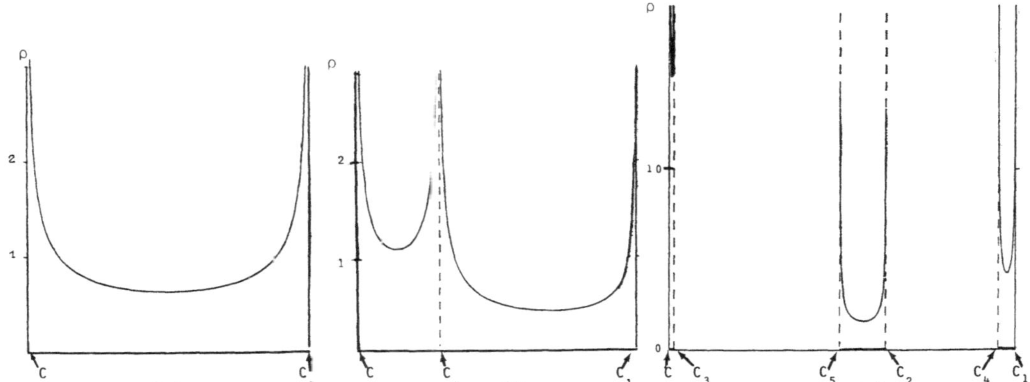

(a) $\lambda = \lambda_1^*$

For a quadratic function $f(x,\lambda)$ $\rho(x)$ is symmetrical. For other symmetrical functions, gererally $\rho(x)$ is unsymmetrical.
Only the case (B) occurs .

(b) $\lambda = \lambda_2^{*1}$

Cases (A) and (B) occur .

(c) $\lambda = \lambda_3^{*1}$

Only the case (B) occurs .

FIG. 4.11 - Classical situations of invariant measures .

curve $\rho = \rho(x)$ is piecewise continuous, and possesses vertical asymptotic straight lines having the abscissae of critical points C_i, $i = 0,1,2,\ldots$, $C_o \equiv C$. The corresponding situations are called *classical*, and two cases are possible : $\rho \to \infty$ on both sides of C_i (*case (A)*), $\rho \to \infty$ on one side of C_i and $\rho = 0$ (or is not defined) on the other side (*case (B)*). So for $f(x,\lambda) \equiv x^2-\lambda$, $\lambda = \lambda_1^* = 2$, the general solution of (4.10) is $\rho(x) = A/[1-x^2/4]^{\frac{1}{2}}$, A = arbitrary constant, $C = -2$, $C_1 = 2$, $-2 \leqslant x \leqslant 2$; $\rho = 0$ if $|x| > 2$ (case (B)). If $\lambda = \lambda_3^{*1}$, on each of the 3 disjoint intervals (cyclic chaotic segment of order 3) $[C,C_3]$, $[C_5,C_2]$, $[C_4,C_1]$, $\rho(x)$ has an equivalent shape to that obtained for λ_1^* (but without the symmetry of this case), with $\min\rho(\overline{CC_3}) > \min\rho(\overline{C_4C_1}) > \min\rho(\overline{C_5C_2})$ (case (B) also). If $\lambda = \lambda_{2^1}^*$, the two intervals (cyclic chaotic segment of order 2) $[C,C_2]$ $[C_3,C_1]$ are such that $C_2 \equiv C_3 \equiv q_2$. On each of these two intervals, the invariant density $\rho(x)$ has an equivalent shape to that obtained for λ_1^* (without symmetry) with $\min\rho(\overline{CC_2}) > \min\rho(\overline{C_3C_1})$. So on both sides of $C_2 \equiv C_3$, the case (A) is obtained, and for C, C_1 it is the case (B). It is the same for $\lambda = \lambda_{k2^i}^{*j}$ which gives also the two cases (A) and (B). The first purpose of this paragraph is to show the existence of *non classical* situations for $\hat{\lambda} \neq \lambda_k^{*j}$ ($\hat{\lambda}$ is a limit point of boxes files, without being the boundary of a box), such that $\rho \to \infty$ on one side of C_i and $\rho \to M$, M=constant $< \infty$, on the other side (*case (C)*) (cf. [C 6]). The second purpose is to give a method allowing a numerical evaluation of $\rho(x)$ in the neighbourhood of the critical C_i for the cases (A), (B) and (C).

The simplest non classical situation occurs when $f(x,\lambda) \equiv x^2-\lambda$ is such that the fixed point q_2, with multiplier $S < -1$, merges into the critical point C_3 of rank 4. Then $\hat{\lambda} = 1.89291098791\ldots$, and $\overline{CC_1}$ is a T-invariant chaotic segment. A simple remark, about the iterates of a bilateral neighbourhood V of the extremum, shows that the 3rd iterate of V (fig. 4.12) is located on the left of C_2. This announces for C_2 the case (C), $\overline{CC_1}$ having no sub-interval where $\rho = 0$, because $\overline{CC_1}$ does not contain a cyclic chaotic segment.

For the study of $\rho(x)$, a reduced form of T is considered with $f(x,\lambda) = 1-(x-u)^2/(1-u)^2 = 1-\lambda^2(x-u)^2$, $\lambda = \hat{\lambda}[C_3 \equiv q_2]$, $u = 0.4717131412$ being the abscissa of the maximum of $f(x,\lambda)$. This form is used in order that $\overline{CC_1}$ becomes the interval $[0,1]$. Then one has $C(x = f(u,\hat{\lambda}) = 1)$, C_1

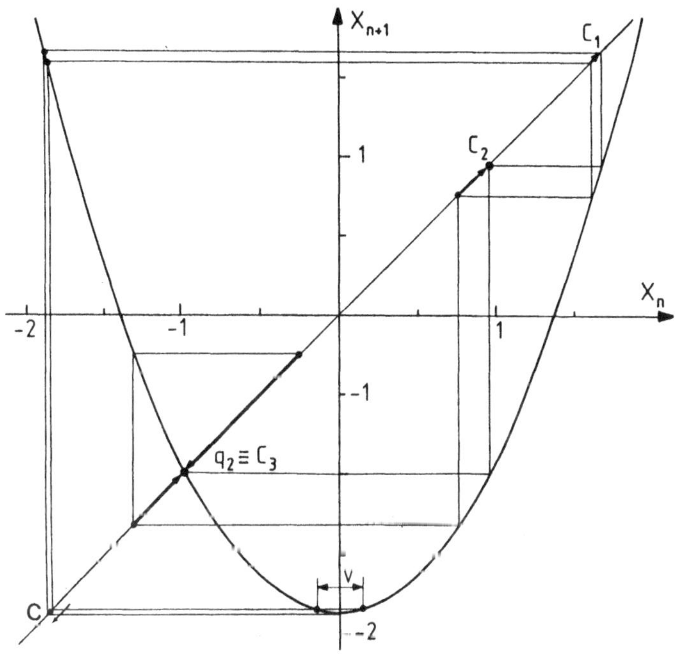

FIG. 4.12

The simplest non classical situation of invariant measure .

The third iterate of the bilateral neighbourhood V of the extremum is located on the left of $C_3 \equiv q_2$. The oriented second iterate indicates that the case (C) of the text occurs in the neighbourhood of the critical point C_2 .

(x = 0). Let ζ = 0.202709968 the abscissa of C_2, ξ = 0.740716314 the abscissa of $C_3 \equiv q_2$. ζ_1 = 2u is the abscissa of C_2^{-1}, the first rank antecedent of C_2, $C_2^{-1} \neq C_1$. The functional equation (4.10) is written:

$$\rho = \frac{d}{dx}\left[\int_{T^{-1}(0,x)} \rho(s)ds\right];\ \text{a.e.}\ \rho(x) = \sum_{y\in T^{-1}(x)} \frac{\rho(y)}{|T'(y)|}, \tag{4.11}$$

$T'(y) = df(x,\lambda)/dy$, a.e = "almost everywhere". Now it is conjectured that $\rho(x)$ is a twice continuously differentiable function except for the points C, C_1, C_2, C_3. This conjecture is justified because it does not lead to theoritical contradictions. Moreover it will be shown that the conjecture leads to estimates which can be independently verified by numerical way with a good accuracy. A limited expansion of $\rho(x)$ is sought in the neighbourhood of x = u (maximum of $f(x,\lambda)$) with the local form : $\rho(u+\varepsilon) = \rho(u) + \rho'(u)\varepsilon + \rho''(u)\varepsilon^2/2 + o(\varepsilon^2)$. This expansion is considered in the interval $\zeta < x < \xi$. In the neighbourhood of ζ_1, one writes : $\rho(\zeta_1+\varepsilon) = \rho(\zeta_1) + \rho'(\zeta_1)\varepsilon + \rho''(\zeta_1)\varepsilon^2/2 + o(\varepsilon^2)$. Then, in C, C_1, C_2, C_3 the only possible compatible local forms, solutions of (4.11) are with $\varepsilon > 0$:

$$\begin{cases} \rho(1-\varepsilon) = A_0\varepsilon^{-1/2} + D_0\varepsilon^{1/2} + o(\varepsilon^{1/2})\ ;\ \rho(\varepsilon) = A_1\varepsilon^{-1/2} + D_1\varepsilon^{1/2} + o(\varepsilon^{1/2}) \\ \rho(\zeta+\varepsilon) = A_2\varepsilon^{-1/2} + B_2 + o(1)\ ;\ \rho(\zeta-\varepsilon) = B_2' + o(1) \\ \rho(\xi+\varepsilon) = A_3\varepsilon^{-1/2} + B_3 + o(1)\ ;\ \rho(\xi-\varepsilon) = A_4\varepsilon^{-1/2} + B_4 + o(1) \end{cases} \tag{4.12}$$

Let $\sigma_0 = |T'(1)|^{-1/2} = 0.513948859$, $\sigma_1 = [T'(0)]^{-1/2} = 0.543895947$, $\sigma = [T'(\zeta)]^{-1/2} = |T'(\xi)|^{-1/2} = 0.720238129$, from (4.12) it is obtained :

$$\begin{cases} A_0 = (1-u)\rho(u)\ ;\ A_1 = \sigma_0 A_0\ ;\ A_2 = \sigma_1 A_1\ ;\ A_3 = (A_2+A_4)\sigma\ ;\ A_4 = A_3\sigma \\ B_2 = \rho(\zeta_1)\sigma_1^2\ ;\quad D_0 = (1-u)^3\rho''(u)/2;\quad D_1 = 3A_0\sigma_0/8 + D_0\sigma_0^3 \\ B_2' = B_2\ ;\quad B_3 = (B_2+B_4)\sigma^2\ ;\quad B_4 = (B_2'+B_3)\sigma^2. \end{cases} \tag{4.13}$$

The relations between the A_i, i = 0,...,4 give the trace of (4.11) on the singularities of ρ. The matrix of the right members of (4.13) is reducible, and $0 < \sigma < 1$, then an iterative process for the determination of the A_i is stable. Besides one has here directly :

$$(4.14) \quad \begin{cases} A_1/A_o = \sigma_o \; ; \quad A_2/A_o = \sigma_o\sigma_1 \; ; \quad A_3/A_o = \sigma_o\sigma_1\sigma/(1-\sigma^2) \; ; \\ A_4/A_o = \sigma_o\sigma_1\sigma^2/(1-\sigma^2) \; ; \quad B_3 = B_4 = B_2\sigma^2/(1-\sigma^2) \end{cases}$$

Now let N_o, N_1, N_2, N_2', N_3, N_4, be the number of iterated points, obtained after N iterations (N being sufficiently large), and located respectively inside an interval of lenght ε begining from $x = 1^-$, $x = 0^+$, $x = \zeta^+$, $x = \zeta^-$, $x = \xi^+$, $x = \xi^-$. By integration of the relations (4.12) inside each interval with the lenght ε, one has : $N_i = 2A_i\varepsilon^{1/2}+2D_i\varepsilon^{3/2}/3+o(\varepsilon^{3/2})$, $N_j = 2A_j\varepsilon^{1/2}+B_j\varepsilon+o(\varepsilon)$, $i = 0,1$; $j = 2,3,4$; $A_1/A_o = N_1/N_o+o(\varepsilon^{1/2})$; $A_2/A_o = (N_2-N_2')/N_o+o(\varepsilon^{1/2})$; $A_m/A_o = (N_m-\alpha N_2')/N_o+o(\varepsilon^{1/2})$, $m = 3,4$, $\alpha = \sigma^2/(1-\sigma^2)$; $B_2 = N_2'/\varepsilon+o(1)$.

Moreover if N(u) is the number of iterated points located inside an interval with the lenght ε, one of its boundary being $x = u$, one has $\rho(u) = N(u)/\varepsilon+O(\varepsilon)$. Then its results $A_o = (1-u)N(u)/\varepsilon+O(\varepsilon)$, $B_2/A_o = N_2'/[(1-u)N(u)]+o(\varepsilon)$. Nevertheless, it is impossible to obtain D_o and D_1 from $A_o\varepsilon^{1/2}$, because it would be necessary to have $N(u) = \rho(u)\varepsilon+O(\varepsilon^3)$. To avoid this difficulty, it is better to consider an interval the center of which is $x = u$. Then:

$A_o = (1-u)N(u)/\varepsilon+O(\varepsilon^2)$, $D_o = (3/2)[N_o-(1-u)N(u)\varepsilon^{-1/2}/2]\varepsilon^{-3/2}+o(1)$

D_1 is also obtained with the error o(1) by (4.13) from A_o and D_o.

In order to justify the conjecture which leads to (4.12), the ratios A_i/A_o, $i = 1,\dots,4$, A_4/A_3 are given from (4.14) with 6 digits, and from numerical tests for given (N,ε) with 4 digits

	By (4.14)	$N=4.10^5$, $\varepsilon=10^{-3}$	$N=10^6$, $\varepsilon=10^{-3}$	$N=10^5$, $\varepsilon=10^{-2}$
A_1/A_o	0.513943	0.5137	0.5164	0.5175
A_2/A_o	0.279535	0.2828	0.2817	0.2762
A_3/A_o	0.418345	0.4309	0.4325	0.4275
A_4/A_o	0.301308	0.3083	0.3114	0.3141
A_4/A_3	0.720238	0.7155	0.7200	0.7347

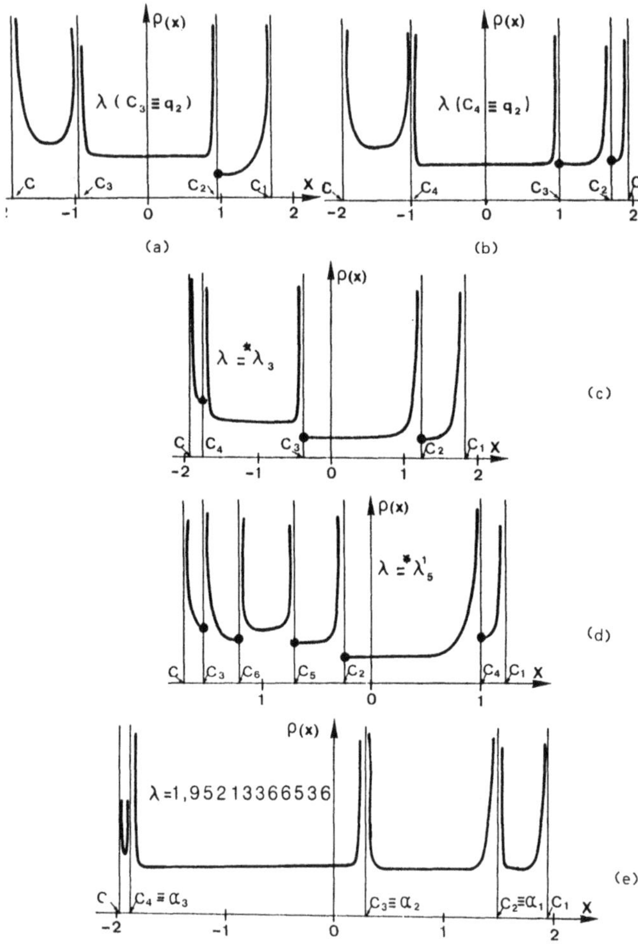

FIG. 4.13

Non classical situations of invariant measures

Moreover with $N = 10^5$, $\varepsilon = 10^{-2}$, $\rho(u)/\rho(\zeta_1) = 0.4507$ and $N(u)\sigma_1^2/(\varepsilon B_2) = 0.4511$, which permits the verification of the equality of these two ratios with an accuracy of 10^{-3}.

Whatever be the quadratic polynomial in the variable x, *the relations* (4.14) *and* $\rho(u)/\rho(\zeta)$ *are the same with the same numerical values.* The method adopted here for $\breve{\lambda}[C_3 \equiv q_2]$ can be used in other situations of accumulation of boxes with $\breve{\lambda} \neq \lambda^{*j}_k$. The fig.4.13 a.e give $\rho(x)$ for $f(x,\lambda) \equiv x^2-\lambda$ [C 6]. Fig.4.13 a, $\lambda = \breve{\lambda}[C_3 \equiv q_2]$ makes appear the case (C) for the critical point C_2, the case (A) in C_3, and the case (B) in C, C_1. Fig.4.13 4.13 b, $\lambda = \breve{\lambda}[C_4 \equiv q_2] = 1.97393204459$ [M34], gives the case (C) in C_2, C_3. Fig.4.13 c, $\lambda = {}^*\lambda_3^1 = 1.92466106319$ [M33] (cf. also § 4.1.5), gives the case (C) in C_2, C_3, C_4 points of the cycle of order 3 with multiplier $S > 1$. Fig 4.13d, $\lambda = {}^*\lambda_5^1 = 1.707364499$ ([M34] and § 4.1.5), gives the case (C) in C_i, $i = 2,\dots,6$, point of the cycle of order 5, with $S > 1$, which is born from $\lambda^1_{(5)_0}$. In fig.4.13 e, C_2, C_3, C_4, are the points α_1, α_2, α_3 of the cycle (3;1) with $S < -1$, giving the case (A) for these points. Though $\breve{\lambda} \neq \lambda^{*j}_k$, this last example corresponds to a classical situation. In all these examples the graphical representation is drawn to the scale, and so takes into account the values of the coefficients A_i, B_i.

Considering the iterates of a bilateral neighbourhood V of the extremum of $f(x,\lambda)$, it is possible to identify which of the cases (A), (B), (C) occurs when the iterated sequence $C, C_1, C_2, \dots$ ends on a repulsive cycle. Indeed the iterates of V give oriented neighbourhoods of $C, C_1, C_2, \dots$, the iterates of the "differentiable" extremum giving $\rho \to \infty$ (representation of an arrow in fig.4.12). So for $\lambda = \breve{\lambda}[C_3 \equiv q_2]$, it appears abscence of arrow on the right of C_2, but one has the arrow on the left of C_2 and on the left of C_1. From this, the case (C) results in C_2. Then taking into account this remark one has :

<u>(P4.10)</u> *For the critical points* C *and* C_1 *the case (B) always occurs. For the critical points, which belong to a repulsive cycle with* $S < -1$, *the case (A) always occurs. For the critical points which belong to a repulsive cycle with* $S > 1$, *only the cases (B) or (C) occur, and if* $\overline{CC_1}$ *does not contain a cyclic chaotic segment, only the case (C) occurs for the critical points different from* C *and* C_1. *For the critical points, different from* C, C_1, *and from the points of a cycle with* $S < -1$, *or* $S > 1$,

which defines the considered $\breve{\lambda}$, *only the case (C) occurs.*

4.7. TOPOLOGICAL ENTROPY.

Among the possible formulations a dynamically somewhat more meaningful relation is the following [M51] :

(4.15) $h(f) = \lim_{k\to\infty} [(1/(k)\ln\alpha_k]$, $\alpha_k - 1$ = extrema number for $f^k(x,\lambda)$

The value of h(f) is therefore related to an asymptotic measure of the number of cycles of the considered recurrence. Consider now the particular recurrence T with $f(x,\lambda) \equiv x^2-\lambda$. If $\lambda = \lambda_1^* = 2$, $\alpha_k = 2^k$, and $h(\lambda_1^*) = \ln 2$. For $\lambda = \lambda_{2^i}^*$, i = 1,2,... (§ 4.4.3), 2^i segments of T^r, $r = 2^i$, having a parabolic shape are in the situation of λ_1^*. Thus the number of intersections of T^r, with $x_{n+1} = x_n$, is 2^{n+i}, and considering T, $\alpha_r(\lambda_{2^i}^*) = 2^{(n+i)/r}$, $h(\lambda_{2^i}^*) = (1/2^i)\ln 2$, $\lim h(\lambda_{2^i}^*) = h(\lambda_{(1)_s}) = 0$ when $i \to \infty$. From [J 3] [G 3] one has:

(P4.11) *Let* $f(x,\lambda) \equiv x^2-\lambda$.

(a) The lower bound of the topological entropy of the cycles, with multiplier S = 0, *is* $h(\lambda_{2^1}^*) = (\ln 2)/2$, *when* $\lambda \in \Gamma_1 \equiv]\lambda_{2^1}^*, \lambda_1^*]$.

(b) $h = h_k^j$ = *constant, for the set of* λ *corresponding to a cycle with* S = 0, *and such that* λ *belongs to a simple (or non-embedded) box (§ 4.4.1)* $\Omega_k^j \in \Gamma_1$.

(c) For every box $\Omega_{2^i k}^{\ell} \in \Gamma_{2^i} \equiv]\lambda_{2^{i+1}}^*, \lambda_{2^i}^*]$, k = 3,4,..., i = 1,2,..., *the topological entropy of the cycles with* S = 0 *is* $h = h_{2^i k}^{\ell} = 2^{-i} h_k^{\ell}$.

Considering that the results, using the notion of kneading invariant ν, are a variant of the Myrberg's ordering law (§ 3.3.4), considering the monotonicity of $h(\nu)$ proved in [M50] [J16] (thus the monotonicity of h([r]), [r] = Myrberg's rotation number) and (P3.8) (cf. § 3.3.3 the remark of the Myrberg's hypothesis H), it is possible to say that $\lambda < \lambda'$ implies $h(\lambda) < h(\lambda')$, for $f(x,\lambda) \equiv x^2-\lambda$. From [J 3] it appears that the [r] of cycles are dense (associated with non singular parameter values), thus the intervals where h = constant are dense. Considering that $h(\lambda)$ is a continuous function [M51] [G3], it results that $h(\lambda)$ is a non-decreasing function in the interval $\Delta_1 = [\lambda_{(1)s}, \lambda_1^*]$, $h(\lambda_{(1)s}) = 0$, $h(\lambda_1^*) = 2$, with h = constant for every simple box $\Omega_k^j \in \Gamma_1$, which verifies the conjecture of [M 6] : $h(\lambda)$ is a piecewise constant function.

From § 4.4.3 it appears that the set of the simple (non-embedded) boxes of Γ_1 has a boxes in files structure. Thus *the graphical representation of* $h(\lambda)$ *is typically that of the devil' staircase. The topological entropy* h *of* $x_{n+1} = x_n^2-\lambda$ *is a continuous, monotonic non-decreasing, piecewise constant function in the interval* Δ_1. *The properties of* h *for* $\lambda \in \Gamma_{2^i}$, $i = 1,2,\ldots$, *are similar to those of* h *for* $\lambda \in \Gamma_1$.

Let Ω_k^j, $\Omega_{k'}^{j'}$ two consecutive simple boxes belonging to a file U_N^α of the degree N, k' = k+m, $k \to \infty$ giving the limit point of U_N^α. Then there exist files $U_{N+1}^{\alpha'}$ of simple boxes with the degree N+1, inside the interval $J = (\lambda^{*j}_k, \lambda^{j'}_{(k')_0})$, and necessarily infinitely many of such boxes have an arbitrarily small length . It is possible to reproduce this remark for an interval $J' \subset J$ separating two consecutive boxes of $U_{N+1}^{\alpha'}$, with an increasing complexity degree, the boundaries of Ω_k^j being limit points for files having different complexity degrees. Thus in the neighbourhood of Ω_k^j, *the definition of the increase of* h *implies the reference to a file of boxes having a given complexity degree* N. Let $[r_\ell]$, λ_ℓ (resp. $[r_{\ell+1}]$, $\lambda_{\ell+1}$) be the Myrberg's rotation sequence, and the λ-value giving the cycle (k;j) (resp. (k';j') with multiplier S = 0. Let h_ℓ (resp. $h_{\ell+1}$) be the h-value in Ω_k^j (resp. $\Omega_{k'}^{j'}$). Writing $\tau_\ell = (h_{\ell+1}-h_\ell)/(\lambda_{\ell+1}-\lambda_\ell)$, one defines the "increase" of h for every file of the complexity degree N by $\tau = \lim \tau_\ell$ when $\ell \to \infty$ [G 4]. The following result was proved in [G 39] :

<u>(P4.12)</u> *Let* $\bar{\lambda}$ *and* $\bar{h}$ *be the limits of the sequences* λ_ℓ *and* h_ℓ, $\ell \to \infty$, *and* $\lim (\lambda_{\ell+1}-\lambda_\ell)/(\lambda_\ell-\lambda_{\ell-1}) = \rho$ *(§ 3.4.2, 3.4.3)*, $\lim (h_{\ell+1}-h_\ell)/(h_\ell-h_{\ell-1}) = s$. *When* $\rho \neq s$, $\tau = 0$ *if* $\rho < s$, *and* $\tau = \infty$ *if* $\rho > s$.

From (P4.12) it results that $\tau = 0$, for $\bar{\lambda}$-values equal to the boundaries $\lambda^j_{(k)_0}, \lambda^{*j}_k$ of a simple box Ω_k^j, $k \neq 1$, $k \neq 2^i$, the convergence speed being higher for λ^{*j}_k as it is shown from numerical experiments. One has also $\tau = \infty$ for λ_1^*, $\lambda^*_{2^i}$, and $\bar{\lambda} \neq \lambda^{*j}_k$. For the $\bar{\lambda} = {}^*\lambda^j_k$, the ratio ρ/s is close to 1, which gives a "critical" case for a conclusion about τ. Considering now the interval $J' \subset J$ corresponding to the complexity degree N+1, one has the same conclusion. Then *it appears that, in an infinitely small neighbourhood of a limit point* $\bar{\lambda}$ *of a file of degree* N *with* $\tau = \tau_N = 0$

or ∞, *there exists infinitely many* λ*-values, corresponding to* $\lambda^{j''}_{(k'')_o}$, or $\tilde{\lambda}$*-values such that* τ_{N+1} *has different values* 0 *or* ∞ [G 4]. This point is the consequence of the fractal properties of the boxes in files bifurcations structure, which is implied by the box-within-a-box bifurcations structure (cf. fig.4.9 for example).

It should be noticed that the monotonicity of $h(\lambda)$ concerns only the case $f(x,\lambda) \equiv x^2-\lambda$. So in the general quadratic case $f(x,\mu) \equiv a(\mu)x^2+2b(\mu)+c(\mu)$, μ = parameter, the equivalence $\lambda = ac-b^2+b$ (§ 3.1) shows that the function $\lambda(\mu)$ is not monotone in general. Thus $h(\mu)$ is not a monotonic function of the parameter μ.

4.8. SOME GENERALIZATIONS.

4.8.1. RECURRENCE DEFINED BY A FUNCTION WITH ONLY ONE EXTREMUM.

The recurrence T(3.1), $x_{n+1} = f(x,\lambda)$, $\lambda_a \leqslant \lambda \leqslant \lambda_b$, is considered. The following hypotheses are formulated :

- (h_1) $f(x,\lambda)$ *is a continuous and differentiable function of* x, *is continuous with respect to the parameter* λ, *and has only one extremum A.*
- (h_2) *One of the hypotheses* (H_1), (H_2), (H_3) *of § 3.1 is satisfied.*
- (h_3) *When* λ *varies monotonically in the interval* (λ_a,λ_b), *there exist a fixed point* q_2 *such that its multiplier* S *decreases monotonically from +1, and a fixed point with* S *which increases monotonicaly from +1.*

Equivalent hypotheses were formulated in [M21] [M25] [G35]. *One can notice that* (H_1) (absence of more than one point of inflexion between two consecutive extrema of T^k, $k = 2,3,\dots$) *is verified when the curve, defined by* $f(x,\lambda)$, *has an only one minimum of radius of curvature* $R(f)$, *which is located at the extremum* [M39]. So let the extremum A be a maximum, and a minimum of $R(f)$ for $x = x_R > x_A$. A λ-value is chosen to obtain T^2 with 3 extrema. The point x_R and its two first rank antecedents x_R^{-1}, $x_R^{\prime -1}$, $x_R^{\prime -1} < x_R^{-1}$, correspond to 3 of the local minima of $R(f)$ for T^2. Thus between the minimum $x = \alpha$ of T^2 and the maximum $x = \beta$ having the largest abscissa, there exist 3 points of inflexion : one due to passage from the minimum to the maximum, and two due to the presence of x_R and x_R^{-1} in the interval

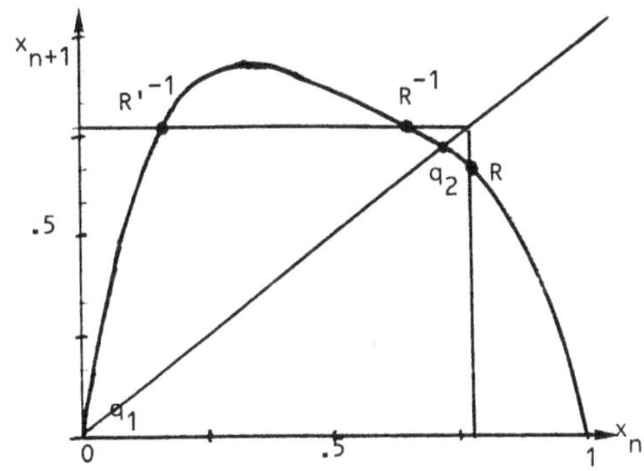

(a)

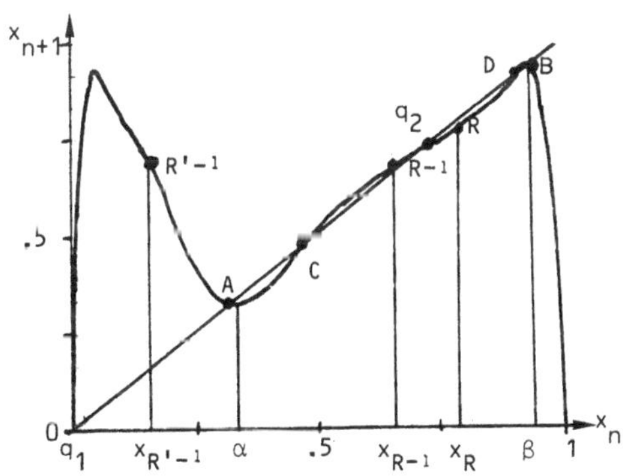

(b)

FIG. 4.14.$\lambda = 1$

T is defined by $f(x,\lambda) = \lambda(7.86x-23.31x^2+28.75x^3-13.3x^4)$ (cf. |S 7|). Fig.(a) represents T . FIG.(b) represents T^2 ,A and B are the points of an attractive cycle of order 2, C and D the points of a repulsive cycle of order 2 . The recurrence T has two attractors: the fixed point q_2 and the cycle A,B .

(α,β). It results that T^2 can intersect $x_{n+1} = x_n$ in 6 points : the fixed points q_1, q_2, and 4 points corresponding to two cycles of order two. For a correct choice of λ, q_2 and one of the two cycles can be simultaneously attractive. By considering $m > 1$ points x_R, $x_R^{(i)} > x_A$, $i = 1,\dots,m$, for T, the number of possible points of inflexion in the interval (α,β) of T^2 becomes 2m+1. So the number of cycles of order two can be larger than 2, and more than two attractors at finite distance are possible. From the above-mentioned considerations it is possible to say :

(P4.13) *Let* T *be a recurrence defined by a function* $f(x,\lambda)$ *having only one extremum and verifying the hypotheses* (h_i), $i = 1,2,3$. *Then* T *has a qualitative behaviour similar to that of* T *with* $f(x,\lambda) \equiv x^2-\lambda$, *has no more than one attractor at finite distance, and has a similar box-within-a-box bifurcations structure.*

This results from the construction of the curves T^k, $k = 2,3,\dots$, in the (x_n,x_{n+1}) plane, which have shapes similar to the ones corresponding to $f(x,\lambda) \equiv x^2-\lambda$, with equivalent geometrical situations with respect to $x_{n+1} = x_n$. Fig. 4.14 represents the case of exitence of two attractors

4.8.2. RECURRENCE DEFINED BY A FUNCTION WITH TWO EXTREMA.

The simplest cases are the symmetrical cubic recurrences :

$$(4.16) \qquad (a)\ x_{n+1} = \lambda x_n + x_n^3,\ \lambda < 0 \ ; \quad (b)\ x_{n+1} = \lambda x_n - x_n^3,\ \lambda > 0.$$

Two attractors at finite distance are now possible (cf. (P2.15) § 2.7.2). The bifurcations organization is more complex than in the quadratic case. It is a box-within-a-box bifurcation structure, which for some intervals of the parameter λ reproduces the structure of the quadratic case,when $f \equiv \lambda x - x^3$, $1 \leqslant \lambda \leqslant \sqrt{27}/2$, considering T, or when $f \equiv \lambda x + x^3$, $-\sqrt{27}/2 \leqslant \lambda \leqslant -2$, considering T^2. Outside these particular λ-intervals,there exist complications described in [M24] (1977) and [G35] (1980). Such complications lead to the existence of a *fuzzy (or chaotic) boundary* (§ 2.8.4) separating the basins of the two attractors at finite distance (cf. p.414 and 418 of [G35]).

The non-symmetrical cubic recurrence :

(4.17) $$x_{n+1} = \lambda x_n + \mu x_n^2 - x_n^3 \quad , \qquad (\lambda,\mu) \in \mathbb{R}^2.$$

implies two parameters λ, μ, and the study of the bifurcations structure may be restricted to $\mu \geqslant 0$, because $f(x,\lambda,\mu) = -f(-x,\lambda,-\mu)$. In (4.16) the bifurcations related to each attractors occur simultaneously. It is not the case here, except for $2\mu^2 + 9\lambda = 9$ which leads to the equivalent of the symmetrical case $f \equiv \lambda x - x^3$. The corresponding box-within-a-box bifurcations structure is described in [F4] and [F5] (1981) in the (λ,μ) plane.

The general non-symmetrical case with two extrema is summarily described p. 142-144 of [G35].

4.9. RECURRENCE DEFINED BY A CONTINUOUS PIECEWISE LINEAR FUNCTION WITH ONLY ONE EXTREMUM.

The first study of the complete bifurcations structure was made in 1977 with [M26] (case of two linear pieces). Let T be the recurrence defined by :

(4.18) $$x_{n+1} = \lambda x_n - \alpha,\ x_n \geqslant 0 \quad ; \quad x_{n+1} = -\beta\lambda x_n - \alpha,\ x_n \leqslant 0$$

where $\lambda > 0$, $\alpha > 0$, $\beta \geqslant 1$. The symmetrical case is given by $\beta = 1$. T has two fixed points q_1 $(x = \alpha/(\lambda-1)$ if $\lambda > 1$, and $q_2(x = -\alpha/(1+\beta\lambda)$. Let C_i be the critical point of rank i+1, i = 0,1,2,..., $C_o \equiv C$, C being the first rank consequent of the extremum x = 0. For $-\alpha/(\lambda-1) < x_o < \alpha/(\lambda-1)$, $x_o \in]q_1^{-1},q_1[$, $\lambda > 1$, x_n tends toward an attractor which is contained inside the interval bounded by C, C_1, if $\lambda < \lambda_1^*$, $\lambda_1^* = (1+\beta)/\beta$ being such that $q_1 \equiv C_1$. (H_3) in § 3.1 permits to assert the existence of only one extremum inside $\overline{CC_1}$. If $0 < \lambda < 1/\beta$, q_2 is an attractive fixed point,the basin of which is the whole x-axis except the point at infinity. If $1/\beta < \lambda < 1/\sqrt{\beta}$, q_2 becomes repulsive and an attractive cycle of order 2 appears. This cycle is repulsive for $\lambda > 1/\sqrt{\beta}$.

Two types of bifurcations occur with T given by (4.18). In the first case two cycles of order k appear from a situation where k extrema of T^k, each of them corresponding to a common point of two segments of straight line, are located on $x_{n+1} = x_n$ in the (x_n, x_{n+1}) plane (§ 2.4.2). Let $\lambda = \lambda^j_{(k)_o}$ be the corresponding λ-value, where the index j characterizes the rotation sequence [r] of the basic cycles (k;j), as in the quadratic case (indeed [r] is defined similarly). For this λ-value one cycle merges into the other, and are separated for $\lambda > \lambda^j_{(k)_o}$. In the second case, for $\lambda = \lambda^j_{(k)_s}$, 2k (or 4k) segments of $\overline{CC_1}$ are made up of points of cycles of order 2k (or 4k). For $\lambda = \lambda^j_{(k)_s}+\varepsilon$, $\varepsilon > 0$, a set of repulsive cycles of order $k.2^i$, $i = 1,2,3,\ldots$, (or $i = 2,3,\ldots$) appears as soon as $\varepsilon = 0^+$. For $k \geqslant 3$ this second type of bifurcation occurs only if β has a sufficiently large value. For k = 1, $\lambda_{(1)_s} = 1/\sqrt{\beta}$, which gives two segments $\overline{C_1C_3}$, $\overline{C_2C_4}$ made up of points of cycles of order 4. When $\lambda > \lambda_{(1)_s} + \varepsilon$, $\varepsilon > 0$, a set of repulsive cycles of order 2^i, $i = 2,3,\ldots$, appears as soon as $\varepsilon = 0^+$.

As for the quadratic case, let $\lambda^*_{2^i}$, $i = 1,2,\ldots$, be the λ-value such that 2^{i-1} critical points C_m, $m = 2^i, 2^i+1, \ldots, 2^i+2^{i-1}-1$, merge into the repulsive cycle of order 2^{i-1} which appears from $\lambda = \lambda_{(1)_s}+\varepsilon$. Considering T^r, $r = 2^i$, the slopes of two segments of straight line, which have a common point in x = 0 in the (x_n, x_{n+1}) plane, are : $(-1)^{i+1}\beta^{\ell}\lambda^r$, $\ell = 2^{i-1}+1$, if $x_n < 0$, and $(-1)^i\beta^{r/2}\lambda^r$ if $x_n > 0$. Writing that C_α, $\alpha = r+1$, merges into the point of the cycle of order 2^{i-1} which is the closest to $x_n = 0$, one has :

$$(4.19) \quad \lambda^*_{2^i} = (1+\beta)^{1/r}\beta^{-(1/2+1/r)}, \quad r = 2^i, \quad i = 1,2,\ldots, \quad \lim_{i\to\infty} \lambda^*_{2^i} = \lambda_{(1)_s} = 1/\sqrt{\beta}$$

comparatively to the quadratic case $\lim \lambda^*_{2^i}$ is now given from an analytical relation.

The attractor of T exists only for $0 < \lambda < \lambda^*_1$. For $\lambda = \lambda^*_1$, $\overline{CC_1} \equiv \overline{q_1^{-1}q_1}$ contains all the repulsive cycles born from all the possible bifurcations occuring in $0 < \lambda < \lambda^*_1$, and T has the properties of § 3.2.4 d. The number of cycles of order k is that of the quadratic case. For $\lambda > \lambda^*_1$ there is not creation of new cycles. For $\lambda^*_{2^1} < \lambda < \lambda^*_1$, the λ^{*j}_k-values, $k = 3,4,\ldots$, cannot be defined if $\beta < \beta^*_1 \simeq 3.07959$, because the k critical points $C_k,\ldots,C_{2k-1}$ are always in the situation for which $\lambda > \lambda^{*j}_k$ in the quadratic case. Thus the boxes Ω^j_k, $k = 3,4,\ldots$, do not exist. Then for $\lambda = \lambda^j_{(k)_o}$,

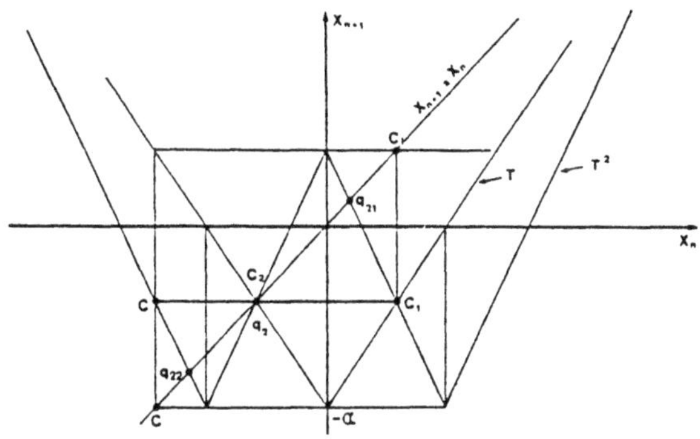

$\lambda = \lambda^*_{(2^1)} = \sqrt{2} \qquad \beta = 1$

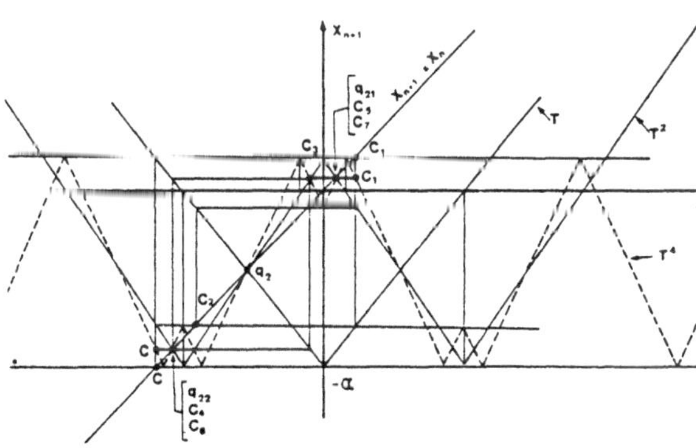

$\lambda = \lambda^*_{(2^2)} = \sqrt[4]{2} \qquad \beta = 1$

FIG. 4.15

T is the piecewise linear recurrence (4.18) .

k points of the x-axis correspond to the merging of a pair of cycles (k;j) and of the whole set of cycles of order mk, m = 2,3,..., which appear from the bifurcations of the box Ω_k^j defined in the quadratic case. All these points become separated when $\lambda = \lambda^j_{(k)_o}+\varepsilon$, $\varepsilon = 0^+$, and are repulsive. On the λ-axis, every box Ω_k^j is degenerate, being reduced to a point.

When $\beta_1^* < \beta < \beta_2^* \simeq 9.95996$, only λ_3^{*1} can be defined, and so the box Ω_3^1 is non-degenerate. For $\lambda = \lambda^1_{(3)_o}+\varepsilon$, $\varepsilon = 0^+$, $\lambda^1_{(3)_o} = 1/(2\beta)+(1+4\beta)^{1/2}/(2\beta)$, a pair of basic cycles (3;1) appears. They are repulsive if $\beta < \beta_c = (\lambda^1_{(3)_o})^{-3} \simeq 4.236065$, but one of them is attractive if $\beta > \beta_c$. These cycles are given by the intersection of $x_{n+1} = x_n$ with T^3. The slopes of the two segments of straight line of T^3, having a common point in $x_n = 0$, are : $\beta^2\lambda^3(x_n \leqslant 0)$, $-\beta\lambda^3(x_n \geqslant 0)$. They give two points of the cycles pair (3;1) which are the closest to $x_n = 0$. When $\beta > \beta_c$, the attractive cycle (3;1) becomes repulsive for $\lambda > \lambda^1_{(3)_s} = \beta^{-1/3}$, and $\lambda^1_{(3)_s} = \lim \lambda^{*1}_{3.2^i}$ if $i \to \infty$. The boundary λ_3^{*1} of Ω_3^1 is defined when C_3, C_4, C_5 merge into the points of the cycle (3;1) with the multiplier $S > 1$, $\lambda_3^{*1} = (1+\beta)^{1/3}\beta^{-2/3}$.

For $\beta > \beta_2^*$, λ_3^{*1} and λ_4^{*1} can be defined, and so the boxes Ω_3^1, Ω_4^1 are non-degenerate. Similarly β_i^* can be defined.

When the boxes Ω_k^j are degenerate, for $\lambda^*_{2^1} < \lambda < \lambda_1^*$, $\overline{CC}_1$ is a chaotic segment. For $\lambda^*_{2^{i+1}} < \lambda < \lambda^*_{2^i}$, $\overline{CC}_1$ contains a cyclic chaotic segment of order 2^i.

The recurrence (4.18) is "generic" in the sense that every piecewise linear recurrence with two segments can be reduced to (4.18). It is the case of :

$$(4.20)\qquad \begin{cases} x_{n+1} = (1-\delta)\gamma^{-1}x_n + \delta, & x_n \leqslant \gamma \\ x_{n+1} = (1-x_n)(1-\gamma)^{-1}, & x_n \geqslant \gamma,\ 0 \leqslant \delta \leqslant 1,\ 1/2 \leqslant \gamma \leqslant 1, \end{cases}$$

for which the correspondance of the parameters is : $\lambda = (1-\delta)\gamma^{-1}$, $\beta = \gamma(1-\delta)^{-1}(1-\gamma)^{-1}$. For (4.20), the criticals points of rank 1 and 2 are $C(x=1)$, $C_1(x=0)$, and a fixed value of γ does not correspond to λ and β fixed ($\gamma = 1-(\beta\lambda)^{-1}$).

Consider the case $\gamma = 1/2$, δ becoming the only parameter of (4.20). With the same notations as for (4.18), one has :

$\delta_1^* = 0, \quad \delta_{(1)_s}^* = 3/4, \quad \delta_{2^1}^* = 2/3, \quad \delta_{2^2}^* \simeq 0.7168$

$\delta_{2^3}^* \simeq 0,7349,\ldots, \quad \delta_{(3)_0}^1 = 1/2, \quad \delta_{(4)_0}^1 = (3-\sqrt{5})/4 \simeq 0,1909$

$\delta_{(5)_0}^1 = 5/8, \quad \delta_{(5)_0}^2 = (3-\sqrt{3})/4, \quad \delta_{(5)_0}^3 \simeq 0,0803.$

For $\delta = \delta_{(3)_0}^1$, $\beta = 2 < \beta_c$, and then the boxes Ω_k^j are degenerate in the interval $0 < \delta < \delta_{2^1}^*$. It is obtained the situation above-mentioned for $\lambda_{(3)_0}^1$. Thus for $0 < \delta < \delta_{2^1}^*, \delta = \delta_{(k)_0}^j$, k points of the x-axis corresponds to the merging of a pair of cycles (k;j) and of the whole set of cycles of order mk, m = 2,3,..., which appear from the bifurcations of the box Ω_k^j defined in the quadratic case. Then the segment $\overline{CC_1} \equiv (0;1)$ is a chaotic segment. For $\delta_{2^i}^* < \delta < \delta_{2^{i+1}}^*$, i = 1,2,..., $\overline{CC_1}$ contains a cyclic chaotic segment of order 2^i.

When $1/2 < \gamma < 1$ non degenerate boxes Ω_k^j become possible. Their existence is determined from the relations between the parameters of (4.20) and those of (4.18).

For a recurrence defined by a continuous piecewise linear function the box-within-a-box bifurcations structure is degenerate.

The results of this paragraph were published in [M26] (1977), and in [G35] (1980). Some of them were rediscovered in [J3] (1981) with a study of the recurrence defined by $f(x,s) \equiv s-1-s|x|$, which is (4.18) with $\alpha = 1-s$, $\lambda = s$, in the particular case $\beta = 1$.

Chapter 5
SOME PROPERTIES OF TWO-DIMENSIONAL RECURRENCES

5.1. INTRODUCTION.

This chapter gives a general view about the properties of two dimensional maps T (or second order autonomous recurrences) having the explicit form :

$$(5.1)\qquad x_{n+1} = f(x_n,y_n,\lambda),\qquad y_{n+1} = g(x_n,y_n,\lambda),\qquad n = 0,1,2,\ldots,$$

where $f(x,y,\lambda)$, $g(x,y,\lambda)$ are reasonably smooth single-valued functions of both the real variables x,y and the real parameter λ. From an initial condition (x_o,y_o), by successive applications of T (5.1) generates an iterated sequence (orbit, or discrete trajectory) :

$$(5.2)\qquad x_n = x(n,x_o,y_o,\lambda),\qquad y_n = y(n,x_o,y_o,\lambda),$$

in the discrete phase plane (x_n,y_n). T can be considered as an implicit definition of the functions x_n, y_n. Such definition of x_n, y_n is theoritically quite satisfactory, but practically it is almost useless, because in general the functions x_n, y_n cannot be expressed explicitly in terms of known elementary and transcendental functions. So, as for one-dimensional recurrences, a meaningful characterization of x_n, y_n consists in the identification of its singularities, and in the description of the latter as λ vary (description of the bifurcations).

The functions x_n, y_n of (5.2) admit, of course, all the singularities of the general solution $x_n = x(n,x_o,\lambda)$ of the autonomous first order recurrence, in addition to some of their own generated by the increased dimensionality. As for the one-dimensional case, T admits cycles of order k as zero-dimensional singularities, given by the fixed points of T^k which verify :

(5.3) $X = T^k X$, $X \neq T^\ell X$, $\ell < k$, $k = 1,2,\ldots,$

where X is the two-dimensional vector with components (x,y), T^k being obtained from $X_{n+k} = F^k(X_n)$. In (5.3) the cycles of order k = 1 are the fixed points of T. Similarly to the one-dimensional case, the fixed points and cycles of T are either attractive, or repulsive, or can correspond to critical cases in the sense of Ljapunov. The higher dimensionality of T permits, however a more varied relation of points singularities with regard to neighbouring points.. This relation is characterized via the definition of one-dimensional singularities which are the *invariant curves* of T. A curve described in the phase plane (x_n,y_n) by $H(x_n,y_n) = C$, C = constant, is called an invariant curve of T, provided the function H satisfies the *functional equation* of automorphic functions :

(5.4) $H[f(x,y,\lambda) , g(x,y,\lambda)] = H(x,y,\lambda)$.

The constant C is determined, for example, by means of an initial condition (x_o,y_o). A similar definition of an invariant curve applies to T^k, and to T^{-k}.

Compared to the one-dimensional case, a new fundamental difficulty arises in the study of (5.1), because very little is known about functional equations of type (5.4). Except in very special case, no existence, uniqueness and continuity, or differentiability, theorems are available. Althought the solutions of (5.4) constitute a fundamental tool for the unraveling of the properties of T, in the most crucial cases they can only be constructed approximately, subject to an independant confirmation of validity, the latter replacing the corresponding unvailable existence, uniqueness and continuity theorems [G36]. The whole approach can be described as a process of symbiotic improvements : a little more understanding of a given recurrence T permits the construction of a better approximation of a particular solution of (5.4), and a more precisely known invariant curve increases the understanding of T, and so forth. In general the corresponding problem is not a simple one. More detailed elements concerning its solution can be found in [G36] p.16-17, 67-71, 76-81, 91-96,

100-103, 208-210, 216-221 and in [G35] p.225-232. These elements are founded on the fact that $H(x,y) = C$, which is not generally single-valued in y inside its x-interval of existence, can be separated into branches $y = \theta(x)$, $\theta(x)$ being a single-valued function, solution of [L 1] :

$$(5.4') \qquad g[x,\theta(x)] = \theta\{f[x,\theta(x)]\} \quad ,$$

each solution describing only one branch of the considered invariant curve. Like in the case of (5.4), the x-intervals of existence of the various $\theta(x)$ are not known a priori. In contrast to (5.4), the solutions of (5.4') are generally unique, i.e. they do not depend on any arbitrary function. This property results not only from the non linearity of (5.4'), but also from the fact that for a given θ, $y_{n+1} = \theta(x_{n+1})$ implies $y_n = \theta(x_n)$ for all x_n for which θ is single valued.

A first part deals with the linear case, i.e. when the solutions of the functional equation (5.4) are obtained from the classical transcendental functions, and different types of fixed points are defined with respect to the corresponding phase portrait of invariant curves. This phase portrait depends on the multipliers, (eigen-values of the matrix of the linear approximation at the fixed point). After that, the autonomous non linear case is considered with the Hadamard-Lattes'results [H 1] [L 1], with the approximate solution of (5.4), and with the intersections of invariant curves at *homoclinic*, or *heteroclinic points*. When the multipliers of a fixed point, or of a cycle, are such as either $|S_1| = 1$, $|S_2| \neq 1$, or $S_{1,2} = \exp(\pm j\phi)$, $j = \sqrt{-1}$, the linear approximation, written at one of the singularities, does not permit the determination of the phase portrait inside its neighbourhood however small it may be. One has a *critical case in the (extended) sense of Ljapunov*, and the phase portrait depends on the non linear terms. Four paragraphs are devoted to these critical cases. One concerns $|S_1| = 1$, $|S_2| \neq 1$. When $S_{1,2} = \exp(\pm j\phi)$, using a method based on a sequence of *Cigala's transformations*, [C 1], two different situations appear. The first, called *non-exceptional case*, is related to the angle ϕ, either non-commensurable with 2π, or, else different from particular commensurable values, for which the Cigala's transformations give rise to the problem of *small denominators*. The second (*excep-*

tional cases) concerns these particular values of ϕ. The Cigala's method consists in obtaining a canonical form of T and in isolating the "dominating non linear terms". The case $\phi = 0$ or π is the subject of a separated paragraph.

Every critical case is associated with a *local bifurcation*, obtained by crossing through this one, when a set of parameters of the recurrence varies. The chapter deals with such bifurcations. In connection with this point, one has to note that *the bifurcation by crossing through* $S_{1,2} = \exp(\pm j\phi)$, $j = \sqrt{-1}$, *is wrongly called Hopf bifurcation*. Indeed the Hopf's contribution (1942) concerns the analytical study of the birth of a limit cycle from a focus, in ordinary differential equations. But even from this point of view, the matter concerns a bifurcation firstly described by Poincaré, what is clearly indicated in the Hopf's paper [H12], and analytically studied by Andronov in 1938 (for this historical aspect cf.also [B14]). Considering recurrences equations, the first contribution is that of Neïmark (1968) [N 2] (cf. also [B 4]), ϕ being not commensurable with 2π, in the particular case for which a complex focus of multiplicity one (or simple complex focus) gives rise to an invariant closed curve. This notion of *multiplicity of a complex focus* ($S_{1,2} = \exp(\pm j\phi)$) is an extension of that used in the ordinary differential equations theory [M 1]. The more general case of any multiplicity was considered in [M14] (1969), [G30] (1974) and [G35]. This chapter also deals with the bifurcations by crossing through exceptional cases, and with two different situations corresponding to $\phi = 0$ (the matrix of the linear approximation can be, or not, reducible to a diagonal form).

The last part of the chapter concerns *global bifurcations* corresponding to the birth of invariant closed curves :

a) from a linear conservative situation perturbed by small non linear dissipative terms ;

b) from a loop constituted by separatrices invariant curves of saddle fixed points.

In the case b), depending on whether homoclinic points exist, or not, two different situations are described from typical examples.

Remark : As above indicated, the analytical existence of the invariant curves (except those given by Hadamard and Lattes' theorems) considered in this chapter is not proved. The series expansions, which are given, must be considered as formal ones. However they are useful from a practical point of view, because one may be assured of this existence by using numerical methods associated with accuracy checks coupled with a certain amount of redundancy. Eventual doubts about the influence of purely numerical phenomena can be resolved for example by recourse to double accuracy.

5.2. AUTONOMOUS LINEAR CASE.

5.2.1. GENERAL SOLUTION.

The simplest autonomous recurrence of second order is a linear one, T :

(5.5) $$x_{n+1} = ax_n + by_n \quad , \qquad y_{n+1} = cx_n + dy_n \quad ,$$

where a,b,c,d, are real constants. As for the first order case, this recurrence is priviledged in the sense that it admits an explicit general solution (5.2), expressible in terms of elementary functions. The availability of such a solution permits to answer all questions pertaining to problems like stability, and existence or uniqueness of invariant curves, in a deductive manner, without any need of intermediate simplifying assumptions.

Let $\Delta = ad-bc$ designate the Jacobian determinant of the right-hand sides of (5.5). T admits a single fixed point $x_n = y_n = 0$ in the finite part of the phase plane, which is characterized by its multipliers S_1, S_2, eigenvalues of the coefficients matrix of (5.5). So S_1, S_2 are the real roots, $S_{1,2} = (a+d)/2 \pm [(a+d)^2/4 - \Delta]^{1/2}$, of the characteristic equation :

(5.6) $$S^2 - (a+d)S + \Delta = 0$$

Consider first $\Delta \neq 0$, which implies $S_1.S_2 \neq 0$. When S_1, S_2 are real and distinct, the general solution of (5.5) is :

$$
(5.7)\quad \begin{cases} x_n = C\,S_1^n + C'S_2^n, \quad y_n = (S_1-a)C\,S_1^n/b + (S_2-a)C'S_2^n/b, & b \neq 0 \\ x_n = C\,S_1^n, \quad y_n = C'S_2^n + C\,S_1^n/(a-d), \quad S_1 = a, \quad S_2 = d, & b = 0, \end{cases}
$$

where C, C' are arbitrary real constants, the values of which are unambiguously defined by the initial conditions $x_n = x_o$, $y_n = y_o$, when $n = 0$ ($C = x_o$, $C(S_1-a) + C'(S_2-a) = by_o$ if $b \neq 0$). The inequality $S_1 \neq S_2$ implies $a \neq d$ so that the second set of equations in (5.7) is never indeterminate. The first set of equations in (5.7) holds also when S_1 and S_2 are complex, with proviso that the constants C and C' are also complex, and so chosen that the dependance between x_n, y_n and x_o, y_o is purely real for all n. This is always possible because when S_1, S_2 are complex, they are necessarily complex conjugate.

When $S_1 = S_2 = S$ is a double root of (5.6), the general solution of (5.5) becomes :

$$
(5.8)\quad \begin{cases} x = C\,S^n + C'nS^n, \quad y_n = (S-a)C\,S^n/b + [(n+1)S-n]C'S^n/b & b \neq 0 \\ x_n = C\,S^n, \quad y_n = C'S^n + CnS^n, \quad S = a = d, & b = 0 \end{cases}
$$

where again C, C' are real constants. When $\Delta = 0$, the right-hand sides of (5.5) are not linearly independant, i.e. they contain a common factor, say $x_{n+1} = ax_n + by_n$, $y_{n+1} = rx_{n+1}$, $n = 0,1,2,\ldots$, r being a known constant. So the recurrence (5.5) degenerates into a first order recurrence with the multiplier $S = a+br$, for $n = 1,2,\ldots$

A stability analysis of the fixed point $x_n = y_n = 0$ can be carried out directly on the basis of the solutions (5.7) and (5.8). By inspection it is obvious that this fixed point is *attractive* (asymptotically stable) when $|S_1| < 1$ and $|S_2| < 1$. It is *repulsive* (unstable when either $|S_1| > 1$, $|S_2| < 1$, or both $|S_1| > 1$ and $|S_2| > 1$. Compared to the information content of (5.7) or (5.8), the knowledge of stability is relatively meager, i.e. it describes only an asymptotic property of the half trajectories (x_n,y_n), $n = 0,1,2,\ldots$ or $n = 0,-1,-2,\ldots$ of T. In fact, from the general solutions (5.7) and (5.8), it is possible to deduce the existence of phase plane curves on which all points of any given half-trajectory, or full trajectory, are located. Two or more half-trajectories are obviously dynamically equivalent if their points are located on the same phase plane curves. Such curves are therefore dynamically more relevant than any discrete set of half-trajectories. This property was already noticed by Poincaré, who called these priviledge curves "analytical" invariant curves, which are particular solutions of (5.4).

5.2.2. INVARIANT CURVES WHEN THE MULTIPLIERS ARE REAL AND DIFFERENT.

The general solutions (5.7) and (5.8) of (5.5) are too compact to be easily used in a discussion of geometrical shapes of these invariant curves. Assume first that the multipliers of the fixed point $(x = y = 0)$ are real and distinct. By means of the linear transformation.

$$(5.9)\quad \begin{cases} x = \alpha X + \beta Y \quad , \qquad y = p_1 \alpha X + p_2 \beta Y \quad ; \qquad \alpha,\beta = \text{constants} \\ p_{1,2} = (S_{1,2}-a)/b = c/(S_{1,2}-d) \quad , \end{cases}$$

the recurrence (5.5) can be brought into the canonical diagonal form :

$$(5.10)\qquad X_{n+1} = S_1 X_n \quad , \qquad Y_{n+1} = S_2 Y_n \quad .$$

The rectangular axes X_n, Y_n are called the principal axes of T, and $p_{1,2}$ the eigen directions of $(x = y = 0)$, given by the eigenvectors of the matrix of the coefficients a,b,c,d. When (5.10) is transformed back to the coordinate frame x_n, y_n, the X_n and Y_n axes have the slopes p_1 and p_2 respectively. The general solution of (5.10) is obviously :

$$(5.11)\qquad X_n = S_1^n X_o \ , \qquad Y_n = S_2^n Y_o \ , \qquad n = 0,\pm 1,\pm 2,\ldots,$$

and no obstacle appears in its extension to continuous n. When $S_1 > 0$ and $S_2 > 0$, n is easily eliminated, and one obtains an explicit representation of the invariant curves covering the phase plane :

(5.12) $Y = Y_o|X/X_o|^\gamma \mathrm{sgn}(X/X_o)$, $\gamma = \ell nS_2/\ell nS_1$, $S_1, S_2 > 0$

When one of the two multipliers is negative, or both are negative, the curves (5.12) are invariant for T^2 for which the two multipliers are now positive, and $\gamma = \ell n|S_2|/\ell n|S_1|$. Then the shapes of invariant curves are unaffected, but the points of an iterated sequence lie now alternately on two distinct invariant curve segments.

When $|S_1| < 1$ and $|S_2| < 1$, or $|S_1| > 1$ and $|S_2| > 1$, the invariant curves resemble parabolas, tangent to $Y = 0$ at $(x = y = 0)$, symmetrical with respect to 0 . By analogy with the trajectories of differential equations the point $(x = y = 0)$ is called a (non-critical) *attractive node* for $|S_i| < 1$, and a *repulsive node* for $|S_i| > 1$, $i = 1,2$. In order to convey the information about the signs of multipliers, a node is said of *type 1* when $S_1 > 0$, $S_2 > 0$, of *type 2* when $S_1.S_2 < 0$, and of *type 3* when $S_1 < 0$, $S_2 < 0$. The corresponding locations of two points $M_n(x_n, y_n)$, $M_{n+1}(x_{n+1}, y_{n+1})$, $n = 0,1,2,\ldots$, of a discrete trajectory are shown fig. 5.1. For the type 1, the vector $\overrightarrow{M_n M_{n+1}}$ does not intersect the principal axes. For the type 2, it intersects one principal axis, and for the type 3, it intersects the two principal axes.

When $|S_1| < 1$, $|S_2| > 1$ (or vice versa) the invariant curves given by (5.12) have a hyperbolic shape, except for the two asymptotes $X = 0$ and $Y = 0$ (fig. 5.2). By analogy with differential equations the fixed point $(x = y = 0)$ is called a *saddle*, of type 1,2 or 3, depending on the signs of S_1, S_2. A saddle is a repulsive fixed point.

5.2.3. INVARIANT CURVES WHEN THE MULTIPLIERS ARE COMPLEX.

Suppose now that the multipliers are complex, $S_{1,2} = \rho\exp(\pm j\phi)$, $j = \sqrt{-1}$, $0 < \phi < \pi$, $\rho > 0$. The equation (5.11) can be written X,Y being complex variables. Let $u = X+Y$, $v = -j(X-Y)$ be the new real variables. Then the diagonal complex-valued recurrence (5.11) is converted to the real-valued canonical form :

(5.13) $u_{n+1} = \rho(u_n\cos\phi - v_n\sin\phi)$, $v_{n+1} = \rho(u_n\sin\phi + v_n\cos\phi)$

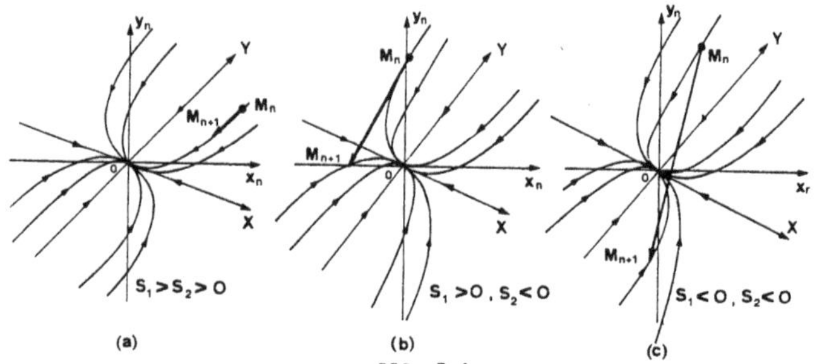

FIG. 5.1

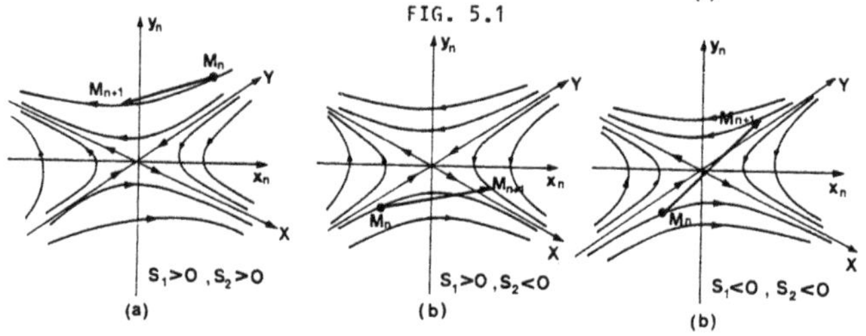

FIG. 5.2

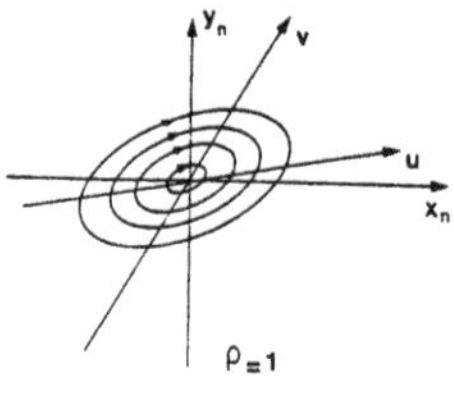

FIG. 5.4

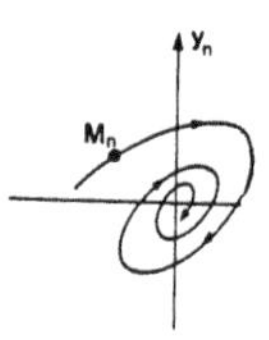

FIG. 5.3

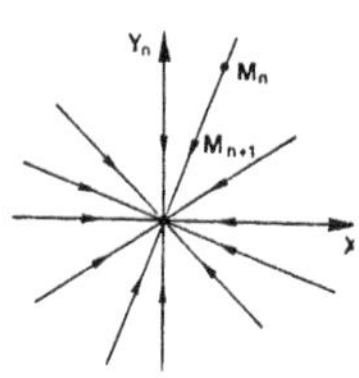

FIG. 5.5

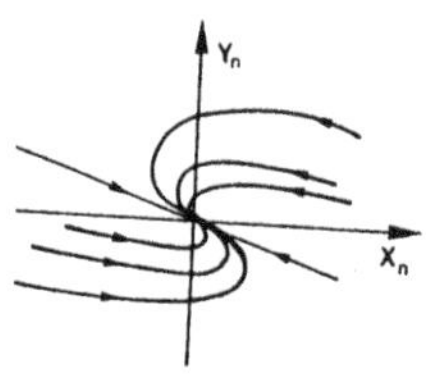

FIG. 5.6

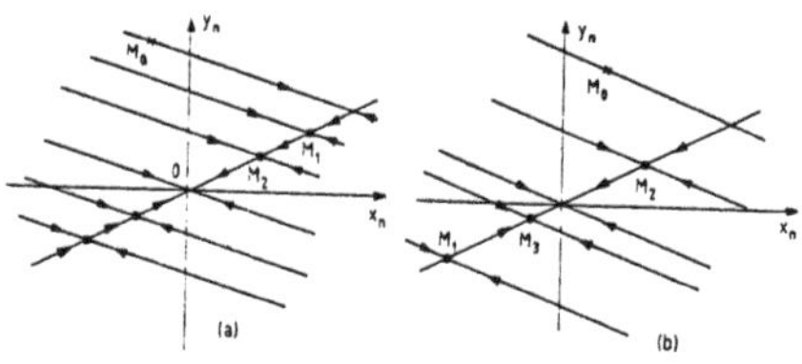

FIG. 5.7

The use of polar coordinates (σ,θ) renders the situation more transparent. If $u = \sigma\cos\theta$, $v = \sigma\sin\theta$, the recurrence (5.13) reduces to :

$$(5.14) \qquad \sigma_{n+1} = \rho\sigma_n \quad , \qquad \theta_{n+1} = \theta_n + \phi \quad ,$$

the solution of which is : $\sigma_n = \rho^n\sigma_o$ $\theta_n = n\phi + \theta_o$, $u_o = \sigma_o \cos\theta_o$, $v_o = \sigma_o \sin\theta_o$. The corresponding invariant curves are described by $\sigma = \rho^{(\theta-\theta_o)/\phi}.\sigma_o$, or :

$$(5.15) \qquad u^2+v^2 = (u_o^2+v_o^2)\rho^{2(\theta-\theta_o)/\phi} \quad , \qquad \theta = \text{arc tg } (u/v)$$

They are therefore logarithmic spirals (fig. 5.3), and the fixed point $(x = y = 0)$ is called an attractive (resp. repulsive) *focus* when $\rho < 1$ (resp. $\rho > 1$).

In the particular case $\rho = 1$, (5.15) indicates that the invariant curves of (5.13) are concentric circles $u^2 + v^2 = u_o^2 + v_o^2$, which correspond to concentric ellipses (fig. 5.4), invariant curves of (5.5). The fixed point $(x = y = 0)$ is called a *centre*, which is stable without being asymptotically stable (attractive). The general solution of (5.13) with $\rho = 1$ is :

$$(5.16) \quad u_n = A\cos(n\phi+\phi_o),\ v_n = A\sin(n\phi+\phi_o),\ u_o = A\cos\phi_o,\ v_o = A\sin\phi_o.$$

Consider an iterated sequence (discrete trajectory) $\{x_n,y_n\}$, $n = 0,\pm1,\pm2,\ldots$ When the rotation angle ϕ is commensurable with 2π, i.e. $\phi = 2\pi k/q$, k,q = positive integers, then $\{x_n,y_n\}$ consists only of k distincts points, although the corresponding invariant curves represent continua. When ϕ is not commensurable with 2π, the points $\{x_n,y_n\}$ fill the whole invariant curve as $n \to \infty$, or $n \to -\infty$. It should be noted that the invariant curves depend neither on the rotation angle ϕ, nor on any auxiliary ordering of points on a discrete trajectory. A centre is an isolated point of non-existence of a solution for the functional equation (5.4). Nodes and saddles are points of non-uniqueness of a solution for (5.4).

5.2.4. CASE OF TWO REAL MULTIPLIERS WITH EQUAL MODULI.

Whatever be the real multipliers S_1, S_2, $|S_1| \neq |S_2|$, from (5.10) the general solution of (5.5) is (5.7). If now $|S_1| = |S_2|$ this solution has also the form (5.7) when the two minors b, c of the coefficients matrix of (5.5) are b = c = 0. Then a = d and the recurrence (5.5) is already in the canonical from (5.10). When $S_1 = S_2 = S > 0$, the invariant curves of T are given by (5.12) with $\gamma = 1$, and they are straight lines passing through the fixed point (fig. 5.5). The fixed point (x = y = 0) is called a *star-node* of type 1. When $S_1 = -S_2$, or $S_1 = S_2 < 0$, considering T^2 leads to the type 1, and (x = y = 0) is a star-node of type 2, or 3. It is attractive when $|S| < 1$ and repulsive when $|S| > 1$.

With real multipliers, $S_1 = S_2 = S$, but b and c do not both vanish, say $b \neq 0$, c = 0, the canonical form of (5.5) (not reducible to a diagonal form), and its general solution are given by :

$$(5.17) \qquad x_{n+1} = Sx_n + by_n , \qquad y_{n+1} = Sy_n$$

$$(5.18) \qquad x_n = S^n(x_o+nby_o/S) , \qquad y_n = by_o[(n+1)S-n]S^{n-1} .$$

When $S_1 = -S_2 = \sqrt{S}$, (5.18) is the solution of the recurrence T^2 defined from (5.17). If n is considered as a continuous parameter and S > 0, (5.18) is the parametric equation of the invariant curve passing through (x_o,y_o). The invariant curves have a parabolic shape, and they are tangent at (x = y = 0) to the line y = px, where p = (S-a)/ b = 0, if $b \neq 0$, $p = c/(S-d) = \infty$ if b = 0. The line y = px is also an asymptotic direction for the invariant curves (fig. 5.6).

5.2.5. DEGENERATE SINGULAR POINTS.

Consider now the recurrence T :

$$(5.19) \qquad x_{n+1} = (a+\mu a')x_n + (b+\mu b')y_n, \quad y_{n+1} = (ar+\mu c)x_n + (br+\mu d)y_n$$

where μ is a parameter, $\mu \to 0$. For $\mu = 0$, the right-hand sides of (5.19) are not linearly independant, $\Delta = 0$ in equation (5.6), and the multipliers are $S_1 = a+br$, $S_2 = 0$. From $n = 1,2,\ldots$, (5.19) is restricted to the straight line $y = rx$, and degenerates into a one-dimensional recurrence $x_{n+1} = (a+br)x_n$. A discrete trajectory obtained from an initial condition (x_o,y_o), $y_o \neq rx_o$, or $y_o = rx_o$ lies entirely on $y = rx$, for $n = 1,2,\ldots$ In spite of this degeneracy the eigendirections of the fixed point ($x = y = 0$) can be determined : $p_1 = r$, $p_2 = -a/b$, for $\mu = 0$. Every point ($x = \alpha, y = r\alpha$) of $y = rx$ has a set of first rank antecedents, defined by $y = -ax/b + \alpha/b$, which can be considered as a degenerate segment of the invariant curve passing through the initial point (x_o,y_o), $y_o = ax_o/b+\alpha/b$. *Then this invariant curve is not smooth and is constituted by the degenerate segment of* $y = -ax/b + \alpha/b$, *and by the "regular" segment* $0 \leqslant x \leqslant \alpha/(a+br)$ *of* $y = rx$. The corresponding fixed point is called a *degenerate node* ($|S_1| < 1$) or a *degenerate saddle* ($|S_1| > 1$), of type 1 ($S_1 > 0$), or of type 2 ($S_1 < 0$) (fig. 5.7).

When $\mu \neq 0$, the invariant curves are smooth, and their limits, when $\mu \to 0$, are the non-smooth invariant curves at $x = \alpha/(a+br)$ obtained for $\mu = 0$. The segment ($\mu \neq 0$) close to the degenerate segment ($\mu = 0$) corresponds to a fast discrete trajectory, with respect to the slow discrete trajectory located on the segment ($\mu \neq 0$) close to the "regular" segment $y = rx$.

Other case of degeneracy occurs when one multiplier, say S_1, is such as $|S_1| = 1$, $|S_2| \neq 1$. From (5.10) it appears that the points of the principal X axis are fixed points if $S_1 = 1$, and points of cycles of order 2 if $S_1 = -1$, attractive when $|S_2| < 1$, repulsive when $|S_2| > 1$. If $|S_1| = |S_2| = 1$ the points of the phase plane are either fixed points, or cycles points of order 2.

5.3. AUTONOMOUS NON LINEAR CASE. THE INVARIANT CURVES.

Considering the singularities of T, the topics of § 2.2 remain valid in the phase plane for fixed points, and cycles, with a two-dimensional interpretation of (2.5). The dimension increase with respect to (2.5) leads to the existence of invariant curves, which are one-dimensional singularities, fixed points and cycles being zero-dimensional ones.

The determination of the phase portrait of the non linear recurrence (5.1) would proceed along the same lines as those the linear recurrence (5.5) if the general solution of the former could be expressed in terms of elementary or known transcendental functions. Unfortunately the functions involved are unknown, and a recourse to approximate methods both analytical and numerical becomes unavoidable. Assume that (0,0) is a fixed point of (5.1) and that $f(x,y,\lambda)$, $g(x,y,\lambda)$ are sufficiently smooth functions with respect x, y. It is possible then to write (5.1) in the form.

$$(5.20)\quad \begin{cases} x_{n+1} = ax_n+by_n + F(x_n,y_n,\lambda), \quad y_{n+1} = cx_n+dy_n+G(x_n,y_n,\lambda) \\ a = f_x(0,0),\ b = f_y(0,0),\ c = g_x(0,0),\ d = g_y(0,0) \end{cases},$$

the subscripts representing partial differentiation with respect to x and y. The functions F, G, $F(0,0,\lambda) = G(0,0,\lambda) = 0$, contain only terms of a degree higher than one.

By a continuity argument it is to be expected that when the phase portrait of the linear approximation of (5.20) is a node, a focus or a saddle, ($|S_i| \neq 1$, $i = 1,2,$), then this phase portrait remains at least locally the same in the presence of F and G, i.e. it is inert with respect to the structural perturbations F and G. This conjecture is confirmed by an analytical argument, which is simple in principle but rather cumbersome in practice as it appears in [H 1] and [L 1]. In these papers it is proved :

<u>(P5.1)</u> *Let* S_1, S_2 *be the multipliers of the fixed point* (0,0) *of* (5.20), *such that* $S_1 \neq S_2$.

- (a) *If* $|S_2| < 1 < |S_1|$, *only two invariant curves* (C_1), (C_2) *pass through the fixed point, each of them being tangent to the eigendirections of* (0,0). *When* F *and* G *are analytical in* x,y, *these two invariant curves are analytical. When* F *and* G *are differentiable in* x,y, *till an order* M-1, (C_1) (C_2) *are differentiable till the order* M.

- (b) *If* $|S_2| < |S_1| < 1$, *or* $|S_2| > |S_1| > 1$, *infinitely many invariant curves pass through* (0,0). *One of them* (C_1) *is tangent to one of the eigendirections. All the others are tangent to the second eigendirection. When* F *and* G *are analytical,* (C_1) *is analytical, as one of the invariant cur-*

ves (C_2) *tangent to the second eigendirection. The other invariant curves are non-analytical. When* F *and* G *are differentiable in* x, y *till an order* M-1, (C_1) *and* (C_2) *are differentiable till the order* M, *but it is not the case of the other invariant curves.*

When $|S_1| = 1$, $|S_2| \neq 1$, or $|S_1| = |S_2| = 1$, however small the neighbourhood of the fixed point (0,0) may be, inside this one the phase portrait depends on the non-linear terms F, G. One has a *critical case in the sense of Ljapunov,* and (5.20) is not inert (or it is structurally unstable). This occurs in particular when the linear approximation of (5.20) gives a centre. The corresponding phase portrait is generally not inert, because F and G may involve sources of damping or anti-damping, which affect the invariant curves at an arbitrary small distance from (0,0). Even if the Jacobian determinant :

$$(5.21) \qquad J(x,y) = (a+F_x)(d+G_y) - (b+F_y)(c+G_x)$$

of (5.20) is identically equal to unity (conservative case), the non-linear terms F and G are usually dominant. Similarly to second order differential equations, the survival of a centre of the recurrence (5.20) requires some rather special circumstances.

When (0,0) is a saddle, or a node for $F = G \equiv 0$, then the known solutions (5.12), of the functional equation (5.14) can be used to define small segments of the invariant curves of (5.10) with any prescribed accuracy, provided the initial point (x_o,y_o) is sufficiently close to (0,0) and these segments are sufficiently short. A continuation by means of numerically computed consequents or antecedents is possible, provided (5.20) and its inverse (or inverses) are iterated N times. The maximally permissible value of N is rather small, because the initial error grows usually rather rapidly with N. Invariant curve segments of appreciable length require therefore highly accurate initial segments. For conciseness the latter are called *germs* of the corresponding invariant curves.

If F and G are analytical in x, y, when the eigenvalues of (0,0) are real and distinct, two germs can be determined in the form of a Mc Laurin development :

$$(5.22) \qquad y = \theta(x) = \beta_1 x + \beta_2 x^2 + \dots \quad , \qquad \theta(0) = 0 \quad ,$$

where the β_i, $i = 1,2,\dots$ are real constants. Inserting (5.22) into the functional equation (5.4') yields a recursive system of linear algebraic equations for the coefficients β_i. The equation defining β_1 coincides with that of the eigendirections p_1 and p_2 at (0,0). Hence, one may choose either $\beta_1 = p_1$ or $\beta_1 = p_2$ (cf. eq.(5.9)). With β_1 fixed, the determination of β_i, $i > 1$, is straightfoward and unambiguous [G36] :

$$(5.23) \quad \begin{cases} \beta_2 = \beta_2(\beta_1) = \frac{1}{2}[(F_{xx}+\beta_1^2 F_{yy}+2\beta_1 F_{xy})\beta_1 - (G_{xx}+\beta_1^2 G_{yy}+2\beta_1 G_{xy})]/ \\ \qquad\qquad\qquad [c-\beta_1 b-(a+\beta_1 b)^2] \quad , \\ \beta_3 = \beta_3(\beta_1,\beta_2) = \dots\dots \quad , \qquad \beta_4 = \beta_4(\beta_1,\beta_2,\beta_3) = \dots \quad , \end{cases}$$

where all derivatives of F and G are evaluated at (0,0). The explicit expressions of β_i increase very rapidly in notational bulk as i increases, and for this reason they are indicated only symbolically. The convergence of the series (5.22) has been proven analytically for sufficiently small $|x|$ by constructing a majorating series [L1]. Since the method of proof in non constructive, the radius of convergence is unknown. Carefully controlled high-accuracy computations have shown that this radius, although sometimes quite small, is sufficiently large to render the development (5.22) highly useful. An iterative method of determining the coefficients β_i, $i > 1$, is also known ([H 1], [P 1], see also the discussion in [M57]). Since the accuracy of the β_i so determined increases with the number of iterations, practically usefull results can only be obtained by means of a computer ([G35] p.226-228 and [G5]) .

The same reasoning applies when (0,0) is not a fixed point but one point of a cycle of order k. It is sufficient to replace the recurrence (5.1) by its k-th iterate and the functional equation (5.4') by :

$$(5.24) \qquad g_k(x,\theta(x)) = \theta(h_k(x,\theta(x))) \qquad .$$

The determination of the iterated functions f_k, g_k is usually quite tedious without the use of a computer.

One germ of the form (5.22) can also be obtained in the case of the fig. 5.6. Because the proof of convergence is also of the majorizing type [L3], the radius of convergence is unknown. Like in the case of a saddle or a simple node, it is small, but usually not unusable.

The singular invariant curves generated by the preceding analytical germs are simply the axes X = 0 and Y = 0 (for the case of fig. 5.6, only X = 0) of the linear canonical forms, distorted by the non linear terms F and G in (5.20) (fig. 5.8). Distortions of other invariant curves of the linear approximation cannot be determined in the same manner, because their logarithmic dependence on S and Y precludes the existence of a Mc Laurin series $\theta(x)$ at x = 0. Corrections of the linear approximation, independent of the existence of a Mc Laurin series at x = 0, can be obtained by means of a small parameter development. The corresponding details are given in [G36] p.78-81.

Exceptionally the invariant curves can be expressed from simple analytical forms solutions of the functional equation (5.4), i.e. its solutions H(x,y) = C are known explicitly. This is the case, for example, of the recurrences [G36] :

$$(5.25) \qquad x_{n+1} = y_n \,, \quad y_{n+1} = -x_n+2y_n/(a^2-y_n^2) \,, \qquad a \neq \pm 1$$

$$(5.26) \qquad x_{n+1} = y_n \,, \quad y_{n+1} = -x_n+2ay_n/(1+y_n^2) \,, \qquad a \neq +1 \quad ,$$

the Jacobian determinant (5.21) of which is unity (except at $y = \pm a$ in (5.25), where it is undefined). They are conservative maps. The corresponding invariant curves are described explicitly by H(x,y) = c, c = positive "integration" constant, where $H(x,y) = x^2y^2-a^2(x^2+y^2) + 2xy$ for (5.25) and $H(x,y) = x^2y^2 + x^2 + y^2 - 2axy$ for (5.26), which can be easily verified by substitution in (5.4). In these exceptional cases the considered recurrence is related to a two-dimensional autonomous ordinary differential equation the phase trajectories of which are the invariant curves of the recurrence (cf. [G36] p. 96-104).

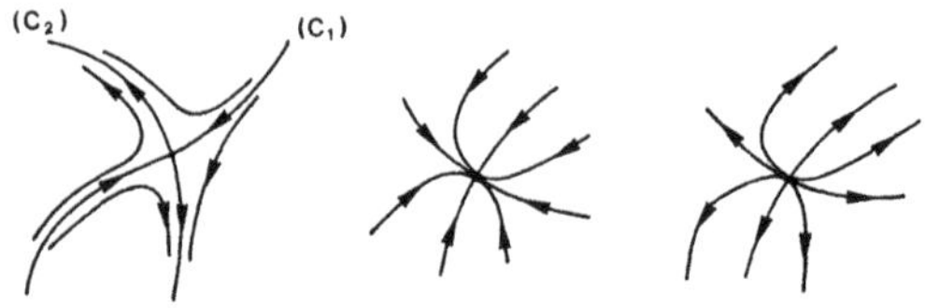

FIG. 5.8 - Invariant curves passing through non linear saddle and node

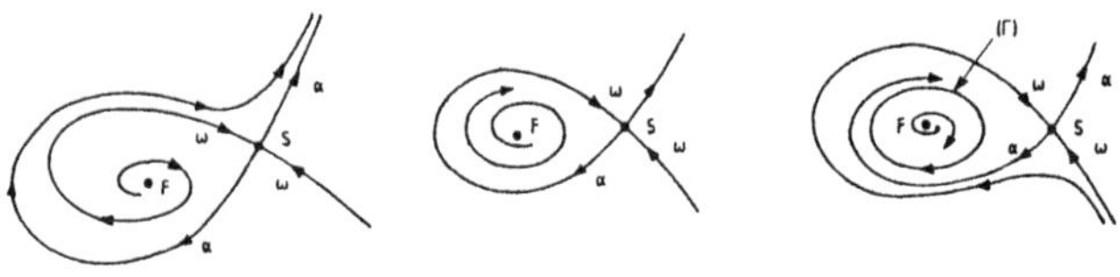

FIG. 5.9 - Invariant curves of a saddle . Situation without homoclinic point .

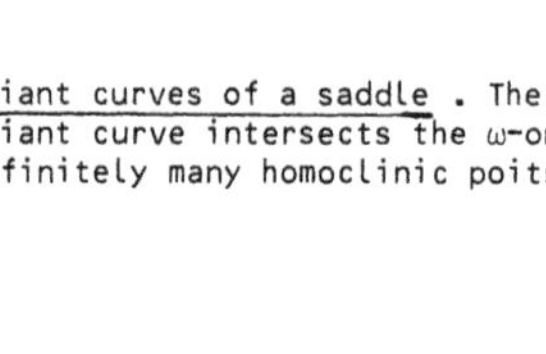

FIG. 5.10 - Invariant curves of a saddle . The α-invariant curve intersects the ω-one at infinitely many homoclinic poits .

5.4. AUTONOMOUS NON LINEAR CASE. HOMOCLINIC AND HETEROCLINIC POINTS.

Consider a saddle fixed point. Call ω-invariant curve that which is associated with the multiplier $|S_1| < 1$, and α-invariant curve that corresponding to $|S_2| > 1$. The two invariant curves passing through the saddle fixed point either do not cross each other, giving a classical situation encountered with the phase trajectories of two-dimensional autonomous ordinary differential equations (fig. 5.9), or cross each other at infinitely many points called by Poincaré *homoclinic points* (fig. 5.10). A α (resp. ω) invariant curve passing through each point of a saddle cycle of order k, either does not intersect a ω (resp. α) invariant curve passing through a different point of the cycle (classical situation of the two dimensional autonomous differential equations), or intersects the latter at infinitely many points called also homoclinic points (fig. 5.10). When a α (resp. ω) invariant curve of a saddle cycle intersects a ω (resp. α) invariant curve belonging to an other cycle at infinitely many points, they are called *heteroclinic points* (Poincaré's terminology). Birkhoff gave the important following result [B13] :

(P5.2) *In the general analytic case, an arbitrarily small neighbourhood of a homoclinic point contains infinitely many repulsive cycles and a homoclinic point.*

Without the hypothesis of analycity for f(x,y), g(x,y) in (5.1) this result also occurs for the most part of cases. As for the one-dimensional case, the presence of infinitely many repulsive periodic states is again related to the existence either of a stable chaos (strange attractor in the non-conservative case), or of an unstable chaos (strange repeller giving a fuzzy (chaotic) basins boundaries or transient chaos). Therefore *the presence of homoclinic points is a sign of chaotic behaviours* for two-dimensional maps.

Particular values of the parameter λ in (5.1) give a limit situation for which a α-invariant curve have infinitely many points of tangency with a ω-one (*homoclinic* or *heteroclinic tangency*). Such a situation can coexist with homoclinic, or heteroclinic intersections, for other neighbouring cycles. Homoclinic, and heteroclinic tangencies correspond to *global bifurcations. The most important one is that which finally separates the existence of homoclinic and heteroclinic points, in a whole region of the phase plane, from the non-existence of any such points.* In general this last global bifurcation is an accumulation of other homoclinic or heteroclinic tangencies coexisting with homoclinic or heteroclinic intersections for neighbouring cycles, and characterizes a transition order ‡ chaos in the considered region of the phase plane.

5.5. CRITICAL CASE $|S_1| = 1$, $|S_2| \neq 1$.

5.5.1. PARTICULAR CASE, THE MAP HAS A WHOLE CURVE MADE UP OF FIXED POINTS, OR OF POINTS OF ORDER TWO CYCLES.

Let T be the recurrence :

$$x_{n+1} = x_n + h(x_n,y_n)X_1(x_n,y_n) \;;\; y_{n+1} = y_n + h(x_n,y_n)Y_1(x_n,y_n) \,, \tag{5.27}$$

where X_1, Y_1 are functions of the real variables x, y the series expansion of which can begin by terms of degree zero, $h(x,y) = 0$ being the equation of a curve (C) in the phase plane (x,y). *Every point of this curve, made up of fixed points, has a multiplier* $S_1 = +1$, *the second being* S_2 *such that* $|S_2| \neq 1$. Indeed if (C) exists, the Jacobian determinant $\partial[(x_{n+1}-x_n),(y_{n+1}-y_n)]/\partial[x_n,y_n]$ is zero for every point of (C), and this is equivalent to the existence of a multiplier $S_1 = +1$ solution of the equation of the eigenvalues.

When h, X_1, Y_1 are analytical functions, it was shown by Lattes [L1] that an analytical invariant curve different from (C), associated with the multiplier S_2, $|S_2| \neq 1$, passes through every fixed point of (C), and the following proposition results :

(P5.3) *Let* $\widehat{AB}$ *be a segment of the curve of fixed points for which* $S_1 = 1$, $|S_1| < 1$. *On the both sides of* $\widehat{AB}$ *their exists a region of the* (x,y) *plane such that the iterated sequence, generated by an initial point of this region, tends toward a point of* $\widehat{AB}$, *and remains on the same analytical invariant curve when the data of* (5.27) *are analytical. When* $|S_2| > 1$, *an equivalent result is obtained, but the iterated sequence moves off from* $\widehat{AB}$.

Considering T^k, an equivalent situation occurs, for k segments made up of points of cycles of order k. Examples when T is an endomorphism are given in [G35] p.162, 252-260.

An example when T is a diffeomorphism is given by :

$$(5.28)\qquad x_{n+1} = x_n-(x_n+y_n)(1/4+x_n+y_n), \qquad y_{n+1} = y_n(x_n+y_n)(x_n+y_n-1/4) ,$$

for which (C) is the line y = -x made up of attractive fixed points with $S_1 = 1$, $S_2 = 1/2$. Another example is :

$$(5.29)\qquad x_{n+1} = x_n+(x_n+y_n)(1-x_n-y_n), \qquad y_{n+1} = y_n(x_n+y_n)(1+x_n+y_n) ,$$

for which (C) is the line y = -x made up of repulsive fixed points with $S_1 = 1$, $S_2 = 2$.

5.5.2. GENERAL CASE.

By means of the linear transformation (5.9), the recurrence (5.20) can be brought into the diagonal form T :

$$(5.30)\qquad x_{n+1} = S_1x_n+ X(x_n,y_n), \qquad y_{n+1} = S_2y_n+ Y(x_n,y_n), \qquad S_1 = +1$$

where the series expansions of X, Y contain only terms of degree > 1. The fixed point (0,0) gives here a critical case in the sense of Ljapunov, $|S_2| \neq 1$, and it is supposed that this point is isolated, i.e. T has not the form (5.27). The stability of (0,0), with respect to the coefficients of the non linear terms X(x,y), Y(x,y), can be determinated using a Ljapunov's function (extension of this notion to recurrence) ([M 9] and [G35] p.162-166).

The aspect of the invariant curves (phase portrait), in a sufficiently small neighbourhood of (0,0), appears, when using polar coordinates, via $x = (r+s)/2$, $y = -j(r-s)/2$, $j = \sqrt{-1}$, $r = \rho e^{j\theta}$, $s = \rho e^{-j\theta}$, it is obtained :

$$(5.31)\quad \begin{cases} \rho_{n+1} e^{j\theta_{n+1}} = \rho_n\left[(1+S_2)e^{j\theta_n} + (1-S_2)e^{-j\theta_n}\right]/2 \\ \qquad + X(\rho_n \cos\theta_n, \rho_n \sin\theta_n) + jY(\rho_n \cos\theta_n, \rho_n \sin\theta_n) \\ \rho_{n+1} e^{-j\theta_{n+1}} = \rho_n\left[(1-S_2)e^{j\theta_n} + (1+S_2)e^{-j\theta_n}\right]/2 \\ \qquad + X(\rho_n \cos\theta_n, \rho_n \sin\theta_n) - jY(\rho_n \cos\theta_n, \rho_n \sin\theta_n) \end{cases}$$

$$(5.32)\quad \begin{cases} \rho_{n+1}^2 = \rho_n^2(\cos^2\theta_n + S_2^2 \sin\theta_n) + \cdots \\ \theta_{n+1} = \operatorname{arc\,tg}\left[\rho_n S_2 \sin\theta_n + Y(\rho_n \cos\theta_n, \rho_n \sin\theta_n)\right]/ \\ \qquad \left[\rho_n \cos\theta_n + X(\rho_n \cos\theta_n, \rho_n \sin\theta_n)\right] \end{cases}$$

When $\rho_n \to 0$ the slopes of the invariant curves passing through the fixed point $0(0,0)$ are such that $\theta_{n+1} = \theta_n = \bar{\theta}$. They are given from the second relation (5.32) : $\bar{\theta}_1 = 0, \pi$, $\bar{\theta}_2 = \pm\pi/2$, and the corresponding multiplier is $S_\theta = [d\theta_{n+1}/d\theta_n]_{\theta=\bar{\theta}} = S_2[(1+tg^2\theta)/(1+S_2^2 tg^2\theta)]_{\theta=\bar{\theta}}$. When $\bar{\theta} = 0, \pi$, $S_\theta = S_2$. Then $\bar{\theta}$ is attractive with respect to θ if $|S_2| < 1$, and repulsive if $|S_2| > 1$. When $\bar{\theta} = \pm\pi/2$, $S_\theta = 1/S_2$, and the inverse situation is obtained. When X, Y, of (5.30) are two analytical functions, let Γ_1, Γ_2 be the two analytical invariant curves, with the slopes 0, $\pi/2$ at 0. They limit four regions of the (x,y) plane, corresponding to invariant curves having the same shape for sufficiently small ρ-values in every one of them. The curve Γ_1 tangent to the y-axis can be written $x = h(y)$, solution of the functional equation :

$$(5.33)\qquad h(y) + X[h(y),y] = h\{S_2 y + Y[h(y),y]\} \quad .$$

The first terms of the h(y) expansion are given by $h(y) = a_2 y^2 + a_3 y^3 + \ldots$ Considering (5.33) and its successive derivatives, the coefficients a_2, $a_3, \ldots$ can be determinated. If ℓ is the smallest degree of $X(0,y)$, $X(0,y) = \gamma_\ell y^\ell + \ldots$, then the germ of Γ_1 is $x = \gamma_\ell y^\ell/(S_2^\ell - 1) + \ldots$ Considering now (5.30) restricted to the points of Γ_1, and sufficiently small ρ-values, when $|S_2| < 1$ the fixed point $\rho = 0$ is attractive. It is repulsive when $|S_2| > 1$.

The invariant curve Γ_2 tangent to the x-axis is obtained similarly. Let β_o be the coefficient of the smallest degree q of $Y(x,0)$. Then the germ of Γ_2 is $y = \beta_o x^q/(1-S_2) + \ldots$ If the convergence properties of the iterated sequence are known on the invariant curve Γ_2 with those of θ_n,,

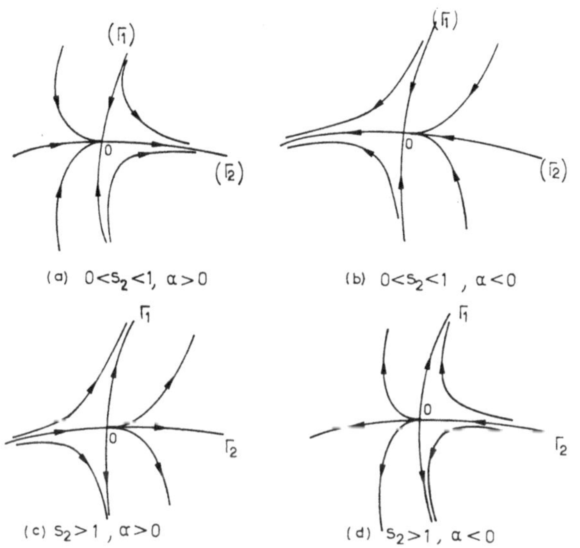

FIG. 5.11 - Critical case, node-saddle fixed point .

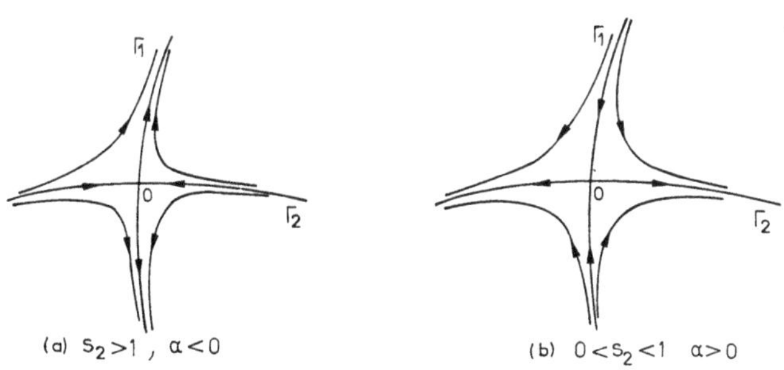

FIG. 5.12 - Critical case, saddle fixed point .

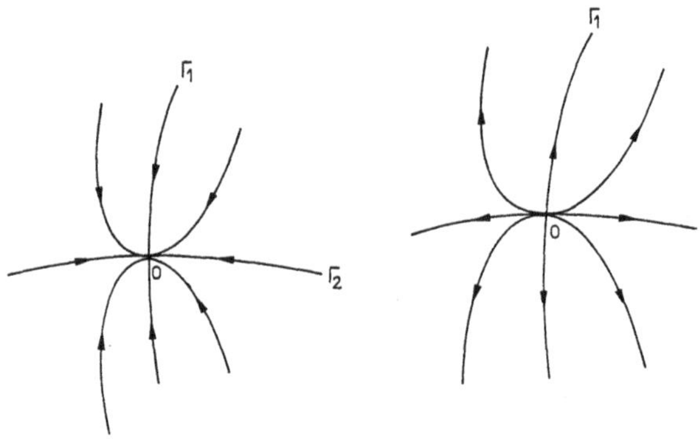

(a) $0<S_2<1$, $\alpha<0$ (b) $S>1$, $\alpha>0$

FIG. 5.13 - Critical case, node fixed point .

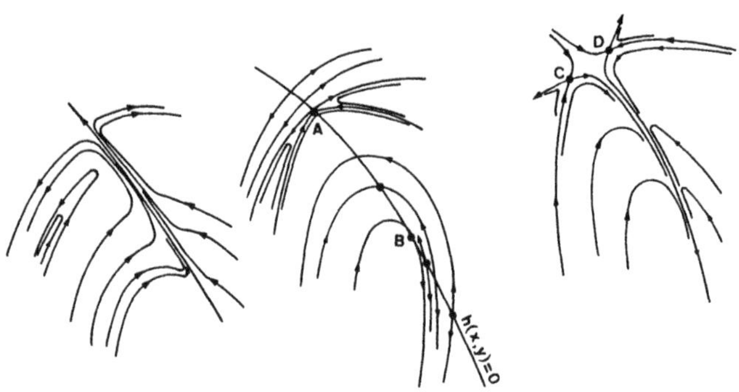

(a) $\mu<\mu_0$ (b) $\mu=\mu_0$ (c) $\mu>\mu_0$

FIG. 5.14 - Non classical local bifurcation .
Birth of a cycle of order 2 from a fixed point A ($S_1=-S_2=+1$), located on a curve $h(x,y)=0$ made up of fixed points . The invariant curves of T^2 are represented in (a),(b),(c). In fig. (b),the fixed point B has the multiplier $S_2=1$. All the points of $h(x,y)=0$ have $S_1=1$ .On the segment $\widehat{AB}$: $|S_2|<1$. For $\mu>\mu_0$ the cycle (C,D) of order 2 appears without existence of any fixed point in its neighbourhood .$\mu \equiv \lambda$ of the text .

and those on Γ_1, in a sufficiently small neighbourhood of 0 the shape of the invariant curves can be determined inside the four sectors limited by Γ_1, Γ_2. With increasing degrees, the functions $X(x,y)$, $Y(x,y)$ are written :

$$(5.34) \quad \begin{cases} X(x,y) = \sum_{j=o}^{m-1} \alpha_j x^{m-j} y^j + \ldots + \sum_{h=o}^{\ell} \gamma_h x^{\ell-h} y^h + \ldots \\ Y(x,y) = \sum_{r=1}^{p} \delta_r x^{p-r} y^r + \ldots + \sum_{k=o}^{q} \beta_k x^{q-k} y^k + \ldots \end{cases} ,$$

where $m,p,q,\ell \geqslant 2$, m is the smallest degree of $X(x,0)$, p is the smallest degree of $Y(x,y)$, $\gamma_\ell \neq 0$, ℓ is smallest degree of $X(0,y)$, $\ell \geqslant m$, $\beta_o \neq 0$, q is the smallest degree of $Y(x,0)$, $q \geqslant p$. Inside a sufficiently small neighbourhood of 0, the iterated sequences restricted to Γ_2 are defined from the one-dimensional recurrence :

$$(5.35) \qquad x_{n+1} = x_n + X[x_n, \beta_o x_n^q/(1-S_2)] \quad ,$$

i.e. from (P2.1). Finally one has

(P5.4) *Let* αx^m *be the monomial of* $X[x, \beta_o x^q/(1-S_2)]$ *which has the smallest degree. If* m *is even, the fixed point* 0 *is a node-saddle with the phase portraits of* fig. 5.11. *If* m *is odd, with* $0 < S_2 < 1$, $\alpha > 0$, *or* $S_2 > 1$, $\alpha > 0$, *the fixed point* 0 *is a critical saddle* (fig. 5.12). *If* m *is odd, with* $0 < S_2 < 1$, $\alpha < 0$, 0 *is a critical attractive node* (fig. 5.13a). *If* m *is odd, with* $S_2 > 1$, $\alpha > 0$, 0 *is a repulsive node* (fig. 5.13b). *A similar result is obtained for a cycle of order* k *by considering* T^k.

When $S_1 = +1$, $S_2 < 0$ the fixed point has the type 2. It is also the case when $S_1 = -1$, $S_2 > 0$. Fixed points of type 3 are obtained when $S_1 = -1$, $S_2 < 0$.

5.6. BIFURCATIONS BY PASSING THROUGH THE MULTIPLIER $S_1 = \pm 1$.

5.6.1. PASSING THROUGH THE PARTICULAR CRITICAL CASE OF § 5.5.1.

The recurrence (5.1) is written in the following form T :

$$(5.36)\quad \begin{cases} x_{n+1} = x_n + h(x,y)X_1(x,y) + (\lambda-\lambda_o)\bar{X}(x,y,\lambda) \\ y_{n+1} = y_n + h(x,y)Y_1(x,y) + (\lambda-\lambda_o)\bar{Y}(x,y,\lambda) \end{cases} ,$$

where X_1, Y_1, $\bar{X}$, $\bar{Y}$ are sufficiently smooth functions the expansions of which can begin with monomials of degree zero. For $\lambda = \lambda_o$, the curve $C(\lambda_o)$ such as $h(x,y) = 0$ is made of fixed points. When $\lambda \neq \lambda_o$, the fixed points are given by :

$$h(x,y)/(\lambda-\lambda_o) = \bar{X}(x,y,\lambda)/X_1(x,y) = \bar{Y}(x,y,\lambda)/Y_1(x,y) \quad .$$

When $\lambda = \lambda_o + \varepsilon$, $|\varepsilon|$ having a sufficiently small value, the curve $C(\lambda_o)$ gives rise to isolated fixed points from the particular points of $C(\lambda_o)$ which verify :

$$(5.37)\qquad h(x,y) = 0, \qquad R(x,y) = \bar{X}(x,y,\lambda_o).Y_1(x,y) - \bar{Y}(x,y,\lambda_o).X_1(x,y) = 0 ,$$

with the Jacobian determinant $\partial(h,R)/\partial(x,y) \neq 0$. These fixed points are located on one, or several invariant curves, which tend toward $C(\lambda_o)$ when $\lambda \to \lambda_o$, and one of their two multipliers tends toward $S_1 = +1$. Considering T^k equivalent result is obtained for k curves made up of points of cycles of order k.

Generally at least one segment of $C(\lambda_o)$ is made up of fixed points with $|S_2| < 1$ ($S_1 = 1$), the complementary part being such that $|S_2| > 1$. These segments are bounded by fixed points with $S_2 = \pm 1$. *A point of* $C(\lambda_o)$ *such that* $S_2 = -1$ *and* $R(x,y) \neq 0$ *can give rise to a cycle of order two, which is not associated with a fixed point when* $\lambda = \lambda_o + \varepsilon$*, as it is the case for the classical bifurcations* (2.24) (2.25), i.e. *the neighbourhood of this cycle does not contain a fixed point, because the fixed point with* $S_2 = -1$ *for* $\lambda = \lambda_o$*, disappears when* $\lambda \neq \lambda_o$. The fig. 5.14 represents the invariant curves phase portrait describing this *non-classical bifurcation*. From the considerations of this paragraph, it results :

(P5.5) *Consider the recurrence* (5.36) *which has a curve* $C(\lambda_o)$ *of fixed points when* $\lambda = \lambda_o$*. For* $\lambda \neq \lambda_o$*, isolated fixed points appear from the* *points of* $C(\lambda_o)$ *given by* (5.37). *A fixed point of* $C(\lambda_o)$ *with* $S_2 = -1$ *can give rise to a cycle of order two. When* $\lambda \neq \lambda_o$, *this cycle exist without the existence of a fixed point in its neighbourhood.*

An example illustrating (P5.5) is [G22] [G5] [G35] :

$$(5.38) \qquad x_{n+1} = y_n \quad , \qquad y_{n+1} = 0.5x_n + 0.1x_n^2 + \lambda x_n y_n \quad .$$

Whatever λ may be $F(x = y = (0.2+2\lambda)^{-1})$ is a fixed point and $E_1(0;5)$, $E_2(5;0)$ are the two points of a cycle of order two and $O(0;0)$ is an attrative star node of type 2. F is a repulsive node of type 2 is $\lambda < 0.1$, and a saddle if $\lambda > 0.1$. (E_1,E_2) is a saddle cycle of type 1 if $\lambda < 0.1$, and a repulsive node of type 1 if $\lambda > 0.1$. From (5.38) T is obtained :

$$(5.39) \quad \begin{cases} x_{n+2} = x_n + 0.1(x_n+y_n-5) + \mu x_n y_n \\ y_{n+2} = y_n + 0.1(x_n+y_n-5)(0.1x_n+1)y_n + \mu x_n y_n [0.5+0.1x_n+(0.2+\mu)y_n] \end{cases} ,$$

with $\mu = \lambda - 0.1$. When $\mu = 0 (\lambda = \lambda_o = 0.1)$, (5.39) has the form (5.27) and the straight line Δ, $h(x,y) \equiv x+y-5 = 0$ is made up of points of order two cycles with $S_1 = +1$ for (5.35). Using (5.37), it appears that the fixed point F and the cycles (E_1,E_2) are born of the corresponding points of Δ.

Reducing to zero the Jacobian determinant of (5.38) gives the curve (LC_{-1}) $0.2x + \lambda y = 0.5$, having a first order consequent $(\ell.c)$ $(0.5+\lambda x)^2 + 0.4y = 0$, called a *critical curve*, which is the two-dimensional equivalent of the critical point in the sense of Julia-Fatou. Thus (5.38) is a two dimensional endomorphism. The fig.5.15, 5.16, 5.17, represent the basin boundary (F) of the fixed point O. The index "-i", i = 1,2,..., associated with the points F, E_1, E_2, A, B, C, α, β, indicates the rank i-antecedents of these points. Here α, β are the intersections of $(\ell.c)$ with (F). Fig. 5.15, 5.16, 5.17 represent only the $y > 0$ part of this boundary. The parameter value $\lambda_o = 0.1$ corresponds to a *non-classical global bifurcation.* Indeed crossing through λ_o causes a discontinuous increase of the basin area, the origin of which is the existence of a basin related to the segment $\overline{AB}$ of Δ made up of points of attractive order two cycles, A(x = -13.508), B(x = -5), and its first rank consequent $\overline{A'B'}$, A'(x = 18.508),

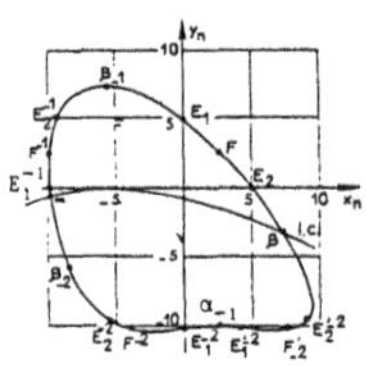

a) λ = 0,09

FIG. 5.15

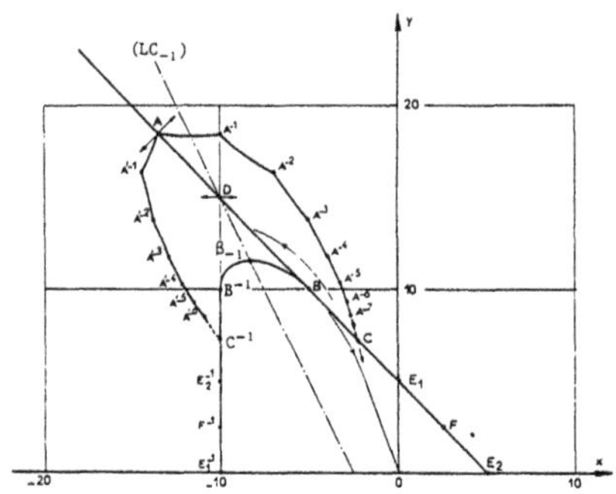

b) λ = 0,1

FIG. 5.16

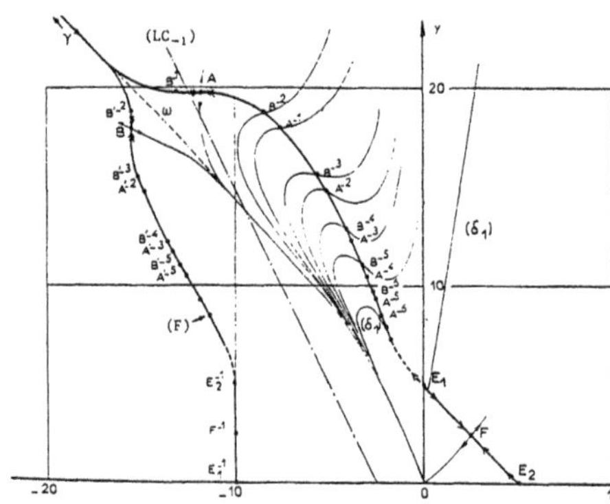

FIG. 5.17

c) λ = 0,101

FIG. 5.15, 5.16, 5.17 - Non classical global bifurcation.

For $\lambda = \lambda_0 + \varepsilon$, $\varepsilon > 0$ being small, the basin of the fixed point O abruptly expands by addition of the basin of the segments $\overline{AB}$, A'B' made up of cycles of order 2, obtained for $\lambda = \lambda_0$.

$B'(x = 10)$. One has $S_1(A,A') = S_1(B,B') = S_2(B,B') = +1$, $S_2(A,A') = -1$, the multiplier S_2 is $|S_2| \leqslant 1$ for the points of AB, $\overline{A'B'}$, and $|S_2| > 1$ for the complementary part of Δ. This basin, which exists only when $\lambda = \lambda_o$, is bounded by the non-smooth curve $\widehat{AA^{-1}}$, $\widehat{A^{-i}A^{-i-1}}$, $\widehat{AA'^{-i}}$, $\widehat{A'^{-i}A'^{-i-1}}$, $i = 1,2,\ldots$, $\lim A^{-i} = C$, $\lim A'^{-i} = C^{-1}$, when $i \to \infty$. Here C, C^{-1} are two points of the basin boundary of O, on which are located the points E_1, E_2, F, C, B, $\beta_{-1}, C_{-1}, E_1^{-1}, E_2^{-1}, F^{-1}$ and their consequents (fig.5.16). In this figure two segments of invariant curves $\widehat{OB}$, $\widehat{\beta_{-1}B}$, $\widehat{A'_{-1}A}$, $\widehat{A_{-1}A}$ are indicated.

For $\lambda = \lambda_o + \varepsilon$, $\varepsilon > 0$ having a sufficiently small value, the basin of the fixed point O abruptly expands with the addition of the basin of the segments $\overline{AB}$, $\overline{A'B'}$ obtained only for $\lambda = \lambda_o$. The basin boundary of O has now three points at infinity : Υ located on $x+y = 5$, and its two first rank antecedents. Moreover the order two cycle (A,A'), $S_1 = +1$, $S_2 = -1$, obtained for $\lambda = \lambda_o$, gives rise to an order four saddle cycle (A,B,A',B'), for $\lambda = \lambda_o + \varepsilon$, which appears without the existence of an order two cycle in its neighbourhood. It results a *non-classical local bifurcation* given by (P5.5). Corresponding invariant curves for T^4 and their antecedents are represented in fig.5.17. Considering the recurrence restricted to the basin boundary (F) (one-dimensional endomorphism), when λ increases from λ_o, the saddle cycle of order four undergoes a bifurcations set having the box-within-a-box structure.

5.6.2. GENERAL CASE WITH ONE MULTIPLIER $S_1 = +1$.

The following recurrence is now considered :

$$(5.40)\quad \begin{cases} x_{n+1} = A(\lambda) + [S_1+\varepsilon_1(\lambda)]x_n + B(\lambda)y_n + X(x_n,y_n) + \bar{X}(x_n,y_n,\lambda) \\[2ex] y_{n+1} = C(\lambda) + [S_2+\varepsilon_2(\lambda)]y_n + D(\lambda)x_n + Y(x_n,y_n) + \bar{Y}(x_n,y_n,\lambda) \end{cases},$$

where X, Y are given by (5.34), $\bar{X}$, $\bar{Y}$, are sufficiently smooth functions of x, y, $S_1 = +1$. $A,B,C,D,\bar{X},\bar{Y}$, ε_1, ε_2 are continuous with respect to the parameter λ in the neighbourhood of λ_o, $A(\lambda_o) = B(\lambda_o) = C(\lambda_o) = D(\lambda_o) = X(x,y,\lambda_o) = Y(x,y,\lambda_o) = \varepsilon_1(\lambda_o) = \varepsilon_2(\lambda_o) = 0$. So for $\lambda = \lambda_o$, (5.40) becomes the recurrence (5.30) having a critical case. The fixed points of

(5.40) are solutions of :

$$(5.41)\qquad \begin{cases} A(\lambda) + \varepsilon_1(\lambda)x + B(\lambda)y + X(x,y) + \bar{X}(x,y,\lambda) = 0 \\ C(\lambda) + [S_2-1 + \varepsilon_2(\lambda)]y + D(\lambda) + Y(x,y) + \bar{Y}(x,y,\lambda) = 0 \end{cases} .$$

Considering (5.34), the theorem of implicit functions ($S_2 \neq 1$) permits to obtain in a sufficiently small neighbourhood of O(0;0) :

$$(5.42)\qquad y = u(x,\lambda) = \beta_o x^q/(1-S_2) + W(x) + \bar{W}(x,\lambda) \quad ,$$

from the second relation of (5.41). Here $W(x)$ is made up of degrees higher than q, and $\bar{W}(x,\lambda_o) = 0$. The curve $y = u(x,\lambda_o)$ on which are located the fixed points has the same germ than that of the curve Γ_2 (cf. § 5.5.2). With (5.42) the first relation (5.40) becomes the one dimensional recurrence (2.14) :

$$(5.43)\qquad x_{n+1} = A_o(\lambda) + [1+\varepsilon_1^*(\lambda)]x_n + F_1(x_n) + F_2(x_n,\lambda) \quad ,$$

$A_o(\lambda_o) = \varepsilon_1^*(\lambda_o) = 0$, $F_2(x,\lambda_o) = 0$, $F_1(x)$, $F_2(x,\lambda)$ being two functions defined for (2.14). Then (P2.9) gives the fixed points on (5.42) and their stability. The nature of every fixed point is defined by two multipliers $S_1(\lambda)$, $S_2(\lambda)$, $S_1(\lambda_o) = +1$, $S_2(\lambda_o) = S_2$. When λ is sufficiently close to λ_o, $|y|$ being sufficiently small, $S_2(\lambda) \# S_2 \neq 1$ characterizes the behaviour of the iterated sequences on the invariant curve, passing through the considered fixed point, and which merges into Γ_1 (cf. § 5.5.2) when $\lambda \to \lambda_o$. From these considerations it results :

<u>(P5.6)</u> *Let* $T(\lambda)$ *be the recurrence* (5.40), *with* $S_1 = +1$, *which becomes* (5.30) *when* $\lambda = \lambda_o$. *Let* $y = u(x,\lambda)$ *be the curve* (5.42), *and the associated one-dimensional recurrence* (5.43) *which is the recurrence* (2.14) *of* § 2.4.1. *The fixed points, resulting from the bifurcation by crossing through* λ_o, *are located on* $y = u(x,\lambda)$ *and are the solutions of* (5.41).

a) *When the coefficients* $b_j(\lambda)$ *of the relation* (2.14) *are not simultaneously equal to zero, the number of fixed points obtained for* $\lambda = \lambda\pm\varepsilon$, $|\varepsilon|$ = *sufficiently small value, is given by* (P2.9). *Two invariant curves*

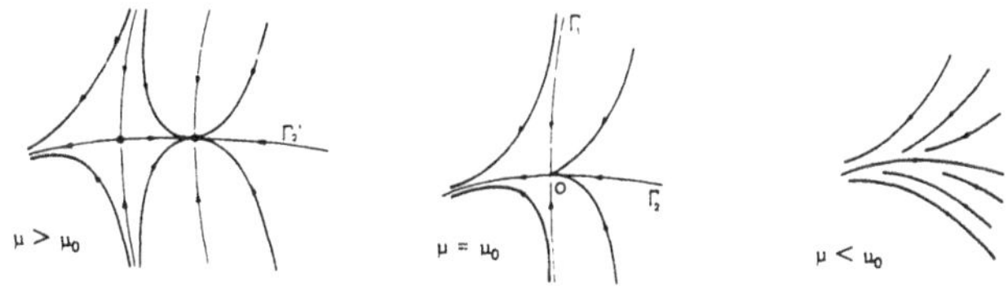

FIG. 5.18

Bifurcation through the critical case $S_1=1$, m is even, $p_1=2$, $p_2=0$.

$\mu \equiv \lambda$

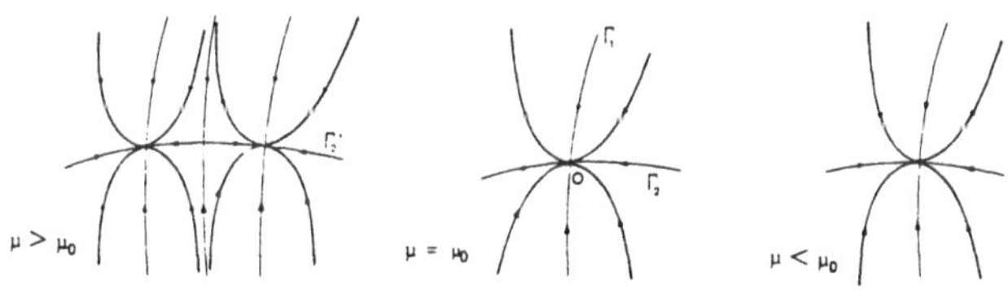

FIG. 5.19

Bifurcation through the critical case $S_1=1$, m is odd, $\alpha<0$, $q_1=3$, $q_2=1$.

$\mu \equiv \lambda$

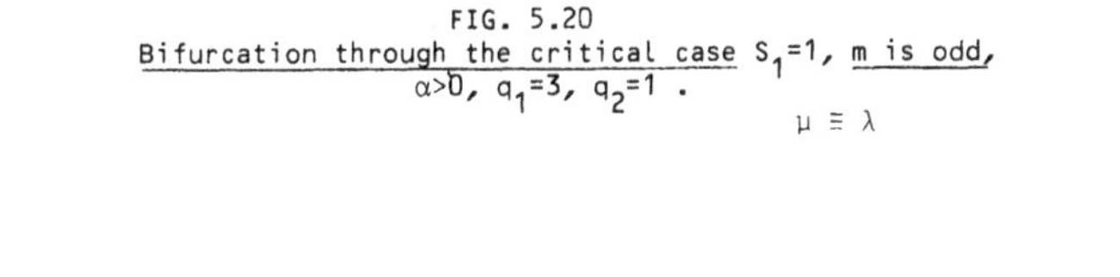

FIG. 5.20

Bifurcation through the critical case $S_1=1$, m is odd, $\alpha>0$, $q_1=3$, $q_2=1$.

$\mu \equiv \lambda$

(C_1), (C_2) *defined in* (P5.1) *cross through every fixed point. These invariant curves merge into the curves* Γ_1, Γ_2 *defined in* § 5.5.2 *when* $\lambda = \lambda_o$.

b) *The nature of the two extreme fixed points on* $y = u(x,\lambda)$ *is determinated by the monomial* αx^m *having the smallest degree in* $X[x, \beta_o x^q/(1-S_2)]$, *and the multiplier* S_2. *These extreme points are two nodes, or two saddles when* m *is odd, and a node and a saddle when* m *is even. All other fixed points on* $y = u(x,\lambda)$ *are such that nodes and saddles alternate.*

So when m is even, $0 < S_2 < 1$, and $\alpha > 0$ (resp. $\alpha < 0$), the extreme points are an attractive node on the left (resp. on the right), a saddle on the right (resp. on the left), and are of type one (cf. the simplest case with fig. 5.18 where $p_1 = 2$, $p_2 = 0$, are given by (P2.9)). When $-1 < S_2 < 0$, the node and the saddle are of type 2, with an oscillation of the iterated sequence on both sides of the invariant curve tending toward Γ_2 if $\lambda \to \lambda_o$. When $|S_2| > 1$, the extreme fixed points are a saddle on the left (resp. on the right), and a repulsive node on the right (resp. on the left), of type 1 ($S_2 > 0$) or of type 2($S_2 < 0$). When m is odd, $0 < S_2 < 1$, the exteme fixed points are two attractive nodes if $\alpha < 0$ (simplest case with fig. 5.19, $q_1 = 3$, $q_2 = 1$) or two saddles if $\alpha > 0$ (fig. 5.20). When $S_2 > 1$, they become two repulsive nodes if $\alpha > 0$, and two saddles if $\alpha < 0$. When $S_2 < 0$ these fixed points are of type 2.

5.6.3. GENERAL CASE WITH ONE MULTIPLIER $S_1 = -1$.

The considered recurrence T is (5.40) with $S_1 = -1$. Operating as in § 5.7.1, it appears that the fixed points are given by (2.17). Then in a sufficiently small neighbourhood of O(0;0), with $\lambda \neq \lambda_o$, only one fixed point exists tending toward O when $\lambda \to \lambda_o$. However, as in the one-dimensional case, passing through λ_o gives rise to cycles of order 2. These cycles are determinated by the consideration of T^2, which leads to the situation of § 5.6.1. The results of (P5.6) occur for T^2, but now the two extreme points have the same nature. The fixed point of T is either of type 2, or of type 3.

Let $F_1(x) = a_o x^m + a_1 x^{m+1} + \ldots$, $m > 1$, be the function appearing in (2.17) from $X[x, \beta_o x^q/(1-S_2)]$ written for $T(\lambda_o)$. When m is odd, $0 < S_2 < 1$, the extreme cycle is an attractive node of type 1 if $a_o > 0$, and a saddle of type 1 if $a_o < 0$. With $S_2 > 1$, the extreme cycle is a repulsive node of type 1, if $a_o < 0$, and a saddle of type 1 if $a_o > 0$. When m is even, $0 < S_2 < 1$, the extreme cycle is an attractive node of type 1 if $a_1 > 0$ $(m > 2)$, $a_1 + 2a_o^2 > 0$ $(m = 2)$, and a saddle of type one in the opposite case. With $S_2 > 1$, the extreme cycle is a repulsive node of type 1, if $a_1 < 0$ $(m > 2)$, $a_1 + 2a_o^2 < 0$ $(m = 2)$, and a saddle of type 1 in the opposite case. When $S_2 < 0$, the extreme cycle becomes of type 2, it is the same for the other cycles, and the fixed point is of type 3.

5.7. CRITICAL CASE WITH MULTIPLIERS $S_{1,2} = \exp(\pm j\phi)$, $j = \sqrt{-1}$. THE NON-EXCEPTIONAL CASE.

The recurrence, or map T, has a fixed point with these multipliers. By means of a linear transformation, T can be brought into :

$$(5.44) \quad r_{n+1} = r_n \cos\phi - s_n \sin\phi + R(r_n, s_n) \ ; \ s_{n+1} = r_n \sin\phi + s_n \cos\phi + S(r_n, s_n),\ \phi \neq 0, \pm\pi,$$

where R and S are two sufficiently smooth functions, the expansions of which begin with monomial of degree larger than one, $0(r = s = 0)$ being now the considered fixed point. In a sufficiently small neighbourhood of 0, the absence of homoclinic, or heteroclinic points is supposed. Let $R_i(r,s)$, $S_i(r,s)$ the monomials of degree i, i = 2,3,4,...

$$(5.45) \qquad R_i(r,s) = \sum_{p=0}^{p=i} A_{ip}\, r^{i-p} s^p, \quad S_i(r,s) = \sum_{p=0}^{p=i} B_{ip}\, r^{i-p} s^p$$

By means of a complex linear transformation $r = (x+y)/2$, $s = j(y-x)/2$, (5.44) is brought into the recurrence with complex variables :

$$(5.46) \quad \left\{ \begin{array}{l} x_{n+1} = e^{j\phi} x_n + \sum_{i=2}^{M} P_i(x_n, y_n), \quad y_{n+1} = e^{-j\phi} y_n + \sum_{i=2}^{M} Q_i(x_n, y_n) \\ P_i(x,y) = \sum_{p=0}^{p=i} C_{ip}\, x^{i-p} y^p, \quad Q_i(x,y) = \sum_{p=0}^{p=i} C^*_{ip}\, x^p y^{i-p} \end{array} \right.$$

C_{ip} and C^*_{ip} being two complex conjugate coefficients. (5.46) can also be directly obtained via (5.9). It is supposed that T is a diffeomorphism, and that ϕ is different from certain values, commensurable with 2π, corresponding to what it is called *exceptional cases* (cf.5.8).

This critical case, and the exceptional cases can be studied by means of a local method based on a reduction to certain canonical forms [M15]-[M17] [G35]. This method was used by Cigala for solving the problem of stability in the case of two-dimensional conservative diffeomorphisms [C1]. However this approach, called here *Cigala's method*, has more general implications. The essence of the method of reduction to a canonical form consists in identifying and isolating "*dominating terms*" of the non linear parts. This objective is obtained by means of a sequence of non linear transformations, which successively remove non-dominating terms. Practically the method of successive transformations is only successful when dominating terms happen to be of a sufficiently low order, because in general the complexity as well as the notational bulk of the successive transformations increases very rapidly. In what follows, because of the continuity and differentiability assumptions already made, the "order" or "degree" m will be defined as the degree i of $P_i(x,y)$, $Q_i(x,y)$ in (5.46), which are called an "i-th order form".

The succesive transformations permitting dominating terms to be identified are of the almost-identity type :

$$(5.47)\qquad \left\{\begin{array}{l} \xi = x + \phi_m(x,y),\ \eta = y + \Psi_m(x,y) \\ \phi_m(\xi,\eta) = \sum\limits_{p=0}^{m} \alpha_p\, \xi^{m-p}\, \eta^p,\ \Psi_m(\xi,\eta) = \sum\limits_{p=0}^{m} \alpha_p^*\, \eta^{m-p}\, \xi^p, \end{array}\right.$$

$m \geqslant 2$ being the lowest degree of the P_i and Q_i, α^* being the complex conjugate of α. This almost-identity transformation permits the conservation of the invariant curves shape inside a sufficiently small neighbourhood (Δ) of the fixed point O. The free undetermined constants α are so chosen that in the new variables ξ, η the recurrence (5.46) contains as few terms as possible in the corresponding m-th order forms. Inside (Δ) from (5.47) it is obtained :

$$(5.48)\qquad \left\{\begin{array}{l} x = \xi - \phi_m(\xi,\eta) + \phi_m(\xi,\eta)\, \partial\phi_m/\partial\xi + \Psi_m(\xi,\eta)\, \partial\phi_m/\partial\eta \ \ldots \\ y = \eta - \Psi_m(\xi,\eta) + \Psi_m(\xi,\eta)\, \partial\Psi_m/\partial\eta + \phi_m(\xi,\eta)\, \partial\Psi_m/\partial\xi \ \ldots \end{array}\right.$$

$$
(5.49)\begin{cases} \xi_{n+1} = e^{j\phi}\,\xi_n - e^{j\phi}\,\phi_m(\xi_n,\eta_n) + P_m(\xi_n,\eta_n) + \phi_m(e^{j\phi}\xi_n, e^{-j\phi}\eta_n) + \cdots = e^{j\phi}\,\xi_n + U_m + \cdots \\ \eta_{n+1} = e^{-j\phi}\eta_n - e^{-j\phi}\psi_m(\xi_n,\eta_n) + Q_m(\xi_n,\eta_n) + \psi_m(e^{j\phi}\xi_n, e^{-j\phi}\eta_n) + \cdots = e^{-j\phi}\eta_n + V_m + \cdots \end{cases}
$$

where U_m, V_m, are m-order forms. Choosing the α_p such that :

$$
(5.50) \qquad [e^{j\phi(m-2p-1)} - 1]\alpha_p = -e^{-j\phi} c_{mp}
$$

$p = 0,1,\ldots,m$, the corresponding terms of the m-th order form are eliminated in (5.49). But *this process will succeed if does not occur* $(m-2p-1)\phi = 2k'\pi$, $k' = 0,\pm1,\pm2,\pm3,\ldots$

<u>Definition</u>. *Let* ϕ *be commensurable with* 2π, $\phi = 2k\pi/q$, k *and* q *being two incommensurable integers . A necessary condition in order that the considered critical case may be an exceptional case is* $(m-2p-1)k/q = h$, $h = \pm1,\pm2,\ldots$

It is supposed now that the angle ϕ is either incommensurable with 2π, or, if it is commensurable, it is such as $(m-2p-1)k/q \neq h$.

a) When m is an even number, $m = 2\mu$, the relation (5.50) shows that the m-order form can be eliminated and (5.49) becomes :

$$
(5.51) \qquad \xi_{n+1} = e^{j\phi}\xi_n + \sum_{i=2\mu+1}^{\infty} P_i'(\xi_n,\eta_n) \; ; \; \eta_{n+1} = e^{-j\phi}\eta_n + \sum_{i=2\mu+1}^{\infty} Q_i'(\xi_n,\eta_n)
$$

Which leads to the consideration of m being an odd number.

b) When m is an odd number, $m = 2\mu+1$, the relation (5.50) shows that only the terms of the m-order form, such as $p \neq \mu$, can be eliminated. Then with an arbitrary choice of α_μ, the recurrence (5.49) is brought into :

$$
(5.52) \qquad \begin{cases} \xi_{n+1} = e^{j\phi}\,\xi_n + \gamma_{2\mu+1,\mu}\,\xi_n^{\mu+1}\eta_n^{\mu} + \cdots \\ \eta_{n+1} = e^{-j\phi}\eta_n + \gamma^*_{2\mu+1,\mu}\,\eta_n^{\mu+1}\xi_n^{\mu} + \cdots, \end{cases}
$$

where the dotted line corresponds to forms of order higher than $2\mu+1$. When the lowest order form in (5.46) is odd, $m = 2\mu+1$, then $\gamma_{2\mu+1,\mu} = c_{2\mu+1,\mu}$. When it is even, $m = 2\mu$, a first transformation permits to obtain (5.51),a and a second one, leads to (5.52) with $\gamma_{2\mu+1,\mu} \neq c_{2\mu+1,\mu}$. For convenience sake the variables ξ, η are designated by the same characters, though they are different in (5.51), (5.52), the latter resulting from the application of two successive Cigala's transformations. Suppose $\gamma_{2\mu+1,\mu} \neq 0$ using the polar coordinates $\xi = \rho e^{j\theta}$, $\eta = \rho e^{-j\theta}$ in (5.52), with ρ having a sufficiently small value, one has :

$$\text{(5.53)}\quad \begin{cases} \rho^2_{n+1} = \rho^2_n + G_\mu \rho_n^{2\mu+2} + \ldots \\ G_\mu = e^{-j\phi}\gamma_{2\mu+1,\mu} + e^{j\phi}\gamma^*_{2\mu+1,\mu} = 2|\gamma_{2\mu+1,\mu}|\cos(\phi+\tau_\mu),\ \tau_\mu = \arg\gamma^*_{2\mu+1,\mu} \end{cases}$$

The case $G_\mu = 0$ with $\gamma_{2\mu+1,\mu} \neq 0$ is a special one, not considered here. If $\gamma_{2\mu+1,\mu} = 0$, let $\gamma_{2\ell+1,\ell}$, $\ell \geqslant \mu$, be the first coefficient different from zero which is obtained after a sequence of Cigala's transformations. It is supposed that $(i-2p-1)\phi \neq 2h\pi$, $i = 2\mu, 2\mu+1, \ldots, 2\ell, 2\ell+1$, h being a positive or negative integer. The recurrence is now :

$$\text{(5.54)}\quad \begin{cases} \rho^2_{n+1} = \rho^2_n + G_\ell \rho_n^{2\ell+2} + \ldots \\ \theta_{n+1} = \theta_n + \phi - |\gamma_{2\ell+1,\ell}|\rho^{2\ell}\sin(\phi+\tau_\ell) + \ldots \\ \tau_\ell = \arg\gamma^*_{2\ell+1,\ell},\ G_\ell = e^{-j\phi}\gamma_{2\ell+1,\ell} + e^{j\phi}\gamma^*_{2\ell+1,\ell}, \end{cases}$$

From this equation one has the following proposition.

<u>(P5.7)</u> *Let* (5.46) *be the complex recurrence, obtained by means of a linear transformation, and which has* $O(x = y = 0)$ *as a fixed point. It is supposed the absence of homoclinic, or heteroclinic points, or other chaotic solutions in a sufficiently small neighbourhood* (Δ) *of* 0. *Let the angle* ϕ *be such that* $(i-2p-1)\phi \neq 2h\pi$, $i = 2\mu, 2\mu+1, \ldots, 2\ell, 2\ell+1$, h = *positive, or negative integer. Let* $\gamma_{2\ell+1,\ell}$ *be the first coefficient different from zero, which is obtained after* $2(\ell-\mu+1)$ *successive Cigala's transformations, and* G_ℓ *given by* (5.54). *Then, in the neighbourhood* (Δ) *of* 0, *the phase portrait of invariant curves is that of a focus, which is attractive if, and only if,* $G_\ell < 0$, *and repulsive if, and only if,* $G_\ell > 0$. *Such a fixed point, in a critical case, is called a complex focus of multiplicity* ℓ.

The presence of homoclinic, or heteroclinic points does not permit a conclusion of stability of 0, from the sign of G_ℓ, because the terms of degree higher than $(2\ell+1)$ can give invariant curves, coming from other singular points outside of (Δ), which cross through (Δ).

The coefficients G obtained via the Cigala's method are the same as those obtained via the use of a Ljapunov's function extended to recurrence [B 2]. For the determination of G_1, the index $\mu = 1$ in (5.51) and, $m = 2$ in (5.48). So the recurrence (5.49) is :

$$(5.55)\quad \begin{cases} \xi_{n+1} = e^{j\phi}\,\xi_n - U_2 + \phi_2(\partial U_2/\partial\xi_n) + \Psi_2(\partial U_2/\partial\eta_n) + P_2\left[\partial\phi_2(e^{j\phi}\xi_n, e^{-j\phi}\eta_n)/\partial(e^{j\phi}\,\xi_n)\right] \\ \qquad + Q_2\left[\partial\phi_2(e^{j\phi}\xi_n, e^{-j\phi}\eta_n)/\partial(e^{-j\phi}\,\eta_n)\right] + P_3 + \cdots \\ \eta_{n+1} = e^{-j\phi}\,\eta_n - V_2 + \Psi_2(\partial V_2/\partial\eta_n) + \phi_2(\partial V_2/\partial\xi_n) + Q_2\left[\partial\Psi_2(e^{j\phi}\xi_n, e^{-j\phi}\eta_n)/\partial(e^{-j\phi}\,\eta_n)\right] \\ \qquad + P_2\left[\partial\Psi_2(e^{j\phi}\xi_n, e^{-j\phi}\eta_n)/\partial(e^{j\phi}\,\xi_n)\right] + Q_3 + \cdots \end{cases}$$

U_2 and V_2 being the 2nd-order forms. The forms ϕ_2 and Ψ_2 are determinated in order to eliminate the terms of the second degree. So (5.55) becomes :

$$(5.56)\quad \begin{cases} \xi_{n+1} = e^{j\phi}\,\xi_n - P_2\left[2C_{20}\,\xi_n/(e^{j\phi}-1) + e^{-2j\phi}C_{21}\,\eta_n/(e^{-j\phi}-1)\right] \\ \qquad - Q_2\left[C_{21}\,\xi_n/(e^{-j\phi}-1) + 2e^{-2j\phi}C_{22}\,\eta_n/(e^{-3j\phi}-1)\right] + P_3 + \cdots \\ \eta_{n+1} = e^{-j\phi}\,\eta_n - Q_2\left[2C^*_{20}\,\eta_n/(e^{-j\phi}-1) + e^{2j\phi}\,C^*_{21}\,\xi_n/(e^{j\phi}-1)\right] \\ \qquad - P_2\left[C^*_{21}\,\eta_n/(e^{j\phi}-1) + 2e^{2j\phi}\,C^*_{22}\,\xi_n/(e^{3j\phi}-1)\right] + Q_3 + \cdots \end{cases}$$

$\gamma_{3,1}$ is the coefficient of the term $\xi^2\eta$ in ξ_{n+1}

$$(5.57)\quad \begin{aligned} \gamma_{3,1} = &-\left[2/(e^{j\phi}-1) + e^{-2j\phi}/(e^{-j\phi}-1)\right]C_{20}\,C_{21} \\ &- C_{21}\,C^*_{21}/(e^{-j\phi}-1) - 2e^{-2j\phi}\,C_{22}\,C^*_{22}/(e^{-3j\phi}-1) + C_{31} \end{aligned}$$

and G_1, given by (5.53) is the real expression :

$$(5.58)\quad \begin{cases} G_1 = \mathcal{R}e(c_{20}\,c_{21})\cdot(\cos 3\phi - 3\cos 2\phi + 2\cos\phi)/(1-\cos\phi) \\ \qquad + \mathcal{J}m(c_{20}\,c_{21})\cdot(\sin 3\phi - 3\sin 2\phi + 2\sin\phi)/(1-\sin\phi) + 2\mathcal{R}e(c_{31})\cos\phi \\ \qquad + 2\mathcal{J}m(C_{31})\sin\phi - [\mathcal{R}e(C_{21})]^2 - [\mathcal{J}m(C_{21})]^2 - 2[\mathcal{R}e(C_{22})]^2 - 2[\mathcal{J}m(C_{22})]^2, \\ \phi \neq 2k\pi/3,\ k = 1,2,\ \phi \neq 2k\pi/4,\ k = 0,1,2,3,\ \phi \neq 0,\ \phi \neq \pi \end{cases}$$

Here $\mathcal{R}e(C)$ means "real part of c" and $\mathcal{J}m(C)$ "imaginary part of c"

Remarks :

In some particular cases the coefficient G_ℓ can be defined even if $(i-2p-1)\phi = 2h\pi$, $i = 2\mu,\ldots,2\ell+1$. They occur when certain coefficients C_{jp}, or γ_{jp}, obtained after the Cigala's transformations, are equal to zero, which appears from (5.50). So G_1 can be defined with $\phi = \pm 2\pi/3$ when $C_{22} = 0$, or with $\phi = \pm 2\pi/4$ when $C_{31} = C_{33} = 0$. The same occurs for $\phi = \pi$ when $C_{30} = C_{32} = C_{33} = 0$. When these coefficients are not equal to zero, true exceptional case are obtained.

The relations between the coefficients A_{ip} of (5.45) and the coefficients c_{ip} of (5.46) are given in annex.

5.8. THE EXCEPTIONAL CASE WITH $\phi = 2k\pi/q$, q = ODD INTEGER.

The integers k and q are incommensurable, and it is supposed that one has a true exceptional case. A first study [B4] (1968) was conducted in the particular case q = 3 without using the Cigala's results, the general one being treated in [M15] (1969) and [M17] (1970). From the relation $(i-2p-1)\phi = 2h\pi$, $i = 2\mu,\ldots,2\ell+1$, $h \neq 0$, two different situations appear.

5.8.1. FIRST SITUATION.

It corresponds to the following conditions : $q = \pm(2\ell+1)$, $\gamma_{3,1} = \gamma_{5,2} = \ldots = \gamma_{2\ell-1,\ell-1} = 0$, $\ell > 1$, $\gamma_{2\ell+1,\ell}$ cannot be determinated for obtaining (5.52) because one has $(i-2p-1)\phi = 2h$ for $i = 2\ell$, and so one of the coefficients $\gamma_{2\ell,j}$ of the 2ℓ-th order form cannot be eliminated by a Cigala's transformation.

The integers k and $(2\ell+1)$ being incommensurable, the relation $(i-2p-1)\phi = 2h\pi$ occurs only for $p = 2\ell$, from which it results $h = -k$. With a Cigala's transformation all the terms of the degree 2 can be eliminated, except the term $\eta^{2\ell}$ in the relation giving ξ_{n+1}, and $\xi^{2\ell}$ in the relation giving η_{n+1}. Then the new recurrence T is :

$$\xi_{n+1} = e^{j\phi}\xi_n + \gamma_{2\ell,2\ell}\eta_n^{2\ell} + \cdots,\quad \eta_{n+1} = e^{-j\phi}\eta_n + \gamma^*_{2\ell,2\ell}\xi_n^{2\ell} + \cdots,$$

where the dotted line corresponds to the terms of degree higher than 2ℓ. Using polar coordinates (ρ,θ), when ρ has a sufficiently small value, one has :

$$(5.59)\qquad \begin{cases} \rho^2_{n+1} = \rho^2_n + 2|\gamma_{2\ell,2\ell}|\rho^{2\ell+1}\cos\left[(2\ell+1)\theta_n + \phi + \tau_{2\ell}\right] + \cdots \\ \theta_{n+1} = \theta_n + \phi - |\gamma_{2\ell,2\ell}|\rho^{2\ell-1}\sin\left[(2\ell+1)\theta_n + \phi + \tau_{2\ell}\right] + \cdots \end{cases}$$

where $\tau_{2\ell} = \arg \gamma^*_{2\ell,2\ell}$. For a sufficiently small values of ρ, the directions $\bar{\theta}$ such that $\theta_{n+1} = \theta_n + \phi$, given by $\sin[q\theta+\phi+\tau_{2\ell}] = 0$, are invariant by applications of T^q, the equation of which is :

$$(5.60)\qquad \begin{cases} \rho^2_{n+q} = \rho^2_n + 2q|\gamma_{2\ell,2\ell}|\rho_n^{2\ell+1}\cos(q\theta_n+\phi+\tau_{2\ell}) + \cdots \\ \theta_{n+q} = \theta_n - q|\gamma_{2\ell,2\ell}|\rho_n^{2\ell-1}\sin(q\theta_n+\phi+\tau_{2\ell}) + \cdots \end{cases}$$

2q invariant directions exist, separated by angles $\Phi/2$. They are given by :

$$(5.61)\qquad q\bar{\theta} = k\pi - \phi - \tau_{2\ell}$$

Considering the derivative :

$$(5.62)\qquad d\theta_{n+q}/d\theta_n = 1 - q^2|\gamma_{2\ell,2\ell}|\rho_n^{2\ell-1}\cos(q\theta_n+\phi+\tau_{2\ell}) + \cdots$$

evaluated for $\bar{\theta} = \theta$, it plays the role of the multiplier $S_{\bar{\theta}}$ of the fixed point $\bar{\theta}$ of the recurrence defined by the second relation (5.60), ρ having a fixed value. The coordinate ρ having a sufficiently small value, the value of this derivative shows that the solutions $\bar{\theta}$ of (5.61) are alternately attractive and repulsive with respect to the coordinate θ, when n increases. Furthermore considering now the first relation (5.60) it appears that ρ_n increases with n on an attractive direction ($0 < S_{\bar{\theta}} < 1$), and that ρ_n decreases on a repulsive direction ($S_{\bar{\theta}} > 1$). The study of (5.60), for sufficiently small values of ρ, shows that the 2q invariant directions are the asymptotes of the invariant curves, and that ρ has 2q minima separated by angles equal to $\Phi/2$. The fig. 5.21 give the phase portrait of invariant curves in a neighbourhood of the fixed point 0, for q = 3 and q = 5. The corresponding fixed point will be called a *complex saddle*.

5.8.2. SECOND SITUATION.

It corresponds to the following conditions : $q = \pm(2\delta+1)$, $\gamma_{3,1} = \gamma_{5,2} = \cdots = \gamma_{2\ell-1,\ell-1} = 0$, $\delta < \ell$, $\gamma_{2\delta,2\delta} = 0$, $\gamma_{2\ell+1,\ell}$ cannot be deter-

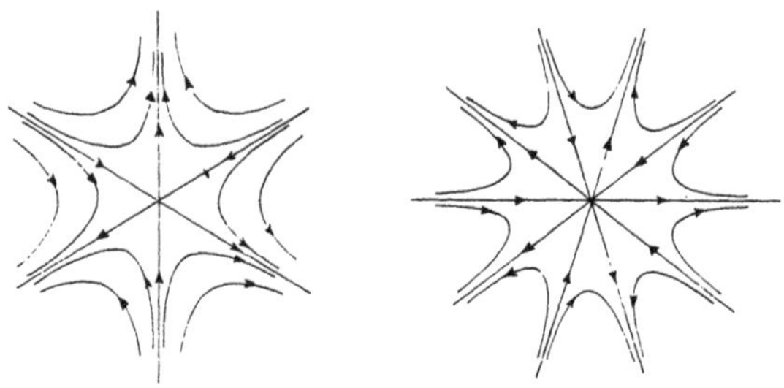

q= 3 q = 5

FIG. 5.21

Exceptional case with q=odd integer

Complex saddle .

$q=3 \quad \gamma_{3,1}=0$

FIG. 5.22

Exeptional case with q=odd integer.

Flower type fixed point .

minated for obtaining (5.52), because certain terms of the degree 2ℓ cannot be eliminated with the Cigala's transformation. This point means that $(i-2p-1)\phi \neq 2h\pi$, $h \neq 0$, when $i < 2\ell$, or if $(i-2p-1)\phi = 2h\pi$, $i < 2\ell$, the corresponding coefficient $\gamma_{jp} = 0$. This last relation is verified when $i = 2\ell$. Furthermore it is supposed that $(2\delta+1)$ is not divisor of $(2\ell-2p-1)$, $p = 0,1,\dots,2\ell$. It appears that $(i-2p-1)\phi = 2h\pi$ is only satisfied for two values of p :

$p = \ell+\delta$ with $h = -k$ and $p = \ell-\delta-1$, with $h = k$. Then it is supposed that at least one of the coefficient $\gamma_{2\ell,\ell+\delta}$, $\gamma_{2\ell,\ell-\delta-1}$ is different from zero, so that a new Cigala's transformation does not permit to define $\gamma_{2\ell+1,\ell}$. Thus a sequence of Cigala's transformations, the last being the one with the order $m = 2\ell$, leads to the recurrence T :

$$(5.63)\qquad \begin{cases} \xi_{n+1} = e^{j\phi}\,\xi_n + \gamma_{2\ell,\ell-\delta-1}\,\xi_n^{\ell+\delta+1}\,\eta_n^{\ell-\delta-1} + \gamma_{2\ell,\ell+\delta}\,\xi_n^{\ell-\delta}\,\eta_n^{\ell+\delta} + \cdots \\ \eta_{n+1} = e^{-j\phi}\,\eta_n + \gamma^*_{2\ell,\ell-\delta-1}\,\xi_n^{\ell-\delta-1}\,\eta_n^{\ell+\delta+1} + \gamma^*_{2\ell,\ell+\delta}\,\xi_n^{\ell+\delta}\,\eta_n^{\ell-\delta} + \cdots, \end{cases}$$

becoming in polar coordinates :

$$(5.64)\qquad \begin{cases} \rho^2_{n+1} = \rho^2_n + 2\rho_n^{2\ell+1}\Big[|\gamma_{2\ell,\ell-\delta-1}|\cos(q\,\theta_n - \phi - \tau_{\ell-\delta-1}) \\ \qquad + |\gamma_{2\ell,\ell+\delta}|\cos(q\,\theta_n + \phi + \tau_{\ell+\delta})\Big] + \cdots \\ \theta_{n+1} = \theta_n + \phi + \rho_n^{2\ell-1}\Big[|\gamma_{2\ell,\ell-\delta-1}|\sin(q\,\theta_n - \phi - \tau_{\ell-\delta-1}) \\ \qquad - |\gamma_{2\ell,\ell+\delta}|\sin(q\,\theta_n + \phi + \tau_{\ell+\delta})\Big] + \cdots \end{cases}$$

with $\tau_{\ell-\delta-1} = \arg \gamma^*_{2\ell,\ell-\delta-1}$ and $\tau_{\ell+\delta} = \arg \gamma^*_{2\ell,\ell+\delta}$. The recurrence T^q :

$$(5.65)\qquad \begin{cases} \rho^2_{n+q} = \rho^2_n + 2q\,\rho_n^{2\ell+1}\Big[|\gamma_{2\ell,\ell-\delta-1}|\cos(q\,\theta_n - \phi - \tau_{\ell-\delta-1}) \\ \qquad + |\gamma_{2\ell,\ell+\delta}|\cos(q\,\theta_n + \phi + \tau_{\ell+\delta})\Big] + \cdots \\ \theta_{n+q} = \theta_n + q\,\rho_n^{2\ell-1}\Big[|\gamma_{2\ell,\ell-\delta-1}|\sin(q\,\theta_n - \phi - \tau_{\ell-\delta-1}) \\ \qquad - |\gamma_{2\ell,\ell+\delta}|\sin(q\,\theta_n + \phi + \tau_{\ell+\delta})\Big] + \cdots \end{cases}$$

allows the determination of invariant curves when ρ has sufficiently small values. Indeed T^q has 2q invariant directions $\bar{\theta}$, separated by angles $\Phi/2$, and which are the solutions of :

$$|\gamma_{2\ell,\ell-\delta-1}|\sin(q\,\bar{\theta} - \phi - \tau_{\ell-\delta-1}) - |\gamma_{2\ell,\ell+\delta}|\sin(q\,\bar{\theta} + \phi + \tau_{\ell+\delta}) = 0$$

They are given by :

$$\operatorname{tg} q\bar{\theta} = \frac{\left[|\gamma_{2\ell,\ell-\delta-1}|\sin(\phi+\tau_{\ell-\delta-1}) + |\gamma_{2\ell,\ell+\delta}|\sin(\phi+\tau_{\ell+\delta})\right]}{\left[|\gamma_{2\ell,\ell-\delta-1}|\cos(\phi+\tau_{\ell-\delta-1}) - |\gamma_{2\ell,\ell+\delta}|\cos(\phi+\tau_{\ell+\delta})\right]}$$

A direction $\bar{\theta}$ is attractive for the coordinate θ, if the multiplier $S_{\bar{\theta}}$ given by :

$$\left[d\theta_{n+q}/d\theta_n\right]_{\theta=\bar{\theta}} = 1 + q^2\rho_n^{2\ell-1}\Big[|\gamma_{2\ell,\ell-\delta-1}|\cos(q\bar{\theta}-\phi-\tau_{\ell-\delta-1}) - |\gamma_{2\ell,\ell+\delta}|\cos(q\bar{\theta}+\phi+\tau_{\ell+\delta})\Big] + \dots,$$

is such that $0 < S_{\bar{\theta}} < 1$, ρ having a fixed value. It is repulsion when $S_{\bar{\theta}} < 1$. In other words, for sufficiently small values of ρ, the sign of the expression :

$$H = |\gamma_{2\ell,\ell-\delta-1}|\cos(q\bar{\theta}-\phi-\tau_{\ell-\delta-1}) - |\gamma_{2\ell,\ell+\delta}|\cos(q\bar{\theta}+\phi+\tau_{\ell+\delta}) \tag{5.66}$$

defines the behaviour θ in the neighbourhood of $\bar{\theta}$. The behaviour of ρ is obtained from the sign of the expression

$$K = |\gamma_{2\ell,\ell-\delta-1}|\cos(q\bar{\theta}-\phi-\tau_{\ell-\delta-1}) + |\gamma_{2\ell,\ell+\delta}|\cos(q\bar{\theta}+\phi+\tau_{\ell+\delta}) \tag{5.67}$$

The invariant directions are alternately attractive and repulsive for the coordinate θ_n. *In a sufficiently small neighbourhood of the fixed point* 0, *from* (5.66), (5.67) *it results that the invariant curves of* T^q *have the shape of a complex saddle* (fig.5.21) *the arrows indicate the sense of the increasing index* n *when the signs of* H *and* K *are opposite. When these signs are identical, the invariant curves are those given by the* fig.5.22. *The corresponding fixed point will be called of flower type.*

For example, the fig.5.22 is obtained with the recurrence (5.44), with $R(r,s) \equiv r^4+s^4$; $S(r,s) \equiv 0$.

5.9. THE EXCEPTIONAL CASE WITH $\Phi = 2k\pi/q$, q = EVEN INTEGER.

The integers k and q are incommensurable. The angle ϕ verifies $(i-2p-1)\phi = 2h\pi$, $h \neq 0$, and the corresponding coefficients C_{ip}, γ_{ip}, given by (5.50), are not equal to zero. A first study [B 4] (1968) was conducted in the particular case q = 4. The general one was treated in [M16]

[M17] (1970). As for the case q = odd integer, two different situations appear.

5.9.1. FIRST SITUATION.

It corresponds to the following conditions : $q = \pm 2(\ell+1)$, $\gamma_{3,1} = \gamma_{5,2} = \dots = \gamma_{2\ell-1,\ell-1} = 0$ for $\ell \geqslant 1$, $\gamma_{2\ell+1,\ell} = 0$ or $\neq 0$ for $\ell \geqslant 1$. In other words a sequence of Cigala's transformations permits the elimination of the terms of degree $2,3,\dots 2\ell$. Then the relation $(i-2p-1)\phi = 2h\pi$, $i = 2\ell+1$, is verified for $p = \ell$ $(h = 0)$, and $p = 2\ell+1$ $(h = -k)$. Thus a Cigala's transformation eliminates all the terms of the degree $(2\ell+1)$ except the terms having the coefficients $\gamma_{2\ell+1,\ell}$ and $\gamma_{2\ell+1,2\ell+1}$, and the following recurrence T is obtained :

$$(5.68)\qquad \begin{cases} \xi_{n+1} = e^{j\phi}\,\xi_n + \gamma_{2\ell+1,\ell}\,\xi_n^{\ell+1}\,\eta_n^{\ell} + \gamma_{2\ell+1,2\ell+1}\,\eta_n^{2\ell+1} + \dots \\ \eta_{n+1} = e^{-j\phi}\,\eta_n + \gamma^*_{2\ell+1,\ell}\,\eta_n^{\ell+1}\,\xi_n^{\ell} + \gamma^*_{2\ell+1,2\ell+1}\,\xi_n^{2\ell+1} + \dots \end{cases}$$

where the dotted line corresponds to the terms of a degree higher than $2\ell+1$. Using the new coordinates $\xi = \rho e^{j\theta}$, $\eta = \rho e^{-j\theta}$, $j = \sqrt{-1}$, when ρ has a sufficiently small value one has

$$(5.69)\qquad \begin{cases} \rho^2_{n+1} = \rho^2_n + 2\rho_n^{2\ell+2}\Big[|\gamma_{2\ell+1,\ell}|\cos(\phi+\tau_\ell) \\ \qquad\qquad + |\gamma_{2\ell+1,2\ell+1}|\cos(q\theta_n+\phi+\tau_{2\ell+1})\Big] + \dots \\ \theta_{n+1} = \theta_n + \phi - \rho_n^{2\ell}\Big[|\gamma_{2\ell+1,\ell}|\sin(\phi+\tau_\ell) \\ \qquad\qquad + |\gamma_{2\ell+1,2\ell+1}|\sin(q\theta_n+\phi+\tau_{2\ell+1})\Big] + \dots \end{cases}$$

with $\tau_\ell = \arg \gamma^*_{2\ell+1,\ell}$, $\tau_{2\ell+1} = \arg \gamma^*_{2\ell+1,\ell+1}$, $G_\ell = 2|\gamma_{2\ell+1,\ell}|\cos(\phi+\tau_\ell)$. It is supposed here that $(\phi+\tau_\ell) \neq \pm\pi/2$. The directions such $\theta_{n+i} = \theta_n + \phi$, are the invariant directions of T^q given by :

$$(5.70)\qquad \begin{cases} \rho^2_{n+q} = \rho^2_n + 2q\,\rho_n^{2\ell+2}\Big[|\gamma_{2\ell+1,\ell}|\cos(\phi+\tau_\ell) \\ \qquad\qquad + |\gamma_{2\ell+1,2\ell+1}|\cos(q\theta_n+\phi+\tau_{2\ell+1})\Big] + \dots \\ \theta_{n+q} = \theta_n - q\,\rho_n^{2\ell}\Big[|\gamma_{2\ell+1,\ell}|\sin(\phi+\tau_\ell) \\ \qquad\qquad + |\gamma_{2\ell+1,2\ell+1}|\sin(q\theta_n+\phi+\tau_{2\ell+1})\Big] + \dots, \end{cases}$$

These invariant directions $\overline{\theta}$ are the solutions of :

$$(5.71)\qquad \sin(q\overline{\theta}+\phi+\tau_{2\ell+1}) = -|\gamma_{2\ell+1,\ell}|\cdot|\gamma_{2\ell+1,2\ell+1}|^{-1}\cdot\sin(\phi+\tau_\ell) = a$$

and three qualitative different situations occur :

- $|a| > 1$, real invariant directions do not exist.

- $|a| < 1$, 2q invariant directions exist. Considering (5.71) it appears that when a = o they are separated by angles $\Phi/2$, but it is no longer the case when $a \neq 0$ for which two adjacent angles are not equal. Starting from two such adjacent angles, the others are obtained by q successive rotations through an angle Φ.

- $|a| = 1$, q double invariant directions exists.

When $|a| \leqslant 1$, considering the multiplier $S_{\bar{\theta}}$:

$$(5.72) \qquad \left[d\theta_{n+q}/d\theta_n\right]_{\theta=\bar{\theta}} = 1 - q^2|\gamma_{2\ell+1,2\ell+1}|\rho_n^{2\ell}\cos(q\bar{\theta}+\phi+\tau_{2\ell+1}) + \cdots,$$

ρ having a fixed value, $0 < S_{\bar{\theta}} < 1$ gives an attractive invariant direction $\bar{\theta}$, and $S_{\bar{\theta}} > 1$ a repulsive one. When $|a| < 1$, these invariant directions are alternately attractive and repulsive.

(a) *Phase portrait for* $|a| > 1$.

When ρ has sufficiently small values, from (5.70) it appears that the invariant curves have a focus shape which is deformed by the non-linearities (fig. 5.23). The resulting distortions correspond to the existence of q local maxima, and q local minima for ρ when θ rotates through an angle equal to 2π. The difference between a minimum and a consecutive maximum is the more important as $|\gamma_{2\ell+1,2\ell+1}|$ is large with $|a| > 1$. Thus the fixed point 0 is a *complex focus of multiplicity* ℓ. From (5.70) one has :

<u>(P5.8)</u> *When a given by* (5.71) *is such* $|a| > 1$, *the necessary and sufficient condition for* 0, *complex focus of multiplicity* ℓ, *being attractive* (*resp. repulsive*) *is* $\cos(\Phi+\tau_\ell) > 0$ (resp $\cos(\Phi+\tau_\ell) < 0$).

(b) *Phase portrait for* $|a| < 1$, $|\gamma_{2\ell+1,\ell}| < |\gamma_{2\ell+1,2\ell+1}|$

Consider the case $G_\ell < 0$, and let $\bar{\theta}_r$ be a repulsive invariant direction. The first relation (5.70) shows that in the neighbourhood of $\bar{\theta}_r$, ρ_n decreases when n increases, because $\cos(\Phi+\tau_\ell) < 0$ and $\cos(q\bar{\theta}_r+\Phi+\tau_{2\ell+1}) < 0$. In the neighbourhood of an attractive invariant direction $\bar{\theta}_a$, $\cos(q\bar{\theta}_a+\Phi+\tau_{2\ell+1}) > 0$. From (5.71) one has $|\sin(q\bar{\theta}_a+\Phi+\tau_{2\ell+1})| < |\sin(\Phi+\tau_\ell)|$. Then $|\cos(q\bar{\theta}_a+\Phi+\tau_{2\ell+1})| > |\cos(\Phi+\tau_\ell)|$, and :

$$(5.73) \qquad |\gamma_{2\ell+1,\ell}|\cdot|\cos(\phi+\tau_\ell)| < |\gamma_{2\ell+1,2\ell+1}|\cos(q\bar{\theta}_a+\phi+\tau_{2\ell+1}),$$

FIG. 5.23
|a|>1, complex focus of multiplicity l.

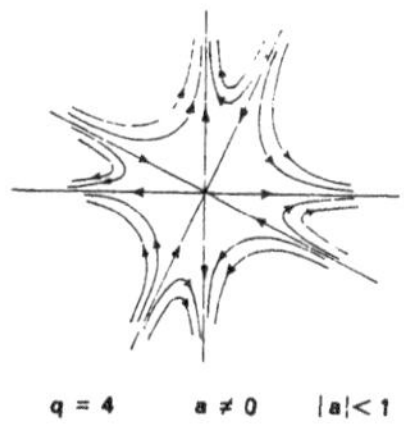

FIG. 5.24
Complex disymmetric saddle .

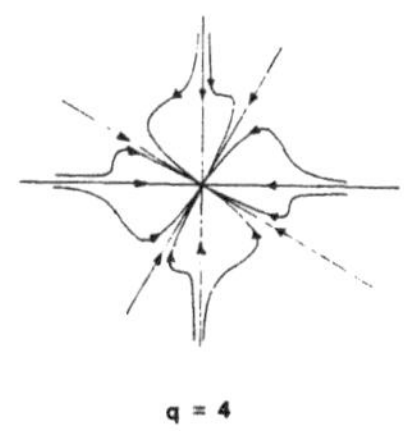

FIG. 5.25
Complex disymmetric node |a|<1

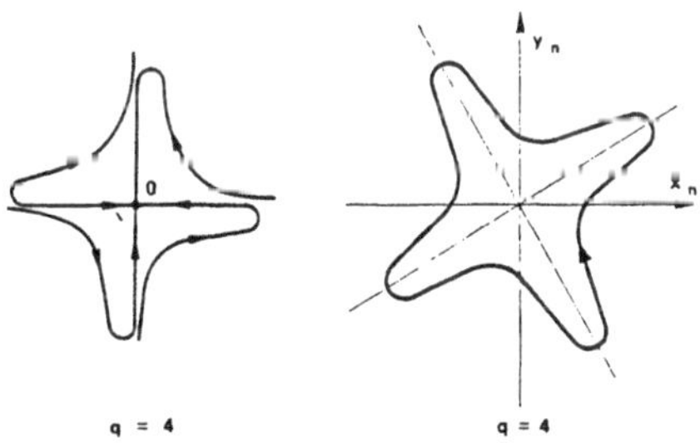

(a) (b)

FIG. 5.26
|a|=1

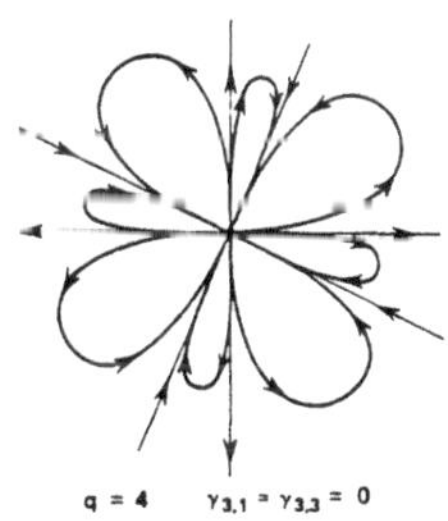

FIG. 5.27

This inequality associated with the first relation (5.70) shows that ρ_n is an increasing function of n, in a sufficiently small neighbourhood of $\bar{\theta}_a$.

Consider now the case $G_\ell > 0$, i.e. $\cos(\phi+\tau_\ell) > 0$. Let $\bar{\theta}_a$ be an attractive invariant direction, i.e. $\cos(q\bar{\theta}_a+\phi+\tau_{2\ell+1}) > 0$. The first relation (5.70) shows that ρ_n is an increasing function of n, in a sufficiently small neighbourhood of $\bar{\theta}_a$. Let $\bar{\theta}_r$ be a repulsive invariant direction, i.e. $\cos(q\bar{\theta}_r+\phi+\tau_{2\ell+1}) < 0$. From (5.71) $|\cos(q\bar{\theta}_r+\phi+\tau_{2\ell+1})| > |\cos(\phi+\tau_{2\ell})|$ and

(5.74) $$|\gamma_{2\ell+1,2\ell+1}|\ |\cos(q\bar{\theta}_r+\phi+\tau_{2\ell+1})| > |\gamma_{2\ell+1,\ell}|\ |\cos(\phi+\tau_\ell)|$$

This inequality associated with the first relation (5.70), shows that ρ_n is a decreasing function of n, in a sufficiently small neighbourhood of $\bar{\theta}_r$. Moreover, it appears that ρ has 2q minima when the variation of θ is 2π (cf. (5.69)). It results :

Whatever be the sign of G_ℓ*, the* 2q *invariant directions are the asymptotes of the invariant curves of* T^q*, in a sufficiently small neighbourhood of the fixed point* 0. *The corresponding phase portrait is given by the fig. 5.24, and is that of a complex saddle.*

(c) *Phase portrait for* $|a| < 1$, $|\gamma_{2\ell+1,\ell}| > |\gamma_{2\ell+1,2\ell+1}|$

Consider the case $G_\ell < 0$. Similarly to the case (b), in a sufficiently small neighbourhood of a repulsive invariant direction $\bar{\theta}_r$, ρ_n is a decreasing function of n. For an attractive invariant direction $\bar{\theta}_a$, the sense of the inequality (5.73) is now reversed, and ρ_n is a decreasing function of n. The invariant curves are tangent to the q directions $\bar{\theta}_a$. The fixed point is attractive. The corresponding phase portrait is given fig. 5.25 for q = 4, ρ having a sufficiently small value.

When $G_\ell > 0$, symmetrical results are obtained, and the fixed point is repulsive.

(d) *Phase portrait for* $|a| = 1$

The case $|a| = 1$ is a boundary between the two qualitative different situations of a complex focus and of a complex saddle. Now the coefficient of $\rho^{2\ell}$, in the expression of the multiplier $S_{\bar{\theta}}$ given by (5.72), is equal to zero. For sufficiently small values of ρ, only q invariant directions $\bar{\theta}$ exist. They give simultaneously the common tangent to invariant curves at the point 0 and their asymptotes. Every $\bar{\theta}$ is attractive on one side and repulsive on the other side. When $G_\ell < 0$, on $\bar\theta$ ρ_n is a decreasing function of n, and the fixed point 0 is attractive, the corresponding phase portrait is given by the fig. 5.26a. When $G_\ell > 0$ a similar phase portrait is obtained but with a repulsive fixed point. When T is a conservative map, in the absence of homoclinic or heteroclinic points, the phase portrait is given by the fig. 5.26b.

5.9.2. SECOND SITUATION.

It corresponds to the following conditions : $q = \pm 2(\delta+1)$, $\gamma_{3,1} = \gamma_{5,2} = \dots = \gamma_{2\ell-1,\ell-1} = 0$. Moreover $\gamma_{2\ell+1,\ell} = 0$, or $\neq 0$, for $\ell > 1$, $\gamma_{2\delta+1,2\delta+1} = 0$, $\delta < \ell$. Otherwise, either the relation $(i-2p-1)\phi = 2h\pi$ is not satisfied for $i < 2\ell+1$, or if it is, the corresponding coefficient $\gamma_{ip} = 0$. This relation is verified for $i = 2\ell+1$. Furthermore it is supposed that $(\delta+1)$ is not a divisor of $(\ell-p)$, $p = 0,1,2,\dots,2\ell+1$. It appears that the relation $(i-2p-\phi) = 2h\pi$, is not satisfied for three values of p :
$p = \ell (h = \ell)$; $p_1 = \ell+\delta+1$ $(h = -k)$; $p_2 = \ell-\delta-1$ $(h = k)$
At least one of the three corresponding coefficients is supposed to be different from zero. After a sequence of Cigala's transformations, the recurrence T becomes with cartesian coordinates and polar ones :

$$(5.75)\quad \begin{cases} \xi_{n+1} = e^{j\phi}\,\xi_n + \gamma_{2\ell+1,p_1}\,\xi_n^{\ell-\delta}\,\eta_n^{p_1} + \gamma_{2\ell+1,p_2}\,\xi_n^{\ell+\delta+2}\,\eta_n^{p_2} + \gamma_{2\ell+1,\ell}\,\xi_n^{\ell+1}\,\eta_n^{\ell} + \cdots \\ \eta_{n+1} = e^{-j\phi}\,\eta_n + \gamma^*_{2\ell+1,p_1}\,\eta_n^{\ell-\delta}\,\xi_n^{p_1} + \gamma^*_{2\ell+1,p_2}\,\eta_n^{\ell+\delta+2}\,\xi_n^{p_2} + \gamma^*_{2\ell+1,\ell}\,\eta_n^{\ell+1}\,\xi_n^{\ell} + \cdots \end{cases}$$

$$(5.76)\quad \begin{cases} \rho^2_{n+1} = \rho^2_n + 2\rho_n^{2\ell+2}\Big[|\gamma_{2\ell+1,p_1}|\cos(q\,\theta_n + \phi + \tau_{p_1}) + |\gamma_{2\ell+1,p_2}|\cos(q\,\theta_n - \phi - \tau_{p_2}) \\ \qquad + |\gamma_{2\ell+1,\ell}|\cos(\phi + \tau_\ell)\Big] + \cdots \\ \theta_{n+1} = \theta_n + \phi - \rho_n^{2\ell}\Big[|\gamma_{2\ell+1,p_1}|\sin(q\,\theta_n + \phi + \tau_{p_1}) - |\gamma_{2\ell+1,p_2}|\sin(q\,\theta_n - \phi - \tau_{p_2}) \\ \qquad + |\gamma_{2\ell+1,\ell}|\sin(\phi + \tau_\ell)\Big] + \cdots \\ \tau_{p_1} = \arg\gamma^*_{2\ell+1,p_1},\quad \tau_{p_2} = \arg\gamma^*_{2\ell+1,p_2},\quad \tau_\ell = \arg\gamma^*_{2\ell+1,\ell} \end{cases}$$

and the recurrence T^q :

$$(5.77)\quad \begin{cases} \rho^2_{n+q} = \rho^2_n + 2q\,\rho_n^{2\ell+2}\Big[|\gamma_{2\ell+1,p_1}|\cos(q\,\theta_n + \phi + \tau_{p_1}) + |\gamma_{2\ell+1,p_2}|\cos(q\,\theta_n - \phi - \tau_{p_2}) \\ \qquad + |\gamma_{2\ell+1,\ell}|\cos(\phi + \tau_\ell)\Big] + \cdots \\ \theta_{n+q} = \theta_n - q\,\rho_n^{2\ell}\Big[|\gamma_{2\ell+1,p_1}|\sin(q\,\theta_n + \phi + \tau_{p_1}) - |\gamma_{2\ell+1,p_2}|\sin(q\,\theta_n - \phi - \tau_{p_2}) \\ \qquad + |\gamma_{2\ell+1,\ell}|\sin(\phi + \tau_\ell)\Big] + \cdots \end{cases}$$

can give the phase portrait in a sufficiently small neighbourhood of the fixed point 0. Indeed the invariant directions $\bar{\theta}$ of T^q are the solutions of :

$$(5.78)\quad \begin{cases} A\sin q\bar{\theta} + B\cos q\bar{\theta} = C,\ A = |\gamma_{2\ell+1,p_1}|\cos(\phi+\tau_{p_1}) - |\gamma_{2\ell+1,p_2}|\cos(\phi+\tau_{p_2}) \\ B = |\gamma_{2\ell+1,p_1}|\sin(\phi+\tau_{p_1}) + |\gamma_{2\ell+1,p_2}|\sin(\phi+\tau_{p_2}),\ C = -|\gamma_{2\ell+1,\ell}|\sin(\phi+\tau_\ell) \end{cases}$$

and, supposing $\phi+\tau_\ell \neq \pi/2+k\pi$, two cases must be considered

- (a) $C^2 > A^2 + B^2$. The invariant directions $\bar{\theta}$ are not real. From (5.77) it appears that the invariant curves have a focus shape. When θ rotates through an angle equal to 2π, ρ presents q local maxima, and q local minima. The difference between a minimum and a consecutive maximum is the more important as C^2 is close to (A^2+B^2). The fixed point is a *complex focus of multiplicity* ℓ.

- (b) $C^2 < A^2 + B^2$. 2q invariant directions $\bar{\theta}$ exist now. Every $\bar{\theta}$ is associated with a multiplier $S_{\bar{\theta}}$ given by :

$$(5.79)\quad \left[d\theta_{n+q}/d\theta_n\right]_{\theta=\bar{\theta}} = 1 - q^2\rho_n^{2\ell}\Big[|\gamma_{2\ell+1,p_1}|\cos(q\bar{\theta}+\phi+\tau_{p_1}) - |\gamma_{2\ell+1,p_2}|\cos(q\bar{\theta}-\phi-\tau_{p_2})\Big] + \cdots$$

when ρ has a fixed value. So these invariant directions are alternately attractive and repulsive with respect to the coordinate θ. Contrary to the case q = odd integer, these 2q invariant directions are such that two adjacent angles are not equal, except when C = 0.

Considering (5.77) and (5.79), ρ having sufficiently small values, it appears that the sign of the expression :

$$(5.80)\quad D = |\gamma_{2\ell+1,p_1}|\cos(q\bar{\theta}+\phi+\tau_{p_1}) - |\gamma_{2\ell+1,p_2}|\cos(q\bar{\theta}-\phi-\tau_{p_2})$$

determinates the behaviour of the variable θ in the neighbourhood of $\bar{\theta}$. The behaviour of the variable ρ is determinated by the sign of the expression.

$$(5.81)\quad \begin{aligned} E = {} & |\gamma_{2\ell+1,p_1}|\cos(q\bar{\theta}+\phi+\tau_{p_1}) + |\gamma_{2\ell+1,p_2}|\cos(q\bar{\theta}-\phi-\tau_{p_2}) \\ & + |\gamma_{2\ell+1,\ell}|\cos(\phi+\tau_\ell) \end{aligned}$$

Let $D(\bar{\theta}_a) > 0$ and $E(\bar{\theta}_a)$ be the values of D and E for an attractive invariant direction, and $D(\bar{\theta}_r) < 0$, $E(\bar{\theta}_r)$ these values for a repulsive direction. The different combinations of the signs of these expressions lead to the four following situations :

- $D(\bar{\theta}_a) > 0$, $E(\bar{\theta}_a) < 0$, $D(\bar{\theta}_r) < 0$, $E(\bar{\theta}_r) > 0$, *the phase portrait is that of a flower type fixed point* (fig.5.22).

- $D(\bar{\theta}_a) > 0$, $E(\bar{\theta}_a) < 0$, $D(\bar{\theta}_r) < 0$, $E(\bar{\theta}_r) < 0$, *the phase portrait is that of the fig.5.25.*

- $D(\bar{\theta}_a) > 0$, $E(\bar{\theta}_a) > 0$, $D(\bar{\theta}_r) < 0$, $E(\bar{\theta}_r) > 0$, *the phase portrait is that of the fig 5.25, 0 being a repulsive fixed point.*

- $D(\bar{\theta}_a) > 0$, $E(\bar{\theta}_a) > 0$, $D(\bar{\theta}_r) < 0$, $E(\bar{\theta}_r) < 0$, *the phase portrait is that of a complex saddle* (fig.5.24).

Remark

For q = 4, the expressions of the coefficients $\gamma_{3,1}, \gamma_{3,3}$, determined from a Cigala's transformation are :

$$\gamma_{3,1} = \left[2C^*_{2,2}C_{2,2} - 4C_{2,1}C_{2,0}\right]/(j-1) + \left[4C^*_{2,1}C_{2,1} - 2C_{2,1}C_{2,0}\right]/(j-1) + 3C_{3,1}$$

$$\gamma_{3,3} = C_{2,2}C_{2,1}(j-1) - C^*_{2,0}C_{2,2}(j+1) + C_{3,3}$$

$G_1 = e^{-j\phi}\gamma_{3,1} + e^{j\phi}\gamma^*_{3,1}$, is given by (5.58).

5.10. CRITICAL CASES WITH MULTIPLIERS $S = \exp(\pm j\phi)$, $\phi = 0$, or π.

At first consider the case $\phi = 0$, i.e. $S_1 = S_2 = 1$, for the recurrence (5.44). This case was studied in |M19|. In a sufficiently small neighbourhood of the fixed point 0(r = s = o) the absence of homoclinic, and heteroclinic points is supposed. By means of the complex linear transformation $r = (x+y)/2$, $s = j(y-x)/2$, $j = \sqrt{-1}$, (5.44) is brought into the recurrence with complex variables :

$$(5.82) \quad \begin{cases} x_{n+1} = x_n + X^{(m)}(x_n,y_n) + \Phi(x_n,y_n) = x_n + \sum_{i=0}^{m} C_{mi}\, x^{m-i} y^i + \Phi(x_n,y_n) \\ y_{n+1} = y_n + Y^{(m)}(x_n,y_n) + \psi(x_n,y_n) = y_n + \sum_{i=0}^{m} C^*_{mi}\, y^{m-i} x^i + \psi(x_n,y_n) \end{cases}$$

where $m \geq 2$, C^* is the complex conjugate of c, Φ and Ψ represent the set of the terms of degree higher than m. Using now polar coordinates $x = \rho e^{j\theta}$, $y = \rho e^{-j\theta}$, (5.82) becomes :

$$(5.83) \quad \rho^2_{n+1} = \rho^2_n + \rho^{m+1}_n P(\theta_n) + \cdots, \quad \theta_{n+1} = \theta_n + \rho^{m-1}_n Q(\theta_n) + \cdots,$$

$$P(\theta) = \sum_{i=0}^{k-1} 2|C_{mi}|\cos[(m-2i-1)\theta - \tau_i] + \sum_{i=-1}^{\ell} 2|C_{m,m-i-1}|\cos[(m-2i-1)\theta + \tau_{m-i-1}],$$

$$Q(\theta) = \sum_{i=0}^{k-1} |C_{mi}|\sin[(m-2i-1)\theta - \tau_i] - \sum_{i=-1}^{\ell} |C_{m,m-i-1}|\sin[(m-2i-1)\theta + \tau_{m-i-1}],$$

where $\tau_i = \arg C^{\star}_{mi}$; $k = (m-1)/2$, $\ell = k$ when m = odd integer ; $k = m/2$, $\ell = k-1$ when m = even integer. The dotted line corresponds to terms of degree higher than m. When m = odd integer, $P(\theta)$ and $Q(\theta)$ contain a constant term corresponding to $|C_{mk}|$. It is supposed now that $Q(\theta)$ is not identically equal to zero, $Q(\theta) \not\equiv 0$. The two following cases are considered :

- (a) $Q(\theta) = 0$ has not real roots :

The invariant curves are those of a *complex focus* in the general case, and those of a center in the conservative case. This situation occurs only when m = odd integer, and the complex focus is attractive if $|C_{mk}| \cos \tau_k < 0$, and repulsive in the opposite case. The winding sense of the focus is obtained from the sign of $\sin \tau_k$.

- (b) $Q(\theta) = 0$ has real different roots.

In the interval $0 \leqslant \theta \leqslant \pi$, $Q(\theta) = 0$ cannot have more than m+1 real roots which correspond to the invariant directions of (5.83), for sufficiently small values of ρ. Let $\bar{\theta}_i$, $i = 1,2,\dots, s < m+1$, be these roots, then all the roots of $Q(\theta) = 0$ are $\bar{\theta}_{i\alpha} = \bar{\theta}_i + \alpha\pi$, α = positive, or negative integer. Every $\bar{\theta}$ is associated with a multiplier $S_{\bar{\theta}}$:

$$(5.84) \qquad S_{\bar{\theta}_i} = (d\theta_{n+1}/d\theta_n)_{\bar{\theta}_i} = 1+\rho_n^{m-1}\left[dQ(\theta_n)/d\theta_n\right]_{\bar{\theta}_i}+\dots = 1+\rho_n^{m-1}Q'(\bar{\theta}_i)+\dots$$

The invariant direction $\bar{\theta}_i$ is attractive when $Q'(\bar{\theta}_i) < 0$, and repulsive when $Q'(\bar{\theta}_i) > 0$. It is supposed that $Q'(\bar{\theta}_i) \neq 0$ and $P(\bar{\theta}_i) \neq 0$. In a sufficiently small neighbourhood of $\bar{\theta}_i$, four different behaviours of the invariant curves appear :

- (1) $Q'(\bar{\theta}_i) < 0$, $P(\bar{\theta}_i) < 0$: ρ_n is a decreasing function of n, and $\bar{\theta}_i$ is attractive (fig.5.28a). The behaviour of $\bar{\theta}_i$ is called "of attractive node type" (AN).

- (2) $Q'(\bar{\theta}_i) > 0$, $P(\bar{\theta}_i) > 0$: ρ_n is an increasing function of n, and $\bar{\theta}_i$ is repulsive (fig.5.28b). The behaviour of $\bar{\theta}_i$ is called "of repulsive node type" (RN).

- (3) $Q'(\bar{\theta}_i) > 0$, $P(\bar{\theta}_i) < 0$, ρ_n is a decreasing function of n, $\bar{\theta}_i$ is repulsive (fig.5.28c). The behaviour of $\bar{\theta}_i$ is called "of A saddle type" (AS).

- (4) $Q'(\bar{\theta}_i) < 0$, $P(\bar{\theta}_i) > 0$, ρ_n is an increasing function of n, $\bar{\theta}_i$

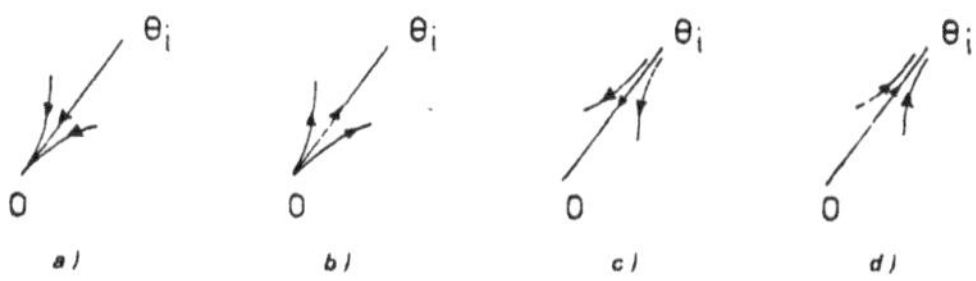

FIG. 5.28

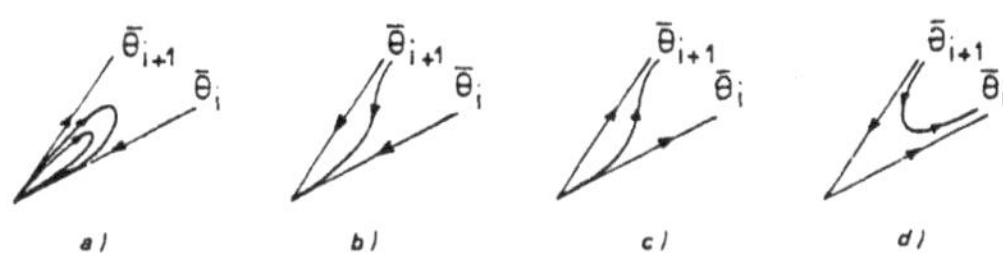

FIG. 5.29

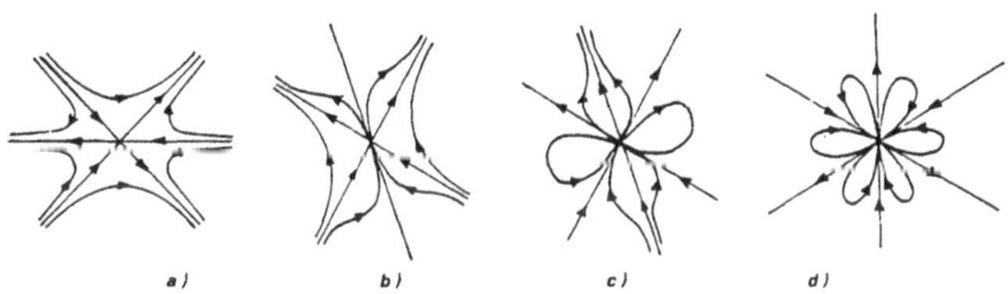

FIG. 5.30

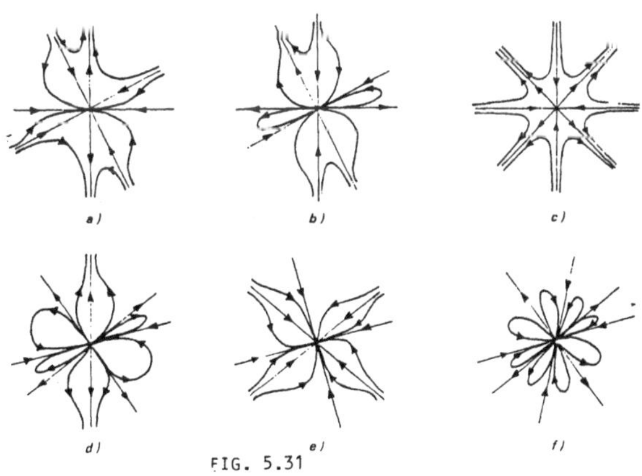

FIG. 5.31

is attractive (fig.5.28d). The behaviour of $\bar{\theta}_i$ is called "of R saddle type" (RS).

Considering now two consecutive invariant directions $\bar{\theta}_i$, $\bar{\theta}_{i+1}$, they may give rise to four different types of sectors : the *elliptic sector* (one $\bar{\theta}$ is AN, the other is RN, fig.5.29a), the *attractive parabolic sector* (one $\bar{\theta}$ is AS, the other is AN, fig.5.29b), the *repulsive parabolic sector* (one $\bar{\theta}$ is RS, the other RN, fig.5.29c), the *hyperbolic sector* (one $\bar{\theta}$ is AS, other RS, fig.5.29d).

The maximum number of elliptic, or hyperbolic, sectors is $(2m+2)$. The maximum number of parabolic sectors is $2m$ when m is even, and $(2m+2)$ when m is odd. When $m = 2$, and $Q(\theta) = 0$ has three different real roots, the phase portraits of fig.5.30 are possible. When $m = 3$, $Q(\theta)$ having four real zeros, the fig.5.31 are obtained, and the results of § 5.8 and 5.9 appear as particular cases if T^q, with $S_1 = S_2 = +1$, is considered.

When $Q(\theta) = 0$ has multiple roots $\bar{\theta}_k$, $P(\bar{\theta}_k) \neq 0$, one has $Q'(\bar{\theta}_k) = 0$, and in (5.84) the coefficients of ρ_n^a, $a > m-1$, define now the attractive, or repulsive character of $\bar{\theta}_k$. In particular, if a root $\bar{\theta}_k$ is double, $\bar{\theta}_k$ can be attractice on one side, and repulsive on the other side. Then the possible phase portraits can be obtained from fig.5.30, 5.31 when a pair of two consecutive invariant directions $\bar{\theta}_i$, $\bar{\theta}_{i+1}$ are such $\bar{\theta}_i \to \bar{\theta}_{i+1}$.

When $\phi = \pi$, $S_1 = S_2 = -1$, the phase portraits are obtained by consideration of T^2, which leads to $\phi = 0$.

5.11. BIFURCATION BY CROSSING THROUGH THE CRITICAL CASE $S = \exp(\pm j\phi)$.

5.11.1 CROSSING THROUGH A NON-EXCEPTIONAL CASE.

Consider the dynamical system defined by the recurrence, or map, T, with the real variables :

$$(5.85) \qquad r_{n+1} = f(r_n, s_n, \mu_{i'}) \ , \quad s_{n+1} = g(r_n, s_n, \mu_{i'}) \ , \quad n = 0,1,2,\ldots,$$

where $\mu_{i'}$ are real parameters, $i' = 1,2,\ldots,a$, which gives a non-exceptional critical case when $\mu_{i'} = \mu_{i'_o}$. Then the multipliers of the considered fixed point are $S_{1,2} = e^{\pm j\phi_o}$, $j = \sqrt{-1}$. It is supposed that the functions f and g are either analytical, or at least differentiable up to and including an order N, with respect the variables r,s (functions of class N) in a sufficiently large domain G of the phase plane (r,s), which contains the fixed point. It is also supposed that f and g are least twice differentiable functions with respect the parameters $\mu_{i'}$ in a neighbourhood of $\mu_{i'}$, $\mu_{i'_1} \leqslant \mu_{i'_o} \leqslant \mu_{i'_2}$. As indicated in the introduction, *the corresponding bifurcation in wrongly attributed to Hopf* (for the correct references cf the § 5.1). By means of a linear transformation, T is brought into a form, such that the fixed point is located at $O(r = s = 0)$ for $\mu_{i'} = \mu_{i'_o}$:

$$(5.86) \quad \begin{cases} r_{n+1} = r_n\cos\phi_o - s_n\sin\phi_o + R(r_n, s_n, \mu_{i'}) \\ s_{n+1} = r_n\sin\phi_o + s_n\cos\phi_o + S(r_n, s_n, \mu_{i'}) \end{cases} ,$$

R, S being two functions of class N, the lowest degree terms of which are higher than one. It is also supposed that the critical case of (5.86) is not an exceptional one, and that in the under-mentioned neighbourhood of the fixed point, there is neither homoclinic nor heteroclinic points. For $\mu_{i'} = \mu_{i'_o}$, the fixed point $(r = s = 0)$ is a complex focus of multiplicity ℓ, without chaotic behaviour in its neighbourhood, and sufficiently far from cycles points which do not result of the described bifurcation. The purpose of this paragraph is to show that, depending on the non linear terms of T, the bifurcation gives rise to a certain number of *invariant closed curves*, which are either *regular* in the sense that they are not made up of segments of invariant curves passing through pairs of cycles (saddle + node or focus), or *irregular* because they are constituted by such segments. This last case results from the first one, via a "boxes in files" bifurcations structure of the map of the circle onto itself, related to an invariant closed curve.

In order to simplify the results presentation, it is supposed that the fixed point stays in $O(r = s = 0)$ when the parameters $\mu_{i'}$ vary. With the complex linear transformation $r = (x+y)/2$, $s = j(y-x)/2$, $j = \sqrt{-1}$, (5.85) becomes :

$$(5.87)\quad \begin{cases} x_{n+1} = \left[1+\varepsilon(\mu_i)\right]e^{j\phi(\mu_i)}\, x_n + \sum_{i=2}^{\infty} P_i(x_n,y_n,\mu_i), & P_i = \sum_{p=0}^{p=i} C_{ip}(\mu_i)\, x_n^{i-p}\, y_n^p, \\ y_{n+1} = \left[1+\varepsilon(\mu_i)\right]e^{-j\phi(\mu_i)}\, y_n + \sum_{i=2}^{\infty} Q_i(x_n,y_n,\mu_i), & Q_i = \sum_{p=0}^{p=i} C^*_{ip}(\mu_i)\, x_n^{i}\, y_n^{i-p}, \end{cases}$$

when the functions f, g are analytical, c* and c being complex conjugate, $\varepsilon(\mu_{i'_o}) = 0$. A Cigala's transformation (5.48) (5.49) is applied to (5.87), and the coefficients α_p are chosen such that :

$$(1+\varepsilon)\alpha_p\left[(1+\varepsilon)^{m-1}\, e^{j\phi(m-2p)} - e^{j\phi}\right] = -C_{mp},$$

Considering all the possible values of the coefficients, after a sequence of Cigala's transformations for m = 2,4,6,...2ℓ, all the terms of even degree can be eliminated, and a sequence for m = 3,5,...,2ℓ+1 eliminates the terms of odd degree $m = 2\alpha+1$, $\alpha \leqslant \ell$, $p \neq \alpha$. The recurrence (5.87) is now written :

$$(5.88)\quad \begin{cases} \xi_{n+1} = \left[1+\varepsilon(\mu_i)\right]e^{j\phi(\mu_i)}\xi_n + \gamma_{31}(\mu_i)\xi_n^2\eta_n \\ \qquad + \gamma_{52}(\mu_i)\xi_n^3\eta_n^2 + \cdots + \gamma_{2\ell+1,\ell}(\mu_i)\xi_n^{\ell+1}\eta_n^{\ell} + \cdots, \\ \eta_{n+1} = \left[1+\varepsilon(\mu_i)\right]e^{-j\phi(\mu_i)}\eta_n + \gamma^*_{31}(\mu_i)\xi_n\eta_n^2 \\ \qquad + \gamma_{52}(\mu_i)\xi_n^2\eta_n^3 + \cdots + \gamma_{2\ell+1,\ell}(\mu_i)\xi_n^{\ell}\eta_n^{\ell+1} + \cdots, \end{cases}$$

where the dotted lines represent the terms having a degree higher than (2ℓ+1). Here $\gamma_{2\beta+1,\beta}(\mu_{i'_o}) = 0$, $\lim \gamma_{2\beta+1,\beta}(\mu_{i'}) \to 0$ when $\mu_{i'} \to \mu_{i'_o}$, for $\beta < \ell$, and $\gamma_{2\ell+1,\ell}(\mu_{i'_o}) \neq 0$, O(0;0) being a complex focus of multiplicity ℓ. Passing to polar coordinates $\xi = \rho e^{j\theta}$, $\eta = \rho e^{-j\theta}$, for sufficiently small values of ρ, on has :

$$(5.89)\quad \begin{cases} \rho_{n+1} = F(\rho_n) = \left[1+\varepsilon(\mu_i)\right]\rho_n + 0{,}5\left[G_1(\mu_i)\rho_n^3 + G_2(\mu_i)\rho_n^5 + \cdots + G_\ell(\mu_i)\rho_n^{2\ell+1}\right] + \cdots, \\ G_\beta(\mu_i) = \left[1+\varepsilon(\mu_i)\right]\left[e^{-j\phi(\mu_i)}\gamma_{2\beta+1,\beta}(\mu_i) + e^{j\phi(\mu_i)}\gamma^*_{2\beta+1,\beta}(\mu_i)\right], \\ G_\beta(\mu_{i_o}) = 0 \ \text{si}\ \beta < \ell \\ \theta_{n+1} = \theta_n + \phi(\mu_i) + \rho^2 g(\theta,\rho,\mu_i)\ ;\ g(\theta,\rho,\mu_{i_o}) = -|\gamma_{2\ell+1,\ell}|\rho^{2\ell-2}\sin(\phi+\tau_\ell) + \cdots \end{cases}$$

where the dotted line in the expression of ρ_{n+1} represents the terms having a degree higher than $(2\ell+2)$, depending on ρ and θ, and also on the terms having a degree $d \geq 2$ of smallness with the respect to μ_i (such as the products $\gamma_{2\beta+1,\beta}\ \gamma_{2\beta'+1,\beta'}$; $\beta, \beta' < \ell$). When the functions f, g of (5.85) are no longer analytical, but only of class $N \geq 2\ell+1$ inside a domain G containing the fixed point 0, the coefficients $\gamma_{2\alpha+1,\alpha}$ of (5.88) and G_α of (5.89) are again defined for $\alpha \leq \ell$.

In a sufficiently small neighbourhood U_ε of the fixed point 0, and for sufficiently small values of $|\mu_i - \mu_{i_0}|$, considering $F(\rho)$, the terms of a degree higher than $(2\ell+2)$, which are the coefficients of quasi periodical functions of θ, and thus of n, are disregarded. The reasons of this approximation lead to a bulky argumentation which is equivalent to that of the differential equations with periodical coefficients (cf. the §67 of [M 1]). Then the fixed points of the approximated first relation (5.89), $\rho_{n+1} = F(\rho_n)$, are given by the positive real roots of :

$$(5.90) \qquad \varepsilon(\mu_i) + 0{,}5\left[G_i(\mu_i)R + \cdots + G_\ell(\mu_i)R^\ell\right] = 0, \qquad R = \rho^2$$

These fixed points corresponds to s, $1 \leq s \leq \ell$, invariant closed curves by T, located inside U_ε if $(d\varepsilon/d\mu_i) \neq 0$. Let $\bar{\rho}_k$ be such a fixed point, and $S_k = (dF/d\rho)_{\bar{\rho}_k}$ its multiplier which defines also the stability of the corresponding invariant closed curve (Γ_k). Here S_k is always positive. If the S_k are different from +1, the results of the chapter 2, about one-dimensional recurrences permit to conclude to the alternance of repulsive $(S_k > 1)$ and attractive $(0 < S_k < 1)$ fixed points, and therefore to the alternance of repulsive and attractive invariant closed curves. The curve (Γ_{max}) corresponding to the largest $\bar{\rho}_k$ has its stability which is defined from the sign of G_ℓ of (5.54). Indeed this curve degenerates into the fixed point 0, surrounding all the other (Γ_k), and the stability of 0 is defined by G_ℓ. So (Γ_{max}) is attractive if $G_\ell < 0$, and repulsive if $G_\ell > 0$.

When $S_k = +1$, $\bar{\rho}_k$ is a multiple root corresponding to two, or more Γ_k merging into one another. The resulting invariant closed curve (Γ) is attractive on one side, repulsive on the other (semi-stable curve) when the multiplicity is even. It is either attractive, or repulsive when the multiplicity is odd. In the both cases, the alternance of stability of the (Γ_k) permits to define the property of (Γ). From these results one has [M14] [G30] :

(P5.9) *Let* T *be the recurrence* (5.85), *verifying the hypotheses of* § 5.11.1 *and such that when* $\mu_{i'} = \mu_{i'_o}$, T *becomes* (5.86) *the fixed point* 0 *of which is a complex focus of multiplicity* ℓ. *Crossing through the value* $\mu_{i'_o}$, *the point* 0 *gives rise to* s *invariant closed curves*, $0 \leqslant s \leqslant \ell$. *When* $S = (dF/d\rho)_{\bar{\rho}_k} \neq 1$, $F(\rho)$ *being given by* (5.89), $\bar{\rho}_k$ *being a root of* (5.90), *the invariant closed curves are alternately attractive and repulsive. The curve having the largest amplitude is attractive if* $G_\ell < 0$. *It is repulsive if* $G_\ell > 0$. *When* $S = 1$ *for a* $\bar{\rho}_k$, *the corresponding invariant closed curve is multiple.*

The bifurcation described by (P5.9) also occurs for a cycle of order k'. Then the consideration of the map $T^{k'}$ leads to the situation of k' fixed points for every one of which one has (P5.9). So every point of the considered cycle gives rise to s, $0 \leqslant s \leqslant \ell$, invariant closed curves by $T^{k'}$. Before the bifurcation the solution is periodical of order k'. After the bifurcation, s quasi-periodical solutions appear, when the bifurcation gives rise to regular closed curves, or s sub-harmonic solutions (period = multiple of k') appear in the case of regular closed curves. When $G_\ell > 0$, the basin of the attractive cycle of order k', appearing after the bifurcation, is bounded by the k' closed invariant curves having the smallest amplitude.

Considering every invariant closed curve $(\Gamma_{\bar{k}})$ generated by the bifurcation, the map restricted to such a curve is a diffeomorphism of the circle onto itself, when T is a diffeomorphism. Then, when the parameters $\mu_{i'}$ vary, bifurcations of *boxes in files type* occur on every curve $(\Gamma_{\bar{k}})$. The boundaries of one of these boxes correspond to a bifurcation having the type described by fig.5.18, and giving a cycle node-saddle for (5.85). The invariant closed curve $(\Gamma_{\bar{k}})$ is made up of segments of invariant curves passing through saddle cycles and ending at node, or focus, cycles of the same order. When $(\Gamma_{\bar{k}})$ is a regular closed curve one has the situation of an irrational rotation number.

5.11.2. CROSSING THROUGH AN EXCEPTIONAL CASE.

The variety of exceptional cases gives rise to the same diversity of such bifurcations. Depending on whether ϕ is constant, or a function $\phi(\mu_{i'})$ with $\phi(\mu_{i'_o}) = 2k\pi/q$, the bifurcation makes appear a saddle cycle of order q, or an invariant closed curve with ρ having q local maxima, and q local minima.

The following examples given in [B4] illustrate the first situation. Let T be the recurrence (5.85), such that in (5.86), ϕ does not depend on the parameters μ_i, $\phi = 2\pi/3$. Using the notations of (5.60) with $S_{1,2}(\mu) = (1+\mu)\exp(\pm j2\pi/3)$, by consideration of only one parameter μ, $\mu_o = 0$ giving the critical case, one has :

$$(5.91) \quad \begin{cases} \rho_{n+3} = (1+3\mu)\rho_n + 3|\gamma_{2,2}|\rho_n^2 \cos(3\theta_n + 2\pi/3 + \tau_{2\ell}) + \cdots \\ \theta_{n+3} = \theta_n - 3|\gamma_{2,2}|\rho_n \sin(3\theta_n + 2\pi/3 + \tau_{2\ell}) + \cdots \end{cases}$$

With $|\mu|$ having a sufficiently small value, for $\mu > 0$, or < 0, a saddle cycle of order 3 appears from the exceptional case $\mu = 0$. The three points of this cycle are located on three invariant directions θ, which are not the same for $\mu > 0$ and $\mu < 0$ (cf the invariant curves of T^3 fig. 5.32a). When ρ is sufficiently small, the coordinate ρ of the points of the cycle is $\rho \# \mu/|\gamma_{2,2}|$. Considering now the same problem for $\phi = \pi/2$, and using the notations of (5.70), a = 0, $S_{1,2}(\mu) = (1+\mu)\exp(\pm j\pi/2)$, one has :

$$(5.92) \quad \begin{cases} \rho_{n+4}^2 = (1+8\mu)\rho_n^2 + 8\rho_n^4\left[|\gamma_{3,1}|\cos(\pi/2 + \tau_1) + |\gamma_{3,3}|\cos(4\theta_n + \pi/2 + \tau_3)\right] + \cdots \\ \theta_{n+4} = \theta_n - 4\rho_n^2\left[|\gamma_{3,1}|\sin(\pi/2 + \tau_1) + |\gamma_{3,3}|\sin(4\theta_n + \pi/2 + \tau_3)\right] + \cdots \end{cases}$$

Here a saddle cycle of order 4 appears from the exceptional case $\mu = 0$ (fig. 5.32b).

When ϕ depends on the parameters $\mu_{i'}$, generally the exceptional case $\phi = 2k\pi/q$ gives rise to one, or more, invariant closed curves such that the coordinate ρ presents q local maxima and q local minima.

5.12. CRITICAL CASE WITH TWO MULTIPLIERS $S_1 = S_2 = +1$ WHEN THE LINEAR APPROXIMATION MATRIX IS NOT REDUCIBLE TO A DIAGONAL FORM.

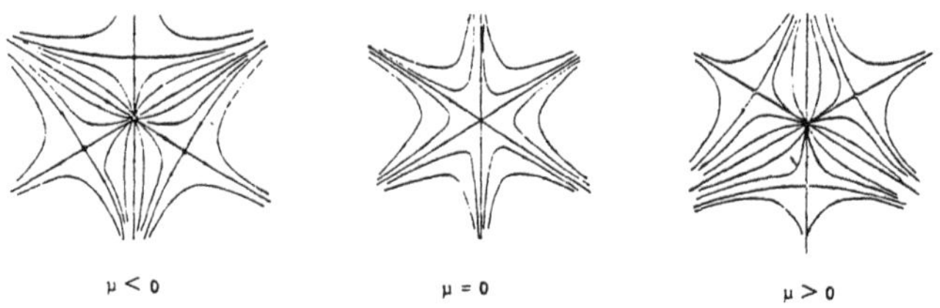

(a) q = 3 .

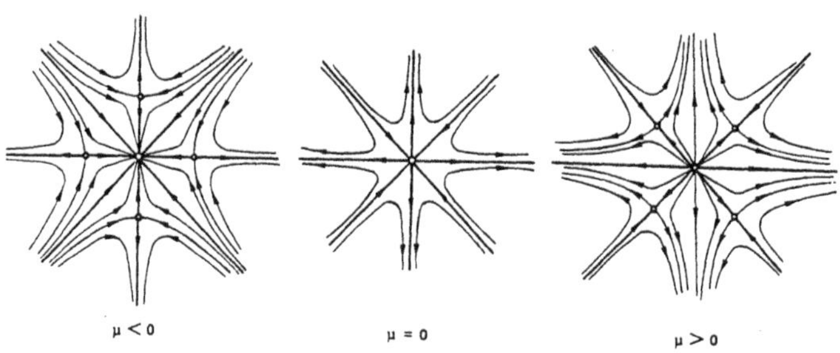

(b) q = 4 .

FIG. 5.32 - Bifurcation through an exceptional case . $\mu \equiv \lambda$ of the text .

The critical case with $S_{1,2} = \exp(\pm j\phi)$, $j = \sqrt{-1}$, $\phi = 0$, leads to $S_1 = S_2 = +1$, but in a situation for which a linear transformation gives a diagonal form to the matrix of the linear approximation written at the considered fixed point. In another situation this is not possible and such a transformation leads to the following form :

$$(5.93) \qquad r_{n+1} = r_n + s_n + R(r_n, s_n), \qquad s_{n+1} = s_n + S(r_n, s_n) \quad .$$

It is supposed here that R, S are two analytical functions, the expansions of which begin with terms of a degree higher than one. The origin 0 of the phase plane is a fixed point (r = s = 0). The results of this paragraph are given with the hypothesis of the absence of homoclinic, or heteroclinic points in the considered neighbourhoods of 0. The proofs and some details are omited. They can be found in [L18] [L19] and [G35].

Via transformations equivalent to Cigala's ones, a maximum of terms of degree $m \geqslant 2$ are eliminated in (5.93). Two final forms with real variables can be obtained. The one, which is used in the demonstrations, is called *normal form* :

$$(5.94) \qquad x_{n+1} = x_n + y_n + \sum_{j=k}^{M} \phi_{j0} x_n^j + \cdots, \quad y_{n+1} = y_n + \sum_{j=\ell}^{M} \psi_{j0} x^j + \cdots, \quad k, \ell \geq 2$$

where the dotted lines represent terms of degrees higher than M. Particular invariant curves, locally given by y = f(x), play a fundamental role. They are called *principal invariant curves*. In a sufficiently small neighbourhood of the fixed point 0, they are given by a fractional,or non-fractional, power series expansion :

$$(5.95) \qquad y = f(x) = p_o x^{\alpha} + \sum_{i=1}^{\infty} p_i x^{\alpha+i/2} \qquad , \alpha > 1 \quad ,$$

where α is either a positive fraction the denominator of which is 2, or a positive integer, i is a positive integer. It is an integer power series expansion when α is an integer, and $p_i = 0$ for i = odd number. Considering (5.94) and the functional equation of the invariant curves y = f(x), according to the respective values of the powers k and ℓ, three different cases appear :

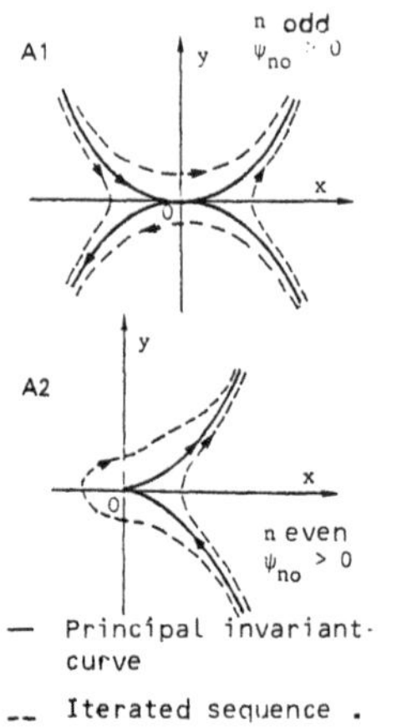

FIG. 5.33 - Critical case with two multipliers equal to one, when the matrix of the linear approximation is not reducible to a diagonal form . Case A .

Here m corresponds to the index k of the text, and n corresponds to ℓ .

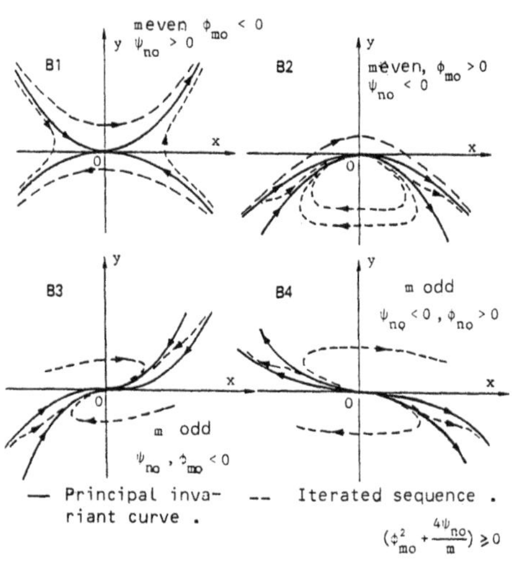

FIG.5.34 - Idem as fig. 5.33, but case B .

m corresponds to the index k, and n to ℓ.

- case A ($\ell < 2k-1$) : $\alpha = (\ell+1)/2$, $p_o^2 = 2\Psi_{\ell o}/(\ell+1)$
- case B ($\ell = 2k-1$) : $\alpha = k$, $p_o = (-\phi_{k_o} \pm \sqrt{\phi_{k_o}^2 + 4\Psi_{\ell o}/k})/2$
- case C ($\ell > 2k-1$) : $\alpha = \ell-k+1$ with $p_o = \Psi_{\ell_o}/[(\ell-k+1)\phi_{k_o}]$, or $\alpha=k$ with $p_o = -\phi_{k_o}$.

The coefficients p_i ($i = 1,2,...$) can be determinated from the coefficients of (5.93), and from the p_j ($j < i$). In the case B, it is supposed that $(2k+q)p_o \neq -(k+q)\phi_{k_o}$, q = positive integer, (if this does not occur cf. [L 18] and [G 35] p.430). When p_o is real, all the p_i are also real, and, except in the case A with ℓ = even integer, they are equal to zero for i = odd number ($f(x)$ is an integer power series expansion). The existence of the principal invariant curves can be proved in the cases A,B,C, the convergence of the series being a consequence of the results of [D 5]. So in the case A, ℓ being an even, or odd integer, and with $\Psi_{\ell_o} > 0$, two principal invariant curves exist (the p_o and p_i are real). They are given from the two determinations of p_o, $p_o^2 = 2\Psi_{\ell_o}/(\ell+1)$. In the case B, α is always an integer. The coefficient p_o is real, if $\phi_{k_o}^2 + 4\Psi_{\ell_o}/k \geqslant 0$, and a principal invariant curve corresponds to each of the two determinations of p_o, or a double curve when the equality occurs. In the case C, since α has two possible integer values, and p_o has two real determinations, two real principal invariant curves always exist. For the non-principal invariant curves, the equation of which is given in a parametric form, $x = F(t)$, $y = G(t)$, their existence results from the existence of the principal invariant curves.

The fig. 5.33, 5.34 and 5.35 give the different possible phase portraits in the cases A,B,C when the principal invariant curves are real. Each invariant curve can be divided into two segments, $x > 0$ and $x < 0$, except when it is not defined (case A, with ℓ = even integer). The arrow on an invariant curve indicates the positive direction of an iterated sequence when n increases : segment of type ω (resp α) when it tends toward (resp. moves off) 0 if $n \to \infty$. The determination of this direction is obtained remarking that, from (5.94), the difference $(y_{n+1}-y_n)$ has the same sign as $\Psi_{\ell_o}x^{\ell}$, on $y = f(x)$, when $|x|$ has a sufficiently small value. Then it is possible to obtain the conditions of instability of the fixed point 0 (0 is repulsive) from the existence of at least a segment α.

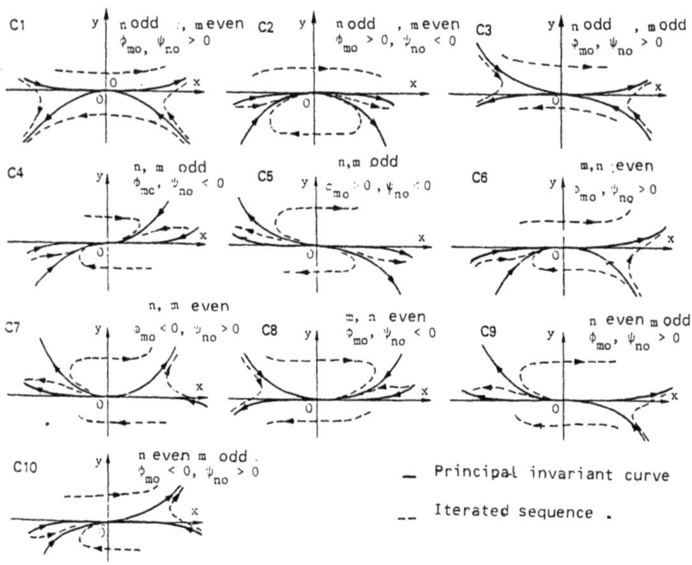

FIG. 5.35

Critical case with two multipliers equal to one, when the matrix of the linear approximation is not reducible to a diagonal form . Case C .

Here m corresponds to the index k of the text, and n corresponds to the index ℓ.

(P5.10) *The fixed point* 0 *of* (5.94) *is repulsive when one of the following condition is satisfied* : (1) ℓ *is even* ; (2) ℓ *is odd with* $\Psi_{\ell_o} > 0$; (3) ℓ *is odd*, k *is even*, $\Psi_{\ell_o} < 0$, $\ell > 2k-1$; (4) ℓ *and* k *are odd*, $\Psi_{\ell_o} < 0$, $\Phi_{k_o} > 0$, $\ell > 2k-1$; (5) $\ell = 2k-1$, $\Psi_{\ell_o} < 0$, $\Phi_{k_o} > 0$, $\Phi_{k_o}^2 + 4\Psi_{\ell_o}/k \geqslant 0$. The remaining cases correspond either to the absence of real principal invariant curves, or to real principal invariant curves made up of segments of type ω.

Using a method based on an associated differential equation [L19] [G35], and the corresponding Ljapunov's results [L20], one has :

(P5.11) *The fixed point* 0 *of* (5.94) *is attractive when one of the following conditions occurs :* (a) $k = \ell$, ℓ *is odd*, $\Phi_{k_o} < \Psi_{\ell_o}$; (b) $k < \ell$, k *and* ℓ *being odd*, $\Psi_{\ell_o} < 0$, $\Phi_{k_o} < 0$.

The fixed point 0 *is repulsive when one of the following conditions occurs* : (c) $k = \ell$, ℓ *is odd*, $\Psi_{\ell_o} < 0$ $\Phi_{k_o} > \Psi_{\ell_o}$; (d) $k < \ell$, k *and* ℓ *being odd*, $\Psi_{\ell_o} < 0$ $\Psi_{k_o} > 0$; (e) $k > \ell$, ℓ *is odd*, $\Psi_{\ell_o} < 0$.

When the principal invariant curves are not real, the phase portrait is that of a *complex focus* which can present an important dissymmetry (in the neighbourhood of y = 0, the variation of y is very different for y > 0, and y < 0). In the conservative case it can be that of a center. The *complex focus* is *attractive* with (P5.11a), or (P5.11b) when $\ell = 2k-1$ and $\Upsilon = \Phi_{k_o}^2 + 4\Psi_{\ell_o}/k < 0$. It is *repulsive* in (P5.11c), or (P5.11d) with $\ell = 2k-1$ and $\Upsilon < 0$. The other cases of (P5.11) correspond to *complex nodes*. The invariant curves of complex focus, or center type, can be represented analytically from the *periodical transcendental functions* Csθ, Snθ, *of Ljapunov* (cf. [G35] p.436 and [L18] [L19]).

The other remaining cases may be solved from the use of the associated differential equation ([G35] p.432-436).

5.13. BIFURCATION BY CROSSING THE CRITICAL CASE OF THE § 5.12.

The considered recurrence T is (5.1) for which the functions $f(x,y,\lambda)$ $g(x,y,\lambda)$ are supposed to be analytical with respect to the real variables x,y, and continuous with respect to the parameter λ. Using transformations equivalent to the Cigala's ones, it is supposed that T becomes the recurrence (5.94) for $\lambda = \lambda_o$. Let $\varepsilon = \lambda - \lambda_o$ be the new parameter, and T be the recurrence :

$$(5.96) \quad \begin{cases} x_{n+1} = a(\varepsilon)x_n + b(\varepsilon)y_n + X(x_n, y_n, \varepsilon) + \phi(\varepsilon)x_n^k + \phi'(\varepsilon)x_n^{k+1} + \cdots \\ y_{n+1} = \varepsilon x_n + d(\varepsilon)y_n + Y(x_n, y_n, \varepsilon) + \psi(\varepsilon)x_n^{\ell} + \psi'(\varepsilon)x_n^{\ell+1} + \cdots, \end{cases}$$

Here X (resp Y) is a polynomial of the degree k+1 (resp ℓ+1) which does not contain neither linear terms in x,y, nor terms in x^k, x^{k+1} (resp x^{ℓ}, $x^{\ell+1}$). It is also supposed that $X(x,y,0) = Y(x,y,0) = 0$, $a(0) = b(0) = d(0) = 1$, $\phi(0) = \phi_{k_o} \neq 0$, $\phi'(0) = \phi_{k+1,o}$, $\Psi(0) = \Psi_{\ell_o} \neq 0$, $\Psi'(0) = \Psi_{\ell+1,o}$, and that the dotted lines corresponds to terms of degree highers than k+1, for the first relation, and to terms of degrees higher than ℓ+1, for the second one. The results [G35] of this paragraph are obtained from the determination of the fixed points of (5.96), and their multipliers for $\varepsilon > 0$, $\varepsilon < 0$, and also from the results of § 5.12. They are limited to $k \leqslant 3$, $\ell \leqslant 7$, without claiming that all the possible cases are described even with these limitations. The existence of invariant closed curves, appearing from the fixed point 0 (x = y = 0) for $\varepsilon = 0$, is numerically established taking the precautions of use which are implied by such a method. The analytical proof of existence appears to be on a too high level of difficulty in the general case. The case for which T is conservative when $\varepsilon = 0$ is not considered here.

The crossing through the cases A,B,C of the § 5.12 (existence of two real principal invariant curves), and through the case for which such curves do not exist, are considered. Only the simplest situations, limited to the merging of a maximum of three singularities for $\varepsilon = 0$, are given. The following notations are used : AN = attractive node, RN = repulsive node, AF = attractive focus, RF = repulsive focus, Γ_a = attractive invariant closed curve surrounding a repulsive focus RF, Γ_r = repulsive invariant closed curve surrounding an attractive focus AF. The saddles, nodes and focus are supposed to have the type one. The details proving the described bifurcations are not given here for reasons of analytical bulk. However as exercise the reader can obtained the proof by considering the simplest case of :

$$(5.97) \quad x_{n+1} = (1\pm\varepsilon^a)x_n+y_n+\phi_{ko}x_n^k \ ; \ y_{n+1} = \pm\varepsilon^b x_n+y_n+\Psi_{\ell_o}x_n^{\ell}, \quad a,b > 0,$$

a,b are integers, or fractions. With such an example, the determination of the fixed points and their nature is not a difficult problem.

5.13.1. CROSSING THROUGH CRITICAL CASES WITH REAL PRINCIPAL INVARIANT CURVES.

The corresponding bifurcations are characterized by the diagrams (5.98) - (5.103) and are related to the cases marked a),b),...m) which are defined by k,ℓ, and the signs of ϕ, Ψ. The first diagrams are of the type :

(5.98)
- AN (resp RN) + saddle ⇆ saddle + AN (resp RN) ; or nothing
- AF (resp RF) + saddle ⇆ saddle + AF (resp RF) ; or nothing .

Here "nothing" indicates the absence of fixed point after the crossing through $\varepsilon = 0$. The bifurcations (5.98) (fig. 5.36a) occur for :

- (a) $\ell < 2k-1$, ℓ = even integer, with the fig. 5.33 A2 when $\varepsilon = 0$. The simplest example is obtained with (5.97), $\ell = k = 2$.

The second type of bifurcation diagrams (fig. 5.36.b) is :

(5.99)
- Saddle + AF (or RF) + saddle ⇆ saddle
- Saddle + AN (or RN) + saddle ⇆ saddle .

It occurs for the every of the four following cases :

- (b) $\ell < 2k-1$, ℓ = odd integer, $\Psi_{\ell_o} > 0$ (fig. 5.33 A1 for $\varepsilon = 0$).
- (c) $\ell = 2k-1$, $\gamma > 0$, k = even integer, $\Psi_{\ell_o} > 0$ (fig. 5.34 B1 for $\varepsilon = 0$).
- (d) $\ell > 2k-1$, ℓ = odd integer, k = even integer, $\phi_{k_o} > 0$, $\Psi_{\ell_o} > 0$ (fig. 5.35 C1, $\varepsilon = 0$).
- (e) $\ell > 2k-1$, ℓ and k = odd integers, $\phi_{k_o} > 0$, $\Psi_{\ell_o} > 0$ (fig. 5.35 C3, $\varepsilon = 0$).

The third type of bifurcation diagram (fig. 5.36 c and d) is :

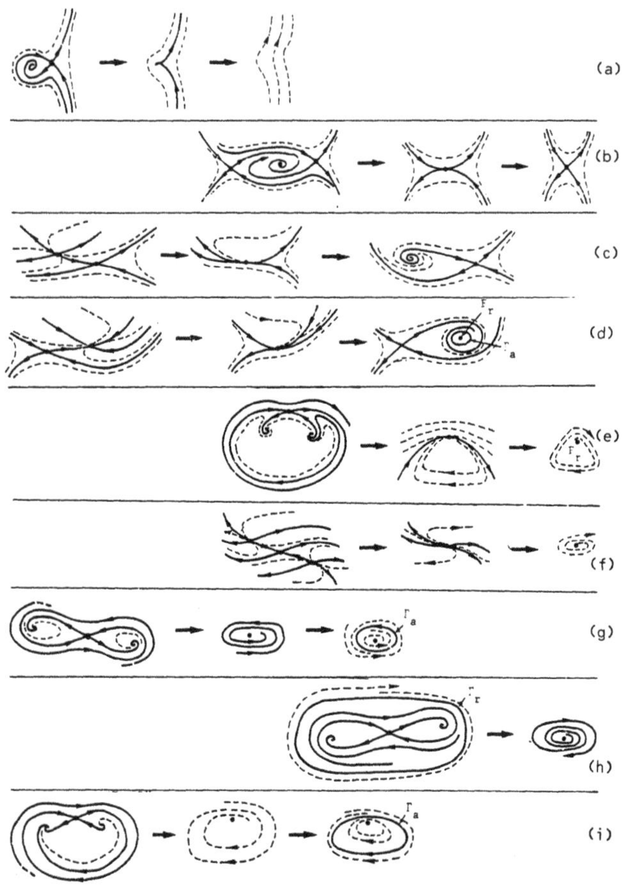

FIG. 5.36

Simplest bifurcations by passing through the critical case with two multipliers equal to one, when the matrix of the linear approximation is not reducible to a diagonal form .

(5.100) Node + saddle → saddle + Focus ; or → saddle + Focus + Γ_r (or Γ_a), the node is either attractive or repulsive, it is the same for the focus. It occurs for each of the four following cases.

- (f) $\ell > 2k-1$, ℓ and k = even integers, $\phi_{k_o} < 0$ or > 0, $\Psi_{\ell_o} > 0$ (fig. 5.35 C7, $\varepsilon = 0$).

- (g) $\ell > 2k-1$, ℓ = even integer, k = odd one, ϕ_{k_o} and $\Psi_{\ell_o} > 0$ (fig. 5.35 C9, $\varepsilon = 0$).

- (h) $\ell > 2k-1$, ℓ and k are even with $\phi_{k_o} > 0$, $\Psi_{\ell_o} > 0$, or $\phi_{k_o} < 0$, $\Psi_{\ell_o} < 0$ (fig. 5.35 C6, C8, $\varepsilon = 0$).

- (i) $\ell > 2k-1$, ℓ is even, k is odd with $\phi_{k_o} < 0$, $\Psi_{\ell_o} > 0$ (fig. 5.35 C10, $\varepsilon = 0$).

The 4th type of bifurcation diagram (fig. 5.36 e) is :

$$(5.101)\quad \begin{cases} \text{RN + saddle + AN} \to \text{RF (or AF) ; or} \to \text{RF} + \Gamma_a \text{ ; or} \to \text{AF} + \Gamma_r \\ \text{RF + saddle + AF} \to \text{RF (or AF).} \end{cases}$$

It occurs for the following cases :

- (j) $\ell = 2k-1$, $\gamma \geqslant 0$, k is even, $\Psi_{\ell_o} < 0$ (fig. 5.34 B2), $\varepsilon = 0$). When $\gamma = 0$, $\phi_{k+1,o}$, $\Psi_{\ell+1,o}$, respectively coefficients of x^{k+1} in x_{n+1}, and of $x^{\ell+1}$ in y_{n+1}, are necessary for the determination of the phase portrait.
- (k) $\ell > 2k-1$, k is even, ℓ is odd, $\Psi_{\ell_o} < 0$ (fig. 5.35 C2, $\varepsilon = 0$).

The 5th type of bifurcation diagram (fig. 5.36 f) is :

$$(5.102)\quad \begin{cases} \text{RN + saddle + RN} \to \text{RF ; or} \to \text{AF} + \Gamma_r \text{ ; or} \to \text{RN} \\ \text{AN + saddle + AN} \to \text{AF ; or} \to \text{RF} + \Gamma_a \text{ ; or} \to \text{AN} \end{cases}$$

It occurs for the following cases

- (ℓ) $\ell > 2k-1$, ℓ and k are odd, ϕ_{k_o} and $\Psi_{\ell_o} < 0$, or $\phi_{k_o} > 0$, $\Psi_{\ell_o} < 0$ (fig. 5.35 C4 and C5, $\varepsilon = 0$).

- (m) $\ell = 2k-1$, k is odd, ϕ_{k_o} and $\Psi_{\ell_o} < 0$, or $\phi_{k_o} > 0$, $\Psi_{\ell_o} < 0$ (fig. 5.34 B3 and B4, $\varepsilon = 0$).

5.13.2. CROSSING THROUGH CRITICAL CASES WITHOUT REAL PRINCIPAL INVARIANT CURVES.

The details about the corresponding situations for $\varepsilon = 0$ are given in the appendix D of [G35]. When $\varepsilon = 0$, these situations lead to invariant curves of the complex focus type in the dissipative case, and of the ordinary center type, or of the "Birkhoff's unstable center" type (cf.[G35]). The so obtained focus, or center, for $\varepsilon = 0$ is either of the *symmetric type* (when the polar angle θ varies from 0 to π, the maximum variation of y, Δy_1, has an order equivalent to that of the maximum variation of y, Δy_2, when θ varies from π to 2π), or of the *dissymmetric type* (when θ varies from 0 to 2π, Δy_1 and Δy_2 are quantities having very different orders).

The 6th type of bifurcation diagram (fig. 5.36 g,h) is :

$$(5.103)\quad \begin{cases} \text{AF (resp RF) + saddle + AF (resp RF)} \to \text{AF (resp RF) ; or} \to \\ \text{RF} + \Gamma_a (\text{resp AF} + \Gamma_r) . \end{cases}$$

It occurs in the following *symmetric cases* :

- (n) $\ell < 2k-1$, $\Psi_{\ell_o} < 0$, ℓ is even, $k \leqslant \ell$.
- (o) $\ell = 2k-1$, $\phi^2_{k_o} + 4\Psi_{\ell_o}/k < 0$, $\Psi_{\ell_o} < 0$, k is odd.

These two cases lead also to a variant of (5.103) :

$$(5.104)\quad [\text{AF (resp RF) + saddle + AF (resp RF)}] + \Gamma_r\ (\text{resp } \Gamma_a) \to \text{RF (resp AF)}$$

where Γ_r (resp Γ_a) surrounds the three fixed points which are indicated between square brackets.

The 7th type of bifurcation diagram (fig. 5.36 i) is :

$$(5.105)\quad \text{AF + saddle + RF} \to \text{RF} + \Gamma_a \text{ ; or} \to \text{AF} + \Gamma_r \text{ ; or} \to \text{RF (resp AF)}.$$

It occurs in the following *dissymetric cases* :

- (p) $\ell = 2k-1$, $\phi^2_{k_o} + 4\Psi_{\ell_o}/k < 0$, $\Psi_{\ell_o} < 0$, ℓ is odd and k is even.
- (q) $\ell < 2k-1$, $k < \ell$, $\Psi_{\ell_o} < 0$, ℓ is odd.

Remarks

The bifurcations (5.98) - (5.105) concern the simplest cases with two or three fixed points which merge when $\varepsilon = 0$. In the general case a larger number of these points are concerned by these bifurcations.

In order to obtain the zero order approximation of the invariant closed curve Γ_a, or Γ_r, in (5.105), it is better to use the form $x = \rho cs\theta$, $y = -\rho^{\gamma} Sn\theta$, $\gamma = (\ell+1)/2$ (case (p)), as the first term of the series expansion of Γ_a, or Γ_r. Here $Cs\theta$, $Sn\theta$ are the Ljapunov's periodical functions (cf. [L20] and [G35] p.436).

All the bifurcations figures 5.36 are obtained with the hypothesis of absence of homoclinic, or heteroclinic, points. The presence of such points gives an "oscillation" of the considered invariant curves.

5.14. BIRTH OF INVARIANT CLOSED CURVES FROM A LINEAR CONSERVATIVE CASE PERTURBED BY SMALL NON LINEAR DISSIPATIVE TERMS.

The question is that of a global bifurcation, which occurs in a reccurence T, by crossing through parameters values giving a center phase portrait made up of ellipses. This phase portrait is structurally unstable, and after the crossing it can give rise to a certain number of invariant closed curves which are alternately attractive and repulsive. The first study was made in 1973 [G37] [M20] [G35], and the simplest way is to consider the reccurence (or map) T with the form :

$$(5.106) \quad \begin{cases} r_{n+1} = r_n \cos\phi - s_n \sin\phi + \alpha\, p(r_n, s_n, \alpha) \\ s_{n+1} = r_n \sin\phi + s_n \cos\phi + \alpha\, q(r_n, s_n, \alpha), \quad p(0,0,\alpha) = q(0,0,\alpha) = 0 \end{cases}$$

where α, and ϕ are two parameters, $\phi \neq 2k\pi$, $k = 0, \pm 1, \pm 2, \ldots$, p and q are two functions, each one having an expansion with respect to the real variables r, s, which will be limited to a polynomial of degree N, N having a sufficiently large value. It is supposed also that p and q are sufficiently smooth with respect to the parameter α. When $\alpha = 0$, T is a linear conservative recurrence . The corresponding phase portrait is of the center type, and is constituted by the set of the circles (Γ_o) having 0 (r = s = 0) as the centre. In the more general case a linear transformation permits to obtain (5.106).

For $\alpha = \pm\varepsilon$, $\varepsilon > 0$ having a sufficiently small value, ϕ being different from certain values commensurable with 2π, generally the center phase portrait disappears. Numerical simulations, associated with adequate accuracy checks and redundancy, make appear the existence of isolated invariant closed curves (Γ) which are alternately attractive and repulsive. The analytical existence of such invariant closed curves is not proved here. The purpose of this paragraph is only to obtain an approximate formulation of the curves Γ, by using a method equivalent to that of the "Poincare's small parameter" in the non-linear differential equations theory, and by considering some circles (Γ_o) as generating solutions. It is supposed that ε is sufficiently small in order to obtain the non-existence of homoclinic, or heteroclinic points, in the considered regions of the phase plane (r,s).

After the linear transformation $r = (x+y)/2$, $s = j(y-x)/2$, $j = \sqrt{-1}$, (5.106) becomes a recurrence with complex conjugate variables :

$$(5.107)\quad \begin{cases} x_{n+1} = e^{j\phi} x_n + \alpha R(x_n,y_n,\alpha) = e^{j\phi} x_n + \alpha \sum_{i=1}^{N} R_i(x_n,y_n,\alpha) \\ y_{n+1} = e^{-j\phi} y_n + \alpha R^*(x_n,y_n,\alpha) = e^{-j\phi} y_n + \alpha \sum_{i=1}^{N} R_i^*(x_n,y_n,\alpha) \\ R_i(x,y,\alpha) = \sum_{p=0}^{p=i} C_{ip}(\alpha) x^{i-p} y^p ,\ R_i^*(x,y,\alpha) = \sum_{p=0}^{p=i} C_{ip}^*(\alpha) x^p y^{i-p}, \end{cases}$$

C_{ip} and C_{ip}^* being complex conjugate.
With $x = \rho e^{j\theta}$, $y = \rho e^{-j\theta}$, (5.107) has the polar form :

$$(5.108)\begin{cases} \rho_{n+1} = \rho_n + 0{,}5\alpha\left[R(\rho_n,\theta_n,\alpha)e^{-j(\phi+\theta_n)} + R^*(\rho_n,\theta_n,\alpha)e^{j(\phi+\theta_n)}\right] + \cdots \\ \theta_{n+1} = \theta_n + \phi + \arg\left[1 + \alpha\,\rho_n^{-1}\, e^{-j(\theta_n+\phi)}\, R(\rho_n,\theta_n,\alpha)\right], \end{cases}$$

where the dotted line indicates the powers of α higher than one. Let θ_o, ρ_o be an initial condition for $n = 0$; $\rho = f(\theta;\theta_o,\rho_o,\alpha)$, the equation of the invariant curve passing through (θ_o,ρ_o). Without a loss of generality it is possible to choose $\theta_o = 0$. Supposing that $|\rho-\rho_o|$ and $|\alpha|$ have sufficiently small values, it is possible formally to write :

$$\rho = \rho_0 + \alpha\psi(\theta;\rho_0,\alpha) = \rho_0 + \alpha\psi(\theta;\rho_0;0) + \alpha^2\psi'_\alpha(\theta;\rho_0,0) + \cdots \tag{5.109}$$

$\Psi(\theta,\rho_o,0)$ being a sufficiently smooth periodical function of θ. If an invariant closed curve (Γ) exists when $\alpha \neq 0$, it is generated from the circle $\rho = \rho_o$, invariant closed curve for $\alpha = 0$. The equation of (Γ) is (5.109), and $\Psi(2\pi,\rho_o,\alpha) = 0$. A solution $\rho = f(\theta;0,\rho_o,\alpha)$ must be such that $\lim \rho = \rho_o$ when $\alpha \to 0$, and ρ_o appears as a generating solution in the sense of the Poincaré's method of the small parameter.

By consideration of (5.109) and (5.108), writing the invariance of (Γ), and identifying the terms coefficients of α, lead to the difference equation :

$$\begin{aligned}\psi(\theta_n+\phi,\rho_0,0) - \psi(\theta_n,\rho_0,0) = 0{,}5\Big[R(\theta_n,\rho_0,0)e^{-j(\phi+\theta_n)} \\ + R^*(\theta_n,\rho_0,0)e^{j(\phi+\theta_n)}\Big] = F(\theta_n,\rho_0,\phi),\end{aligned} \tag{5.110}$$

where the right hand side is made up of terms having the form :

$$0{,}5\rho_0^i \left\{ \sum_{p=0}^{i} (C_{ip}e^{j[\theta(i-2p-1)-\phi]} + C_{ip}^* e^{-j[\theta(i-2p-1)-\phi]} \right\} \tag{5.111}$$

The periodicity of Ψ implies that the secular terms in the sense of the above mentioned Poincaré's method are equal to zero, i.e. $\int_o^{2\pi} F(\theta_n,\rho_o\phi)d\theta_n = 0$ or :

$$P = \sum_{i=1}^{N} \rho_0^i \Big[C_{i,(i-1)/2}\, e^{-j\phi} + C^*_{i,(i-1)/2}\, e^{+j\phi}\Big] = 0 \tag{5.112}$$

So P = 0 corresponds to the terms i-2p-1 = 0. Then the solution of (5.110) is :

$$\left\{\begin{aligned} &\psi = 0{,}5 \sum_{i=1}^{N} \rho_0^i \sum_{p=0}^{i} \overline{C}_{ip} e^{j\theta(i-2p-1)} + \overline{C}^*_{ip} e^{-j\theta(i-2p-1)} \\ &i - 2p - 1 \neq 0,\ \overline{C}_{ip} = e^{-j\phi(i-2p)}\Big[1-e^{-j\phi(i-2p-1)}\Big]^{-1} \cdot C_{ip}\end{aligned}\right. \tag{5.113}$$

From (5.113) it appears that some values of ϕ, which are close to commensurable with 2π values, such as $\phi(i-2p-1) = 2h\pi$, $h > 0$, or $\leqslant 0$, being an integer, give rise to *small denominators* in the solution. It is supposed that angle ϕ does not have these values. The general solution of (5.110) without right hand side is an arbitrary periodical function of period ϕ. In order to satisfy the periodicity condition of Ψ, this arbitrary periodical function is chosen equal to a constant A, and ρ_o is chosen such as $P = 0$. Then every solution ρ_o^* of the equation $P = 0$, $\partial P/\partial\rho_o \neq 0$, corresponds to a circle which is in (5.109) the zero order approximation ($\alpha = 0$) of one of the invariant closed curves (Γ) of (5.106) when $\alpha \neq 0$. The constant A is determinated in order to satisfy $\Psi(2\pi,\rho_o,0) = 0$, and Ψ gives the first order approximation of (Γ). The following approximation of (Γ) is obtained by the same way, after identification of the terms coefficient of α^2. The convergence of (5.109) is not analytically proved. It is conjectured that it occurs for a whole class of functions p and q, corresponding to polynomials of degree $N \leqslant 5$, because the results obtained from the first terms of the expansion (5.109) are compatible with numerical checks.

When ϕ is far from values giving a small divisor in the solution expression, the corrective term $\alpha\Psi(\theta;\rho_o,0)$ of the first approximation is small, if $|\alpha|$ is small. A curve (Γ) is the more close to the circle defined by ρ_o^* as $|\alpha|$ is small. When ϕ approaches the values of an interval I, $\phi_c-\delta\phi_1 < \phi < \phi_c+\delta\phi_2$, $\phi_c(i-2p-1) = 2h\pi$, *the corrective term $\alpha\Psi(\theta;\rho_o,0)$ is no longer small, and the closed curve (Γ) is such that ρ presents* t *minima, and* t *maxima,* $t = i-2p-1$, *when θ rotates through an angle equal to* 2π. *When $\phi \in$ I, the effect of the small divisor is such that the invariant closed curve exists no longer. This occurs in presence of at least a saddle cycle and a focus (or node) cycle of order* t. *The larger $|\alpha|$ is, the larger* I *is.*

In order to determinate the stability of the invariant curves (Γ) (the zero order approximation of which is given by (5.112)), the variational equation, verified by the variation $\Delta\rho$ with respect to a given (Γ), is written. To this end, the expression :

$$\rho = \rho_o + \alpha\Psi(\theta;\rho_o,0) + \Delta\rho + \ldots$$

is used into (5.108), and the variational equation is obtained from the linear terms with respect to $\Delta\rho$:

$$\Delta\rho_{n+1} = \Delta\rho_n + 0,5\alpha\left[\left(\frac{\partial P}{\partial\rho_0}\right)^{*}+G(\theta_n)\right]\Delta\rho\ , \tag{5.114}$$

where $(\partial P/\partial\rho_o)^*$ is defined for the considered root ρ_o^* of $P = 0$. $G(\theta_n)$ is a function of θ_n made up of terms of the form $A_k\cos(k\theta_n+\gamma)$, k=integer≠0. With respect to the index n of iteration, G is a almost periodic function G_n, with :

$$\lim_{h\to\infty} \frac{1}{h} \sum_{n=1}^{h} G_n = 0$$

Then a Halanay's theorem (cf. theorem 3 [H14] p.86) permits to formulate the following proposition.

(P5.12) *The invariant closed curve* (Γ), *the zero order approximation of which is the circle with the radius* ρ_o^* *solution of* (5.112), *is attractive when* $(\partial P/\partial\rho_o)^* < 0$, *and repulsive when* $(\partial P/\partial\rho_o)^* > 0$.

The case $(\partial P/\partial\rho_o)^* = 0$ corresponds to the existence of a multiple root of (5.112), i.e. to the merging of several curves (Γ). A critical case in the Liapunov sense occurs. Then the determination of the stability requires the knowledge of non linear terms in (5.114). The case $(\partial P/\partial\rho\ )\equiv 0$, is related to $P \equiv 0$, and it occurs when the equation (5.106) is a non linear conservative map.

5.15. BIRTH OF INVARIANT CLOSED CURVES IN SOME OTHER CASES OF GLOBAL BIFURCATIONS.

Equivalent situations to those of limit cycles of two-dimensional autonomous ordinary differential equations [A 3] also occur for the invariant closed curves of autonomous two-dimensional recurrences. They belong to the following cases :

- (a) *birth of m invariant closed curves* (Γ) *from a loop constituted by a* α*-branch and* ω*-branch of a saddle fixed point* (§ 5.4). The fig.5.37a

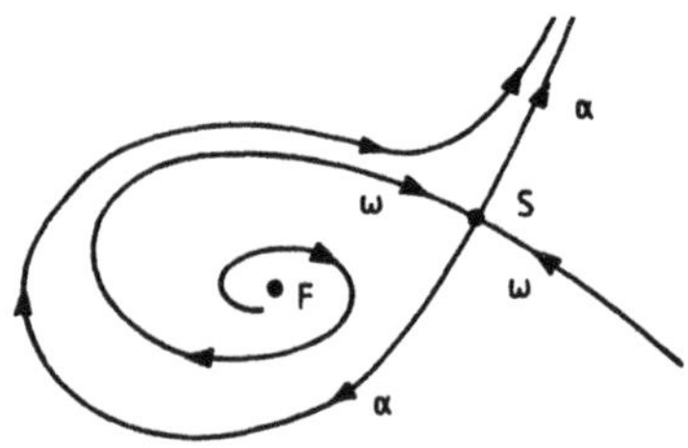

$\lambda < \lambda_0$

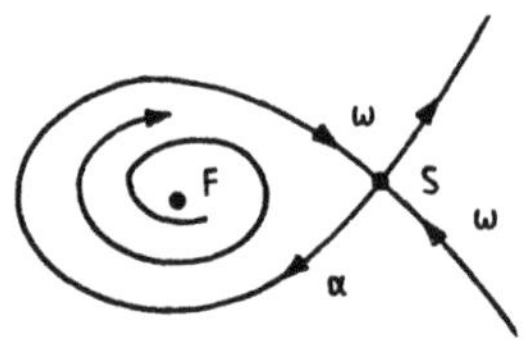

$\lambda = \lambda_0$
Bifurcation value

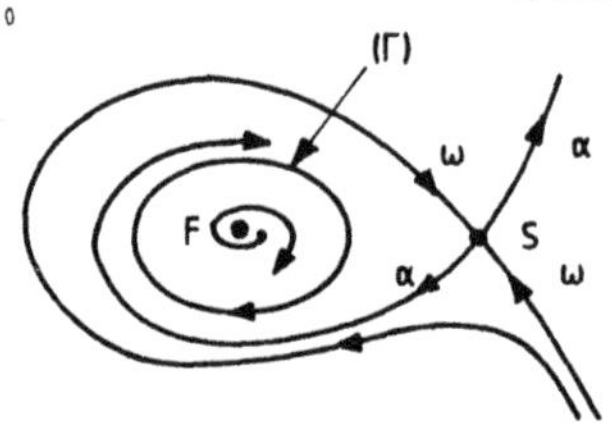

$\lambda > \lambda_0$

Birth of an invariant closed curve (Γ) from the saddle loop obtained for $\lambda = \lambda_0$

(a) m=1

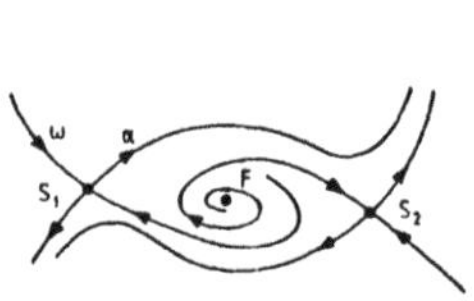

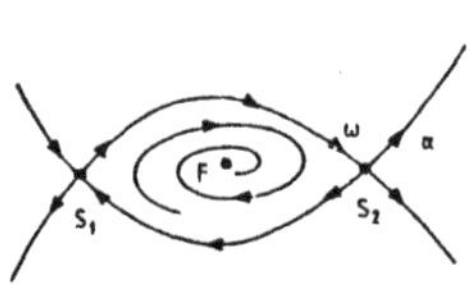

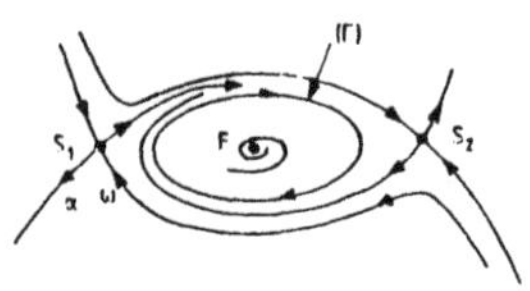

(b) Variant of the case of fig. (a)
m = 1, k = 2

FIG. 5.37

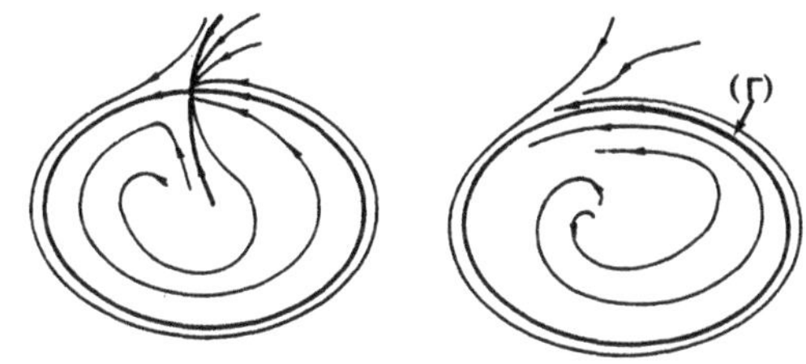

$\lambda = \lambda_0$ $\lambda > \lambda_0$

FIG. 5.38

Birth of an invariant closed curve from the loop of a node-saddle fixed point obtained for $\lambda = \lambda_0$. m = 1 .

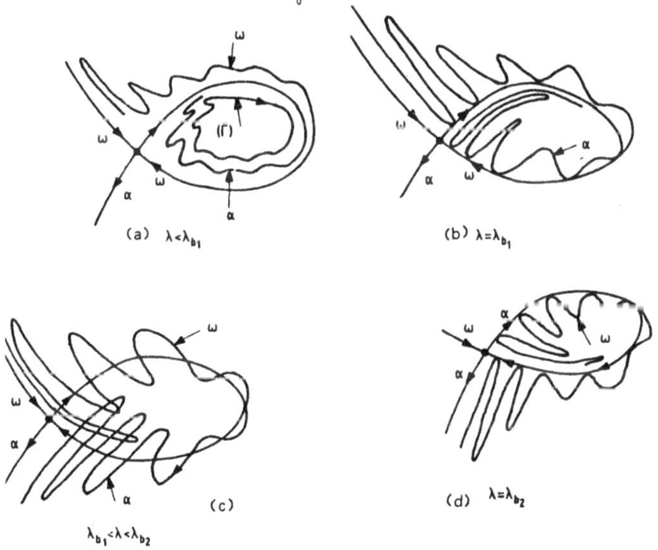

FIG. 5.39

Vanishing of an invariant closed curve after a homoclinic tangency (fig.b) . Fig. c gives the homoclinic situation . In fig. d a second bifurcation by homoclinic tangency occurs .

represents this bifurcation in the simplest case $m = 1$. A variant of this bifurcation is the birth of m invariant closed curves from a "critical" invariant closed curve, made up of the merging of k α-branches and k ω-branches of a saddle cycle of order k, or of k saddles fixed points (cf. fig. 5.37b for $m = 1$).

- (b) *birth of m invariant closed curves from a loop constituted by the invariant curve passing through a saddle-node fixed point* (fig. 5.38); and equivalent situation for a saddle-node cycle.

- (c) *birth of m invariant closed curves from a complex invariant closed curve* (Γ) *of multiplicity* $p \geqslant m$. The Ljapunov exponents of (Γ) (cf. the extension of this notion to recurrences in [C 8], or [G18] p.283) correspond to a critical case.

An analytical study of such bifurcations presents a high level of difficulties, till now it seems that only numerical methods permits to make appear these bifurcations. Another problem makes particularly hard such a study in the cases (a) (b). It is related to the fact that the more often the bifurcations of fig. 5.37 - 38 do not occur exactly. This is due to the effect of non-linearities, which produces homoclinic points from α and ω-branches of the saddles, for parameter values $\lambda_{b_1} \leqslant \lambda \leqslant \lambda_{b_2}$. Here λ_{b_1}, λ_{b_2} corresponds to two values giving two diffent situations of homoclinic tangencies (§ 5.4) as indicated fig. 5.39, and the limit existence of the invariant closed curve (Γ) corresponds to $\lambda = \lambda_{b_1}$. Before this bifurcation, (Γ) fits to the "oscillations" of the α (or ω)-branch of the saddle, which announce the proximity of a situation giving homoclinic points. So a part of (Γ) presents an oscillatory shape [G35]. This phenomenon makes harder an analytical study.

5.16. FIRST EXAMPLE OF THE BIFURCATIONS OF THE TWO PRECEDENT PARAGRAPHS.

Consider the recurrence T (5.106) such as $p(r,s) \equiv 0$ and $q(r,s) \equiv as + bs^2 + cs^3 + ds^4 + es^5$ [M20] [G37]. The point O(0,0) is an attractive focus when $\alpha a \cos\phi < 0$, and a repulsive focus when $\alpha a \cos\phi > 0$. The equation of the secular terms (5.112), $P = 0$, is :

$$\rho_o \cos\phi[a + 3c\rho_o^2/4 + 5e\rho_o^4/8] = 0 \ .$$

Writing $\rho_o^2 = R$, $3c/4 = A$, $5e/8 = B$, the solutions ρ^* are obtained from the positive roots of $a + AR + BR^2 = 0$. Thus zero, one (double), or two invariant closed curves (Γ) can exist. When $A^2 - 4aB > 0$, two curves (Γ_1), (Γ_2) are obtained, if the two roots R are positive. Let (Γ_1), (resp (Γ_2)) be the curve associated with the smallest (resp largest) ρ_o^*. (Γ_1) is attractive when $\alpha a \cos\phi > 0$, and repulsive when $\alpha a \cos\phi < 0$. The curve (Γ_2) has an opposite stability. When $A^2 - 4aB = 0$, with $AB < 0$, then $(\partial P/\partial\rho_o)^* = 0$, and one complex invariant closed curve (Γ) of multiplicity two (§ 5.15b) is obtained. It is the result of the merging of (Γ_1) and (Γ_2), and (Γ) is attractive on one side, repulsive on the other side. When $A^2 - 4aB < 0$, invariant closed curve does not exist.

The existence of the curves (Γ) implies ϕ to be outside the interval I, $\phi_c - \delta\phi_{c_1} < \phi < \phi_c + \delta\phi_{c_2}$, $\phi_c(i-2p-1) = 2h\pi$. Here the values $\phi_c = \pm 2\pi/3$ $(b \neq 0)$, $\phi_c = \pm 2\pi/4$ $(c \neq 0)$, $\phi_c = 2k\pi/5$ (k = integer for $d \neq 0$), $\phi_c = 2k\pi/6$ $(e \neq o)$, give rise to small divisors for the solution of (5.110). When $\phi \in I$, invariant closed curves (Γ) do not exist. When ϕ approachs the values of I, the shape of (Γ) is perturbed by $t = i-2p-1$ minima of ρ, and t maxima of ρ.

Consider now the more particular case $p \equiv 0$, $q \equiv -0.3s_n + 4s_n^3$ [M20]. The solution of (5.112) is $\rho_o^* = \sqrt{10}/10$, and (Γ) is attractive if $\alpha\cos\phi < 0$, repulsive if $\alpha\cos\phi > 0$. The stability of the focus fixed point O(0,0) is opposite to that of (Γ). When $\phi = \pm\pi/2$. The recurrence T is conservative, the jacobian determinant of the right hand side being equal to 1. T is a a diffeomorphism when $Q = (0.3\alpha\cos\phi - 1)\alpha\cos\phi < 0$, and an endomorphism when $Q > 0$. T is such as $b = d = e = 0$, thus only $\phi_c = \pm 2\pi/4$ gives rise to a small denominator in (5.113). When ϕ is close to the values of the interval I, the invariant closed curve (Γ) presents 4 minima of ρ, and 4 maxima of ρ for a 2π variation of θ, and (Γ) becomes close to the invariant curves passing through a saddle cycle of order 4. This cycle coexists with a focus cycle of order 4.

The parameter $\alpha > 0$ having a sufficiently small fixed value, two values of ϕ, $\phi = \phi_c - \delta\phi_{c_1}$, $\phi = \phi + \delta\phi_{c_2}$, are such that the bifurcation situation of § 5.15(a) is obtained. I.e. four α-branches of the saddle cycle of order 4 merge into four ω-branches of the same cycle, and give rise to an invariant closed curve (Υ) passing through the four points of

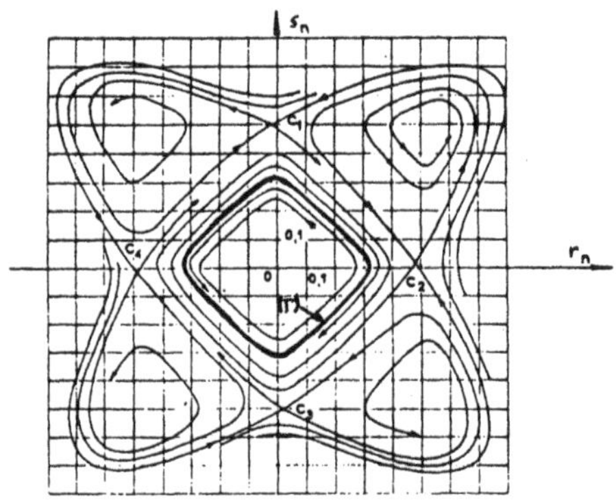

$\varphi \simeq -1{,}616$, $\alpha = 0{,}132$

(a)

The invariant closed curve (Γ) is attractive .

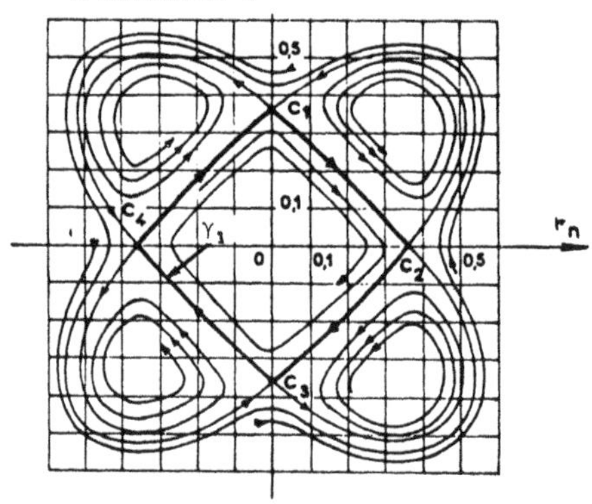

$\varphi = \varphi_{b_2} \simeq -1{,}5868\,\text{rd}$ $\alpha = 0{,}132$

(b)

Bifurcation situation
(Γ)merges into (γ_1)

FIG . 5.40 (First part)

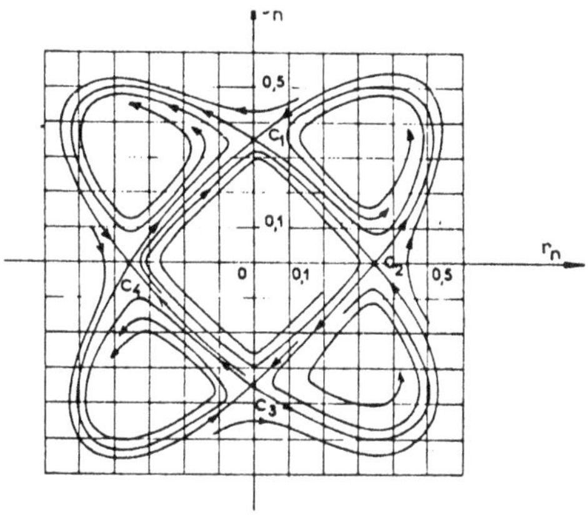

$\varphi_{b_1} < \varphi < -\frac{\pi}{2}$ $\alpha = 0{,}132$
(c)

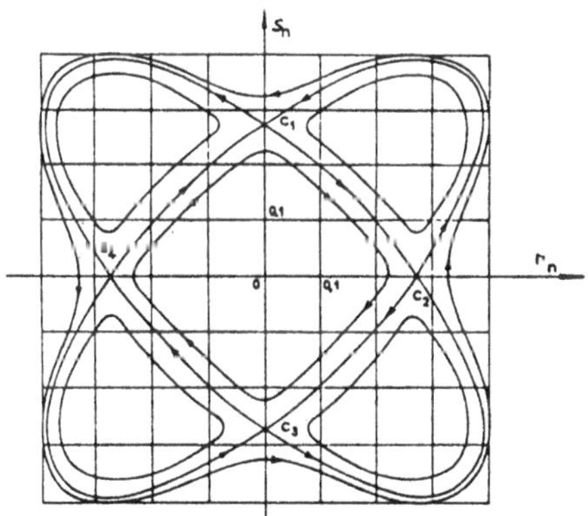

$\varphi = -\frac{\pi}{2}$ $\alpha = 0{,}132$
(d)

Conservative case

FIG. 5.40 (Second part)

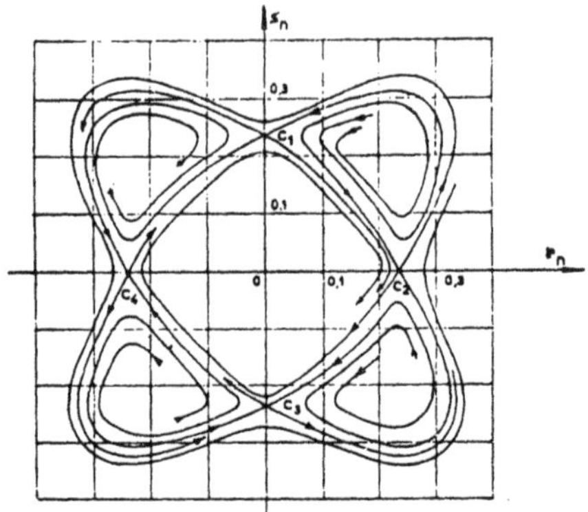

$-\frac{\pi}{2} > \varphi > \varphi_{b_2}$, $\alpha = 0{,}132$, $\varphi = 1{,}5680$ rd
(e)

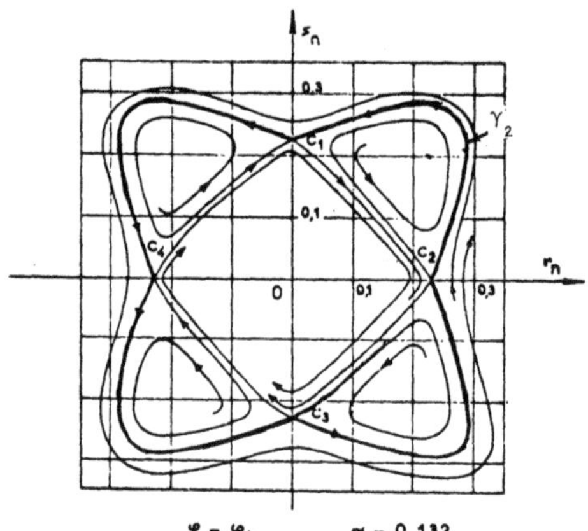

$\varphi = \varphi_{b_2}$ $\alpha = 0{,}132$
(f)

Second situation of bifurcation
The closed curve (γ_2) gives rise to
a new invariant closed curve (Γ)

FIG. 5.40 (3rd part)

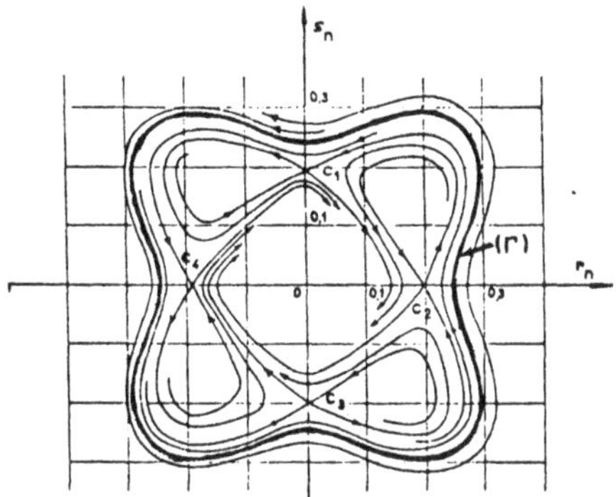

φ = −1,5608rd α = 0,132

(g)

The invariant closed curve (Γ) is now repulsive .

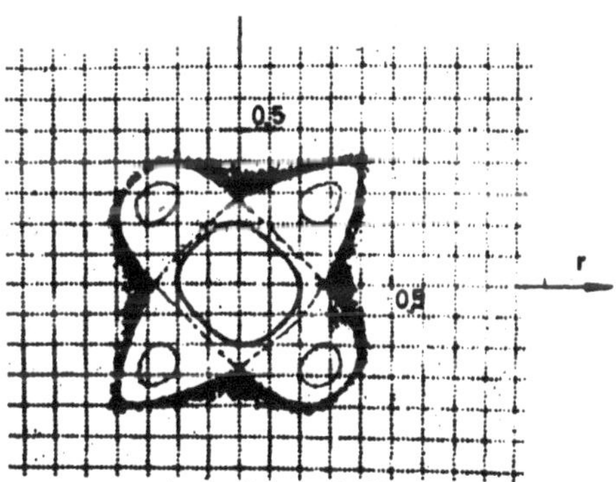

$\phi = \pi/2$, $\alpha = 0.332$

(h)

With such an increase of non-linearity homoclinic points appear .

FIG. 4.50 (4th part)

the saddle. The fig. 5.40 give the different phase portraits when ϕ varies from a value $< \phi_{c_1}$, to a value $> \phi_{c_2}$. For $\phi = -\pi/2 - \delta\phi_{c_1} \simeq -90.92°$, a closed curve (γ_1) is obtained, and for $\phi = -\pi/2 + \delta\phi_{c_2} \simeq -89.84°$ it is a curve (γ_2) (fig. 5.40f). In these two global bifurcation cases, the invariant closed curve (Γ) merges into (γ_i), $i = 1,2$, and (Γ) exists no longer when $\phi \in I$.

In order to make appear the equivalent phenomenon with $\phi_c = \pm 2\pi/3$, quadratic non linearities are necessary, as it results from (5.113). A simple example consists in choosing : $p \equiv 0$, $q \equiv -0.3s + 2s^2 + 4s^3$ [M20] $\phi_c = \pm 2\pi/3$. Fig. 5.41 give the different phase portraits when Φ varies passing through the interval I. As for the example of fig. 5.40, *on one side of* I *the pair of cycles of order* $t = i-2p-1$ *is inside the finite domain bounded by* (Γ), *on the other side of* I *these cycles are outside this domain.*

The value $|\alpha|$ *being sufficiently small, when* Φ *is close to values commensurable with* 2π *but different from* Φ_c, *the shape perturbation of* (Γ) *is small and it occurs a boxes in files bifurcation structure* for the recurrence constrained to (Γ). The boundaries of a box of this structure correspond to a bifurcation resulting from the merging of a saddle cycle with a node cycle of the same order, one of their multipliers becoming equal to +1. Inside a box the Poincaré rotation number is given from the characteristies of the cycles. It is worth of noting that *the crossing through such values* $\Phi \neq \Phi_c$ *is not accompanied by the disappearance of the invariant closed curve* (Γ), *which occurs for the values* Φ_c, *when* $\Phi \in I$.

When $|\alpha|$ is not sufficiently small, the bifurcations of § 5.15 (a) (b) do not exactly occur as indicated in this paragraph. So considering (5.106) and the crossing through the interval I, when $\phi_c = \pm\pi/2$ in presence of cubic non linearities, the phase portraits of the fig. 5.42 are now obtained. Here two points C of the saddle cycle of order 4, the corresponding α and ω-branches of these points, a segment of the invariant closed curve (Γ), and one point F of the focus cycle of order 4 are represented. Let λ be one of the parameters of p, q in (5.106). Fig. 5.42a corresponds to $\lambda < \lambda_{b_1}$, the shape of (Γ) follows the "oscillations" of the α-branches of the four points of the saddle cycle. Fig. 5.42b corres-

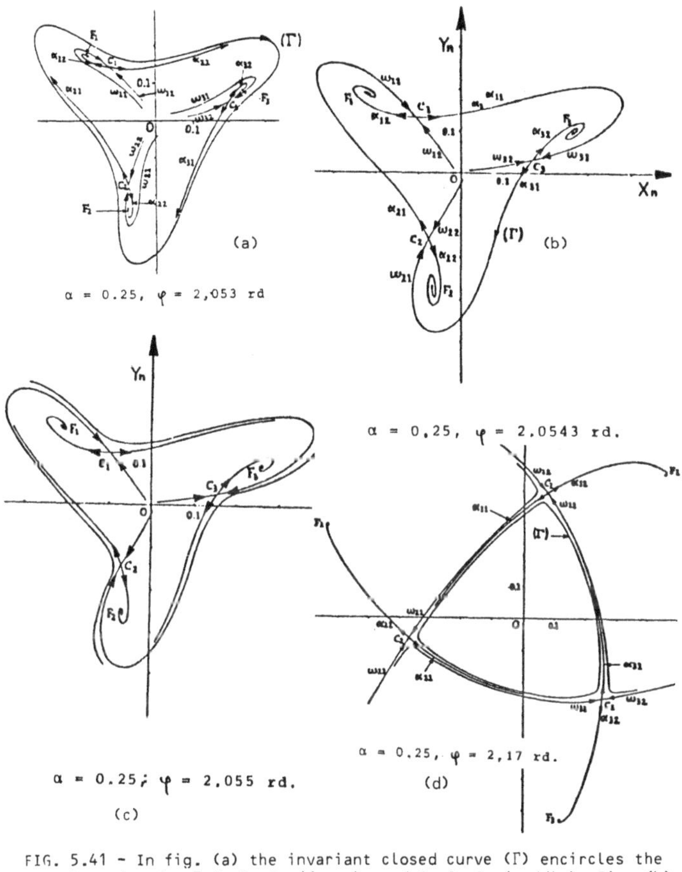

FIG. 5.41 - In fig. (a) the invariant closed curve (Γ) encircles the two cycles of order 3 F_1, F_2, F_3 (focus), and C_1, C_2, C_3 (saddle). Fig. (b) gives a first bifurcation, limit of existence of (Γ). In fig. (c) (Γ) has disapeared. A new bifurcation, when complementary segments of inva-curves of the saddle cycle form a closed curve, occurs. From this closed curve a new curve (Γ) appears, but now the two cycles are outside the domain encircled by (Γ) (fig. (d)).

ponds to $\lambda = \lambda_{b_1}$, bifurcation value giving an homoclinic tangency. Homoclinic points are obtained for $\lambda_{b_1} < \lambda < \lambda_{b_2}$ (fig. 5.42c). A second bifurcation of homoclinic tangency type occurs for $\lambda = \lambda_{b_2}$ (fig. 5.42d). This tangency is obtained with the α and ω-branches different from those corresponding to $\lambda = \lambda_{b_1}$. For $\lambda > \lambda_{b_2}$, it can be obtained an invariant close curve (Γ), which surrounds the pair of cycles of order four, and follows the oscillations of the α-branches of fig. 5.42d.

5.17. SECOND EXAMPLE. GLOBAL BIFURCATIONS IN PRESENCE OF CHAOTIC BEHAVIOUR.

The considered recurrence, or map T is [R4]

$$(5.115) \qquad x_{n+1} = y_n + \alpha[ax_n + bx_n^2 + cx_n^2 y_n(1-x_n^2)] \; ; \; y_{n+1} = (1+\alpha e)x_n$$

The jacobian determinant is $J = (1+\alpha e)[1-\alpha c x^2(1-x^2)]$. When $0 < \alpha c < 4$ and $\alpha \neq -1/e$, T is a diffeomorphism. By means of a linear transformation (5.115) can be brought into the form (5.108) and the equation of the secular terms (5.112) is $P \equiv \rho_o^4 + 2\rho_o^2 + 8e/c = 0$. When $\alpha^2 a^2 < 4(1+\alpha e)$, the fixed point O(0,0) is a focus with the multipliers $S_{1,2} = \sqrt{1+\alpha e}\, e^{\pm j\phi}$ $j = \sqrt{-1}$, $tg\phi = [4(1+\alpha e)-\alpha^2 a^2]^{1/2} / (\alpha a)$, $\cos\phi = \alpha / [4(1+\alpha e)]^{1/2}$. It is attractive when $\alpha e < 0$, and repulsive when $\alpha e > 0$. T has at least two other fixed points when $[2 + \alpha(e-a)] / [\alpha c(1+\alpha e)] < 0$ (sufficient condition).

Consider now (5.115) with $a = -0.65$, $b = -0.4$, $c = 0.9$, $e = 0.1$. The equation $P = 0$ has two positive real roots, $\rho_o^{*(1)} \simeq 0.81$, $\rho_o^{*(2)} \simeq 1.15$. Let α be positive and such as $0 < \alpha c < 4$. When α is sufficiently small, two invariant closed curves (Γ_1), (Γ_2) exist. For this example, if $\alpha \to 0$, $\phi \to \pi/2$ which gives rise to a small divisor in (5.113). Then (Γ_1), (Γ_2) do not tend toward circles with radii $\rho_o^{*(i)}$, $i = 1,2$. These radii are only average values of ρ, which has 4 maxima and 4 minima for a 2π-variation of θ. In the case of (5.115), the least order approximation of a curve (Γ) is $\rho = \rho_o + \alpha\Psi(\theta;\rho_o,0)$.

When $0 < \alpha < \bar{\alpha} \simeq 3.586$, O is a repulsive focus, $(\partial P/\partial\rho_o) < 0$ for $\rho = \rho_o^{*(1)}$ thus (Γ_1) is attractive, $(\partial P/\partial\rho_o) > 0$ for $\rho = \rho_o^{*(2)}$ and (Γ_2)

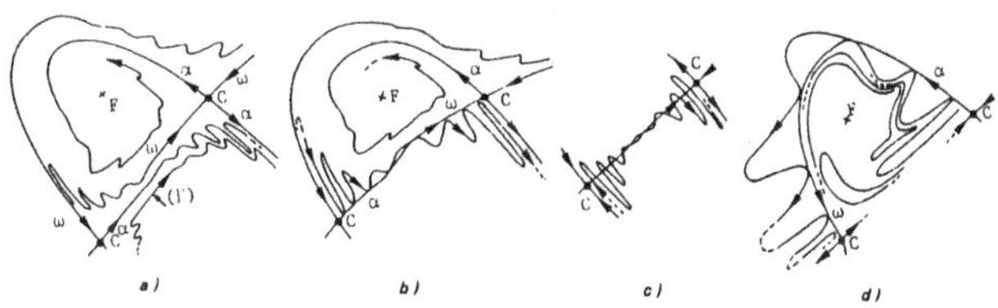

FIG. 5.42

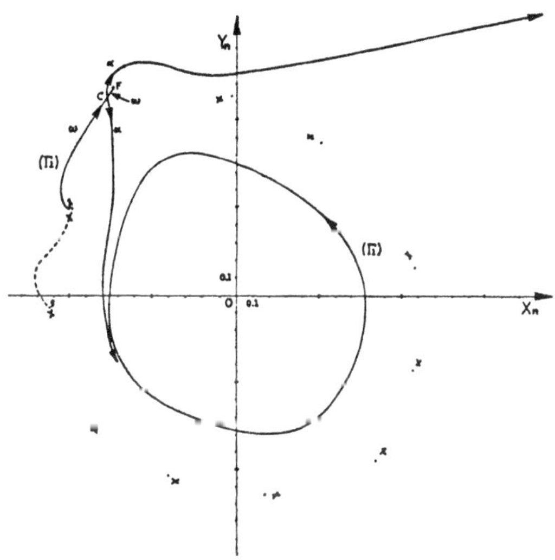

$\alpha = 0.453$

FIG. 5.43

× points of the saddle cycle of order 11
. points of the focus cycle of order 11
(Γ_1) is an attractive invariant closed curve
(Γ_2) repulsive invariant closed curve is made up of the segments
ω passing through the points of the saddle cycle. These segments
"pass through the points of the focus cycle.

is repulsive. Considering T constrained to (Γ_1) or (Γ_2), one has a boxes-in-files bifurcations structure for parameters variations. Fig. 5.43 shows the phase portrait (α = 0.453) corresponding to the box associated with the rotation number $\mu = -3/11$. (Γ_2) is made up of the ω-branches of a saddle cycle of order 11, which pass through a repulsive focus of order 11. This value of α is still sufficiently small to produce a "regular" shape for (Γ_1),(Γ_2). It is no longer the case when α increases, as indicated in Fig. 5.44 (α = 0.67). An important shape perturbation of (Γ_2) occurs with "oscillations". (Γ_1) corresponds to a box with $\mu = 2/7$, defined from a saddle cycle C, and an attractive node cycle of order 7. Between (Γ_1) and (Γ_2) their exists a saddle cycle C', a repulsive focus, and an attractive cyclic invariant closed curve (γ_i) of order 7, i = 1,...7, (Fig. 5.44b). The ω-branches of C' separate the basins of (Γ_1) and (γ_i). The curves (γ_i) disappear from a bifurcation having the type defined § 5.15 (a) for T^7 and from the bifurcation of § 5.11.1. With an increase of α, (Γ_2) becomes a repulsive cyclic "chaotic" segment of order 7 (Fig. 5.45, 5.46). This chaotic segment is made up of the ω-branches of the saddle cycle C of order 7 which are folded in a narrow band with an equivalent shape to that of the α-branches of (5.1), $f \equiv 1 - \lambda x^2 + y$, $g \equiv bx$, (cf. chap.6). The ω-branches of the other saddle cycle C' of order 7 "follow" the shape of the cyclic chaotic segment (fig.5.46).

The bifurcation by homoclinic tangency, described at the end of § 5.15, appears here when α crosses through $\alpha_{b_1} \simeq 1.1969$ (fig. 5.48). The Fig. 5.47 - 5.49 illustrate the variations of the phase portrait. The ω (resp.α)—branches of the saddle cycle C_i of order 3 are designated by ω_{i_1}, ω_{i_2} (resp.α_{i_1},α_{i_2}), i = 1,2,3. For α = 1,196, fig. 5.47 shows (Γ_1) which follows the oscillations of α_{11} (resp.α_{21},α_{31}) in the neighbourhood of C_3 (resp.C_1,C_2), and the ω-branches of C_i which separate the basins of the two attractors : (Γ_1) and the attractive focus cycle of order 3 F_i(the basin of (Γ_1) is the shaded zone). When $\alpha > \alpha_{b_1}$, (Γ_1) disappears with the presence of homoclinic points produced by the α and ω-branches of the saddle cycle C_i, i = 1,2,3.

<u>Bounded chaotic behaviours</u> : <u>strange attractor and strange repeller</u>.

Giving larger values to |a|, |b|, |c|, |d|, |e| in (5.115) incrases the non-linearities effects and makes for chaotic behaviours. Then consi-

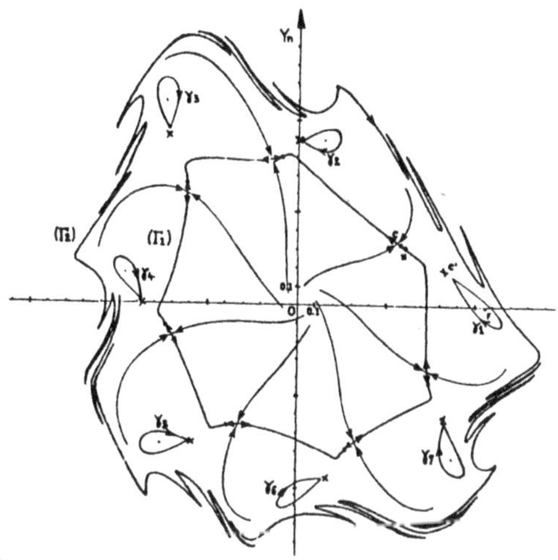

(a)
× saddle cycle of order 7, $\mu=-2/7$
focus cycle of order 7, $\mu=2/7$
Between (Γ_1),(Γ_2) there are 7 cyclic closed curves (γ_i). On (Γ_1) there are a node cycle, and a saddle one of order 7.

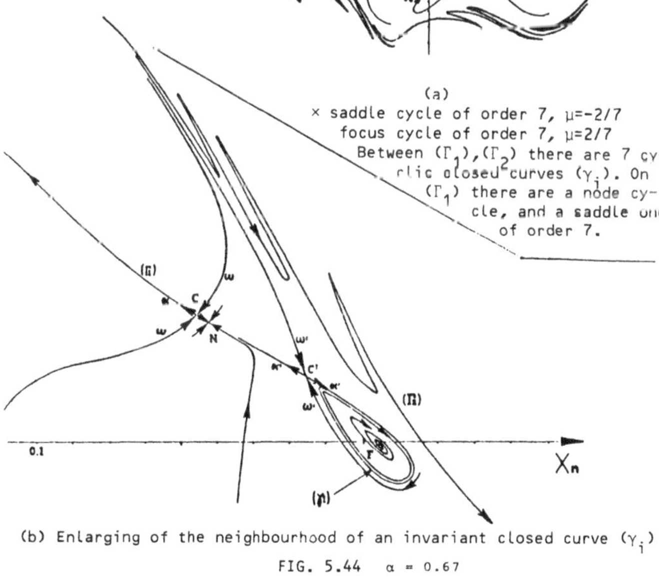

(b) Enlarging of the neighbourhood of an invariant closed curve (γ_i)

FIG. 5.44 $\alpha = 0.67$

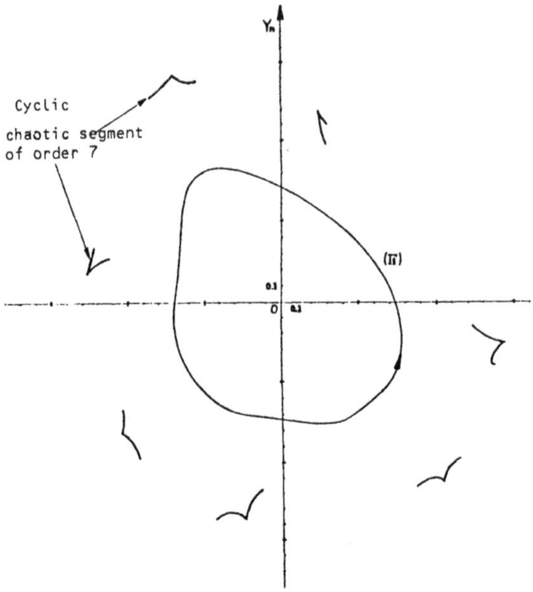

α = 0.80

FIG. 5.45

With an increase of the parameter α the invariant closed curve (Γ_2) turns into a repulsive cyclic chaotic segment of order 7 . Details about this case are given in fig. 5.46 with a higher value of the parameter .

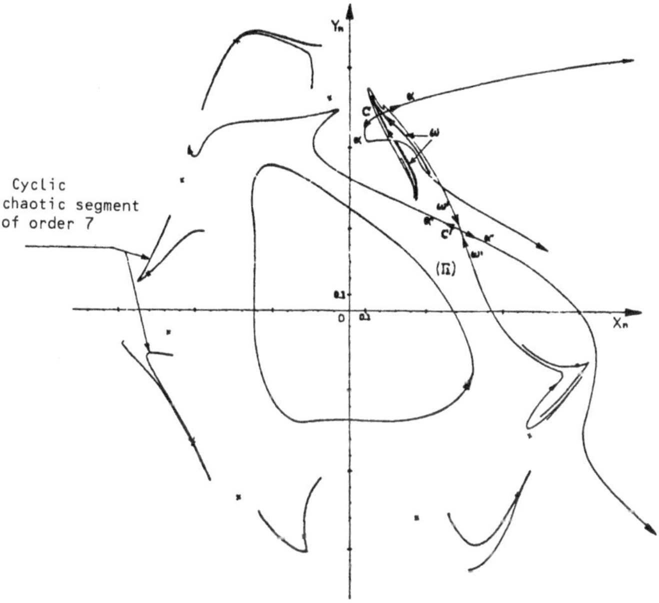

FIG. 5.46

× Points of the saddle cycle (C') of order 7, $\mu=2/7$
•× Points of the saddle cycle (C) of order 7, $\mu=2/7$, every point of this cycle belongs to the cyclic chaotic segment

The cyclic chaotic segment is made up of the ω-branches of the saddle cycle of order 7 (C) which are folded in a narrow band. The invariant curves ω' of the saddle cycle of order 7 (C') tend toward the cyclic chaotic seg ment. The rotation number of (Γ_1) is closed to 1/3.

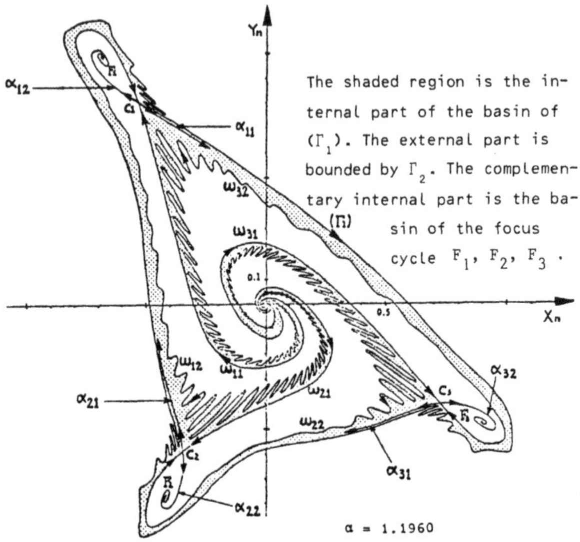

(a)

"Oscillation" of the invariant curve (Γ_1) in the neighbourhood of the saddle cycle C_i,i=1,2,3,associated with the focus cycle F_i,i=1,2,3.

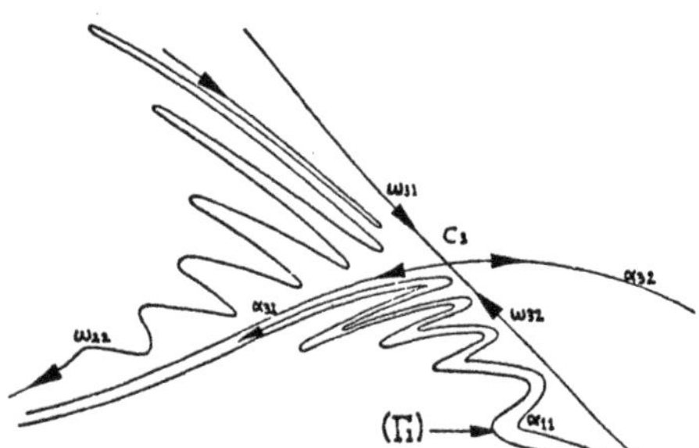

(b)

FIG. 5.47

Enlarging of fig. (a) in the neighbourhood of the saddle point C_3. The invariant curve α_{11} passing through the saddle C_1 follows the "oscillations of (Γ_1)

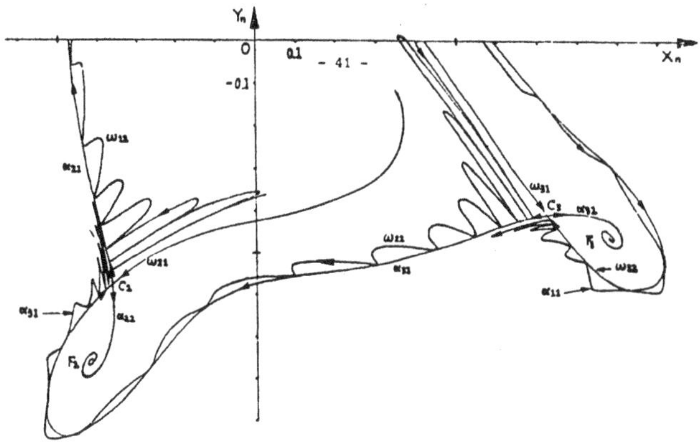

α = 1.1969

FIG. 5.48

First homoclinic tangency: limit of existence of (Γ_1).

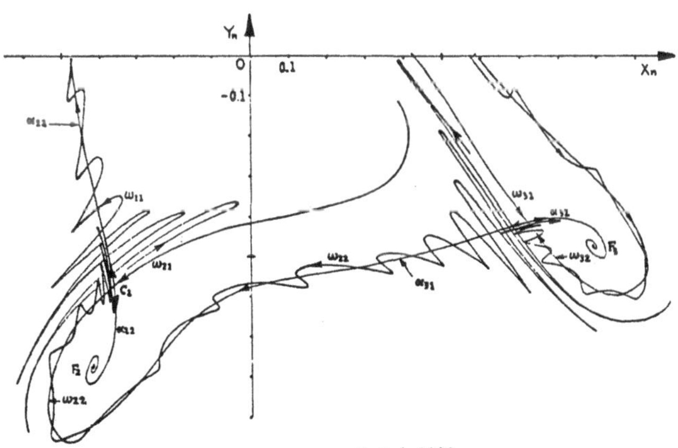

α = 1.1990

FIG. 5.49

Homoclinic points appear and (Γ_1) does not exist .

der (5.115) with a = -0.78, b = -0.28, c = 0.98, e = 0.12. An increase of α from α = 0.78 gives rise to "oscillations" of (Γ_2), and to the formation of a repulsive cyclic chaotic segment of order k = 10 in the same situation as previously with k = 7. When α tends toward α = 1, the chaotic segments, which correspond to the folding of ω-branches of a saddle cycle k = 10, spread in the length and width sense, and give rise to a repulsive *annular chaotic area* (ACA) [B 7]. Fig. 5.50 represents the corresponding situation for α = 1.108. The basin of the attractive invariant close curve (Γ_1) is bounded by the repulsive ACA, which is a *fuzzy (or chaotic) boundary*. More precisely the repulsive ACA contains the ω-branches of two saddle cycles of order 10, C_i of type 1 (positive multipliers), C_i' of type 3 (negative multipliers), i = 1,2,...,10. These ω-branches reproduce the shape of the repulsive ACA (fig.5.51), and give rise to homoclinic and heteroclinic points. A neighbourhood of such points containing other saddle cycles (Birkhoff's theorem), the ω-branches of these cycles cannot intersect other ω-branches (because T is diffeomorphism) and they follows the shape of the ω-branches of the cycles C_i, C_i'. *So the repulsive* ACA *is made up of infinitely many ω-branches of type 1 and 3 saddle cycles of any order k, and of the accumulation curves of these ω-branches. When* $k \to \infty$. *The so-obtained set is called a strange repeller.*

With a further increase of α, the attractive invariant closed curve (Γ_1) undergoes a set of bifurcations similar to that of (Γ_2). For $\alpha \simeq 1.032$, (Γ_2) is made up of the α-branches of a saddle cycle k = 31, ending at an attractive node k = 31. Another increase of α produces an attractive cyclic chaotic segment k = 31, and after an attractive ACA similarly to the above described process, the α-branches playing now the role of the ω-branches. *So the attractive* ACA *is made up of infinitely many α-branches of type 1 and 3 saddle cycles of any order k, and of the accumulation curves of these α-branches when* $k \to \infty$. *The so obtained set is called a strange attractor, if these α-branches are not intersected by ω-branches coming from cycles located outside of the considered* ACA. This point appears fig. 5.52 and 5.53 for α = 1.057. (Γ_1) was transformed into a strange attractor the basin of which is bounded by a fuzzy boundary born of (Γ_2) (fig. 5.52). With the fig. 5.53 it appears that the α-branches of the saddle cycle k = 31 reproduce the shape of the numerically obtained strange

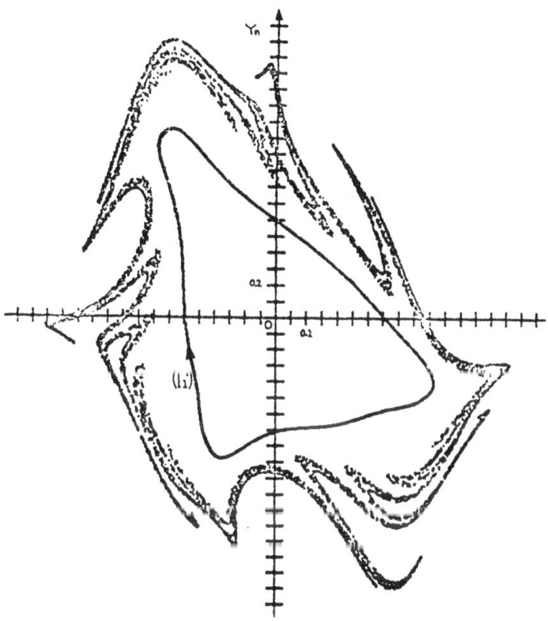

α = 1.018

FIG. 5.50

(Γ_1) is an attractive invariant closed curve the basin of which is bounded by a fuzzy (or chaotic) boundary resulting from the successive bifurcations of (Γ_2). In the fig. this fuzzy boundary encircles (Γ_1), and was obtained by inverse iteration.

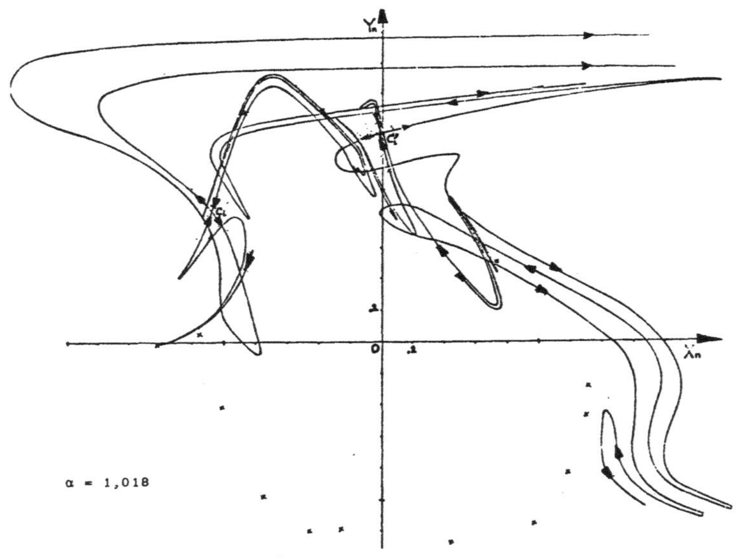

FIG. 5.51

The invariant curves ω of the saddle cycles of order 10 (C_i) and (C'_i), i=1,2,...,10, follow the shape of the fuzzy boudary given by fig. 5.50 .

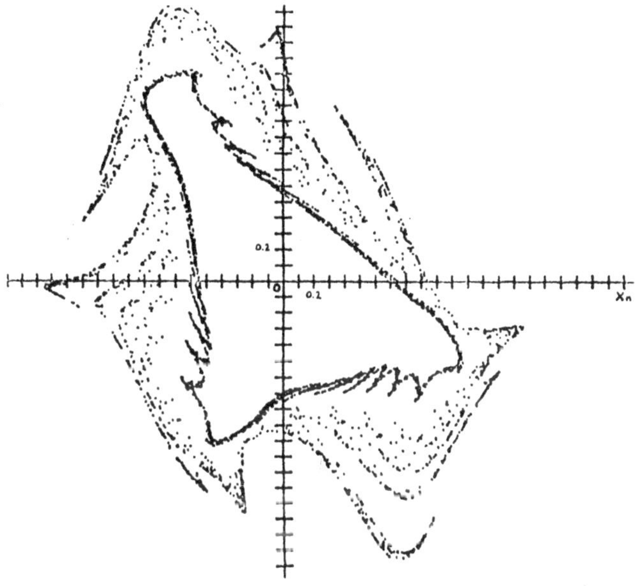

FIG. 5.52

The invariant curve (Γ_1) has turned into an annular strange attractor, the basin of which is limited by the fuzzy boundary appearing in fig. 5.51 for a smaller value of the parameter .

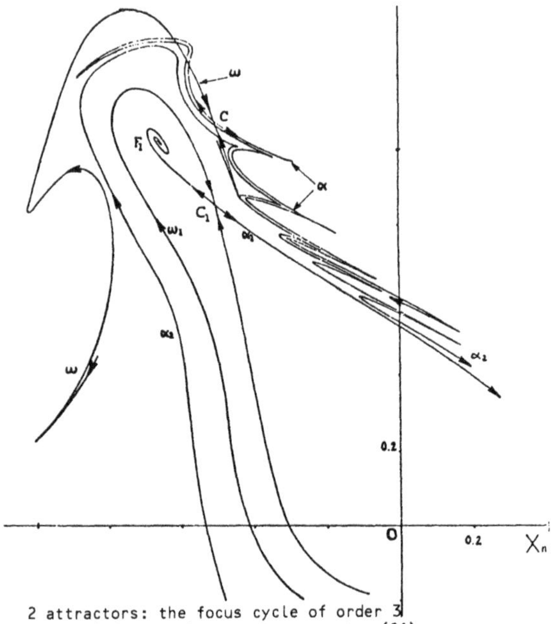

2 attractors: the focus cycle of order 3 F_i, i=1,2,3, and the strange attractor (SA)

$\alpha = 1.057$

FIG. 5.53

The strange attractor of fig. 5.52 is "reproduced" from the α-invariant curves of a saddle cycle of order 31 (C) . α_1, α_2 are α-invariant branches of the saddle cycle of order 3, C_i, i=1,2,3, associated with the focus cycle F_i. The α-branches of C_i, which do not tend toward F_i, tend toward the α-invariant curves of the saddle cycle of order 31 (C). ω-branches of (C_i) constitute the boundary separating the basins of the two attractors (SA) and F_i.

attractor of fig. 5.52. An attractive focus cycle $k = 3$, F_i, $i = 1,2,3$, also exists inside the area limited by the strange attractor. The ω-branches of the corresponding saddle cycle $k = 3$ separate the basins of F_i from that of the strange attractor.

With a further increase of α, for $\alpha = \alpha^* \simeq 1.058$, *an heteroclinic tangency occurs between some ω-branches of the repulsive* ACA *born of* (Γ_2), *and some α-branches of the attractive* ACA *born of* (Γ_1). Such heteroclinic tangency is the result of infinitely many processes of bifurcations accumulations explained in chapter 6. *For* $\alpha > \alpha^*$, *the repulsive* ACA , *and the attractive* ACA *disappears,* $\alpha = \alpha^*$ *gives rise to an important global bifurcation in the field of chaotic dynamics.*

Remark. A whole class of dynamical systems is concerned with the bifurcations appearing from the examples of § 5.16, 5.17. Voluntarily the corresponding results are not presented in an abstract form, because only correctly chosen examples give the best understanding of the complex mechanisms of bifurcations, as it was the case for the one-dimensional endomorphism defined by the quadratic function $x^2-\lambda$.

Fig. 5.43 - 5.53 are extracted from [R4].

Chapter 6

TWO-DIMENSIONAL DIFFEOMORPHISMS AND THE FOLIATED BOX-WITHIN-A-BOX BIFURCATIONS STRUCTURE

6.1. INTRODUCTION.

In this chapter the following family of diffeomorphisms T_b is considered :

(6.1) $$x_{n+1} = f(x_n, a) + y_n \quad ; \quad y_{n+1} = bx_n \qquad n = 0, 1, 2 \ldots,$$

Where x, y are real variables, a, b are real parameters, f(x,a) being a sufficiently smooth function of x and a. T_b has a constant Jacobian determinant J = -b, and its inverse T_b^{-1} is given by :

(6.2) $$x_n = y_{n+1}/b \quad , \quad y_n = x_{n+1} - f(b^{-1}y_{n+1}, a),$$

with J = -1/b . T_b^{-1} being explicitely defined, it is not necessary to use the Hadamard's theorem (§ 1.6) to verify that T_b is a diffeomorphism. As indicated in § 1.7, the interest in studing (6.1) is related to the fact that it is a step in a process of successive simplifications of models. The last step in this process consists in making b = 0 in (6.1), and in considering that T_0 given by :

(6.3) $$x_{n+1} = f(x_n, a) \qquad n = 1,2,\ldots$$

is an endomorphism. For b = 0, n = 1,2,..., the iterated sequence solution of (6.1) is identical to the solution of (6.3). So *the one-dimensional endomorphism* (6.3) *is embedded into the two-dimensional diffeomorphism* (6.1). For $f(x,a) \equiv 1-ax^2$, the first study of T_b was given in [H 7] with a numerical simulation of an iterated sequence, considered for particular values of (a,b). With the finite number of iterations of the simulation, this solution was interpreted as a strange attractor. Afterwards many papers were devoted to this problem, and here only the relevant references, directly related to the object of this chapter, are given.

Consider $f(x,a) \equiv 1 - ax^2$. From § 3.1, it results that, on the parameter a - axis, the bifurcations structure of T_0 is the same as that corresponding to $\bar{f}(x,\lambda) \equiv x^2 - \lambda$, because the relation between the coefficients of a quadratic recurrence and λ gives here $\lambda = a$. Thus for b = 0, the bifurcation structure of (6.1) is a "box-within-a-box" type, described in the chap. 3 , 4. It is worth noting that for $b = \pm 1$, $J = \mp 1$, and that (6.1) is a conservative diffeomorphism. Moreover for b = -1, J = +1, $-1 \leqslant a \leqslant 3$, the most part of the properties of (6.1) are known. Indeed consider (6.1), and a first change of variables (x', y') for obtaining the origin at the fixed point q_2 $(x = [b-1 + \sqrt{(1-b)^2 + 4a}\,]/(2a),\ y = bx)$. From this new form of the recurrence, the non linear transformation (which conserves the qualitative properties):

$$x' = X \quad , \quad y' = Y - g(X)/2 \quad , \quad g(X) = -aX^2 - 2a\, x_{q_2} X$$

leads to the recurrence.

$$(6.4) \qquad X_{n+1} = Y_n + F(X_n) \ , \ Y_{n+1} = bX_n + F(X_{n+1}) \ , \ F(X) = g(X)/2.$$

A relatively detailed study of (6.4) with b = -1, concerning the phase, plane and its bifurcations, was made in [G 25] [G 26] (1972), [G 27] (1972) [G 28] (1973) [G 35] [G 36] (1980) for $F(X) \equiv \mu X + (1-\mu)X^2$, $-1 \leqslant \mu \leqslant 1$. With b = -1, $f(x,a) \equiv 1-ax^2$, it is easy to verify that the relation between the parameter a of (6.1), and μ of (6.4) is :

$$(6.5) \qquad \mu = 1 - \sqrt{a+1} \qquad , \qquad a = \mu^2 - 2\mu$$

So the known properties of (6.4) *for* $-1 \leqslant \mu \leqslant 1$ *can be transfered to* (6.1) *into the interval* $-1 \leqslant a \leqslant 3$. Moreover the properties described in the chapter 4 of [G36], when a small damping term is added to (6.4), permits to obtain informations about the behaviour of (6.1) for $b = -1+\varepsilon$ $\varepsilon > 0$ having a sufficiently small value. The publications [B 7] [G28] [G 35] [G 36] gives the phase plane, and bifurcations properties of (6.4) for various form of F(X)

$$\begin{cases} F(X) \equiv \mu X + (1-\mu)X^3, \\ F(X) \equiv \mu X + (1-\mu)X^2 \exp[(1-X^2)/4] \\ F(X) \equiv \mu X + 2(1-\mu)X^2/(1+X^2) \\ F(X) \equiv X - [\sin(bX + b_0) - \sin b_0](1-\mu)/(b \cos b_0), \ b=\pi-2b_0, \\ \qquad 0 < b_0 < \pi/2, \\ F(X) \equiv X - (1-\mu)\pi^{-1}\sin\pi X. \end{cases}$$

So an equivalent transfer of properties can also be made from (6.4) to (6.1) with $b = -1$, $f(x,a)$ having now a non-quadratic form. Considering T_b (b=0), $n = 0,1,2,\ldots$, as a degenerated two-dimensional map with $J \equiv 0$, it was shown in § 1.8, 1.9 and 5.2.5 that degenerate invariant curves can be defined in the (x, y) plane [M 28] [M 30] (1978) [G 35] (1980). Then a segment of such a degenerate invariant curve is the continuous set of the first rank antecedents of a point of the x_n-axis. Using an expansion of an invariant curve of T_b, $b \neq 0$, passing through a fixed point, it will be shown that limit of such a curve when $b \to 0$ is the degenerate one [G 35] (1980) [M 36] (1982). As long as the analytical expressions of the coordinates of the cycles ($k = 1,2,3$) can be determined, they directly appear for $b = 0$ as continuous functions of b when $b \to 0$. For $k > 3$ this continuity results from that of $f(x,a)$ and that of the solution of the cycle equation $x_{n+k} = x_n = x$, $y_{n+k} = y_n = y$, which is written :

$$g_k(x,y,a,b) = h_k(x,y,a,b) = 0$$

g_k, and h_k being continuous and differentiable with respect to (x,y,a, b). Let S_1, S_2 be the multipliers of a cycle of order k for T_b.

S_1, S_2 are continuous and differentiable functions of (x,y,a,b) if the data of (6.1), i.e. f(x,a) are continuous and sufficiently smooth functions of (x,a). The equation of a bifurcation curve in the (a,b) plane, corresponding to a fold bifurcation $S_1 = 1$, $S_2 \neq 1$ (§ 5.6.2) or to a flip bifurcation $S_1 = -1$, $S_2 \neq 1$ (§ 5.6.3), is given by the parametric(equation with "parameters" (x,y)=coordinates of a cycle point):

$$(6.6) \qquad g_k(a,b\ ;\ x,y) = 0,\ h_k(a,b\ ;\ x,y) = 0,\ S_1(a,b\ ;\ x,y) = \pm 1.$$

Whatever may be the point (x,y) of the considered cycle, (6.6) gives the same bifurcation curve. Except for k = 1,2, the elimination of the parameters (x,y), leading to the implicit equation $G_k(a, b) = 0$, is practically impossible. When $b \to 0$, one has $y \to 0$, and $S_2 \to 0$ because $J = -b \to 0$, and the equation (6.6) tends towards the equation giving a bifurcation value $a^j_{(k)_\circ}$ (S = +1) , a^j_k (S = -1) of T_0 :

$$(6.7) \qquad x = f^k(x,a)\ ,\quad S(a,x) = \pm 1\ ,\quad S \equiv S_1$$

This is due to the continuity properties of g_k, h_k, with respect to their arguments (a,b,x,y), resulting from the continuity of f(x,a), and that of S_1 .

Since [M 28] (1970), the argument of continuity for b = 0, in the phase plane (x,y) and in the parameters plane , has guided the studies which have led to a progressive identification of the bifurcations structure of T_b with $f(x,a) \equiv 1-ax^2$ [G 35] [H 4] [H 5] [M 32] [M 36] [F 6] - [F 9].Consider now g_k (a,b ; x,y) = 0, h_k(a,b ; x,y) = 0, and the points (x,y) of a same cycle of order k, solutions of these equations. It is possible to associate with the cycle a norm $|z|$ characterizing for example an average distance from its k points, to a reference point,or the distance from only one of its point. In the space (a,b, $|z|$), when (a,b) varies,$|z|$ corresponds to a surface sheet of the multi-valued manifold described by $H_k(a,b, |z|) = 0$ obtained from $g_k = h_k = 0$. Let $\Lambda^j_{(k)_\circ}$ be the bifurcation curve (6.6) with $S_1 = +1$ giving the two cycles (k;j) when b = 0, $a = a^j_{(k)_\circ}$. It corresponds to the junction of two sheets along a surface fold. One of the two sheets corresponds to an attractive cycle (node with a multiplier $0 < S < 1$) if $|b| < 1$, the other corresponds to the repulsive saddle cycle ($S_1 > 1$) which merges into the first when $S_1 = +1$. The Myrberg's cascade (or spectrum) ω_k^j by order (or period) doubling, obtained for $b = 0$ (cf chap. 4), gives rise to a *cascade (or spectrum) of bifurcation curves* ($S_1 = -1$) $\Lambda_{k2^i}^{j_i}$, $k = 1,3,4,\ldots$, $i = 0,1,2,\ldots$, $\Lambda_{(k2^0)_0}^{j_0} \equiv \Lambda_{(k)_0}^{j}$, in the parameters (a,b) plane, or to a surfaces sheets organization in the space $(a,b,|z|)$ represented fig. 6.1. The corresponding bifurcations are :

$$(6.8) \qquad \emptyset \rightleftarrows \text{saddle } (k\,;\, j, S_1 > 1) + \text{node } (k;j)$$

with $S_1 = +1$ on $\Lambda_{(k)_0}^{j}$, the node being attractive if $-1 < b < 1$

$$(6.9) \qquad \text{attr.node } (k2^{i-1};\, j_{i-1}) \rightleftarrows \text{saddle } (k2^{i-1};\, j_{i-1}, S_1 < -1) + \text{attr node } (k2^i, j_i) \;.$$

with $S_1 = -1$ on $\Lambda_{(k2^i)_0}^{j_i} \equiv \Lambda_{k2^{i-1}}^{j_i}$, and $-1 < b < 1$, $i = 1,2,\ldots$
The bifurcation(6.8) is the fold one, (6.9) the flip one, and lim $\Lambda_{(k2^i)_0}^{j_i} = \Lambda_{(k)_s}^{j}$ when $i \to \infty$. The representation symbolism $(k;j)$ of cycles is that which is used in the one-dimensional case (cf chap.3 and 4) : a cycle $(k;j)$ is the one which is obtained when $b \to 0$. Fig. 6.1 represents the simplest situation of a Myrberg's spectrum, which gives a "communication" set between the cycles of (6.8) (6.9), i.e., there exists a continuous way on the sheets permiting to pass from a cycle to another. More complex situations will appear with communications between several Myrberg's spectra.

It is worthy of note that after a bifurcation (6.8), or (6.9), when $S_1 = S_2$ the node can be transformed into a focus $S_{1,2} = \alpha \pm j\beta$, $j = \sqrt{-1}$. Starting from a node of type 1 ($S_1 > 0$, $S_2 > 0$), when $S_1 = S_2$ one has $\alpha > 0$. From a focus a continuous parameter variation

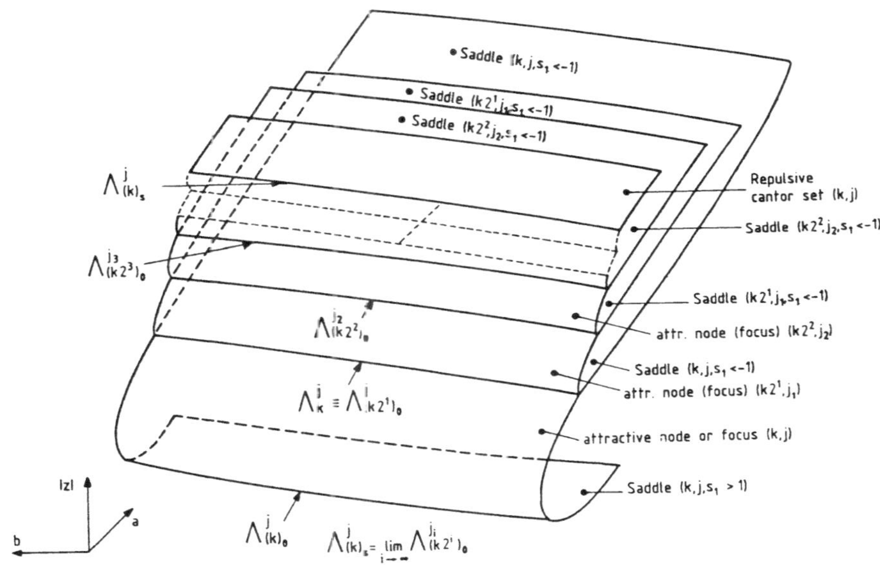

FIG. 6.1 - Surfaces sheets organization of the Myrberg's spectrum ω^j_k (simplest case) .

can lead to $\alpha < 0$ and to $S_1 = S_2$ with $S_1 < 0$, $S_2 < 0$. Then from $S_1 = S_2$ it is obtained a node of type 3 which can undergo the bifurcation (6.9). This behaviour can occur when $b < 0$ with cycles of even, or odd order, and when $b > 0$ with cycles of only even order. Moreover, taking into account the definition of the curves $\Lambda^j_{(k)_0}$ and $\Lambda^j_{(k2^i)_0} \equiv \Lambda^j_{k2^{i-1}}$, a curve $\Lambda^j_{(k2^i)}$ $i = 0,1,2,\ldots$, cannot intersect another curve $\Lambda^j_{(k2^m)_0}$, $m \neq i$. However for $i > 1$ such a curve can form loops with self intersection, two distinct bifurcations occuring simultaneously for a given point of the parameter plane (a,b) as it will be shown in this chapter.

A first part of the chapter is devoted to the behaviour of T_b for small values of $|b|$, and to the degenerate α or ω-invariant curves. Various consequences concerning complex basins are described. A second is a recall of results obtained in [G 25] [G 26] [G 27] [G35] [G 36] about the conservative case $b = -1$ and the almost conservative case. After a paragraph deals with the relation between T_b and the Smale horseshoe, which leads to a symbolic representation of cycles in the two different cases : $b < 0$, $b > 0$. The following paragraphes concern all the consequences of the foliated structure of the bifurcations plane, the problem of communication between the sheets of this structure, and the use of symbolic dynamic to explain the organization of the communications. The bifurcations by homoclinic, or heteroclinic tangency are also the object of a paragraph, but about this problem a more complete report can be found in [H 10]. It is also suggested to consult the very interesting paper [H 11] which gives an original way for treating a part of the problems of this chapter.

6.2. BEHAVIOUR OF THE DIFFEOMORPHISM FOR SMALL VALUES OF $|b|$

6.2.1. DEGENERATE ω-INVARIANT CURVES OF T_b FOR $b = 0$, AND INVARIANT CURVES FOR $b \neq 0$

The results of this paragraph are a consequence of the matter of § 1.7. Consider (6.1), $f(x;a)$ being a continuous function with respect to its arguments, and b=0. Let $A(x = \alpha)$ be a fixed point of T_0 given by (6.3). For T_b, b=0, given by (6.1), A has the coordinates $x = \alpha$, $y = 0$. Let (C_0) be the curve $y = -f(x,a) + \alpha$ in the phase plane (x,y). Using (6.1) with b=0, it is easy to verify that the set of the points of (C_0) has the same point $A(\alpha, 0)$ as first rank consequent. Let α_{-i}, $i = 1,2,\ldots$, be the sequence of rank i antecedents of A, obtained from the inverse mapping T_0^{-1}. When T_0 is an endomorphism, generally this sequence is of "arborescent" type (cf chap. 2,3,4) (for a quadratic map T_0, N_i ($N_i \leq 2^i$) rank i antecedents α_{-i} of A can be obtained). Let (C_i^j) be the curve

$$y = -f(x,a) + \alpha_{-i}^{(j)}\ ,\ i = 1,2,3,\ldots,\ j = 1,2,\ldots N_i (N_i \leq 2^i \text{ if } f(x,a)$$

is quadratic).

The set $\mathcal{C} = \bigcup_{i,j}(C_i^j)$ *made up of the union of all the curves* (C_i^j) *is considered as a degenerate* ω*-invariant curve (stable manifold) passing through the fixed point* A [M 28] [M 30] [G 35]. The degenerate ω-invariant curve is associated with the multiplier $S_2 = 0$. The other multiplier S_1 is that of A for T_0. If A is a repulsive fixed point for T_0, it becomes a *degenerate saddle* for T_b with b=0. Then a *degenerate* α*-invariant curve* (instable manifold) located on y=0 passes through A. In the same way, it is possible to define degenerate α and ω-invariant curves for a saddle cycle of order k.

Consider now that f(x,a) is an analytical function of x, and a sufficiently small neighbourhood (d) of a saddle fixed point, such that inside (d) the ω-invariant curve (C) can be described by a single-valued function $y = \Theta(x)$, solution of functional equation :

$$(6.10) \qquad bx = \Theta\,[f(x,a) + \Theta\,(x)]$$

(cf § 5.1, 5.3). Using a linear transformation of variables, it is assumed that the new coordinates of A are $\bar{x} = \bar{y} = 0$, $\bar{f}\,(\bar{x}\,,a)$ being the transformed function f(x,a). Let $\bar{y} = \Theta(\bar{x}\,) = \sum_{i=1}^{\infty} \Theta_i\,\bar{x}^i$ be the series expansion of the branch of the ω-invariant curve (C) passing through A ($\bar{x}$ = 0, $\bar{y}$ = 0) and located inside the domain of single-valuedness of (C) (cf § 5.1). Substituting $\bar{\Theta}(x)$ into (6.1), written with the new variables, it is obtained

$$(6.11) \qquad \bar{\Theta}(\bar{x}) = -\sum_{i=1}^{\infty} \bar{x}^i [d^i\bar{f}/d\bar{x}^i] / i! + bg_1(\bar{x}) + b^2 g_2(\bar{x}) + \ldots,$$

$$g_j(\bar{x}),\ j = 1,2,\ldots, \text{ being such that } g_j(0)=0.$$

According the Lattes results [L 1], (d) containing A can be chosen sufficiently small in order that $\bar{\Theta}(\bar{x})$ may be a convergent series inside (d). From (6.11) one has lim $\bar{\Theta}(\bar{x}) = -\bar{f}(\bar{x},a)$ when $b \to 0$, or with the variables (x,y) :

$$(6.12) \qquad \lim_{b \to 0} \Theta(x) = -f(x,a) + \alpha$$

which is the equation of the segment (C_0) of the degenerate ω-invariant curve. For $b \neq 0$, (C) is associated with the multiplier S_2, $|S_2| < 1$, lim $S_2 = 0$ when $b \to 0$.

Outside (d), $y = \Theta(x)$ can be continued by iterating the segment (6.11) with T_b^{-1}, in order to obtain (C). With (C) so defined, the following result appears :

$$(6.13) \qquad \lim_{b \to 0} (C) = \bigcup_{i,j} (C_i^j),\ i = 1,2,\ldots,\ j = 1,2,\ldots,N_i \leq 2^i$$

When $b = \varepsilon \neq 0$, $|\varepsilon|$ *having an infinitely small value, the disjoint segments* (C_i^j)*, obtained for* b=0 *, join at infinity in the direction of the y-axis in order to form the continuous curve* (C) *without self intersection because* T_b *is a diffeormorphism. The junctions between the* (C_i^j) *change with the sign of* b*, i.e. it is not the same* (C_i^j) *which are joined for* b>0*, and* b < 0*. When* |b| *is only small,* (C) *is close to the set of the* C_i^j *as long as* $|y| < M$ *,* $M \to \infty$ *if* $\varepsilon \to 0$.
The multipliers of the saddle are such that $S_1(\varepsilon) \to S_1(0)$, $S_2(\varepsilon) \to 0$, when $\varepsilon \to 0$. For $|y| > M$, the radius of curvature of (C) decreases, passes through a minimum for $|y_{(c)}|$ # maximum, $|y|$ decreases till

$|y| < M$ where (C) becomes close to another segment (C_i^j). The analytical proof of this result seems very difficult, but a numerical verification is possible for values of $|b|$ which are not too small. An indirect qualitative justification will be given in § 6.2.2. by considering for T_0 a basin made up of disjoint segments, which by imbedding into T_b must give an all in one block basin. Such a basin can be obtained only with the indicated junctions for $|y| > M$.

6.2.2. CONSEQUENCE : "OSCILLATING" BASINS

Consider T_0 given by (6.3) and f(x,a) corresponding to the curve of fig. 2.16a, $a \equiv \lambda$, for example $f(x,a) \equiv 9ax/4 + 3x^2 + x^3$, $a = 1-\varepsilon'$, $\varepsilon' > 0$ having a sufficiently small value. From § 2.8.2 it appears that the total basin of the attractive fixed point q_2 is made up of disjoint open intervals : the immediate basin (D_0) and its successive antecedents of rank j-1 (D_{j-1}), j = 2,3,..., the limit point of which is q_1 Consider now T_b, b=0, i.e. the degenerate two-dimensional case. For b=0 we call q_i the abscissa of the point q_i. Let (C_0), $y = -f(x,a) + q_3$ be the curve of the (x,y) plane, made up of the set of the first rank antecedents of the repulsive fixed point q_3, which is a degenerate saddle. (C_0) passes through q_3 and its first rank antecedents q_3^{-1}, $q_3'^{-1}$ located on y=0. Let (C_1) be the curve $y = -f(x,a) + q_3^{-1}$, the first rank consequent of which is the point $(x = q_3^{-1}, y=0)$. (C_1) passes through the point q_3^{-2} of y=0. In the (x,y) plane, (C_0) and (C_1) are the boundaries of the immediate basin $(\bar{D}_0)$ of q_2 , $(\bar{D}_0) \cap (y=0) = (D_0)$. Two curves (C_i), $y = -f(x,a) + q_3'^{1-i}$, (C_{i+1}) $y = f(x,a) + q_3^{-i}$, passing through $q_3'^{-i}$, q_3^{-i-1} of y=0, limit a domain (D_{i-1}), i = 2,3..., belonging to the total basin of q_2. When $i \to \infty$, the (C_i) have an accumulation curve (C') $y = -f(x,a) + q_1$. (C') is the ω-invariant curve which passes through the degenerate saddle $(x = q_1, y = 0)$(fig. 6.2).

When $b = \varepsilon$, $|\varepsilon|$ having a sufficiently small value, the fixed point q_2 $(y(q_2) \to 0$ if $\varepsilon \to 0)$ is an attractive node, such that one of its multipliers $S_2 \to 0$ if $\varepsilon \to 0$. T_b is now a diffeomorphism, thus the basin (D) of q_2 must be only all in one block. From the total basin $\bigcup(\bar{D}_j)$, j = 0,1,2,..., of q_2 when b = 0, the only way for the

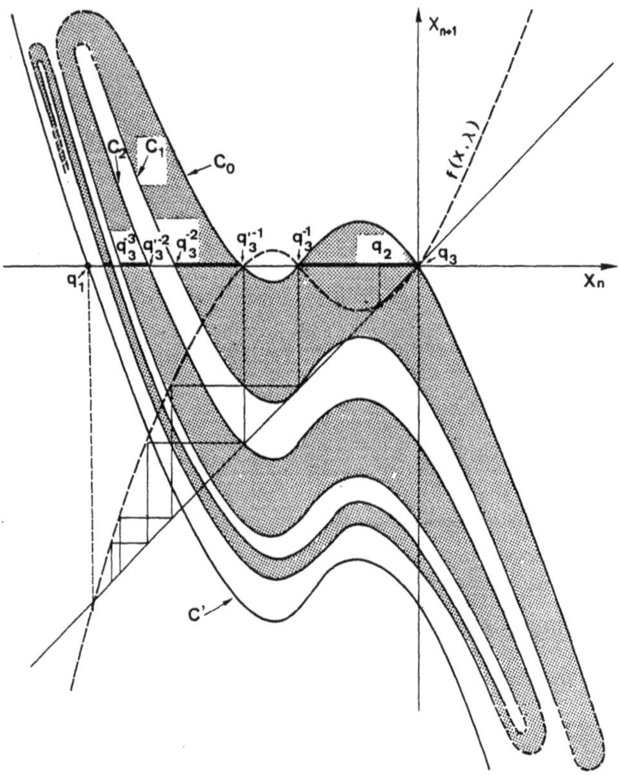

FIG. 6.2
The shaded region is the "oscillating" basin of the attractive fixed point q_2 . The ω-invariant curve of the saddle q_1, called here (C'), is the accumulation curve of these oscillations.

formation of (D) is the junction of the domains $(\bar{D}_j)$ for $|y| > M$, $M \to \infty$ if $b \to 0$, which permits to obtain an all in one block basin by communication of the $(\bar{D}_j)$. The basin (D) of q_2 is the shaded region of fig. 6.2. The boundary (F) of (D) is the ω-invariant curve of the saddle q_3. For $|y| < M$,(F) remain close to the curves (C_i) obtained for $b = 0$. The basin (D) presents itself as an "oscillating ribbon" the oscillations of which have an accumulation curve $(\bar{C}')$, lim $(\bar{C}')$ = (C') when $b \to 0$. $(\bar{C}')$ is the ω-invariant curve passing through the saddle q_1. It is worth of note that the type of junctions between the $(\bar{D}_j)$ for $|y| > M$ changes with the sign of b. So inversing the sign of b in fig. 6.2, the junctions (C_0) with (C_3), (C_1) with (C_2)... occur now for $y < 0$, in a different way.

6.2.3. α-INVARIANT CURVES OF THE MAP T_b, FOR SMALL VALUES OF $|b|$

Consider the case $f(x,a) \equiv 1-ax^2$, and the two fixed points

$$q_1\ (x = [-e-(e^2+4a)^{\frac{1}{2}}]/(2a),\ y = bx),\quad q_2\ (x = [-e+(e^2+4a)^{\frac{1}{2}}]/(2a),\ y=bx),$$

$e = 1-b$, q_1 is a saddle. Suppose a,b such that q_2 is a saddle (hypothesis of this paragraph). The α-invariant curve (unstable manifold) passing through q_2 is given in fig. 6.3 for $b > 0$ ($a=1.4$; $b=0.3$). This α-curve is constituted by two segments : α_1 (continuous line in fig. 6.3), α_2 (broken line). Starting from q_2 on α_2, let C be nearest point corresponding to a minimum of the radius of curvature. C is approximately the point of α_2 having the greatest abscissa. Let C_i, $i = 1,2,3,...$, be the consequent of rank i of C. Let C' be the point of the segment C_1C_3 corresponding a minimum of the radius of curvature, C'_i, its consequent of rank i. Let $y(C_j)$, $y(C'_j)$ be the ordinates of C_j, C'_j, $C_0 \equiv C$, $C'_0 \equiv C'$. The following result was presented in [M28] (1978) and [G35] [M36] :

<u>(P6.1)</u> *When* $b \to 0^+$, *then* $C_j \to C'_j$, $y(C_j) \to 0$, $j = 0,1,2,...$, *and the points* C_j, *which are defined for the bi-dimensional diffeomorphism(6.1), tend toward the critical points of the one-dimensional endomorphism (6.3),* C, C_1, C_2, ... *of rank* 1,2,3,...

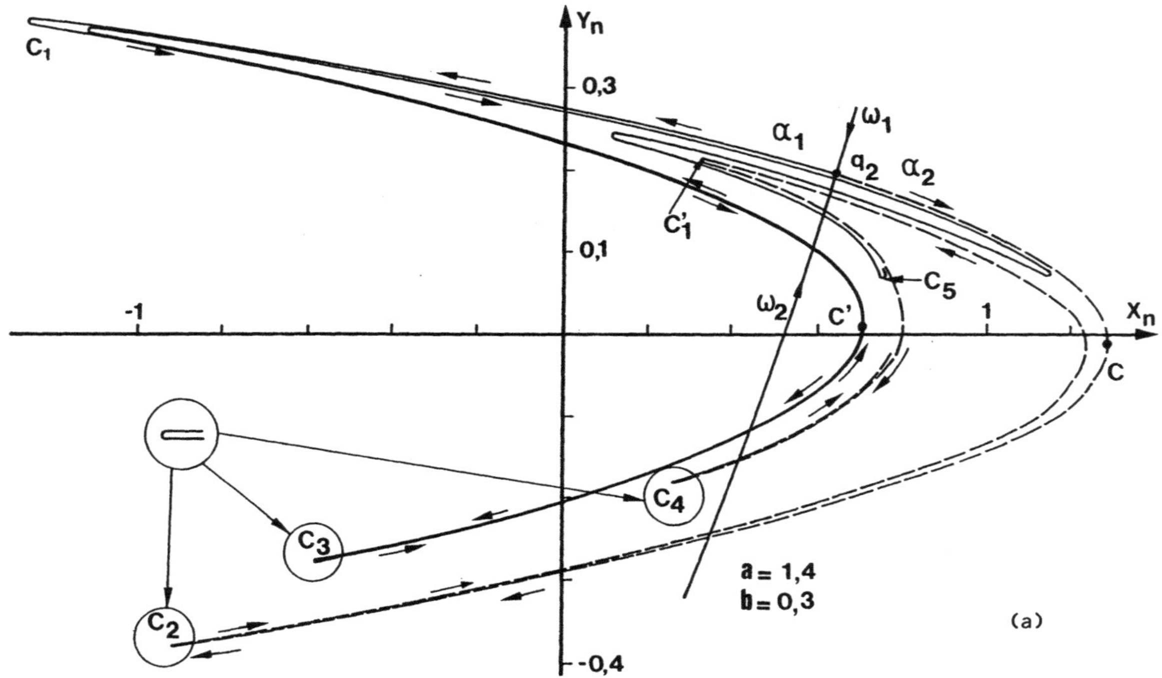

FIG. 6.3 - <u>The α-invariant curve of the saddle</u> q_2.

When $b \to 0$ this curve is crushed on the x-axis, with $\lim_{b\to o} C_i = \lim_{b\to o} C'_i$, $i=0,1,2,\ldots$, $C_o \equiv C$, $C'_o \equiv C'$, the limits of these points being the critical points of the one-dimensional endomorphism.

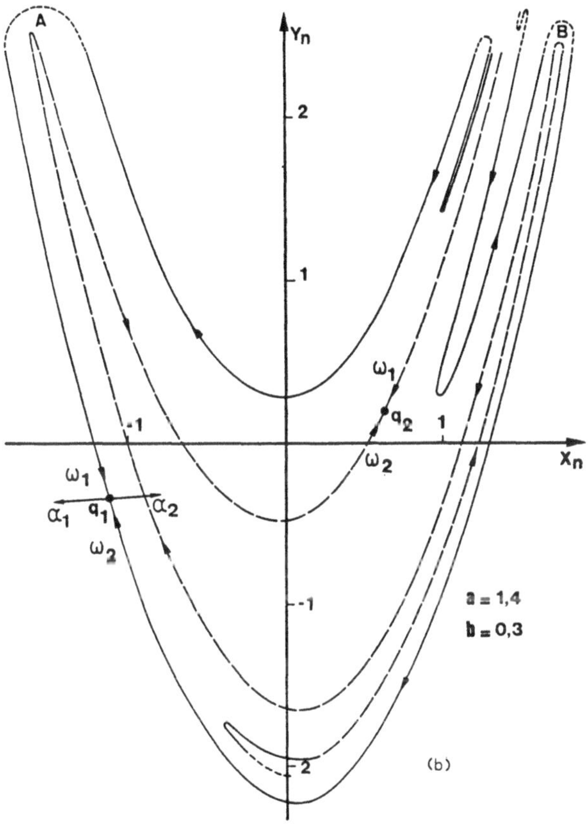

FIG. 6.3.b - The ω-invariant curves of the saddles q_1, q_2 . The parameters are the same as in fig. (a) . These invariant curves emanate from the degenerate ω-invariant curves obtained when b = 0 . This point appears through the parabolic aspect of the curves when b ≠ 0 .

Considering the saddle fixed point q_1, a branch α_1 of the α-invariant curve is such that $x \to -\infty$, the other branch α_2 passes between the branches $\alpha_1, \alpha_2 \in q_2$ without intersecting these ones (for a diffeomorphism a branch α (resp ω) cannot intersect an other branch α(resp ω)). With $b < 0$, the role of $\alpha_1, \alpha_2 \in q_2$ is now played by $\alpha_2 \in q_1$, for the definition of the points C_j, C'_j.

For T_b, $b = 0$, *in the phase plane* (x,y), *the absorptive segment* $\overline{CC_1}$ *on* $y = 0$ *is the degenerate* α*-invariant curve of* q_2 *resulting from the crushing of the* α *curve for* $b \neq 0$ *on* $y = 0$ [M28]. So it appears that the intersections of $\overline{CC_1}$ on $y = 0$ with the degenerate ω-invariant curve of q_2 give the *homoclinic points* related to q_2. The intersections of $\overline{CC_1}$, with the degenerate ω-invariant of q_1 give degenerate *heteroclinic points* . The segment $\overline{q_1C}$ on $y = 0$ belongs to the ω-invariant curve of q_1.

From these considerations it results that *the notions of homoclinic, and heteroclinic points can be extended to the one-dimensional endomorphism* T_0 [G35]. So $\overline{CC_1}$ is the α-set (unstable manifold) of q_2, the interval$)-\infty$, $q_1)$ and $\overline{q_1C}$ constitute that of q_1. The ω-set (stable manifold) of q_2 is the set of points made up of q_2 and all its antecedents of rank 1,2,... The repulsive fixed point q_2 and all repulsive cycles belonging to $\overline{CC_1}$ are limit points for the ω-set of q_2. The *homoclinic trajectory associated with* q_2 is q_2 and the subset of the antecedents of q_2 belonging to $\overline{CC_1}$. *The heteroclinic trajectory related to*(q_1 , q_2) is given by q_1, q_2 and the subset of the antecedents of q_1 belonging to $\overline{CC_1}$. Equivalent definitions can be obtained for α, or ω -sets, homoclinic and heteroclinic trajectories of points of cycles. All these definitions are related to those given by Sharkovskij [S 2] in 1969 without using the notion of degenerate invariant curve. Considering the cycles of order 2^i, $i = 1,2,\ldots$ of the Myrberg's spectrum ω_1, their α and ω sets, it results that T_0 *has homoclinic points if, and only if, it has cycles of order* $\neq 2^i$ [S 2] [S 3].

The points considered in this paragraph have some immediate and important consequences the first basic elements of which appeared in [M28] (1978), [G35] (1980) :

<u>(P6.2)</u> (a) Homoclinic, or heteroclinic, tangencies, generated by the one-dimensional endomorphism T_b*, are obtained for the values* $\hat{a}$ *(which can be of type* $\tilde{a}$, a_k^{*j}, ${}^{*}a_k^{j}$, $a \equiv \lambda$*) defined in § 4.3.4.*

(b) Every homoclinic (resp heteroclinic) point of T_0 *is the limit of an infinite sequence of homoclinic (resp heteroclinic) points of* T_b *when* $b \to 0$.

(c) Every value $\hat{a}$ *(which can be of type* $\tilde{a}$, a_k^{*j}, ${}^{*}a_k^{j}$ *)* $(a \equiv \lambda)$*, defined in § 4.3.4 for* T_0*, and located on the a-axis of the parameter plane* (a,b)*, is the common point(sumit)of a tangled fan* $(\hat{F})$ *made up of an infinite set of bifurcation curves* $\hat{\Lambda}^{(i)}$ *corresponding to "intermediary" homoclinic tangencies. The fan* $(\hat{F})$ *is bounded by two bifurcation curves of homoclinic tangencies. One curve* $\bar{\Lambda}^{*}$*characterizes the limit existence of homoclinic points. The other* Λ^{*}*characterizes the limit situation giving the maximum possible number of homoclinic points for the considered cycle, i.e., crossing through* Λ^{*} *from the interior part of the fan* $(\hat{F})$*, all the possible homoclinic points related to the cycle have been created. Outside* $(\hat{F})$ *from the* Λ^{*}*side, homoclinic tangencies related to the considered cycle are not possible.*

(P6·2a) results from the definition of α and ω-sets of T_0.(P6.2b) appears from the α-curve of fig. 6.3 which is crushed on $y = 0$, when $b \to 0$. (P6.2c) is due to the existence of the double sequence of points C_i, C'_i, $i = 0,1,2,$ $C_0 = C$, $C'_0 = C$, $\lim C_i = \lim C'_i =$ critical point of T_0 , when $b \to 0$. Indeed an infinite number of α-segments with "self-accumulation" passes between C_i, C'_i , $b \neq 0$, and so an ω-invariant curve passing between these points may give infinite possibilities of homoclinic and heteroclinic tangencies. The boundaries $\bar{\Lambda}^{*}$ and $\Lambda^{*}$of the fan $(\hat{F})$ are obtained for a tangency close to the points C'_i, and in another situation for a tangency close to the points C_i. The following result was proved by Holmes & Whitley, [H10] (1984) and gives a fuller information.

<u>(P6.3)</u> In the parameters plane (a,b) *from each of the bifurcation points* $a_{k2^i}^{*j}$ *of the a-axis, grows a "tangled Cantor fan" consisting of uncountably many bifurcation curves of distinct homoclinic tangency which traverse the* (a,b) *plane between* $b = 0$ *and* $b = \pm 1$*. Uncountably*

many bifurcation curves giving quadratic tangencies occur, but there exists only a countable sequence of bifurcation curves giving cubic tangencies.

6.2.4. COMPLEX BASINS : CHAOTIC TRANSIENTS, FUZZY (OR CHAOTIC) BASINS BOUNDARY

Consider (6.1) such that $f(x,a)$ is given by fig. 2.17b (§ 2.8.3). For T_b , $b = 0$, the fixed points q_1, q_2 are degenerate saddles with a multiplier $S_1 > 1$, $S_2 = 0$, and the structure of the basin of the degenerate attractive node A is known in the phase plane (x,y), from the knowledge of the strange repeller $(\bar{F})$ contained inside the segment $\overline{q_1 q_1^{-1}}$ (§ 2.8.3). The α-sets of q_1 and of the repulsive cycles points of $(\bar{F})$ lie on the segment $\overline{AC}$. The ω-sets intersect $y = 0$ only at points of the segment $\overline{q_1 q_1^{-1}}$,thus which countains all the homoclinic and heteroclinic points. In the (x,y) plane the ω-set of a point u of a cycle is made up of the curve $(C_0) y = -f(x,a) + u$ and of the union of all the curves (C_i^j) $y = -f(x,a) + u_{-i}^{(j)}$, $i = 1,2,\ldots$, $j = 1,2,\ldots,2^i$, $u_{-i}^{(j)}$ being one of the rank i antecedents of u. These ω-sets,and their limits when the cycles order $k \to \infty$,constitute the strange repeller$(\bar{F})$ which is an "internal" boundary of the basin of A. The "external" basin boundary is the degenerate ω-invariant curve of the degenerate saddle q_2, which passes through the point q_2^{-1}(without real antecedent) on $y = 0$, with the equation $y = -f(x,a) + q_2$. When $b \neq 0$ the "external" boundary of the basin of A is the ω-invariant curve of the saddle q_2, the two branches of which are without loop, because q_2^{-1} has no real antecedent. Let (D) be the domain limited by this "external" boundary. The "internal" basin boundary is made up of all the ω-invariant curves of all the saddle cycles of order $k = 1,2,\ldots$(on the segment $\overline{q_1 q_1^{-1}}$ when $b = 0$) and of their limit sets when $k \to \infty$. These ω-invariant curves make very complex tangled loops without intersection. The loops are generated by the junctions of the curves (C_i^j)(related to the point u) at infinity when $|b|$ is infinitely small. In particular, it is the case of the ω-curve of q_1 which defines the region of *chaotic transient* inside (D).

Starting from f(x,a) given by fig. 2.17a, *the bifurcation which gives rise to this chaotic transient corresponds to the last homoclinic tangency of the invariant curves of the saddle* q_1 *such that all the possible homoclinic and heteroclinic points, born from* $\overline{q_1 q_1^{-1}}$ *when* $b \neq 0$, *have been created* (curve such as Λ^* in(P6.2c). The chaotic transient inside (D) occurs as long as the α-invariant curve of q_1 does not give heteroclinic points which the ω-invariant curve of q_2.

As soon as such heteroclinic points appear, one has a situation of fuzzy basin boundary which can be explained from that of T_0 (§ 2.8.4) and that of T_b, b = 0. Indeed, in the (x,y) plane with T_b (b = 0), the immediate basin (D_0) of the degenerate node A is bounded by the two degenerate ω-invariant curves $y = -f(x,a) + q_1$, $y = -f(x,a) + q_2$. The successive rank i antecedents of (D_0), i = 1,2,... form an "arborescent" infinite sequence of domains (D_i^j) $j = 1,2,\ldots,2^i$. But with respect to the case of § 6.2.2., the (D_i^j) now have *infinitely many accumulation curves* $(C')y = -f(x,a) + q$, q being a point of a degenerate saddle cycle of order k belonging to $\overline{q_1 q_1^{-1}}$, and the infinite set of the curves (C') have a limit set when $k \to \infty$. Considering the § 2.8.4 it results that the disjoint domains (D_i^j) jumble together in a very complex manner with a complementary infinite sequence of regions belonging to the basin of the point at infinity. This situation characterizes the fuzzy boundary separating the two basins. When $b = \varepsilon$, $|\varepsilon|$ being sufficiently small, the disjoint domains (D_0), (D_i^j) communicate for $|y| > M$, $M \to \infty$ if $\varepsilon \to 0$, and form an all in one block basin (D). This basin has a very complex shape of an "oscillating" ribbon, with infinitely many accumulations of "oscillations"' loops, tangled in a complementary way with the basin of the point at infinity. The region where occurs this tangle of basins is defined by the first loop of the ω-invariant curve of the saddle q_1 (first loop = the monotonic segments on the both sides of the first extremum from q_1). It is a region of homoclinic and heteroclinic points. The resulting tangle of basins leads to a fuzzy boundary separating the two basins, and to the problem of impredictibility discussed in § 2.8.

Such two-dimensional results have been published in [M30] (1978) and after in [G35] [M46] [M47] (for the one-dimensional case cf.

the references given in § 2.8). Since 1983 a set of very interesting studies have been made about this subject with [G6]-[G10][M3][M4][R5][R6]

6.3. BEHAVIOUR OF THE DIFFEOMORPHISM IN THE CONSERVATIVE CASE

6.3.1. QUADRATIC CASE b = -1. PROPERTIES OF THE PHASE PLANE

As indicated in § 6.1 the properties of (6.1) with $f(x,a) \equiv 1-ax^2$, $b = -1$ are obtained via a transformation of variables which leads to the map (6.4), with $F(X) \equiv \mu X + (1-\mu)X^2$, the properties of which are detailed in [G35] [G36]. With $b = -1$ the Jacobian determinant $J = 1$, and the map (6.1) is of orientation preserving type. The fixed points of (6.1) are q_1 $(x = (-1-\sqrt{1+a})/a,\ y = -x)$, $q_2(x = (-1+\sqrt{1+a})/a, y=-x)$, with the multipliers

$$S_i = -ax \pm \sqrt{a^2x^2-1}, \quad i = 1,2, \quad x = x(q_1) \text{ or } x(q_2).$$

For $-1 \leqq a \leqq 3$ the multipliers of q_2 are

$$S_{1,2} = \exp(\pm j\varphi),\ j = \sqrt{-1},\ tg\varphi = -\sqrt{1-a^2x^2}\,/\,(ax),\ \cos\varphi = 1 - \sqrt{a+1}$$

and q_2 is a centre. The fixed point q_1 is a saddle of type 1. With (6.4) q_2 becomes the fixed point $0(X=Y=0)$, centre for $-1 \leqq \mu \leqq 1$, $S_{1,2} = \exp(\pm j\varphi)$, $\cos\varphi = \mu$; q_1 becomes the saddle $(X=1\ ,\ Y=0)$. The invariant curves of (6.4) have $Y=0$ as axis of symmetry. Those of (6.1), $b = -1$, have $y=-x$ as axis of symmetry. Fixed points and cycles are real only when $a \geqq -1$

Consider the k points of a cycle and join the successive points (defined by consequents) by means of straight lines producing thus a polygon with k corners. One of the characteristics of a cycle of order $k > 2$ is its *rotation index* r, equal to the number of turns made around an interior point of the polygon when following successive consequents of (x,y), until one comes back to (x,y). For conciseness, a cycle saddle (or centre) of order k and rotation numer r is designated by saddle (or centre) $k_{(r)}$. It is related to Poincaré's rotation number r/k.

Consider a family of two cycles, born of the same bifurcation, and consisting of a cycle saddle $k_{(r)}$ of type 1 and a cycle centre $k_{(r)}$. By tracing the two invariant curves (four segments) passing through each point of the saddle $k_{(r)}$, one finds that these invariant curves either merge smoothly with the invariant curves passing through other points of the same saddle cycle, or they intersect the latter curves, but remain inside an annulus containing all cycle points $k_{(r)}$.L In the first case, the singularity configuration so obtained is called an *islands structure*, and in the second an *instability ring* in the sense of Birkhoof. In both cases the invariant curve segments emanating from the saddle $k_{(r)}$ surround the points of the cycle centre $k_{(r)}$. According to Birkhoff [B12], the phase plane picture of a conservative recurrence like (6.1) or (6.4), around a reference fixed point centre (φ not giving exceptional cases in the sense of § 5.8, 5.9) is in general as shown in fig. 6.4 The fixed point centre is first surrounded by a set islands structures corresponding to rational Poincaré's rotation number r/k, alternating with smooth closed invariant curves (non-singular in the sense that they do not cross through any saddles of a finite order). These islands structures then are surrounded by instability rings, each bounded by at least two smooth closed and non-singular invariant curves.

It was shown in [G25] [G26] [G27] [G28] [G 35] [G36] that the solutions structure of fig. 6.4 is not sufficient to account all features of (6.1), or (6.4). Oneof the reasons of this situation is the existence of cycles saddles of type 3 for which no place has been provided in the singularity configurations of fig. 6.4. Remark that from an inspection of (6.4), the parameter $\delta = 1-\mu = \sqrt{a+1}$ obviously appears as a sort of measure of the strength of the non-linearity. So in the mentioned references the following results were proved, via a mixed "numerical-qualitative" way :

-(a) Let $P_m\ [k_{(r)}]$ be the m-th point (x_m, y_m) of a saddle cycle of type 1, $k_{(r)}$. Let S_1, S_2 be its multipliers, p_1, p_2 its eigen-directions. Let $B_m\ (\bar{x}, \bar{y})_m$ be the coordinates of the homoclinic points located on the ω-segment bounded by the saddle fixed point q_1 of (6.1) (or (1;0) for (6.4)) and its point of the axis of symmetry the farthest from q_1.

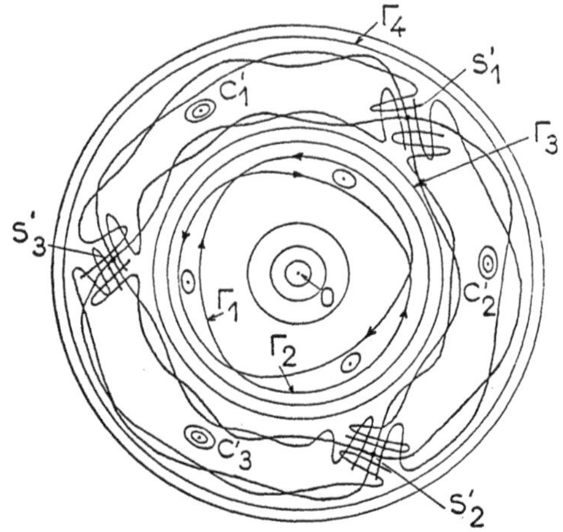

FIG. 6.4 - Phase plane picture according Birkoff

The center 0 is first surrounded by a family of regular invariant curves, then by one or more islands structures defined by the closed singular invariant curves Γ_1, Γ_2 passing through saddles of type 1, and finally by one or more instability rings, defined by the intersecting singular invariant curves emanating from the saddles of type 1 S'_1, S'_2, S'_3. These intersecting invariant curves are limited from below and above by the closed regular invariant curves Γ_3 and Γ_4 respectively. Inside the instability rings there exist closed regular invariant curves surrounding centers C_1, C_2, C_3. (cf also |A5|)

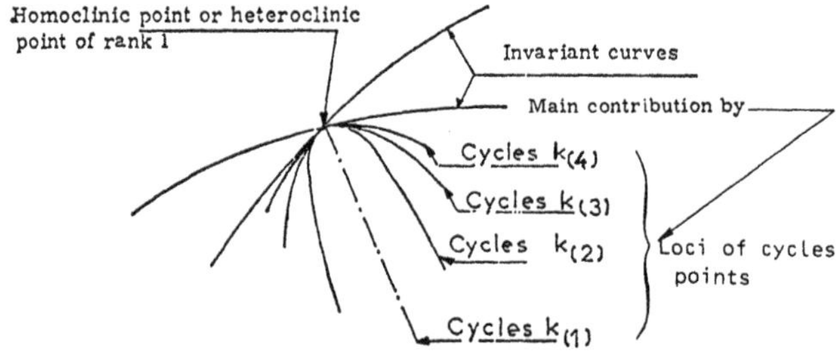

FIG. 6.5

When $k \to \infty$, there exists a set of associated cycles saddles $k_{(r)}$ of type 1, and saddles $k_{(r)}$ of type 3 such that :

(6.14) $r/k \to 0$, $S_1 \to \infty$, $\lim P_m[k_{(r)}] = B_m$, $p_1 - p_2 \to \pi_1 - \pi_2$,

where π_1, π_2 are the slopes of the ω-curve, and the α-curve at B_m. When converging toward B_m, points P_m are aligned on the curves of fig. 6.5. Homoclinic points appear thus as degenerate cycles of infinite order (cf also fig. 6.6)

- (b) When $k \to \infty$, and $r \to \infty$, there exists a set of saddle cycles $k_{(r)}$ of type 1 such that

$$(6.15) \qquad \lim r/k = \bar{\alpha}\ , \quad S_1 \to 1,\ P_m \to \bar{P}_m,\ p_1 - p_2 \to 0,$$

and a set of centres $k_{(r)}$, associated with the first, such that $\varphi \to 0$.

- (c) Considering the phase plane situation for the same strength of the non linearity $\delta = 1-\mu = \sqrt{a+1}$, and different k, and the situation corresponding to the same k and different δ, it appears that an increase of the distance of a cycle from the fixed point centre (q_2 for (6.1)) plays the same role as an increase of δ for a fixed distance.

- (d) The "amplitude" of a "loop" formed by the invariant curves between two consecutive homoclinic points increases rapidly as δ increases, or as the distance from the fixed point centre increases (k increases for the same r). In the same conditions the multiplier $S_1 > 1$, and the angles between the eigenslopes p_1, p_2 ,increase.

- (e) The existence of islands structures and instability rings is limited to cycles having a rotation number r/k such that $r/k > \alpha_1$ for a given parameter a (or μ).

- (f) Let α_2 be the r/k value reached when the loops of the invariant curves of a saddle $k_{(r)}$ have sufficiently large amplitudes to intersect the invariant curves of the fixed point saddle q_1 for (6.1), (1;0)

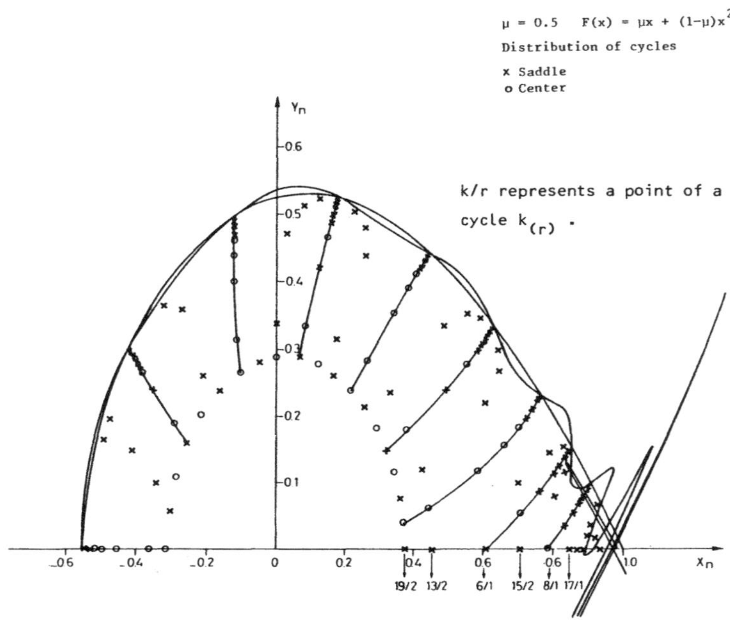

FIG. 6.6.a - Recurrence (6.4) with b = -1 - (|G28||G35|)

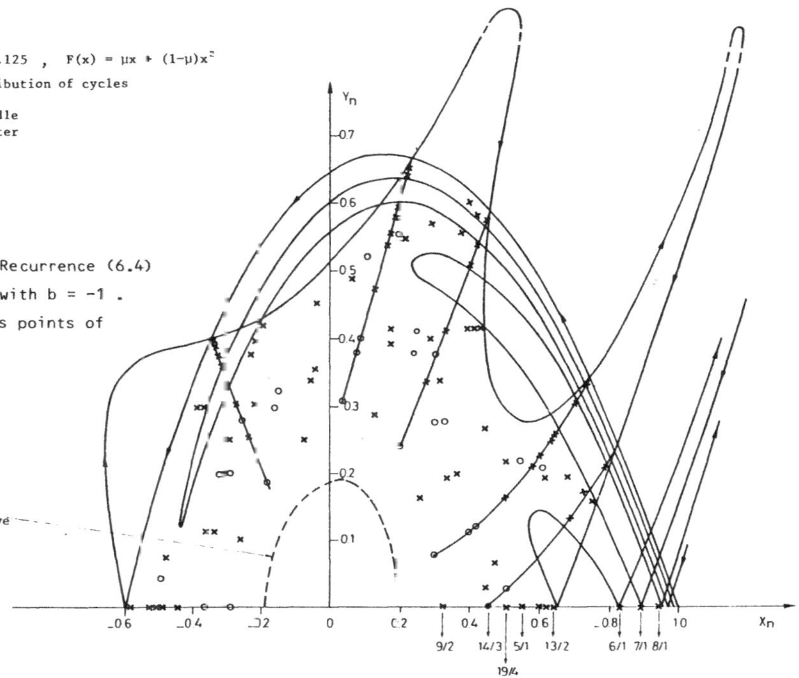

FIG. 6.6.b - Recurrence (6.4)
with $b = -1$.

k/r represents points of cycles $k_{(r)}$.
(|G28||G35|)

for (6.4). For $\alpha_1 < r/k < \alpha_2$, the invariant curves of two cycles, having two different but close r/k, intersect thus giving heteroclinic points. The appearance of heteroclinic points at the smallest possible distance of the fixed point q_2 for (6.1), or (0;0) for (6.4), may be used as an equivalent definition of α_1. In the region defined by $r/k > \alpha_1$, the distance s_n of an iterated sequence, n = 0,1,2,..., from q_2, or (0;0) is bounded. For $\alpha_1 > r/k > \alpha_2$ the distance s_n grows in the mean, but at a less than exponential rate. The corresponding region is called a *region of diffusion*. For $r/k < \alpha_2$ the mean distance s_n grows exponentially. Since the consequents or the antecedents of an initial point (x_0, y_0) form geometrically a complicated sequence, this region is called region of *chaotic instability*. Finally the total region containing intersecting invariant curves, Birkhoff's instability rings included, is a region of chaotic behaviour.

6.3.2. QUADRATIC CASE b = -1. LOCAL BIFURCATIONS

For (6.1) with $-1 \leqq a \leqq 3$, b = -1, the fixed point centre q_2 (or for (6.4) with $-1 \leqq \mu \leqq 1$ the centre (0;0))undergoes the following sequence of bifurcations [G28'] (1972) [G35] [G36] (1980) :

(6.16) centre q_2 $\rightarrow$ centre q_2 + islands structure $k_{(r)}$.

The islands structure $k_{(r)}$ is released from q_2 , each time $\varphi(q_2) = \varphi(r/k) = 2\pi r/k$, r,k being relatively prime integers, and exists only for $\varphi > \varphi(r/k)$, i.e.

$$a > a(r/k) = \cos^2(2\pi r/k) - 2\cos(2\pi r/k), \text{ or } \mu < \cos(2\pi r/k).$$

Let s(x,y) be the "geometrical" distance of one of the two cycles $k_{(r)}$ from q_2, and let s(a) = a(r/k)-a the "parametric" distance. When s(a) increases, the variables s(x,y), $\varphi(k_{(r)})$, $|p_1 - p_2|$ increase from 0, and $S_1(k_{(r)})$ increases from 1 (φ, S_1 are respectively the argument of the multiplier of the cycle centre $k_{(r)}$, and the multiplier greater than one of the cycle saddle $k_{(r)}$, the two cycles constituting the islands structure $k_{(r)}$). Thus it appears that the *bifurcations sequence* (6.16) *has the structure "boxes in files"* described in § 2.9.

Similar bifurcation to that given by (6.16) occurs for a centre cycle defined by r_1/k_1, every point of which releases an islands structure defined by r_2/k_2. The result is a cyclic islands structure defined by $(r_1.r_2)/(k_1.k_2)$. More generally the islands structure $(r_1.r_2...r_m)/(k_1.k_2...k_m)$, m = 1,2,3,..., contained inside the islands structure $(r_1.r_2...r_{m-1})/(k_1.k_2... k_{m-1})$, can be defined, which gives rise to a *"box-within-a-box" structure in the phase plane.*

When a (resp.μ) increases from -1 (resp.1) the angle φ of the cycle centre $k_{(r)}$, born from (6.16), increases from 0 to π, value giving a flip bifurcation by crossing through $\varphi = \pi$, or $S_1 = S_2 = -1$, such that [G28][G35][G 36]

(6.17) centre $k_{(r)} \rightarrow$ saddle $k_{(r)}$ of type 3 + centre $(2k)_{(2r)}$

The invariant curves passing through every point of the saddle $k_{(r)}$ of type 3, surrounds two neighbouring points belonging to the centre $(2k)_{(2r)}$. These curves forms a eight-shaped figure with homoclinic points (fig. 6.7). For (6.1), $f(x,a) \equiv 1 - ax^2$, with q_2 this bifurcation occurs for a = 3 (μ=-1). More generally equivalent bifurcations by doubling order (period) with a Myrberg's cascade occurs :

(6.18) centre $2^i k_{(r_i)}$ $\rightarrow$ type 3 saddle $2^i k_{(r_i)}$+ centre $2^{i+1} k_{(r_{i+1})}$

i = 0,1,2,..., k - 1,3,4,..., for a = $a'(k,r_i)$, and lim $a'(k,r_i)$ = $a'(k,r_i)_s$ when $i \rightarrow \infty$. For k = 1, the cascade (6.18) occurs when $a \geq 3$ ($\mu \leq -1$).

The cycles, which can be represented by $k_{(r)}$, or $(r_1 ... r_n)/(k_1 ... k_m)$, constitute a particular set of cycles generated from the bifurcation (6.16), or from equivalent bifurcations when a cycle centre takes the place of the fixed point q_2. Cycles of order k (cycle (k)), which do not belong to this set, are generated from the following fold bifurcation [G28][G35][G36] :

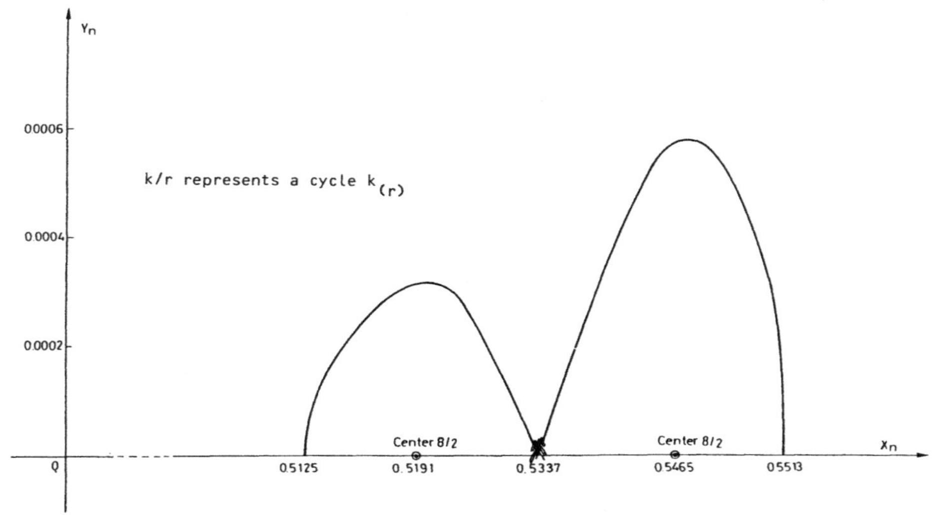

$\mu = -0.014$, $F(x) = \mu x + (1-\mu)x^2$. First loops of invariant curves passing through a 4/1 saddle of type 3 $S_2 = -1.2226$, $p_1 = -0.0545$

FIG. 6.7.a – Recurrence (6.4) with $b = -1$.
(according to |G28||G35||G36|)

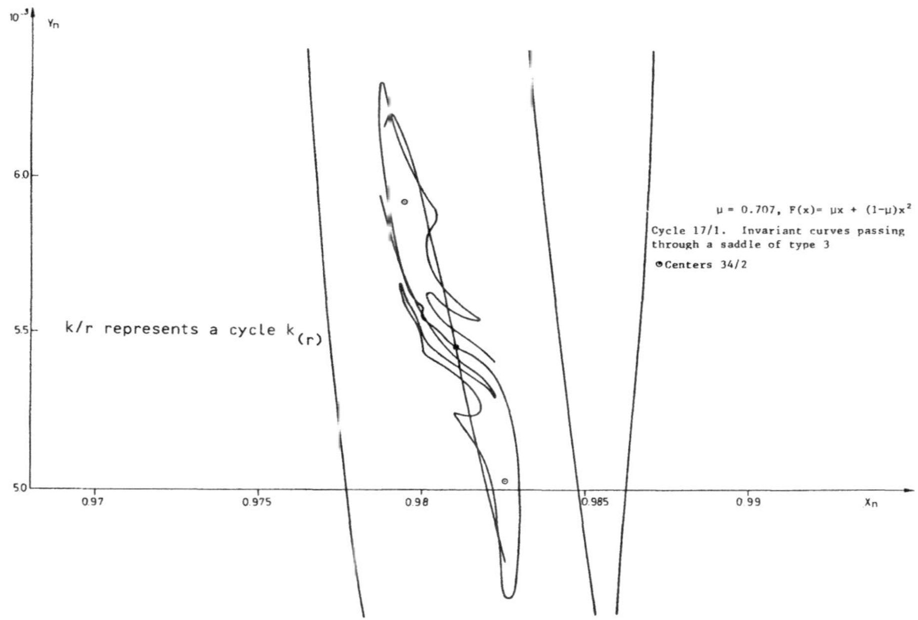

FIG. 6.7.b - Recurrence (6.4) with b = -1 . (|G28||G35||G36|)

(6.19) cusp cycle (k) → centre cycle (k) + type 1 saddle cycle (k)

when the angle φ of the centre is $\varphi = 0$, the multipliers of the saddle being $S_1 = S_2 = +1$. These two cycles merges and gives the cusp cycle at the bifurcation, and the absence of such cycles before the bifurcation.

In the language ofthe differential topology the following bifurcations (6.20), (6.21) are called *co-dimension two bifurcations* :

(6.20) centre cycle (k)→ type 1 saddle cycle (k)+ 2 centre cycles (k)

(6.21) type 1 saddle cycle(k)→ centre cycle(k)+ 2 type 1 saddle cycles (k)

When each bifurcation occurs the angle $\varphi = 0$ for the centres, and $S_1 = S_2 = +1$ for the saddles. For (6.1) with $f(x,a) \equiv 1 - ax^2$, these bifurcations occur with an increasing non linearity strength δ . Such bifurcations are due to typical communications between sheets of the foliated structure of the parameter plane (a,b) (§ 6.9).

Considering cycles $k_{(r)}$ *the crossing through exceptional cases* (§ 5.8, 5.9) corresponds to the following bifurcation diagram [G28'] [G 35] [G36] :

(6.22) saddle $k_{(r)}$ + centre $\bar{k}_{(\bar{r})}$ → centre $\bar{k}_{(\bar{r})}$+ saddle $k_{(r)}$

The bifurcation occurs when the angle φ of the cycle centre $\bar{k}_{(\bar{r})}$ is $\varphi = 2p\,\pi/q$, with the conditions of an exceptional case in the sense of § 5.8, 5.9, and $k = q\bar{k}$. At the bifurcation the k points of the saddle cycle $k_{(r)}$ merge into the $\bar{k}$ points of the centre cycle $\bar{k}_{(\bar{r})}$. Around each point of the centre cycle $\bar{k}_{(\bar{r})}$, the crossing through the bifurcation gives rise to a rotation of angle π/q of the figure formed by the q points of the saddle cycle $k_{(r)}$ which merge into a point of the centre $\bar{k}_{(\bar{r})}$. The merging gives a complex saddle with 2q branches. With respect to other cycles, the saddle cycles $k_{(r)}$ of (6.22) numerically are very sensitive (rapid divergence of consequents in presence of rounding error, cf [G36] p. 136). *With respect to all other cycles* $k_{(r)}$ *the cycles centre* $\bar{k}_{(\bar{r})}$ *of* (6.22) *are not generated by a bifurcation of type*

(6.16), *but by the bifurcation* (6.19). For (6.1) with $f(x,a) \equiv 1-ax^2$, such a bifurcation occurs with $k = 3$, $r = 1$, $\varphi(q_2) = 2\pi/3$, at $a = 5/4$ ($\mu = -1/2$), and the bifurcation (6.19) at $a = 1$ ($\mu = 1-\sqrt{2}$). The cycle centre $3_{(1)}$ ($\bar{k}=3$, $\bar{r}=1$) born from (6.19) undergoes the bifurcation (6.22) with $k=9$, $r=3$. It is conjectured that it is also the case for the saddle cycles $k = 3^i$, $r = 3^{i-1}$, associated with centre cycles $\bar{k} = 3^{i-1}$, $\bar{r} = 3^{i-2}$ $i = 2,3,\ldots$

6.3.3. NON QUADRATIC CASE b = -1

Different results concerning (6.4) have been obtained since 1972 with [G25] [G26] [G27] [G28] [B 7] and after with [G35] [G36] (1980). These results can be transfered to (6.1) via a non linear transformation equivalent to that of § 6.1.

A first set of results concerns the case of a cubic non linearity with $F(X) \equiv \mu X + (1-\mu)X^3$. In the phase plane, results qualitatively equivalent to those of the quadratic case are obtained, but with two fixed points saddle of type 1. The bifurcation (6.16) occurs for an even order $k \geq 4$. But for an odd order $k \geq 3$, this bifurcation becomes :

(6.23) $\quad$ centre $q_2 \rightarrow$ centre q_2 + 2 islands structures $k_{(r)}$.

The two islands structures, with complementary symmetry and the same multipliers , are released from q_2, which is the point (0;0) for (6.4). Bifurcations (6.17) - (6.21) also occur in the cubic case, but (6.22) does not occur for $\varphi = 2\pi/3$. With respect to the quadratic case new bifurcations are encountered :

(6.24) centre $k_{(r)} \rightarrow$ centre $k_{(r)}$ + 2 centres$(2k)_{(2r)}$ +
2 type 1 saddles $(2k)_{(2r)}$

(6.25) type 3 saddle $k_{(r)} \rightarrow$ centre $k_{(r)}$ + type 1 saddle $(2k)_{(2r)}$

which occurs when the angle φ of the centre $k_{(r)}$ is equal to π ($S_1=S_2=-1$)

A second set of results concerns the case of bounded non linearities with $F(X) \equiv \mu X+(1-\mu)X^2 \exp[(1-X^2)/4]$, or $F(X) \equiv \mu X + 2(1-\mu)X^2/(1+X^2)$. For $-1 < \mu < 1$, (0;0) is a fixed point center ; A (1;0) is a type 1 saddle with the exponential non linearity, and a cusp point for the second one (bounded quadratic non linearity). With respect to the quadratic and cubic case, new properties appear in the phase plane. Indeed now there exist not only sequences of cycles, converging toward the homoclinic points of the invariant curves of A, from the interior of the domain (d), bounded by the segments of these curves joining A to their farthest point $x < 0$ on $y = 0$, but also sequences of cycles converging toward these homoclinic points from the exterior of (d). Moreover a third type of cycles appear : the *mixed cycles* i.e. cycles some points of which lie inside (d), others outside (d). The mixed cycles are centres, or type 1 saddles, or type 3 saddles. Other difference concerns the *region of diffusion*. Now the distance s_n (defined in § 6.3.1.f) is bounded. The diffusion is caused by heteroclinic points related to intersections of the invariant curves coming from two, or more type 1 saddle cycles $k_{(r)}$, $k'_{(r')}$,... associated with centre cycles $k_{(r)}$, $k'_{(r')}$. Consider the case of two associations $k_{(r)}$, $k'_{(r')}$. Then the iterated sequence (x_n, y_n) presents series of long chaotic transients in a region of influence of a pair of cycles $k_{(r)}$ or $k'_{(r')}$, separated by short chaotic transients when passing from one influence zone to the other influence zone. Equivalent behaviour occurs in presence of more than two pairs of cycles. Such a region is called a *region of bounded diffusion* and it is due to the existence of mixed cycles.

Another possible singularity is the *unstable centre* (unstable eliptic point in [B10] [B12]) defined by Birkhoff, in the two cases : *with α and ω-branches*, and without *α and ω-branches*.

More details, and the figures corresponding to all the described situations of § 6.3.1. - 6.3.3 can be found in [G35] [G36].

6.3.4. BEHAVIOUR FOR b = +1

The Jacobian determinant of (6.1) now is $J = -1$, and (6.1) becomes an orientation reversing map. But T_b^2 is an orientation preserving map. Thus a part of the results obtained when b=-1 can be transfered to the case b=+1. Now fixed points and cycles exist only for $a \geqq 0$ when $f(x,a) \equiv 1-ax^2$. The fundamental bifurcation (6.16) does not occur for the fixed point q_2 which is a type 3 saddle for $a \geqq 0$, but for all the cycles centre of order $k \geqq 2$. In particular for the cycles of order $k = 2^i$, i = 1,2,..., born from the Myrberg's cascade ω_1. So, consider the cycle centre $k = 2^1$, obtained when $0 \leqq a < 1$, the coordinates of which are $(x = -y = a^{-\frac{1}{2}})$ and $(x = -y = -a^{\frac{1}{2}})$. An island structure, which separately surrounds each of these points, is released from them, each time $\varphi = 2\pi p/(k/2^1)$ when $a = (1-\cos\varphi)/2$.

6.4. BEHAVIOUR OF THE DIFFEOMORPHISM IN THE ALMOST CONSERVATIVE CASE

A set of results for (6.4) with $b = \alpha -1$, $0 < \alpha \ll 1$, are given in [B.7] (1973) [G35] [G36] (1980). Considering the relation between the maps (6.1) and (6.4) such that $J = 1-\alpha$, these results also concern (6.1), and α plays the role of a damping term.

The introduction of this small damping causes very different qualitative behaviours with respect to the case $\alpha = 0$. So the cycles generated from (6.16) or(6.23)do not appear from q_2. But all cycles pairs centre and saddle of type 1 are replaced by cycles pairs focus and saddle of type 1, or cycles pairs node and saddle of type 1, generated from a fold bifurcation (6.8). When the bifurcation occurs, the double cycle node-saddle appears at a distance $d(\alpha)$ from q_2, $\lim d(\alpha) = 0$ if $\alpha \to 0$. With $0 < \alpha << 1$, the mentioned cycles node and focus are attractive, and the map is in situation such that for fixed values of a,b there exist infinitely many attractive cycles (cf also [N 6]). The basins of all these attractors are separated by fuzzy boundaries.

All cycle points $k_{(1)}$ converge to homoclinic points on the invariant curves of $q_1(f(x,a) \equiv 1-ax^2)$ in essentially the same manner as in the conservative case. The points of the cycles $k_{(r)}$, $r > 1$ approach the homoclinic points in a more irregular manner than for $\alpha = 0$. The presence of small damping affects the invariant curves of q_1 very little, producing merely some asymmetry with respect to the y-axis.

Another important qualitative difference occurs for the invariant curves of the cycles saddle of type (1) $k_{(r)}$. Indeed α-segments of these curves can converge toward attractive cycles focus, or toward the attractive fixed point focus q_2, spiraling but also forming many S-loops of the form , which become longer and flatter as they approach the attractor (cf [G36] p.191-195), and so close to the attractor the α-segment becomes practically indistinguishable from a smooth spiral. An example is given fig. 5.47a.

The bifurcation (6.22) does not occur. So for $\bar{k} = 1$, $q = 3$, the distance between q_2 and the saddle $3_{(1)}$ passes through a minimum and then increases.

Non conservative recurrences admit heteroclinic points of an additional type (called N-type in [G 36]) which is qualitatively different from that of conservative recurrences. Indeed in conservative maps, heteroclinic points characterize the overlapping, and thus the destruction, of neighbouring islands structures, whereas the N-type heteroclinic points characterize the tendency of α-invariant curves to terminate at attractive singularities, and ω-segments at repulsive ones. In other words, the existence of N-type heteroclinic points is associated with the existence of "preferential" attractors, like the fixed point q_2. The presence of such attractors constitutes a source of weakly chaotic behaviour for some "ordinary" attractors, because α-invariant curves of distant origin may pass arbitrarily close to the latter, while being effectively on the way to the former.

More details about the almost conservative case can be found in [G36]

6.5. QUADRATIC CASE. FIRST SET OF PROPERTIES OF THE PARAMETER PLANE

Consider (6.1) with $f(x,a) \equiv 1 - ax^2$. T_b has two fixed points $q_1(x=[b-1-(1-2b+b^2+4a)^{\frac{1}{2}}]/(2a),\ y=bx)$, $q_2(x=[b-1+(1-2b+b^2+4a)^{\frac{1}{2}}]/(2a), y=bx)$, the multipliers of which are $S_i = -ax \pm (a^2x^2 + b)^{\frac{1}{2}}$, and the bifurcation curve $\Lambda_{(1)_0}$ $(S_1 = +1)$ $(k = 1$ in (6.8)) is :

$$4a + (1-b)^2 = 0$$

The equation of the curve $\Lambda_1 \equiv \Lambda_{(2^1)_0}$ is obtained by writing the multiplier of q_2 $S_1 = -1$, or the merging of the two points of the cycle of order 2^1, $x = [b-1 \pm \sqrt{4a - 3(b-1)^2}]/(2a)$, $y = b(1-ax^2)/(1-b)$, i.e. :

$$4a - 3(b-1)^2 = 0$$

The multipliers of this cycle are the roots of the equation :

$$[b + 4a^2(1+y)x - 4a^3x^3 - S](b-S) + 2abx[2a^2x^2 - 2a(1+y)] = 0$$

Thus the equation of $\Lambda_{2^1} \equiv \Lambda_{(2^2)_0}$ is obtained for $S = -1$:

$$4a - 5b^2 + 6b - 5 = 0.$$

For $i > 2$ the equation of a curve $\Lambda_{(2^i)_0}$ becomes analytically complicated, and theses curves are obtained numerically via the equation(6.6) In the region $a < a[\Lambda_{(1)_0}]$, the cycles of order 2^i, $i = 0,1,2,\ldots$, belonging to a Myrberg's cascade cannot exist. It is assumed that it is the case for all possible cycles of T_b, i.e. the non-wandering set of T_b is empty at finite distance. Till now this conjecture was not baffled.

Consider now the results of § 6.2.3. The bifurcation curves corresponding to homoclinic, or heteroclinic, tangencies can be obtained numerically (for example via the algorithm of [K 1]). Among these curves, those which are the boundaries $\bar{\Lambda}^*$; $\Lambda^*$of a tangled fan (F) of (P6.2) present a particular interest. When b = 0, chapter 3 has shown that

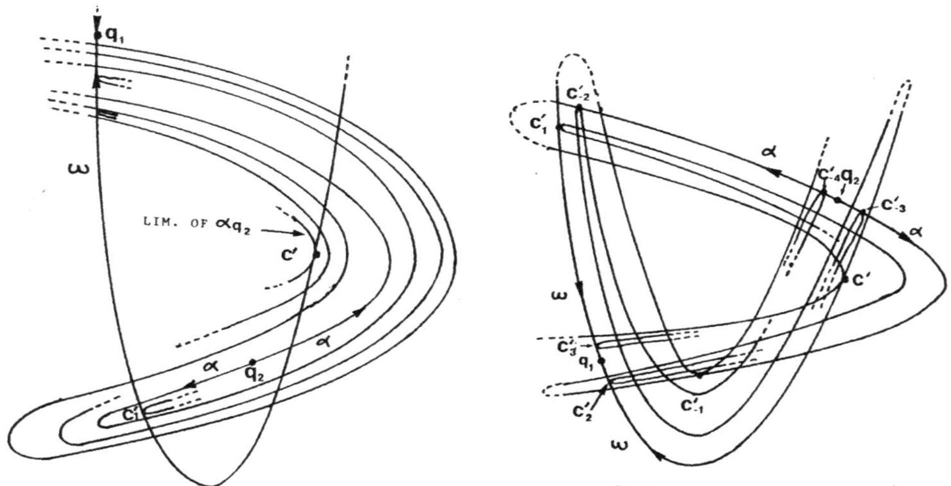

FIG. 6.8 - Last heteroclinic tangencies.

for $a > a_1^* = 2$ the non-wandering set of T_o is complete and made up of repulsive points. Through the point ($a = a_1^*$, $b = 0$) passes the curve Λ_1^*, from which all the possible homoclinic, and heteroclinic points have been created (cf (P6.2)). *Then for* $a > a[\Lambda_1^*]$ *the non-wandering set of* T_b *is complete, and made up of repulsive points.* The curve Λ^* is constituted by two segments $b < 0$, and $b > 0$. For $b < 0$, Λ_1^* corresponds to the last homoclinic tangency between a branch α and a branch ω of the saddle fixed point q_1. The corresponding $\alpha(q_1)$ branch is then a limit curve for the branch $\alpha(q_2)$ of q_2, and thus Λ_1^* is "indirectly" related to the last heteroclinic tangency between $\omega(q_1)$ and $\alpha(q_2)$ branches (fig. 6.8a). For $b > 0$, Λ_1^* corresponds to the last heteroclinic tangency between the $\alpha(q_2)$ branch and the $\omega(q_1)$ branch. But now the corresponding $\alpha(q_2)$ - branch is a limit curve for the branch $\alpha(q_1)$ of q_1, and thus Λ_1^* ($b > 0$) is "indirectly" related to the last homoclinic tangency between $\omega(q_1)$ and $\alpha(q_1)$ (fig. 6.8b).

Fig. 6.6 suggest that for $a > a[\Lambda_1^*]$, T_b is in the situation of the *Smale Horseshoe* [S 8] [G18]. Then it is known that T_b has an invariant (non-wandering) Cantor set Ω, and T_b restricted to Ω is conjugate to the two-sided shift on two-symbols [G 18]. This question will be considered with more details in § 6.6. It is also known that Ω contains countably many cycles, countably many homoclinic and heteroclinic points, and a dense iterated sequence [G18]. The cycles are dense in Ω, and they are of saddle type. Thus for $a > a[\Lambda_1^*]$ all the cycles have been created. From this point, and from the above conjecture, it results that *all the bifurcation curves of* T_b *are located inside the region (box* Ω_1*) defined by* $a \geq a[\Lambda_{(1)_0}]$, $a \leq a[\Lambda_1^*]$ (fig. 6.9).

From the Birkhoff's theorem (P5.2) it results that *the homoclinic points form a part of the non-wandering set* Ω *of* T_b.

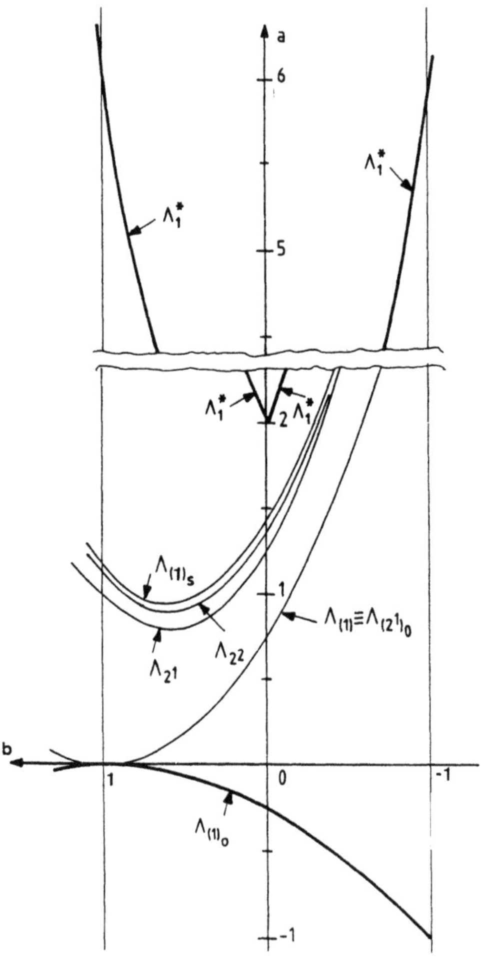

a
6
5
2
1
0
1
-1
-1
b
Λ_1^*
Λ_1^*
Λ_1^*
Λ_1^*
$\Lambda_{(1)_s}$
Λ_{2^2}
Λ_{2^1}
$\Lambda_{(1)} \equiv \Lambda_{(2^1)_0}$
$\Lambda_{(1)_0}$

FIG. 6.9

6.6. THE SMALE HORSESHOE AND THE TWO-DIMENSIONAL QUADRATIC DIFFEOMORPHISM. REPRESENTATION OF THE CYCLES

6.6.1. THE SMALE HORSESHOE

Consider the phase plane (x',y') and a square S with side length equal to one, the centre of which is $(x' = y' = 0)$, the sides being parallel to the axes. The map T leading to the Smale horseshoe is such that $T(S) \cap S$ consists of two components which are mapped rectilinearly by T, according to fig. 6.10a. T can be considered as making a linear horizontal expansion and a vertical contraction of S, by multipliers factors S_1 and S_2 respectively, followed by a folding, the folded portion being outside S, such that restricted to $T(S) \cap S$ (two horizontal bands), T is linear.

By changing the sense of stretching with a folding according fig. 6.10b, $T^{-1}(S) \cap S = T^{-1}[T(S) \cap S]$ is made up of two vertical bands on each of which T has a constant Jacobian [G18] equal to the product $S_1 S_2$, $0 < S_1 < 1/2$, $S_2 > 2$. The points x which remain in S by iterations $T^n x$, $-\infty < n < +\infty$ form a Cantor set Λ. Every point $x \in \Lambda$ is characterized by a bi-infinite binary sequence and one has the following theorem (for the proof cf for example [S8] [S9] [G18]):

(P6.4) The horseshoe map T has an invariant Cantor set Λ which contains : (a) a countable set of cycles of any order ; (b) an uncountable set of bounded non cyclic iterated sequences ; (c) a dense iterated sequence. Moreover T is structurally stable.

Consider now the quadratic diffeomorphism T_b :

$$(6.26) \qquad x_{n+1} = 1 - ax_n^2 + y_n , \qquad y_{n+1} = bx_n,$$

and its inverse T_b^{-1} :

$$(6.27) \qquad x_n = y_{n+1}/b, \qquad y_n = x_{n+1} + ay_{n+1}^2/b^2 - 1.$$

For sufficiently large values of a, $a > a[\Lambda^*_1]$, the square S (ABCD)

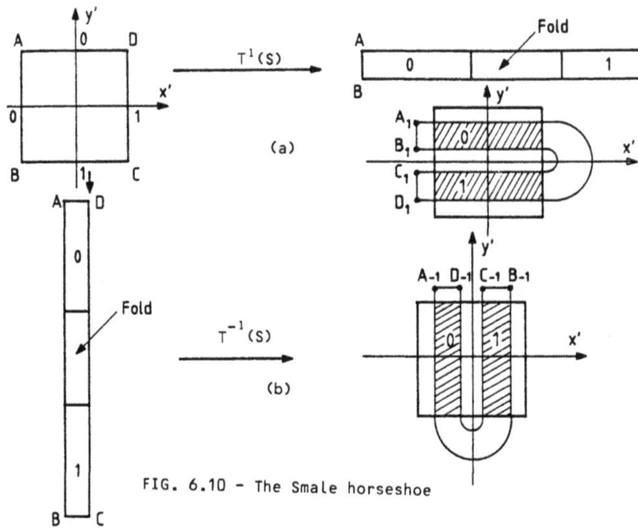

FIG. 6.10 - The Smale horseshoe

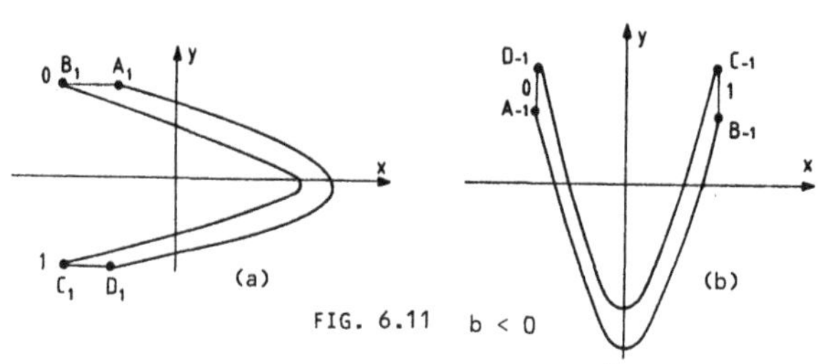

FIG. 6.11 $b < 0$

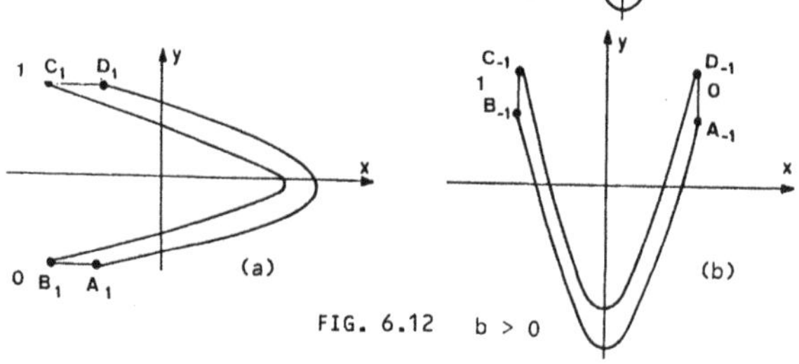

FIG. 6.12 $b > 0$

of fig. 6.10 is mapped by T_b, $b < 0$, so that $T_b(S)$ has a shape (fig. 11a) equivalent to that of $T(S)$ (fig. 10a). With $T_b^{-1}(S)$ a shape (fig. 11b) equivalent to that of $T^{-1}(S)$ (fig. 10b) is obtained. The embedded endomorphism T_0 into T_b is defined by a function $f(x,a)$ with maximum, thus according to the Myrberg's ordering law on the x-axis (P3.9), the $x < 0$ part of S is denoted "0" and the $x > 0$ part of S "1" (here "0" < "1"). So one has $T_b(\overline{AB}) = \overline{A_1B_1}$ associated with "0", and $T_b(\overline{CD}) = \overline{C_1D_1}$ associated with "1", which gives a coding of the y-axis for $b < 0$ (fig. 6.11a).

Considering $T_b^{-1}(S)$, y_n is defined from a function $f(y,a) \equiv ay^2/b^2-1$ with a minimum, thus according to (P3.9) the $y < 0$ part of S is denoted "1" and the $y > 0$ part "0" (here "1" < "0"). So $T_b^{-1}(\overline{BC}) = \overline{B_{-1}C_{-1}}$ is associated with "1", and $T_b^{-1}(\overline{AD}) = \overline{A_{-1}D_{-1}}$ is associated with "0", which gives a coding of the x-axis for $b < 0$ (fig. 6.11.b).

For $b > 0$, T_b is orientation reversing and the equivalent coding is given by fig. 6.12.

Thus it results that for $a > a[\Lambda_1^*]$ T_b given by (6.26) is a horseshoe map (cf [H13] 1984) of two different types : $b < 0$ and $b > 0$.

6.6.2. REPRESENTATION OF THE CYCLES. SYMMETRY.

From § 6.6.1, there is a 1-1 correspondence between the Cantor set Λ of T and the set Ω of the non-wandering points of T_b for $a > a[\Lambda^*_1]$. It was shown in § 3.3 and 3.5 that a cycle $(k;j)$ of T_0 can be represented by a binary word $[q^{(1)j}_k]$, $q^{(1)}$ being the point of minimum (resp maximum) abscissa when T_0 is defined by a continuous function with a unique minimum (resp maximum). Moreover the rotational shifts on the left, and on the right defined in § 3.5 lead to the consideration of a bi-infinite binary sequence $]q^{(1)j}_k[$ associated with the cycle $(k;j)$, which can characterize a cycle point of the horseshoe map T.

a) Case $b < 0$

Associated with the horseshoe map T, a *symbolic phase plane* (x',y') having the symmetry with respect to $y' = -x'$, is defined. The consideration of T^k, $k = \pm 1\ \pm 2,\ \pm 3, \ldots$, leads to more and more foldings when k

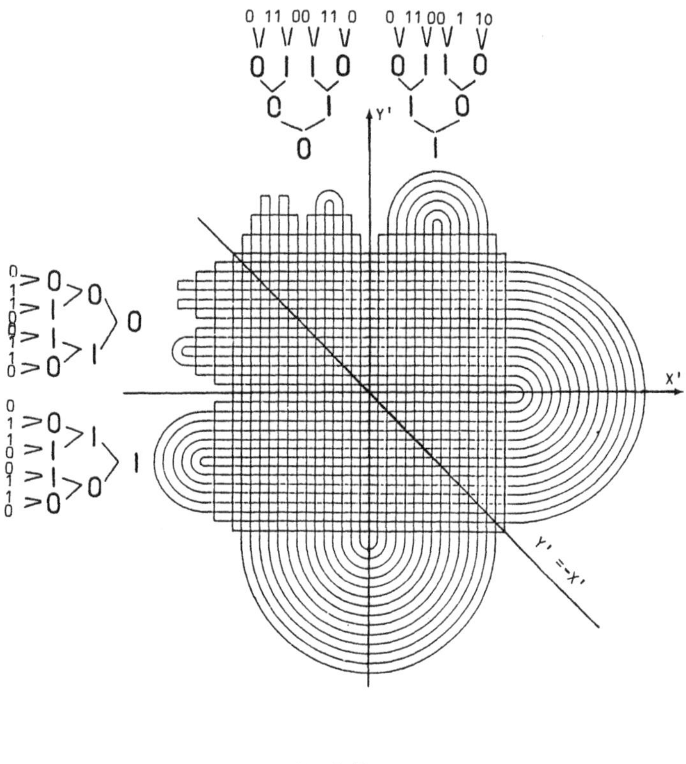

FIG. 6.13

The symbolic phase plane (x', y') when b < 0.

Coding of the x' and y'-axes up to k = 4 .

Folding of the horseshoe by T^k, k = ±1, ±2, ±3

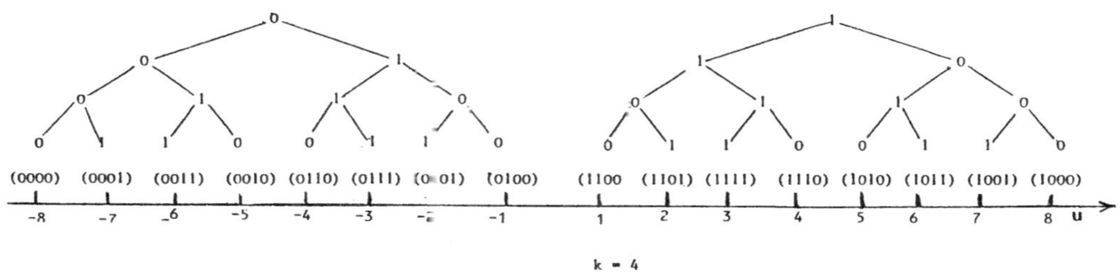

FIG. 6.14 - Numbering of the u-axis from the Myrberg's ordering and the horseshoe coding .

increases, with a coding on the x' and y'-axes which results of these foldings (cf fig. 6.13 for $|k| = 1,2,3$). So on the x'-axis, 2^{k-1} points $x' > 0$, 2^{k-1} points $x' < 0$ equally spaced, are located, and on the y'-axis 2^{k-1} points $y' > 0$, and 2^{k-1} points $y' < 0$. The 2^k points on the x'-axis are numbered from the coding which results from the extension of the schema of fig. 6.14 (inverse Myrberg's order, "0" < "1", because $f(x,a)$ in T_b has a maximum) limited to $k = 4$. Similarly it appears that the 2^k points on the y'-axis are numbered from the inverse coding ("1" < "0"). The numbering gives the points of coordinates $u = -2^{k-1}$, $-2^{k-1}+1$, ..., -1, ..., 2^{k-1} (conventionally 0 is not used) on the x'-axis, and every point is associated with a binary word constituted by k bits. The point 1 of the u-axis corresponds to the word (110^{k-2}), the point -1 to (010^{k-2}), the point $u = 2^{k-1}$ to the word (10^{k-1}), and the point $u = -2^{k-1}$ to the word (0^k). On the y'-axis, $y' = -u$.

Let $[q^{(1)j}_k] = [a_1a_2\dots a_k]$ be the binary word of the point $q^{(1)}$ of a cycle (k;j) of T_0 (b=0 in T_0). In the symbolic phase plane (x',y') this binary word corresponds to the coordinates of the point $q'^{(1)}$ $x' = u\,[a_1a_2\dots a_k]^{-1} \equiv u[a_ka_{k-1}\dots a_1]$ (because the numbering of x' is given from successive applications of T^{-1}), $y' = -u\,[a_1a_2\dots a_k]$ (the numbering of y' is given from successive applications of T). This point $q'^{(1)}$ is in a 1-1 correspondence with the point $q^{(1)}$ of T_b, $b < 0$, in (x,y) plane, $a > a\,(\Lambda_1^*)$. The k-1 other points $q'^{(i)}$, $i = 2,\dots,k$, of this cycle (k;j) are similarly obtained by considering $[q^{(i)}] = [a_i\dots a_ka_1\dots a_{i-1}] = \sigma^i\,[q^{(1)}]$, σ being the rotational shift. So joining by straight line segments the points $q'^{(i)}$ with the order given by $i = 1, 2,\dots,k$, the obtained figure in the (x',y') plane constitutes a representation of the cycle (k;j) (Fig. 6.15) for which the notions of *adjoint cycles* and *self adjoint cycle* of § 3.6 are directly extended.

Consider now the complete non-embedded representation of a cycle defined in § 3.7. From the above indicated construction of a cycle in the symbolic phase plane (x',y') one has the following properties [F 7] [K 5]

<u>(P6.5)</u> When $b < 0$ *for the diffeomorphism* T_b *given by (6.26), every cycle of* T_b *is in a 1-1 correspondence with a cycle of the horseshoe map of fig. 6.10, 6.11.*

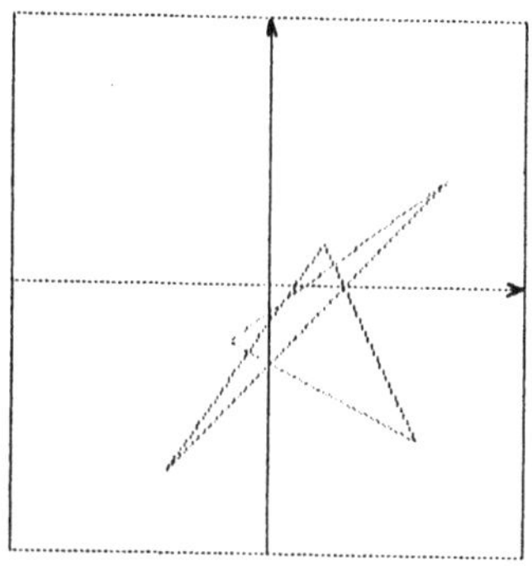

B<0:(1) 1 5 3 2 4 [10110]

FIG. 6.15 - Representation of the cycles of T_b in the symbolic phase plane. $b < 0$. The first number after the sign of b is j', followed by the decimal and the binary rotation sequences for T_o.

Self adjoint cycle (k=5; j'=1; s; n), its points are symmetrical with respect to y' = -x' .

(According to |K 4 |)

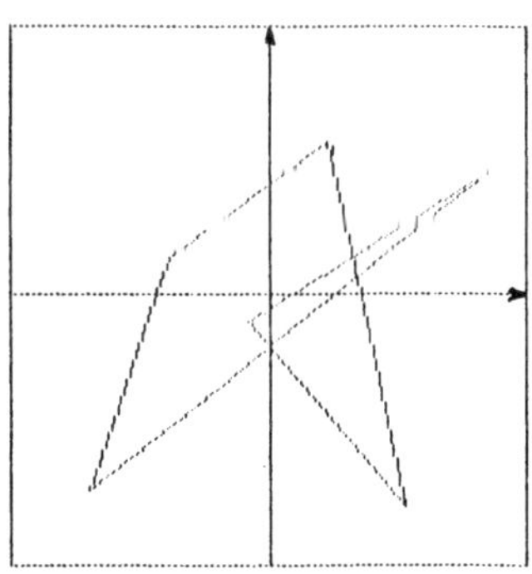

B<0:(2) 1 6 4 2 5 3 [100101]

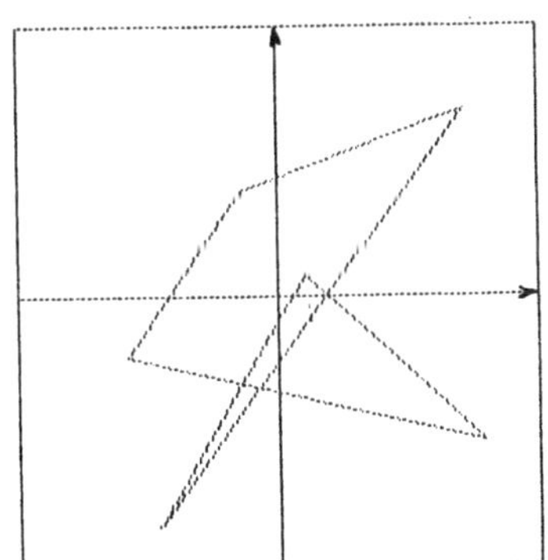

B<0:(3) 1 6 5 3 2 4 [100110]

The two cycles (k=6; j'=2; d; n) and (k=6; j'=3; d; n) are adjoint. The set of their points are symmetrical with respect to y' = -x', but each of these cycle has not this symmetry .

The points of a self adjoint cycle (kj'sp) *or* (kj'sn) *are symmetrical with respect to the axis of symmetry* $y' = -x'$ *(symbolical symmetry).*

The set of the points of two adjoint cycles $(kjd\alpha)$ $(kld\alpha), \alpha = n$ or p *has* $y' = -x'$ *as axis of symmetry, but each isolated cycle is not symmetrical with respect to* $y' = -x'$.

When $b = -1$, *the phase plane* (x,y) *of* T_b *has the same property if* $a > a[\Lambda_1^*]$

For equivalent results, obtained independently via another way cf [H11]

b) <u>Case $b > 0$</u>

The associated symbolic phase plane (x', y') is now different because T_b is orientation reversing, and the horseshoe is symmetrical with respect to $y' = x'$. The inversion of orientation corresponds to a coding of the x'-axis identical to that of the case $b < 0$, for a point $q^{(1)}$ of a cycle (k;j), $x' = u[a_k\ a_{k-1} \dots a_1]$, and to a coding of the y'-axis giving for this cycle $y' = u\ [\bar{a}_1\ \bar{a}_2 \dots \bar{a}_k]$ with $\bar{0} = 1$ and $\bar{1} = 0$.

An important difference occurs for the definition of two adjoint cycles, and a self adjoint cycle. Indeed these notions are now defined from the binary word $[\bar{q}^{(1)j}_{k}] = (\bar{a}_1\ \bar{a}_2 \dots \bar{a}_k)$ $\bar{0}=1$, $\bar{1}=0$. So the cycles $(kjd\alpha)$, $(kld\alpha), \alpha = n$ or p are *adjoint* if a rotational shift σ^α, $\alpha \geq 0$, exists such that $\sigma^\alpha\ [\bar{q}^{(1)}{}^{j}_{k}]^{-1} = [q^{(1)\ell}_{k}]$, $j \neq \ell$. When $j=\ell$ the corresponding cycle is said *self adjoint*. For example the cycle (6;1;d;p) $[q^{(1)}{}^{1}_{6}] = 101110$, $[\bar{q}^{(1)1}_{6}] = 010001$ is the adjoint of the cycle (6;4;d;p) $[q^{(1)4}_{6}] = 100010$ $(\alpha = 0)$. The cycle (6;3;s;n), $[q^{(1)3}_{6}] = 100110$, $[\bar{q}^{(1)3}_{6}] = 011001$ $(\alpha = 0)$ is self adjoint (fig. 6.16). Then the following proprieties occur [F 7] [K 4] :

<u>*(P6.6)*</u> *When* $b > 0$ *for the diffeomorphism* T_b*, every cycle of* T_b *is in a 1-1 correspondence with a cycle of the horseshoe map* T' *of fig. 6.12.*

The points of a self adjoint cycle are symmetrical with respect to $y' = x'$ *(symbolical symmetry).*

The set of the points of two adjoints cycles has $y' = x'$ *as axis of symmetry, but each cycle, separately considered, has not this symmetry.*

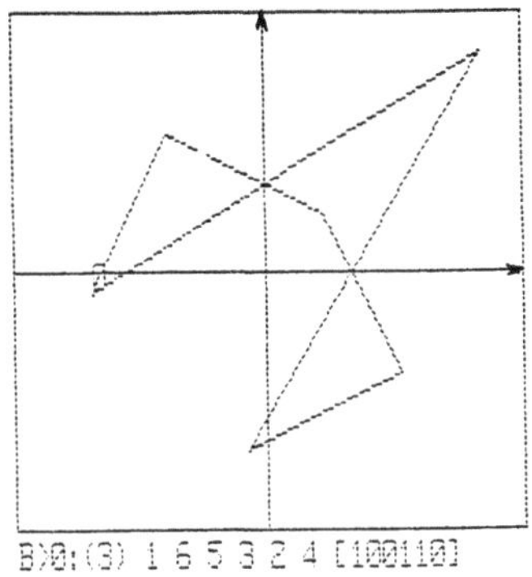

Self adjoint cycle (k=6; j'=3; s; n), its points are symmetrical with respect to y' = x' .

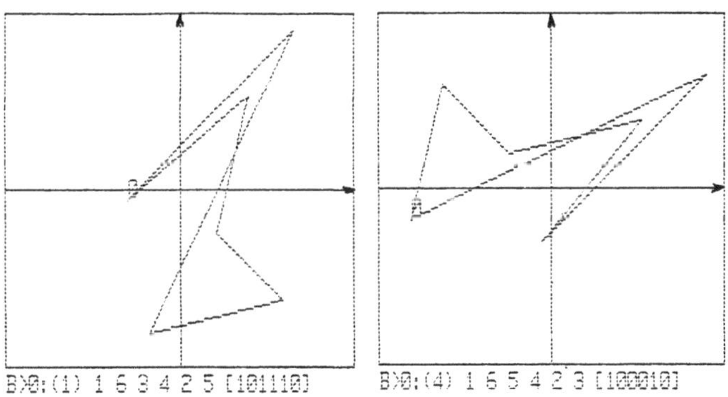

The two cycles (k=6; j'=1; d; p) and (k=6; j'=4; d; p) are adjoint. The set of their points are symmetrical with respect to y' = x', but each isolated cycle has not this symmetry .

FIG. 6.16 - Representation of the cycles of T_b , $b > 0$, in the symbolic phase plane. The first number after the sign of b is j', it is followed by the decimal and the binary rotation sequences of T_o (According to |K 4 |)

When $b = +1$, *the phase plane* (x,y) *of* T_b *has the same property if* $a > a[\Lambda_1^*]$.

c) Remark

From the different points of this paragraph it results that :

(P6.7) The non-embedded representation $(k;j')$ *of a cycle of the one-dimensional endomorphism* T_0, *or its complete form* $(kj'sp)$, $(kj'sn)$, $(kj'dp)$, $(kj'dn)$ (cf §3.7), *or the embedded representation (§ 4.3.2)* $(k;j)$ *of a cycle of* T_0, *defines unambiguously the characteristics of the corresponding cycle when* $b \neq 0$ *has a defined sign. These characteristics (representation in the symbolic phase plane* (x',y')*) are different for* $b < 0$, *and* $b > 0$. *The complete representation gives the properties of symmetry in the conservative case* $b = \pm 1$.

6.7. QUADRATIC CASE. BOX-WITHIN-A BOX FOLIATED BIFURCATION STRUCTURE

6.7.1. ORIGIN OF THE FOLIATION

In [M28] (1978) (cf also [G35]) a first result about the bifurcation structure in the parameters (a,b) plane was indicated. It is given here in the form :

(P6.8) The non-wandering set (including homoclinic points) of T_0 *and* T_b *are identical for* $b = \varepsilon$, *when* $|\varepsilon|$ *is sufficiently small. For* $b = \varepsilon$ *the bifurcations of* T_b *(intersections of the curves* $\Lambda^j_{(k)_0}$, $\Lambda^{ji}_{k2^i}$ *given by* (6.8), (6.9) *with* $b = \varepsilon$*) occur in the same manner as those of* T_0, *and thus the box-within-a-box bifurcations structure defined in chap.4 stretches from* $b = 0$ *to the parameters* (a,b) *plane.*

The "principal" box Ω_1 is bounded by the fold bifurcation curve $\Lambda_{(1)_0}$ and by the "last" curve of homoclinic tangency Λ_1^*. Ω_1 contains an infinite number of boxes of first and second kind (§ 4.3.2, 4.3.3). Every box has the form of ribbon with variable width in the (a,b) plane, its intersection with the axis $b = 0$ giving the corresponding box of T_0. The representation of such a box is that of its intersection with $b = 0$.

A box of first kind Ω_k^j is bounded by the fold bifurcation curve $\Lambda_{(k)_0}^j$ given by (6.8), and by a "last" curve of homoclinic tangency Λ_k^{*j} related to the basic cycles (k;j). For a second kind box $\Omega_{2^i k}^j$, the first boundary $\Lambda_{(2^i k)_0}^j$ is a flip bifurcation curve given by (6.9), the second one being the "last" curve of homoclinic tangency $\Lambda_{2^i k}^{*j}$. All the cycles related to the indexes k,j,... of Λ or Ω are unambiguously defined via (P6.5) - (P6.7). Moreover one has :

(P6.9) Every bifurcation curve starting from $b = 0$ *intersects the two axes* $b = \pm 1$ *of the parameters* (a,b) *plane.*

This assertion results from the first fact that every bifurcation curve is contained in the domain Ω_1, and thus cannot have an asymptote $b = u$, $-1 \leqq u \leqq 1$. The second fact is related either to the impossible existence of what is called a co-dimension two flip bifurcation for $-1 \leqq b \leqq 1$ (cf [H11] p. 62-63), or to horseshoe structure for $b = +1$ $a > a(\Lambda_1^*)$ which implies a 1-1 correspondence between the non-wandering set of T_0 and that of T_b, $b = \pm 1$ (cf § 6.6.2).

In [H4] (1981) [M32] (1981), the following result was proved :

(P6.10) For fixed values of b *which are not small, the bifurcations of* T_b *on the axis* b= *constant occur in sequences which are different from those of* T_0 $(b = 0)$. *In other words when* $b \neq 0$, *in the* (a,b) *plane a box* Ω_k^j *can intersect a box* Ω_m^ℓ, $j \neq \ell$, $k = m$, *or* $j = \ell$, $k \neq m$, *or* $j \neq \ell$, $k \neq m$, *the same occuring for boxes of second kind. So except particular points, every point of the* (a,b) *plane, belonging to the intersection of the two boxes, corresponds to the existence of two attractive cycles (if* $|b| < 1$) *of order* nk *and* pm $(n,p = 1,2,...)$.

The assertion (P6.10) is the consequence of (P6.11) and (P6.12) (below-mentioned) and founded on two fundamental properties already given :

- the first, formulated for $b = 0$, concerns the subset of rotation sequences, directly related to Poincaré's rotation numbers (§ 4.2.5), which are organized in the form of a *boxes in files bifurcations structure*

- the second for $b = \pm 1$ is related to the important bifurcation (6.16), and its equivalent when a centre cycle (k:j) takes the place of the centre fixed point q_2.

Indeed in (6.16) r/k is the rotation number associated with q_2. Every corresponding bifurcation value is the intersection of a fold curve $\Lambda^j_{(k)_0}$ with b = -1, or b = +1. Here the index j is defined from the rotation sequence (b = 0) related to r/k (cf. § 3.3.7 ; 4.2.5 ; 4.3.2 ; and fig. 4.2). This fold bifurcation curve intersects b = 0 at the point $a^j_{(k)_0}$ as a consequence of (P6.9) and § 6.6. Let (k_1; j_1) be a pair of cycles born of the bifurcation (6.16), and giving rise to an island structure, the index j_1 being associated with the rotation number r_1/k_1. According to § 6.3.2, a *box-within-a-box structure in the phase plane* (x,y), characterized by $\mu = (r_1 r_2 \ldots r_m)/(k_1 k_2 \ldots k_m)$, m = 1,2,3,... is generated from the centre cycle (k_1; j_1) via the bifurcation equivalent to (6.16). The curve $\Lambda^j_{(k)_0}$, $k = k_1 k_2 \ldots k_m$, j being defined from μ, belongs to a box Ω^j_k which is embedded into $\Omega^{j_1}_{k_1}$ with the degree m of nesting (§ 4.3.2). Thus it appears that *the box-within-a-box structure in the phase plane, obtained in the conservative case, is directly related to the box-within-a-box bifurcations structure in the parameter plane, and more precisely to the boxes in files structure implied by this latter* (§ 4.2.5 and 4.3.4).

Consider the boxes in files bifurcations structure generated from the centre fixed point q_2 by (6.16). The boundary $\Lambda^j_{(k)_0}$ of every corresponding box intersects b = -1 at the point(b = -1, $a = \bar{a}^j_{(k)_0}$) solution of (cf § 6.3.2) [M32](1981)

$$(6.28) \qquad \bar{a} = \cos^2\varphi - 2\cos\varphi,\ \varphi = 2\pi p/k\ ,\ -1 \leqq \bar{a} \leqq 3,\ p < k/2.$$

One has the intermediary result :

(P6.11) Every angle φ of (6.28) is related to an index j of coding of the box-within-a-box bifurcations structure. Moreover the set of such angles φ is related to a sub-set of the set of simple boxes. These boxes are defined from the boxes in files bifurcations structure Φ which is determinated by (4.7) , (4.9) *and fig. 4.9 when* b = 0.

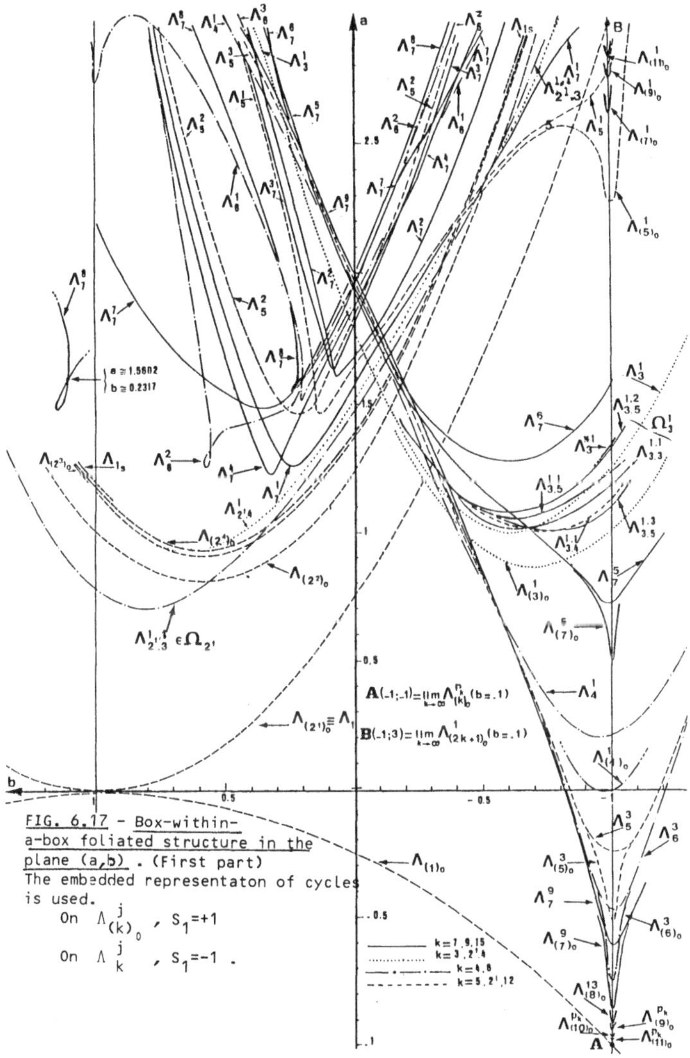

FIG. 6.17 - Box-within-a-box foliated structure in the plane (a,b). (First part)
The embedded representaton of cycles is used.

On $\Lambda^{j}_{(k)_0}$, $S_1=+1$

On Λ^{j}_{k}, $S_1=-1$.

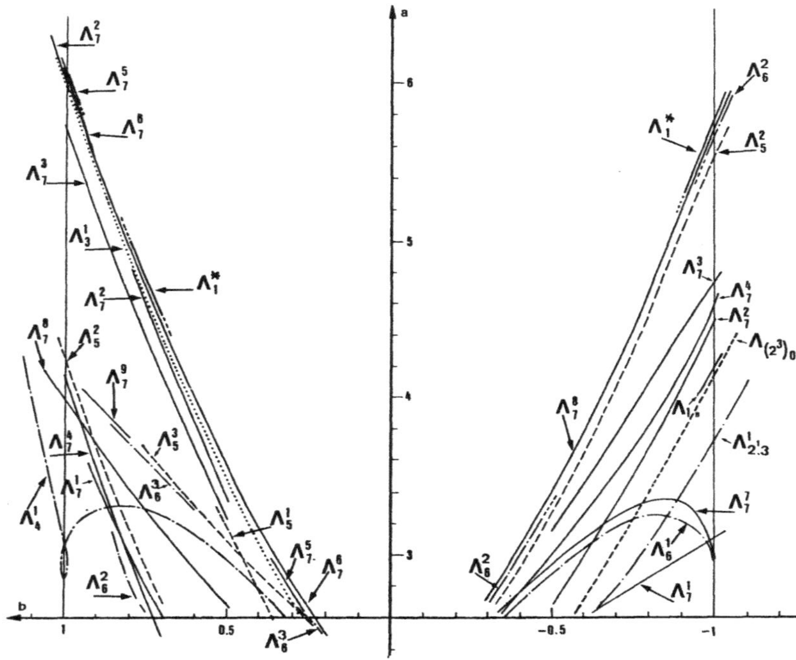

FIG. 6.17 - Box-within a-box foliated structure in the plane (a,b) . (second part)

The embedded representation of cycles is used .

So for $b = -1$, $p = 1$ in $\varphi = 2\pi\ p/k$, $k = 4,5,6,\ldots$ defines the file of complexity degree $D = 1$ of (6.28) which is associated (cf § 4.2.5) with the $\Lambda^{j}_{(k)_\circ}$, $j = p_k$. Here p_k is the index of the last pair of basic cycles of order k (having the rotation sequence $[u] = [1,k,k-1,\ldots,3,2]$ or $[r] = (+)^{k-2}$) in the increasing set of the rotation sequences (or bifurcation values $a^{j}_{(k)_\circ}$ of all the cycles of order k defined for b=0. Thus when k increases the intersections of $\Lambda^{j}_{(k)_\circ}$, $j = p_k$ with $b = 0$ give an *increasing set* of $a^{j}_{(k)_\circ}$, with $\lim a^{j}_{(k)_\circ} = a^{*}_{1} = 2$, when $k\to\infty$. From (6.28) it appears that the intersections of these bifurcation curves with $b = -1$ give a *decreasing set* of $\bar{a}^{j}_{(k)_\circ}$, with $\lim\limits_{k\to\infty} \bar{a}^{j}_{(k)_\circ} =$ $\lim\limits_{\varphi\to 0} a(\varphi) = -1$ (point A of fig. 6.17). It results a first justification of (P6.10), but this property is not limited to the particular set $j = p_k$ (D=1). Indeed considering the files of degrees $D = 2,3,4,\ldots$, (between two consecutive boxes of degree N-1 there are two files of degree N (§ 2.9)), from (6.28) and § 2.9, and from the results of § 3.3.8, 3.3.9, 3.4.2, 3.4.3 for b = 0, it appears that:

<u>*(P6.12)*</u> *The fold bifurcation curves* $\Lambda^{j}_{(k)_\circ}$, j *belonging to the sub-set of indexes related to rotation numbers, give a sequence of intersections with* b = -1 *which is reversed with respect to that obtained with* b = 0.

This result appears in [M32] (cf the relations of limit on the fig. 2 of this paper). Every file has a limit point given for $b = 0$ by § 3.4.2, 3.4.3, and for $b = -1$ by § 2.9. For example the file of boxes of degree $D = 2$, given by μ'_2 of (2.37) with $m = 2$, $\alpha_1 = 1,2,\ldots$, $k = m\,\alpha_1 + m + 1$, corresponds to the index $j = 1$ in the embedded representation, and on $b = -1$, $\varphi \to \pi$ when $\alpha_1 \to \infty$, the set of $\bar{a}$ is increasing $\bar{a}^{1}_{(k)_\circ} \to 3$ when $k \to \infty$ (pt B fig 6.17). For $b = 0$ the corresponding set is decreasing and $\lim a^{1}_{(k)_\circ} = a^{*}_{2^1}$ when $k \to \infty$. B is the intersection of $\Lambda^{1}_{(2^1)_\circ} \equiv \Lambda_1$ with $b = -1$. For $a > a_B$, $b = -1$, B is the limit point for $\Lambda^{1,j}_{(2^1.k)_\circ}$, $j = p_k$=ultimate running number (cf § 4.3.2), when $k \to \infty$

The most part of simple boxes belonging to the intervals Π of b=0, defined by (4.8), (4.9), intersect b =-1 for a > 3. The exceptional simple boxes, intersecting b = -1 for -1 < a < 3, appear from the bifurcation (6.19) (cf for example Λ_7^6 fig 6.17)

REMARKS

(a) In the file D = 1, k = 3, $\varphi = 2\pi/3$ ($\bar{a} = 5/4$) plays a particular role because for b = -1 an *exceptional case* is obtained (cf § 6.3.2), associated with the bifurcation (6.22). The intersection of $\Lambda^1_{(3)_0}$ with b = -1 is $\bar{a} = 1$ ($\mu = 1 - \sqrt{2}$ in (6.5) cf [G35]) which is different from a = 5/4 given by $\varphi = 2\pi/3$ (intersection of Λ_3^1). Thus in a file of any degree, an exceptional case changes the nature of the intersection, it is not the curve $\Lambda^j_{(k)_0}$ which must be considered as intersecting b = ± 1.

(b) All these results recur in a fractal way by means of the box-within-a-box bifurcations structure. More generally, whatever be the sign of b, every box reproduces (with deformation but with conservation of the qualitative properties) what occurs inside the box Ω_1.

(c) It is recalled that *a bifurcation curve belonging to a Myrberg's spectrum (or cascade), given by* (6.9), *cannot intersect another curve belonging to the same spectrum* (§ 6.1). The set of the corresponding sheets of fig. 6.1, associated with this spectrum, is called a spectrum of sheets. Resulting from (P6.10) it appears that *the parameters* (a,b) *plane must be considered as a piling up of spectra of sheets* due to intersections of boxes. It is said that *the box-within-a-box bifurcations structure of* T_b *has a foliated structure.*

(d) The propositions (P6.10) (P6.12) were obtained independantly in [H13] (1984) [H12] (1985) using knot theory. In these papers the bifurcations resulting from (6.28), and the cycles of the boxes in files bifurcations structure are associated with orbits called *resonant torus knots.*

(e) All the precedent results occur for b > 0. So, for b = +1, the curves $\Lambda_{(1)_0}$ and Λ_{2^1} are tangent at C (b = 1, a = 0). The points of the centre cycle of order 2^1 which appears are ($x = -y = \sqrt{a}$) and ($x = -y = -\sqrt{a}$).

This cycle generates a boxes in files bifurcations structure equivalent to that of (6.28), from the relation :

(6.29) $a = (1 - \cos\varphi)/2$, $\varphi = 2\pi p/k'$, $k' = k/2^1$, $0 \leqq a \leqq 1$, $p < k'/2$

The intersection of $\Lambda_{(2^2)_0} \equiv \Lambda_{2^1}$ with $b = +1$ occurs at $a = 1$, which is the limit point ($\varphi = \pi$) of boxes of degree $D = 2$ given by μ'_2 of (2.37) with $m = 2$, $\alpha_1, = 1,2,\ldots$ (cf below fig 6.23).

More generally whatever be $b = +1$ *or* $b = -1$, *every cycle which is a centre, with multipliers* $S_{1,2} = \exp(\pm j\varphi)$, $j = \sqrt{-1}$, *gives rise to a boxes in files bifurcations structure in the interval* $a_1 \leqq a \leqq a_2$ a_1 *corresponding to* $\varphi = 0$ *and* a_2 *to* $\varphi = \pi$. *Equivalent properties to* (P6.12) *occur for the structure generated from each of these cycles.*

(f) With the particular form (6.26) the parameters (a,b) plane presents a cut for $a = 0$, because T_b is then linear. This singularity of the plane does not occur for $f(x,a) \equiv x^2 - a$ in (6.1).

f) Consider a cycle of order k of T_b with the multipliers S_1 S_2. From the expression of the jacobian of T_b^k, it appears that they satisfy the relation:

(6.30) $S_1 S_2 = (-1)^k b^k$

When the multipliers become complex, $S_{1,2} = \rho \exp(\pm j\varphi)$, one has $S_1 S_2 = \rho^2 = (-1)^k b^k$. Thus when $b > 0$, the multipliers of the cycles of odd order are always real.

For $|b| < 1$, $a \in \Omega_1$, all the cycles with complex multipliers are attractive foci .

When $b = +1$, k being an odd integer, $S_1 S_2 = -1$. Then at the fold bifurcation giving rise to a cycle of order k, one has $S_1 = +1$ and thus $S_2 = -1$, from which it results that *the curves* $\Lambda^j_{(k)_0}$ ($S_1 = 1$) *and* Λ^j_k ($S_1 = -1$) *are tangent for* $b = +1$ *if* k *is odd.*

Now if k is even, $S_1 S_2 = 1$ for $b = +1$. It results that the corresponding point of $\Lambda^j_{(k)_0}$ is such that $S_1 = S_2 = 1$ ($\varphi = 0$), and for Λ^j_k $S_1 = S_2 = -1$ ($\varphi = \pi$).

6.7.2. HOMOCLINIC AND HETEROCLINIC SITUATIONS. MORSE-SMALE AREA

The following representations are used :

$\omega(q_i)$, $\alpha(q_i)$, $i = 1,2$, are respectively the ω-invariant curve and the α-one of the saddle fixed point q_1, and of the fixed point q_2 when it is a saddle.

$\bar{\Lambda}_1^*$ is the curve of the first creations of heteroclinic points between $\alpha(q_2)$, $\omega(q_1)$ by heteroclinic tangency (fig. 6.18) [M 28].

Λ_1^* is the curve of the last creations of heteroclinic point between $\alpha(q_2)$, $\omega(q_1)$ by heteroclinic tangency. This curve corresponds also to the last homoclinic tangency of $\alpha(q_1)$, $\omega(q_1)$ (fig 6.8) [M32]

The curves $\bar{\Lambda}_k^{*j}$, Λ_k^{*j} are the equivalent of these two curves for saddle cycles $(k;j)$ (fig. 6.19) [M28]

$\check{\Lambda}_1^*$ is the curve of the first homoclinic tangency of $\omega(q_1)$, $\alpha(q_1)$. It passes through A $(a = b = -1)$ and $(a = 2,\ b = 0)$.

In certain cases the first homoclinic, or heteroclinic, tangency can correspond to very different situations of tangency between the α and ω-invariant curves. This point is illustrated by fig. 6.20, with the invariant curves giving the first homoclinic tangency of $\alpha(q_2)$, $\omega(q_2)$. It results a non-smooth curve $\bar{\Lambda}_{2^1}^*$ represented also in fig.6.20 where the points of non-smoothness have the point C $(a = 1,\ b = 0)$ as limit point [F 6]. Fig.6.21 and 6.22 gives more details about this situation [G35]

In [M32] (1981), on the foundation of asymptotic behaviour of the curves Λ_k^{*j}, the following conjecture was made :

(P6.13) For $b < 0$, *the boundary* Λ_1^* *of the box* Ω_1 *is an accumulation curve for the set of the "last" curve of homoclinic tangency* Λ_k^{*j} *,* $k \to \infty$ *with* $j = p_k - 1$ *, i.e the preultimate running number in the Myrberg ordering.*

For $b > 0$ *the boundary* Λ_1^* *is the envelope of a sub set of curves* Λ_k^{*j} *,* $j > p_k/2$, $k \to \infty$.

The result for $b < 0$ was proved later in [H10] (1984) and [H11].

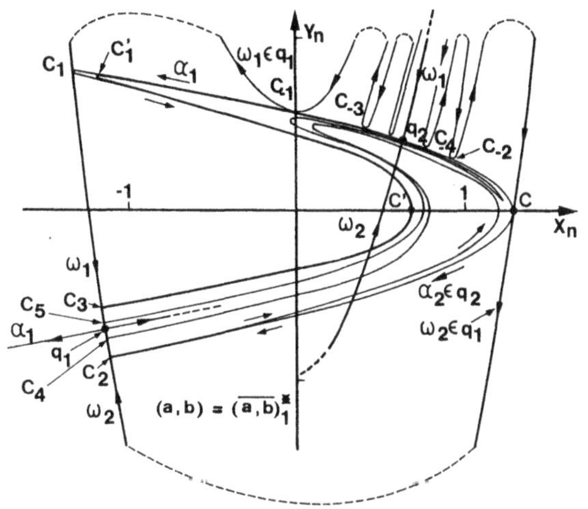

FIG. 6.18 - First heteroclinic tangency $\Lambda = \bar{\Lambda}_1^*$

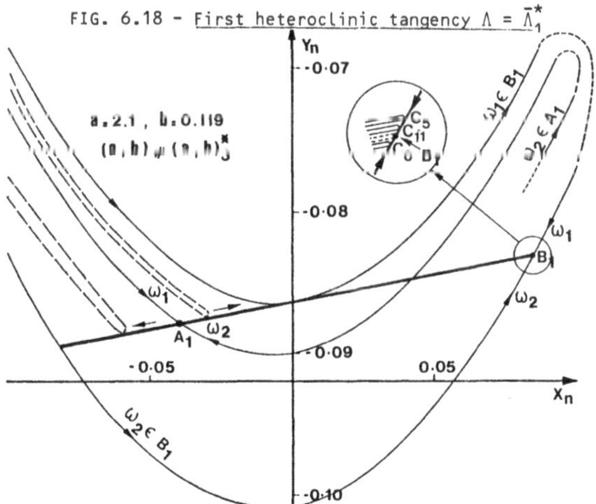

FIG. 6.19 - First heteroclinic tangency $\Lambda = \bar{\Lambda}_3^{*1}$.
A_1, B_1 are points of two saddle cycles of order 3 .

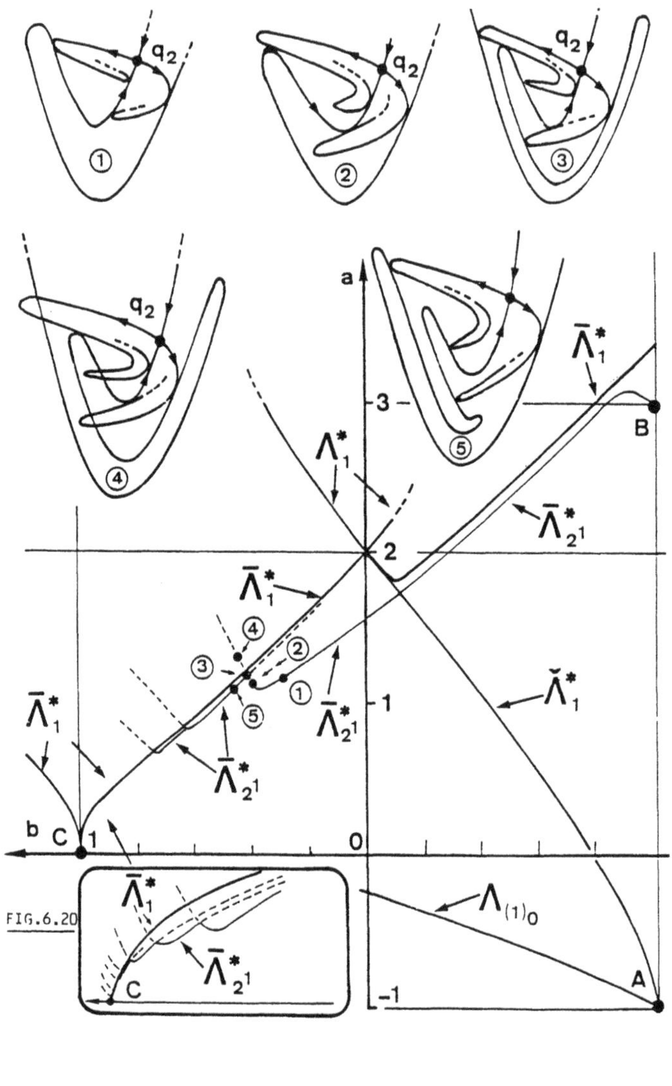
q2
a
b
C
A
B
0
1
2
3
-1
$\bar{\Lambda}^*_1$
Λ^*_1
$\check{\Lambda}^*_1$
$\bar{\Lambda}^*_{2^1}$
$\Lambda_{(1)0}$

FIG.6.20

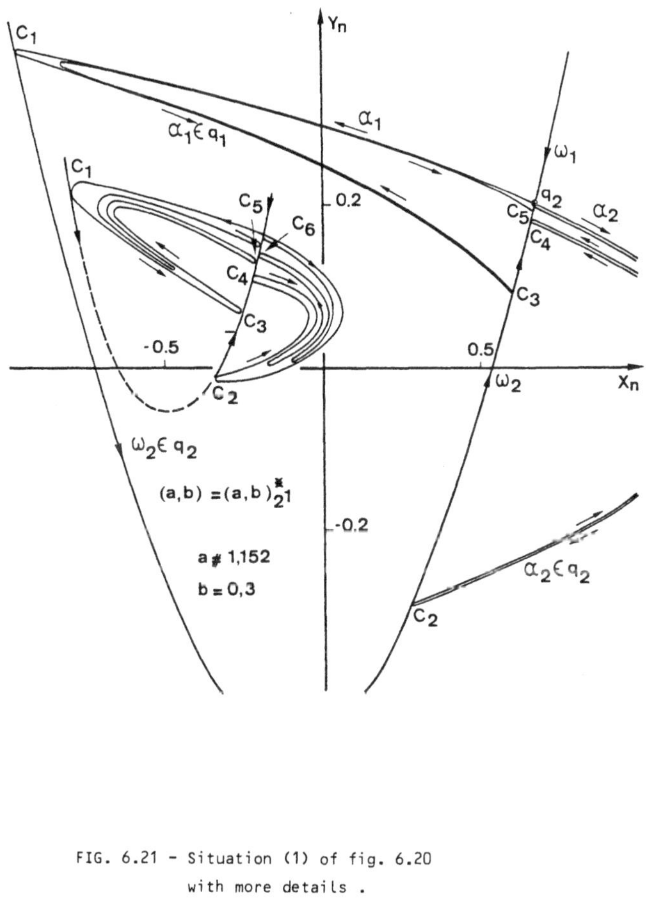

FIG. 6.21 - Situation (1) of fig. 6.20 with more details .

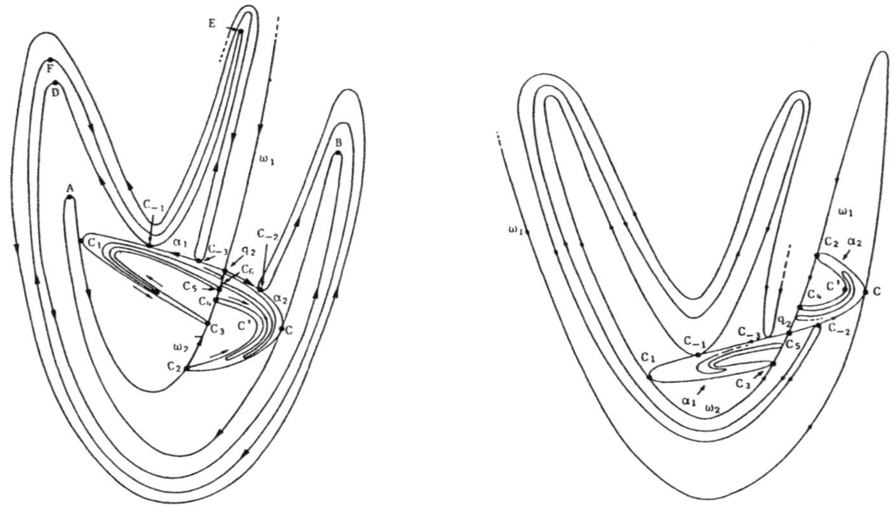

(a) b > 0 (b) b < 0

FIG. 6.22 - First homoclinic tangency $\Lambda = \bar{\Lambda}^{*}_{21}$, related to the saddle q_2 .

The consequents C_i , i = 1, 2,...., of C tend toward q_2 on ω_2. The antecedents C_{-i} of C tend toward q_2 on α_1, α_2, with an alternating convergence. For b < 0 the convergence of the consequents C_i is also of alternating type .

It seems that till now there is not a proof of the case $b > 0$.

To complete the information about homoclinic situations the following proposition, proved in [H10], is given :

(P6.14) On every homoclinic bifurcation curve, within the interior of the primary fan bounded by $\check{\Lambda}_1^*$, Λ_1^* , there is a further countable set of homoclinic bifurcation points. For such parameter values, two distinct non-transverse homoclinic orbits exist simultaneously.

More information about homoclinic situations can be found also in [H10], where in particular an ordering law for certain homoclinic bifurcations is established from a Myrberg's binary coding.

When $b < 0$, the boxes in files structure Φ leads to the definition of a family of curves $\Lambda^j_{(k)_0}$ (first boundary of simple boxes) everyone being associated with the parameter φ, $0 \leq \varphi \leq 2\pi/3$, argument of the multiplier of the centre q_2. Call F_φ the union of this family with the family of the curves $\check{\Lambda}^{*j}_k$, defined as $\check{\Lambda}_1^*$ but for a saddle cycle (k;j) in the place of the saddle fixed point q_1 , and their limits when $k \to \infty$ Consider the set of all the angles φ commensurable with 2π , the irrational rotation numbers with the Cantor set structure defined in (P2.25), and those associated with curves $\check{\Lambda}^{*j}_k$. From (P6.12) it results that the curves of F_φ have an envelope H_1 (fig. 6.23)

When $b > 0$ a countable set of families $F_\varphi^{(2^i)}$, similar to F_φ , $i = 1,2,3...$, is defined. Here φ is the argument of the multipliers $S_{1,2} = \exp(\pm j\varphi)$ of the cycle centre of order 2^i, which appears, for $b = 1$, from a Myrberg's cascade of bifurcations (cf remark(e) of §6.7.1) Every family gives boxes of the second kind, for which the sequence of intersections of their "fold bifurcation" boundary on $b = +1$, is reversed with respect to that on $b = 0$. Besides when i increases these families appear with increasing $\bar{a}$ on $b = 1$, and decreasing $\bar{a}$ on $b = 0$. It results an envelope H_2 (fig. 6.23) defined as for $b = -1$, but associated with the whole set of the families $F_\varphi^{(2^i)}$, $i = 1,2,3,...$ The smoothness of H_2 is not certain, and perhaps isolated points of non-smoothness are possible as for fig. 6.20. They might correspond to the passage from one family, $i = m$, to the following family, $i = m + 1$.

The curves H_1 and H_2 intersect at the point D ($a \simeq 1.54$, $b \simeq -0.095$). At the points a($a = b = -1$) and C($b = 1$, $a = 0$) it seems that $da/db = \infty$

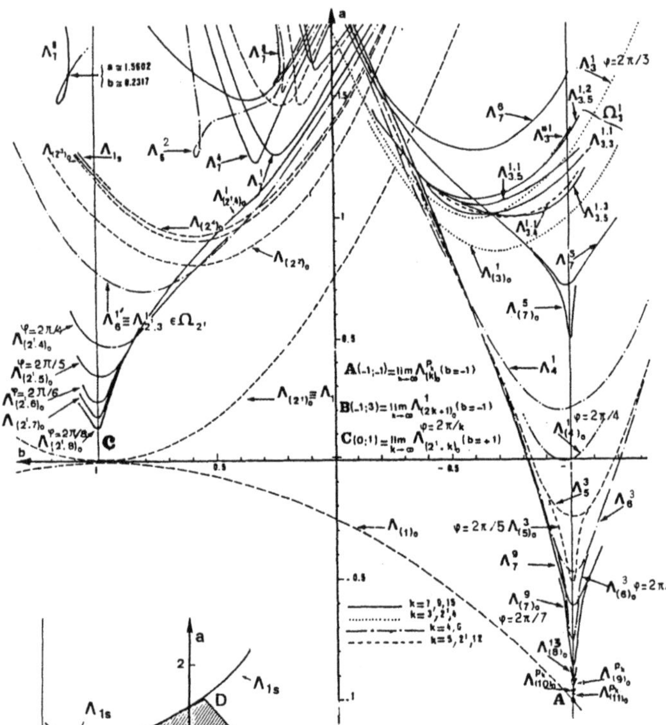

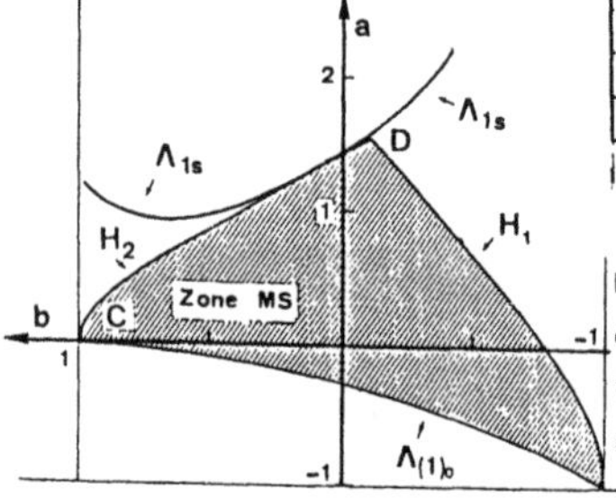

The boundaries H_1, H_2 of the Morse-Smale area are the envelops of families of curves(related to the boxes in files bifucations structure) called F_ϕ and $F_\phi^{(2^i)}$.

FIG. 6.23 - Definition of the Morse-Smale area
(zone MS)

for respectively H_1 , H_2 . H_2 is tangent to $\Lambda_{(1)_S}$ at the point (b = 0, a = $a_{(1)_S}$), and D is a point of non-smoothness.

Consider the domain (MS) bounded by H_1 , H_2 and $\Lambda_{(1)_0}$ [F6] . From many numerical tests it appears that the curves $\Lambda^j_{(k)_0}$, $k \neq 2^i$, i = 1,2,...,do not cross H_1, or H_2 and so do not penetrate into (MS) (fig. 6.23). This property is given an a conjecture. Then (MS) is a region where only segments of the curves Λ_{2^i} , i= 1,2,..., are located.Every point of (MS) corresponds to only one attractor, which is a cycle of order 2^i, i = 0,1,2,... given by one sheet of the foliated structure of fig. 6.1 (k = 1). The region (MS) is characterized by the absence of homoclinic, or heteroclinic points, and inside (MS) the Myrberg's spectrum ω_1 is the only existing (only one sheet of attractive cycles $k = 2^i$). It results that inside (MS) and outside the Λ_{2^i} , T_b is of "*Morse-Smale" type*,i.e. here the number of fixed points and cycles is finite, and each is not in a Ljapunov critical case ($\Lambda \neq \Lambda_{2^i}$), homoclinic and heteroclinic points are absent, the non-wandering set consists of fixed points and cycles alone. The region (MS) is called *Morse-Smale area* [F 6]

6.8. SOME RULES OF FORMATION OF CYCLES REPRESENTED IN THE SYMBOLIC PHASE PLANE

As indicated in § 6.6.2, the binary word $[q^{(1)j}_k]$ of the cycle (k;j), defined for T_0, contains the whole information giving the representation of this cycle, in the symbolic phase plane, when the sign of b is fixed. Thus one has to expect that the properties related to the "structure" of this binary word correspond to properties related to the "shape" of the cycle (k;j) in the symbolic phase plane. The purpose of this paragraph is to show such relations.

From § 3.6 it appears that every binary word $[q^{(1)j}_k]$ is the concatenation of two binary words $[10^m]$ (m+1bits) and [a] (k-m-1bits)

-------- b<0, k = 6, number of rotation sequences [r]=5 ------

1	1.47601	1 6 3 4 2 5	[+---]	10111	0(S-D-3)	1(S-I-3)		
2	1.77289	1 6 4 2 5 3	[++-+]	10010	0(S-D-2)			2
					1(D-I-2)	1 6 5 3 2 4	(3)	2
3	1.90728	1 6 5 3 2 4	[++--]	10011	0(D-I-2)	1 6 4 2 5 3	(2)	3
					1(S-D-2)			3
4	1.96677	1 6 5 4 2 3	[+++-]	10001	0(S-D-2)	1(S-I-2)		
5	1.99637	1 6 5 4 3 2	[++++]	10000	0(S-I-1)	1(S-D-1)		

---- b<0, k = 7, number of rotation sequences [r]=9 ----

1	1.57488	1 7 4 3 5 2 6	[+----]	101111	0(S-I-3)	1(S-D-3)		
2	1.67406	1 7 4 2 6 3 5	[+--+-]	101101	0(S-D-3)	1(S-I-3)		
3	1.83231	1 7 5 2 6 3 4	[++-+-]	100101	0(S-I-3)			3
					1(D-D-3)	1 7 6 3 4 2 5	(4)	3
4	1.88480	1 7 6 3 4 2 5	[++---]	100111	0(D-D-3)	1 7 5 2 6 3 4	(3)	4
					1(S-I-3)			4
5	1.92714	1 7 5 3 2 6 4	[++--+]	100110	0(S-I-2)	1(S-D-2)		
6	1.95370	1 7 6 4 2 5 3	[+++-+]	100010	0(S-D-2)			6
					1(D-I-2)	1 7 6 5 3 2 4	(7)	6
7	1.97718	1 7 6 5 3 2 4	[+++--]	100011	0(D-I-2)	1 7 6 4 2 5 3	(6)	7
					1(S-D-2)			7
8	1.99181	1 7 6 5 4 2 3	[++++-]	100001	0(S-D-2)	1(S-I-2)		
9	1.99909	1 7 6 5 4 3 2	[+++++]	100000	0(S-I-1)	1(S-D-1)		

------- b<0, k = 8, number of rotation sequences [r]=16 ---

1	1.38154	1 8 4 5 2 7 3 6	[+---+-]	1011101	0(S-I-4)	1(S-D-4)		
2	1.52181	1 8 4 5 3 6 2 7	[+-----]	1011111	0(S-D-4)	1(S-I-4)		
3	1.71108	1 8 5 2 7 4 3 6	[+--+--]	1011011	0(S-I-3)	1(S-D-3)		
4	1.81000	1 8 6 2 7 4 3 5	[++-+--]	1001011	0(S-D-3)			4
					1(D-I-3)	1 8 7 4 3 5 2 6	(7)	4
5	1.85173	1 8 6 3 5 2 7 4	[++-+-+]	1001010	0(S-I-3)			5
					1(D-D-3)	1 8 7 4 2 6 3 5	(8)	5
6	1.87000	1 8 6 3 4 2 7 5	[++---+]	1001110	0(S-D-3)			6
					1(D-I-3)	1 8 7 4 2 6 3 5	(8)	6
7	1.89691	1 8 7 4 3 5 2 6	[++----]	1001111	0(D-I-3)	1 8 6 2 7 4 3 5	(4)	7
					1(S-D-3)			7
8	1.91709	1 8 7 4 2 6 3 5	[++--+-]	1001101	0(D-D-3)	1 8 6 3 5 2 7 4	(5)	8
					1(D-I-3)	1 8 6 3 4 2 7 5	(6)	8
9	1.94178	1 8 6 4 2 7 5 3	[+++-++]	1000100	0(S-D-2)			9
					1(D-I-2)	1 8 7 5 3 2 6 4	(12)	9
10	1.96075	1 8 7 5 2 6 3 4	[+++-+-]	1000101	0(S-I-3)			10
					1(D-D-3)	1 8 7 6 3 4 2 5	(11)	10
11	1.9722	1 8 7 6 3 4 2 5	[+++---]	1000111	0(D-D-3)	1 8 7 5 2 6 3 4	(10)	11
					1(S-I-3)			11
12	1.98165	1 8 7 5 3 2 6 4	[+++--+]	1000110	0(D-I-2)	1 8 6 4 2 7 5 3	(9)	12
					1(S-D-2)			12
13	1.98879	1 8 7 6 4 2 5 3	[++++-+]	1000010	0(S-D-2)			13
					1(D-I-2)	1 8 7 6 5 3 2 4	(14)	13
14	1.99433	1 8 7 6 5 3 2 4	[++++--]	1000011	0(D-I-2)	1 8 7 6 4 2 5 3	(13)	14
					1(S-D-2)			14
15	1.99796	1 8 7 6 5 4 2 3	[+++++-]	1000001	0(S-D-2)	1(S-I-2)		
16	1.99977	1 8 7 6 5 4 3 2	[++++++]	1000000	0(S-I-1)	1(S-D-1)		
j'	a	(DRS)	[r]	(BW)	(αβγ)	(DRS)*	j'*	j'

TABLE 1 .

----- b < 0, k = 9, number of rotation sequences [r]=28. --------

j'	a	(DRS)	[r]	(BW)	(αβγ)	(DRS)*	j'*	j'
1	1.55528	1 9 5 4 6 3 7 2 8	[+-------]	10111111	0(S-I-4)	1(S-D-4)		
2	1.59568	1 9 5 4 6 2 8 3 7	[+----+-]	10111101	0(S-D-4)	1(S-I-4)		
3	1.65613	1 9 5 3 7 2 8 4 6	[+--+-+-]	10110101	0(S-I-4)			3
					1(D-D-4)	1 9 6 2 8 4 5 3 7	4	3
4	1.69014	1 9 6 2 8 4 5 3 7	[+--+---]	10110111	0(D-D-4)	1 9 5 3 7 2 8 4 6	3	4
					1(S-I-4)			4
5	1.78586	1 9 6 3 7 4 2 8 5	[++-+--+]	10010110	0(D-D-3)	1 9 7 4 3 6 2 8 5	11	5
					1(D-I-3)	1 9 8 5 2 7 4 3 6	12	5
6	1.82275	1 9 7 2 8 4 5 3 6	[++-+---]	10010111	0(S-I-4)			6
					1(D-D-4)	1 9 8 4 5 3 6 2 7	9	6
7	1.84128	1 9 7 3 6 2 8 4 5	[++-+-+-]	10010101	0(S-D-4)			7
					1(D-I-4)	1 9 8 4 5 2 7 3 6	8	7
8	1.87838	1 9 8 4 5 2 7 3 6	[++---+-]	10011101	0(D-I-4)	1 9 7 3 6 2 8 4 5	7	8
					1(S-D-4)			8
9	1.89077	1 9 8 4 5 3 6 2 7	[++-----]	10011111	0(D-D-4)	1 9 7 2 8 4 5 3 6	6	9
					1(S-I-4)			9
10	1.90311	1 9 7 4 3 5 2 8 6	[++----+]	10011110	0(S-I-3)			10
					1(D-D-3)	1 9 8 5 2 7 4 3 6	12	10
11	1.91144	1 9 7 4 3 6 2 8 5	[++--+-+]	10011010	0(D-D-3)	1 9 6 3 7 4 2 8 5	5	11
					1(S-I-3)			11
12	1.92228	1 9 8 5 2 7 4 3 6	[++--+--]	10011011	0(D-I-3)	1 9 6 3 7 4 2 8 5	5	12
					1(D-D-3)	1 9 7 4 3 5 2 8 6	10	12
13	1.93224	1 9 7 4 2 8 6 3 5	[++--++-]	10011001	0(S-D-3)	1(S-I-3)		
14	1.94957	1 9 7 5 2 8 6 3 4	[+++-++-]	10001001	0(D-I-3)	1 9 8 6 3 5 2 7 4	16	14
					1(D-D-3)	1 9 8 6 3 4 2 7 5	17	14
15	1.95732	1 9 8 6 2 7 4 3 5	[+++-+--]	10001011	0(S-D-3)			15
					1(D-I-3)	1 9 8 7 4 3 5 2 6	18	15
16	1.96402	1 9 8 6 3 5 2 7 4	[+++-+-+]	10001010	0(D-I-3)	1 9 7 5 2 8 6 3 4	14	16
					1(D-D-3)	1 9 8 7 4 2 6 3 5	19	16
17	1.96941	1 9 8 6 3 4 2 7 5	[+++---+]	10001110	0(D-D-3)	1 9 7 5 2 8 6 3 4	14	17
					1(D-I-3)	1 9 8 7 4 2 6 3 5	19	17
18	1.97478	1 9 8 7 4 3 5 2 6	[+++----]	10001111	0(D-I-3)	1 9 8 6 2 7 4 3 5	15	18
					1(S-D-3)			18
19	1.97945	1 9 8 7 4 2 6 3 5	[+++--+-]	10001101	0(D-D-3)	1 9 8 6 3 5 2 7 4	16	19
					1(D-I-3)	1 9 8 6 3 4 2 7 5	17	19
20	1.98381	1 9 7 5 3 2 8 6 4	[+++--++]	10001100	0(S-I-2)	1(S-D-2)		
21	1.98700	1 9 8 6 4 2 7 5 3	[++++-++]	10000100	0(S-D-2)			21
					1(D-I-2)	1 9 8 7 5 3 2 6 4	24	21
22	1.99037	1 9 8 7 5 2 6 3 4	[++++-+-]	10000101	0(S-I-3)			22
					1(D-D-3)	1 9 8 7 6 3 4 2 5	23	22
23	1.99313	1 9 8 7 6 3 4 2 5	[++++---]	10000111	0(D-D-3)	1 9 8 7 5 2 6 3 4	22	23
					1(S-I-3)			23
24	1.99541	1 9 8 7 5 3 2 6 4	[++++--+]	10000110	0(D-I-2)	1 9 8 6 4 2 7 5 3	21	24
					1(S-D-2)			24
25	1.99722	1 9 8 7 6 4 2 5 3	[+++++-+]	10000010	0(S-D-2)			25
					1(D-I-2)	1 9 8 7 6 5 3 2 4	26	25
26	1.99858	1 9 8 7 6 5 3 2 4	[+++++--]	10000011	0(D-I-2)	1 9 8 7 6 4 2 5 3	25	26
					1(S-D-2)			26
27	1.99949	1 9 8 7 6 5 4 2 3	[++++++-]	10000001	0(S-D-2)	1(S-I-2)		
28	1.99994	1 9 8 7 6 5 4 3 2	[+++++++]	10000000	0(S-I-1)	1(S-D-1)		

-TABLE 1-

(Continuation)

$$[q^{(1)j}_{k}] \equiv [10^m a] \ , \ m = k-i, \ i = 1,2,\ldots,k-1, \ a \equiv \Phi \text{ if } i = 1,$$

where [a] begins with the bit "1", Φ indicating the empty set. Here $[10^m]$ is called the *root* of the binary word of the cycle (k;j) and [a] the *stalk*.

The joined tables I (extracted from[K4]) results from the application of the rules of § 3.5, 3.6. They give all the characteristics of the set of cycles of order k, $k \leq 9$, when $b < 0$, using the rules of representation given in § 6.6.2. For $k > 9$ equivalent tables are obtained via the same algorithms for a given sign of b [K4]. From the left to the right these tables give the following informations :

- the non-embedded index j'
- the value of the parameter a
- the decimal rotation sequence (DRS)
- the Myrberg's rotation sequence [r]
- the binary words $[q^{(1)j}_{k}]$ (BW)
- the code (α, β, γ), where α = S (symmetry) indicates a *self-adjoint* cycle, α = D (dissymmetry) indicates a cycle (k;j') which is in *adjoint* relation with another cycle (k;j'*) ; β is related to the *sign of the multiplier* S_1, β = D ($S^1 > 1$, i.e. "p" in the complete representation of § 3.7), β = I ($S_1 < -1$, i.e."n") ; γ = *turns number*, i.e the total rotation of the vector $\overrightarrow{q^{(i)}\,q^{(i+1)}}$, i = 1,2,...k, $q^{(k+1)} \equiv q^{(1)}$, in the (x',y') plane, after k successive iterations (the resulting rotations for every iteration must occur with the same orientation of angles).γ is also called winding number.
- the decimal rotation sequence (DRS)* of (k;j'*), j' $\neq$ j'*, when it exists. When it does not exist for each of the two (self-adjoint) cycles (k;j'), the corresponding line of the column (DRS)*, indicates the (α, β, γ), α = S, of the second self-adjoint cycle
- the index j'*, j' $\neq$ j'* when it exists
- the connexions between adjoint cycles.

These tables permit to see the "evolution" of the binary word $[10^m a]$ in connection with the non-embedded index j', and define families of cycles having the sames roots, every family corresponding to a

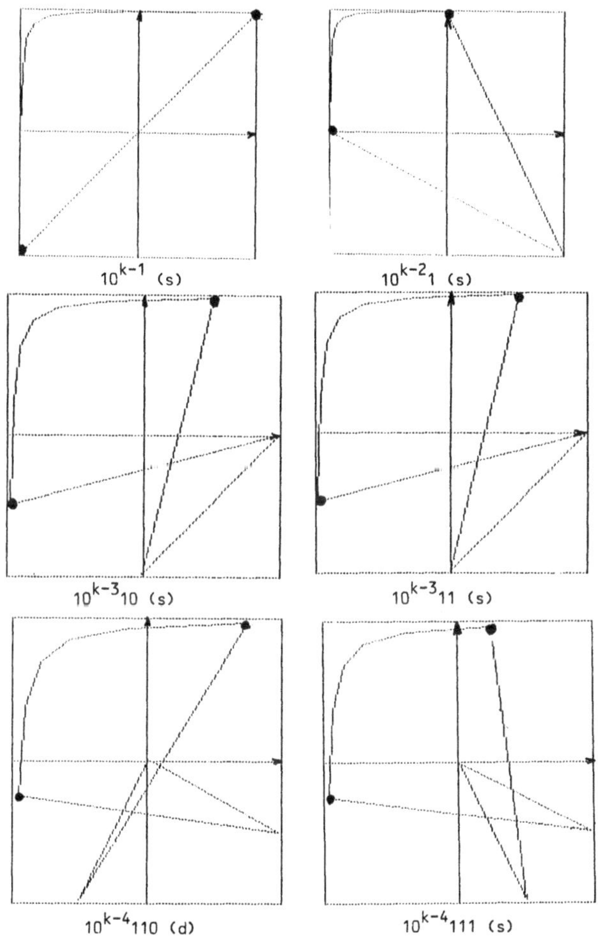

FIG.6.24 - Representation of cycles of order k in the symbolic phase plane (b<0). (s)→self-adjoint cycle; (d)→ adjoint cycle of that symmetric with respect to y'=-x'.

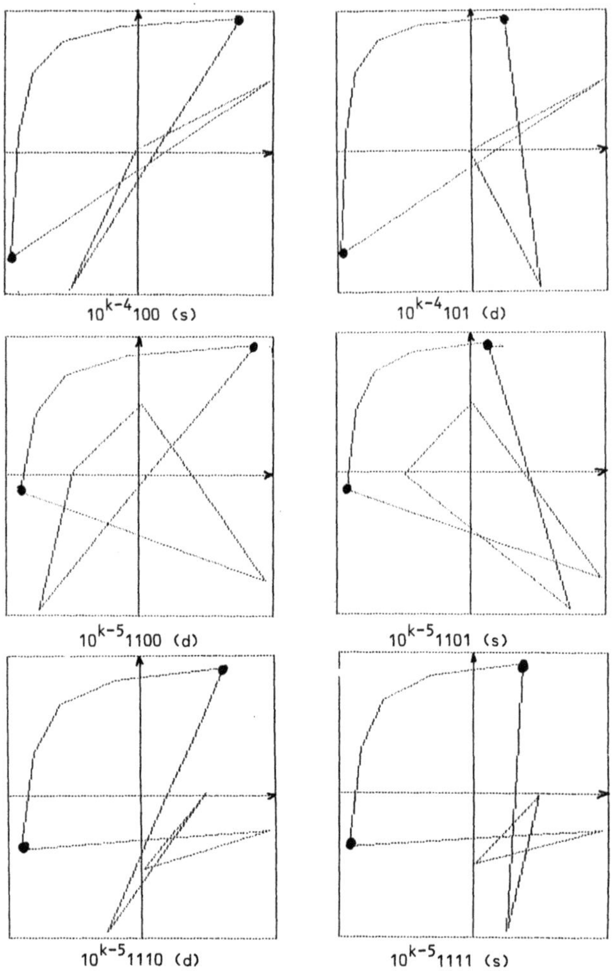

$10^{k-4}100$ (s) $10^{k-4}101$ (d)

$10^{k-5}1100$ (d) $10^{k-5}1101$ (s)

$10^{k-5}1110$ (d) $10^{k-5}1111$ (s)

FIG.6.24 - (Continuation) .

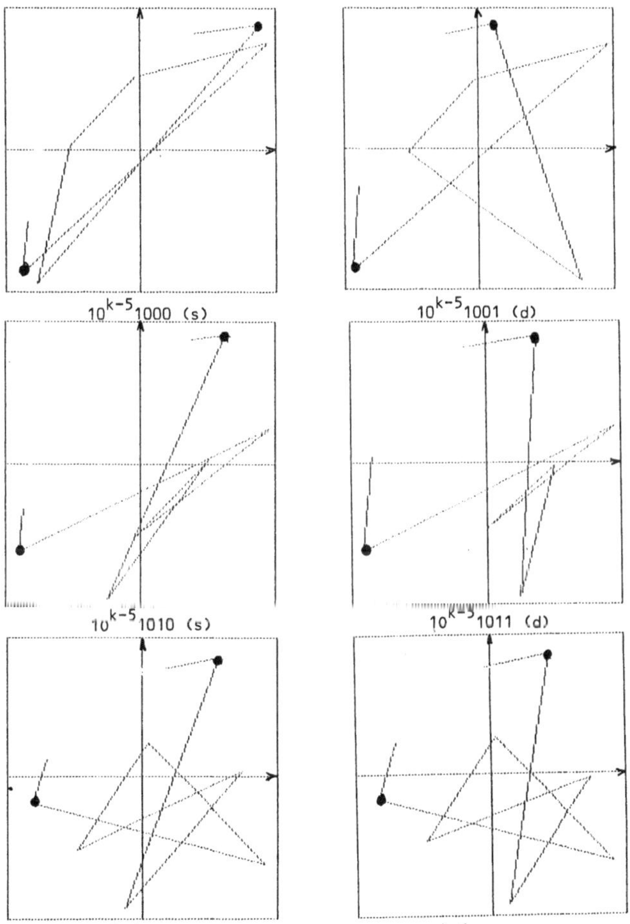

FIG. 6.24 - (Continuation, adapted from |K4|)

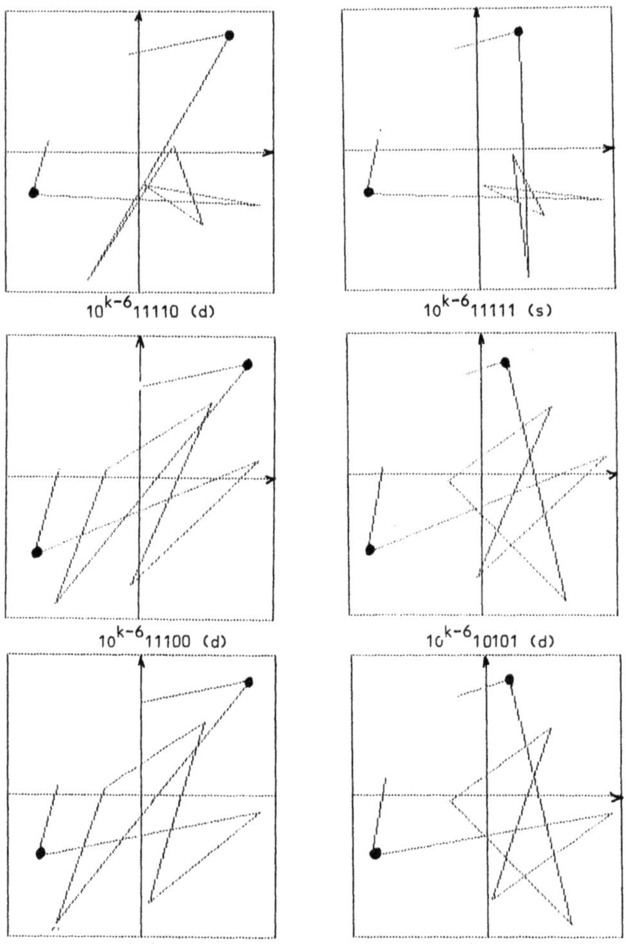

FIG.6.24 - (Continuation).

fixed value of $m = 1,2,..,k-1$, when the parameter k varies.

Representing now the cycles in the symbolic phase plane (x',y') for a fixed sign (< 0 here) of the parameter b, it appears that the root $[10^m]$ corresponds to a segment of convex polygon made up of m sides, the extremities of which are indicated by two small black circles in fig. 6.24 ($b < 0$). The stalk [a] corresponds the remaining part of the representation which generally has a more complex shape with intersections of sides $\overline{q^{(i)}\, q^{(i+1)}}$. The figures 6.24 present an adaptation of the results of [K4]. The multiplier S_1 is > 1, when the number of bits "1" in $[10^m a]$ is even ($\beta = p$). One has $S_1 < -1$, when the number of bits "1" is odd($\beta = n$).Fig. 6.24 also indicates when the cycle is self-adjoint, or symmetrical with respect to $y' = -x'$ (s), or in adjoint relation (d) with another cycle, each isolated cycle being not symmetrical with respect to $y' = -x'$, but the set of the points of the two adjoint cycles having this symmetry.

As adjoint relations in fig. 6.24 one has :

$(10^{k-4}110) = (10^{k-4}101)^*$ $\qquad$ $(10^{k-5}1100) = (10^{k-5}1001)^*$

$(10^{k-5}1110) = (10^{k-5}1011)^*$ $\qquad$ $(10^{k-6}11010) = (10^{k-6}10101)^*$

$(10^{k-6}11011) = (10^{k-6}11101)^*$

6.9. SINGULARITIES OF THE FRACTAL FOLIATED BIFURCATIONS STRUCTURE. COMMUNICATIONS BETWEEN SHEETS OF THIS STRUCTURE.

6.9.1. GENERALITIES

The foliated structure of bifurcations of a Myrberg's spectrum ω_k^j, shown in fig. 6.1, corresponds to the simplest possible situation. Such a spectrum in the $(a, b, |z|)$ representation is isolated with respect to all other spectra, i.e. no communication with sheets belonging to other spectra exists. It will be shown in § 6.11 that *this simplest case occurs when every cycle* $(k2^i\ ;\ j_i')$, $i = 0,1,2,\ldots$, $j_0' \equiv j$, *of* ω_k^j *is self adjoint (or symmetrical) for a given sign of the parameter* b.

The subject matter of the § 6.8 is to make appear three singular situations in the parameters (a, b) plane, which are related to particular communications between sheets of the foliated structure. I.e. it is possible to pass continuously from one sheet to other by continuous variation of parameters. Communications are due to existence of *cusp singularities* on fold bifurcation curves $\Lambda^{j}_{(k)_0}$ (S_1 = +1). For one type of communication, the cusp singularity is associated with a *loop* of a flip bifurcation curve $\Lambda^{j'}_{k}$ (S_1 = -1), j = j' or j ≠ j', giving two distinct bifurcations which occur simultaneously in different places in the phase plane. With the language of differential topology the cusp singularities are related to what is called *co-dimension two bifurcations*.

These three singular situations were called *cross-road area* ("zone échangeur" ZE in french), *spring area* ("zone source" ZS), and *saddle area* ("zone col" ZC) [H5] [H6] (1982) [F7] [F8] [F9]. They firstly appear from the curves described by the parametric equations (6.6), where loops, or cusps give classical singular points in the implicit equations of these curves. The first studies permited the definition of a symbolism of representation of considered areas. From this symbolism the communications are unambiguously identified. It will be also shown in § 6.11 that, the concept of *connections chain* via the adjoint cycles notion, extends the possibilities of the symbolic dynamics, permiting a symbolic description of a complete set of communications (called a *communications cell*) between the sheets. Indeed without drawing the bifurcations curves it is possible to forecast the type of such a set from the knowledge of the corresponding connections chain.

Without the concept of connections chain, the foliated structure in the space (a, b, |z|), associated with each indicated singularity, is determined from bifurcations curves, but with a complementary information given by the evolution of a "bifurcated cycle" and its multipliers, when the point (a, b) of the parameters plane follows a correctly selected path.

It is worth of note that these singular situations lead to communications between different boxes of the box-within-a-box bifurcations structure. More particularly there exists a set of boxes such that starting from a box of the b = 0 axis belonging to this set, and following a correctly selected continuous path in the (a, b) plane, it is possible to get to a different box of b = 0.

In this paragraph the non-embedded representation of cycles (§ 3.3.7 and 3.7), with indexes j', is more often used for spring areas. The reason of this choice is related to the fact that communications occur from sheets obtained via $\Lambda^{j'_i}_{(k2^i)}$ bifurcation curves of a Myrberg's spectrum. However in the figures using non-numerical indexes, given below, the eventual use of "embedded" indexes j_i does not lead to ambiguities because a j_i corresponds to a j'_i, and $j_i = j'_i$ when k is a prime integer. For figures using numerical indexes the nature of the representation (embedded or not) is indicated.

6.9.2. THE CROSS-ROAD AREA

The simplest situation of cross-road area obtained for cycles of order k is illustrated by figure 6.25 [H5] [F9]. Such a communication is designated by the symbolism :

(6.30) $\quad ZF\ k\ (j_1,\ j_2)$ + the sign of b .

The "skeleton" of this cross-road area consists of the bifurcations curves $\Lambda^{j_1}_{(k)_0}$, $\Lambda^{j_1}_{k}$, $\Lambda^{j_2}_{(k)_0}$, $\Lambda^{j_2}_{k}$, $j_2 > j_1$. It is recall that the non-embedded representation use indexes j', the embedded representation using indexes j, but $\Lambda^{j'i}_{(k)_0} \equiv \Lambda^{ji}_{(k)_0}$, $\Lambda^{j'i}_{k} \equiv \Lambda^{ji}_{k}$, i = 1,2, $j_i \leqq j'_i$, $j_i = j'_i$ when k is a prime integer. The curve $\Lambda^{j_2}_{(k)_0}$ $(S_1 = +1)$, $j_2 > j_1$ (or $j'_2 > j'_1$), is that which presents the cusp point, and in a neighbourood of this point there is an exchange of the bifurcation curves Λ^{ji}_{k} $(S_1 = -1)$. This exchange creates two branches, made up of one bifurcation curve with $S_1 = +1$, the other with $S_1 = -1$, having different indexes, and designated by the symbolic representation :

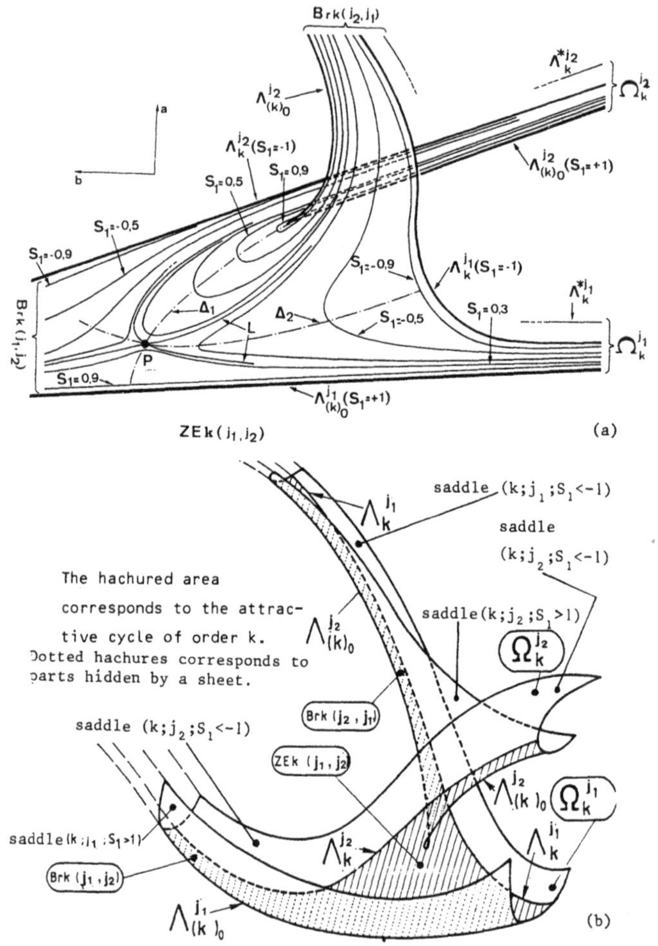

FIG.6.25 - The simplest cross-road area.
Representation in the (a;b) plane and in the (a,b,|z|) space.

(6.31) $Br\ (j_1, j_2)$, $Br\ (j_2, j_1)$

The first index between brackets is always the one associated with the fold bifurcation curve, $S_1 = +1$, and the second is related to the flip bifurcation curve $S_1 = -1$.

Fig. 6.25b represents the sheets structure in the (a, b, |z|) space. Let b be such that $|b| < 1$. A fold $\Lambda^{j}_{(k)_o}$ gives the junction of one sheet corresponding to a saddle cycle (k ; j_i) with $S_1 > 1$, and a second sheet corresponding to an attractive node cycle (k ; j_i) with $|S_1| < 1$, which can become a focus cycle sufficiently far from the junctions $\Lambda^{ji}_{(k_o)}$, Λ^{ji}_{k} of sheets. Crossing through Λ^{ji}_{k} the node becomes a saddle (when $|b| < 1$) with $S_1 < -1$ and gives rise to an attractive node cycle ($k2^1$; j''_i). It results two sheets : one for the saddle, the other for the now node. The representation of fig. 6.25b shows that a continuous path, starting from the branch of the box $\Omega^{j_1}_{k}$ which gives an attractive cycle (k ; j'_i), permits to pass into the equivalent branch of $\Omega^{j_2}_{k}$, or into one of the two branches (6.31), *the cycle remaining attractive* when $|a| < 1$. However the figure of exchange of the cycle points undergoes a distorsion leading to a change of the index j characterizing this cycle, when following such a path. In fig. 6.25a the dotted lines limit, with two bifurcations curves $S_1 = +1$, $S_1 = -1$, a region where every point (a, b) corresponds to two attractive cycles occuring in different places in the phase plane.

Fig. 6.25a represents also the family of curves s = constant, s being the multiplier S_1 which has the highest modulus when S_1 is real, and s = u when $S_{1,2} = \rho \exp(\pm j\varphi) = u \pm jv$ =, $j = \sqrt{-1}$, $v \neq 0$. The structure of the curves of this family is that of contour lines. It gives rise to the existence of a *saddle point* P *in the topographic sense* , which is the intersection of two curves (L) s = constant, and to the existence of *a line of crest* (Δ_1 in fig. 6.25a) associated with a *valley* (Δ_2 in fig. 6.25a).

In the general case (r cusps located on r fold bifurcation curves) a cross-road area can lead to a communication from r boxes of b = 0. Then it is designated by the symbolism :

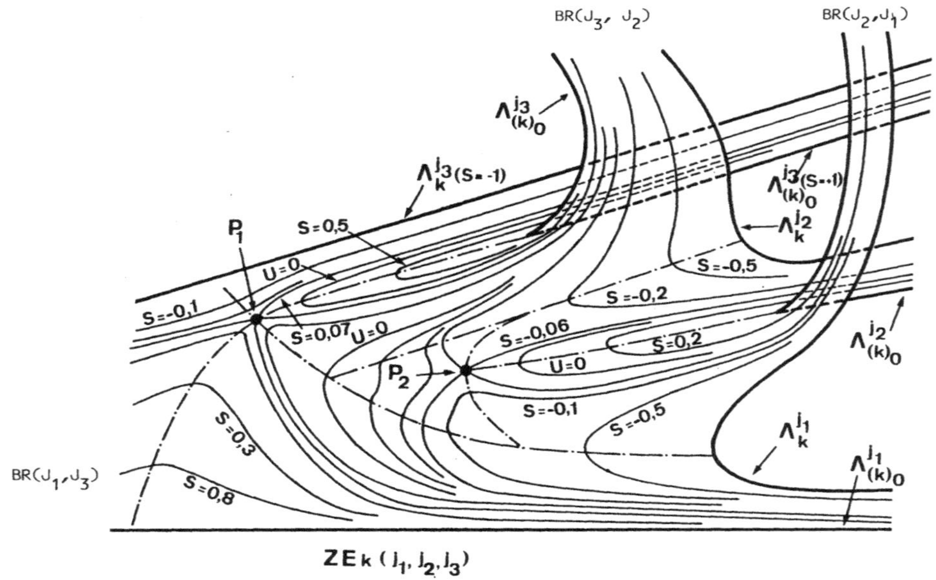

FIG.6.26 - Cross-road area giving rise to three branches .

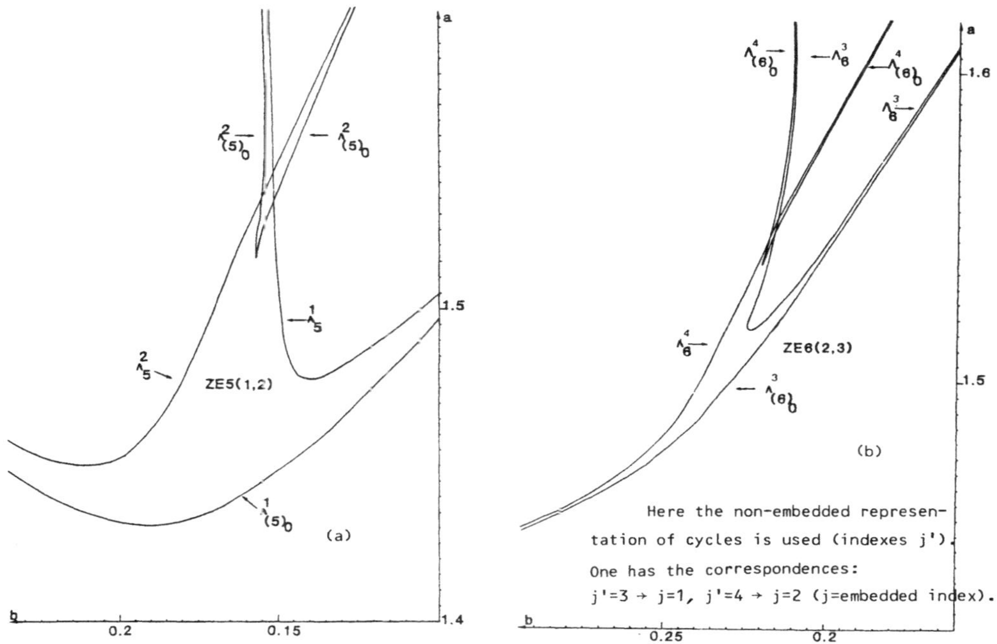

FIG.6.27 - (According to |H6|)

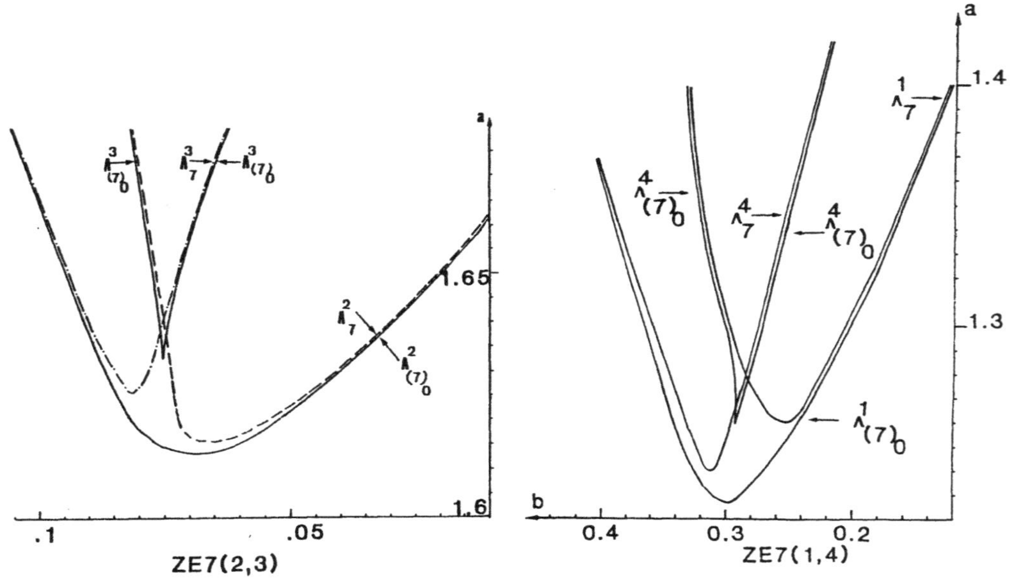

FIG.6.27 - (Continuation)

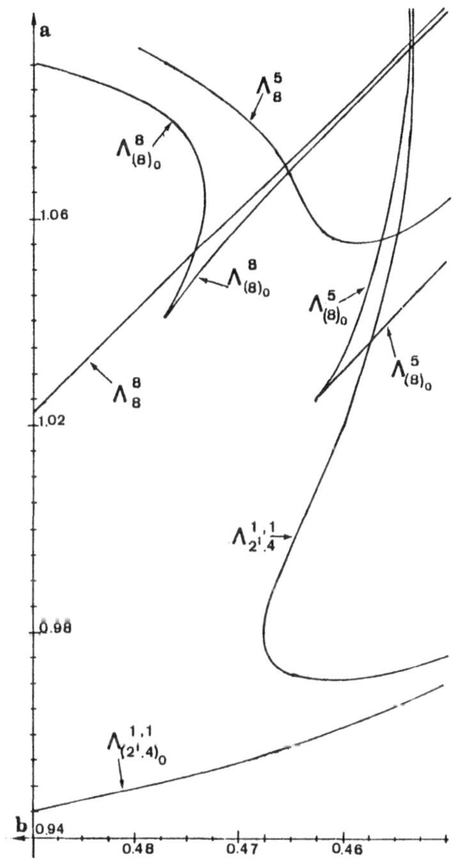

FIG.6.28 - Cross-road area ZE8(1, 5, 8) .

Here the embedded representation of cycles is used. So one has the correspondances between indexes : (j_1=1,j_2=1) (box $\Omega_{2^1.4}$)→ j'=2, j=5→ j'=7, j=8→j'=11, j' being the non-embedded index.

(6.32) $ZE\ k\ (j_1, j_2, \ldots, j_r)$, + the sign of b,

where k is the order of the basic cycles of the communicating boxes, $j_1 < j_2 < \ldots < j_r$ their indexes. It results the creation of r branches :

(6.33) $Br(j_1, j_r), Br(j_2, j_1), Br(j_3, j_2), \ldots, Br(j_r, j_{r-1})$.

Now the family of the curves s = constant presents r - 1 saddle points P_i, i = 1,2,..., r-1. For r = 3, the representation of such a cross road is given fig. 6.26.

For the quadratic diffeomorphism (6.26) the first fold bifurcation curve, corresponding to the index j_1, is that which does not have a cusp point in the cross-road area.

Fig. 6.27, 6.28 represent cross-road areas of T_b given by (6.26), for k ≦ 8, and so complete the knowledge of the foliated bifurcations structure of fig. 6.17.

6.9.3. THE SADDLE AREA

This area implies a foliated structure represented in fig. 6.29. Here the saddle (|b| < 1) cycle (r ; g) (S_1 > 1) appearing from the fold bifurcation $\Lambda^g_{(r)_0}$ (S_1 = +1) is the basic cycle which also defines the fold bifurcation curve $\Lambda^h_{(r)_0}$, g ≠ h, having a cusp point. So following the path *cde* or *cdf*, it appears that the same point (a, b) of the parameters plane can correspond to two different saddle cycles of order r, which are associated with an attractive (|b| < 1) node, or focus, cycle (r ; h) [H6] [F8] [F9]. When b = ±1 the corresponding bifurcation is (6.21). In the simplest case, the flip bifurcation curve (S_1 = -1), associated with $\Lambda^h_{(r)_0}$, is that resulting from the bifurcation (S_1 = -1) of the same basic cycle (r ; h) when b = 0. Then the saddle area is designated by the symbolism :

(6.34) ZC r (g ; h) + sign of b

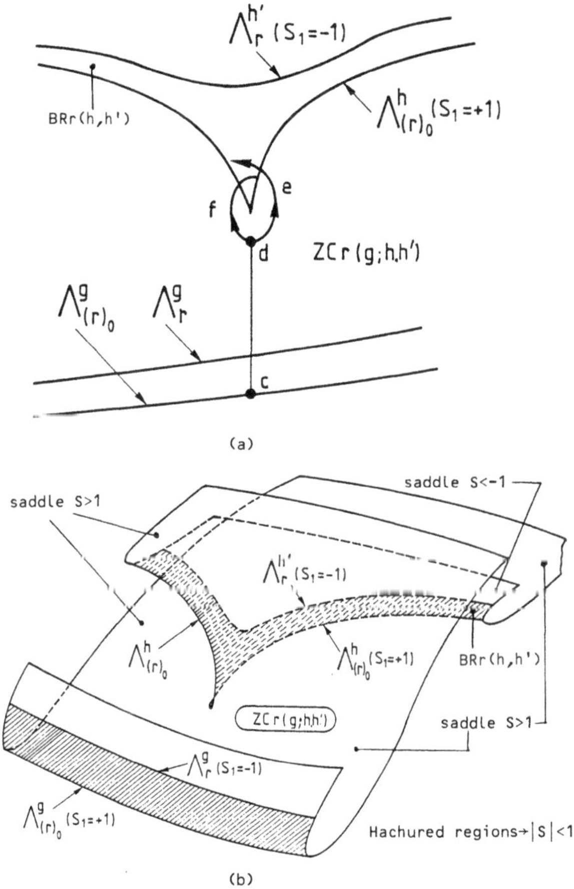

FIG.6.29 - Saddle area.
Representation in the (a,b) plane and in the (a,b,|z|) space.

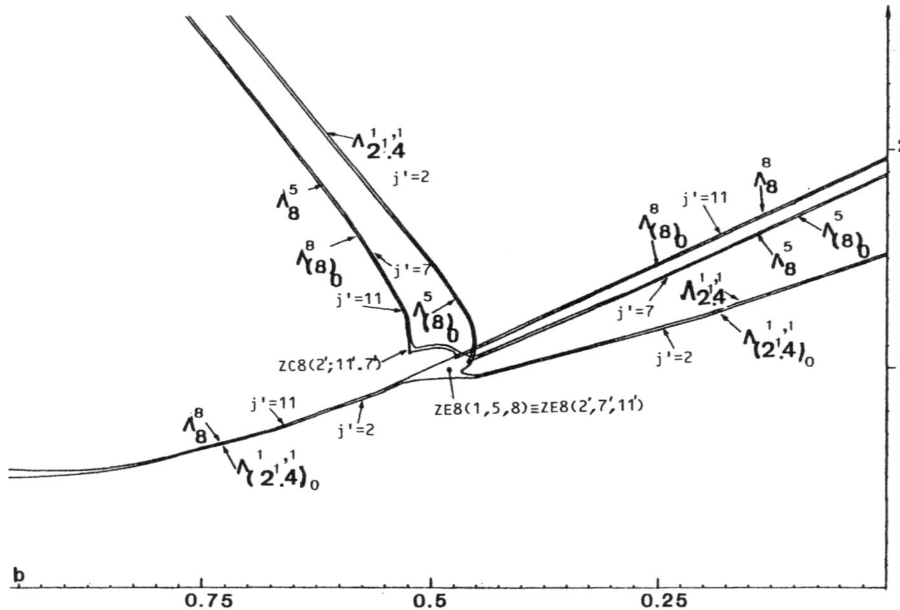

FIG.6.30 - Association of a cross-road area and a saddle area.
Embedded representation and correspondence with the non-embedded one (index j')

However in the saddle area, the curve $S_1 = +1$ with the cusp point, and its associated curve $S_1 = -1$, can correspond to different indexes, i.e these two bifurcation curves form a branch BRr(h,h') coming from a cross-road area. In this case the saddle is represented by the symbolism :

(6.35) ZCr (g ; h.h')

It is this situation which is given fig. 6.29, and also fig. 6.30 related to the diffeormorphism (6.26). So the information about the foliated structure of fig. 6.17 is completed. In fig. 6.29 hachured regios correspond to cycles with $|S_1| < 1$, which are attractive nodes, or foci, when $|b| < 1$. Hidden hachures correspond to a region located below a sheet of the structure.

In the parameter plane, a path passing through the cusp point, and between its two branches,gives rise to the co-dimension two bifurcation ($|b| < 1$) :

(6.36) saddle cycle $(r;h;S_1>1) \to$ attr. node cycle $(r;h)+$ 2 saddles$(r;h;S_1>1)$

6.9.4. THE SPRING AREA. LOCAL AND GLOBAL ASPECTS.

The corresponding foliated structure is given in fig. 6.31. Four bifurcation curves play a fundamental role [H5] [F8]

- Λ_k^j $(S_1 = -1)$ flip bifurcation curve, which also can be $\Lambda_{k2^i}^{j_i}$, one of the curves of a Myrberg's spectrum ω_k^j, $j_\circ \equiv j$,
- $\Lambda_{(k)_\circ}^{\ell_1}$ a fold bifurcation curve having a cusp point,
- $\Lambda_k^{\ell_1}$ $(S_1 = -1)$ the flip bifurcation curve associated with $\Lambda_{(k)_\circ}^{\ell_1}$ (for $b = 0$), presenting a loop which surrounds the cusp point ; or $\Lambda_k^{\ell_1'}$ $\ell_1' \neq \ell_1$, a flip bifurcation curve which forms with $\Lambda_{(k)_\circ}^{\ell_1'}$ a branch BRk (ℓ_1, ℓ'_1) coming from a cross-road area, and presents a loop surrounding the cusp point.

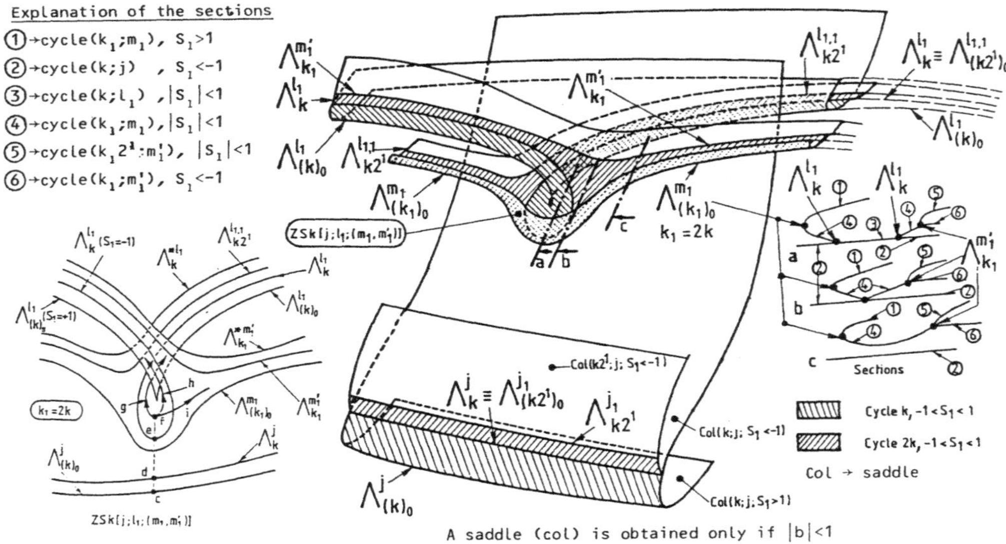

FIG.6.31 - The spring area ZS k $\{j;\ l_1;(m_1,m_1')\}$

Representation in the (a,b) plane and in the (a,b,|z|) space.

- $\Lambda^{m_1}_{(k_1)_o}$, $k_1 = 2k$, a fold bifurcation curve which in the neighbourooh of the spring area is associated with two different flip bifurcation curves $\Lambda^{m'_1}_{k_1}$, $m_1 \neq m'_1$, and $\Lambda^{\ell_1, 1}_{k.2^1}$, according to the schema of fig. 6.31. This last flip bifurcation curve is that obtained by order doubling from $\Lambda^{\ell_1}_{k}$. From this it results the existence of a branch BRk_1 (m_1, m'_1) of a cross-road area.

For the symbolic representation of the spring area only the *non-embedded representation* of cycles must be used to avoid ambiguities due to the "participation" of cycles of order k and k_1, and cycles of a Myrberg's spectrum, to the formation of this foliated structure. According to $\ell_1 \equiv \ell'_1$, or $\ell_1 \neq \ell'_1$, one of the following representations :

(6.36') $ZSk\ [j_i\ ;\ \ell_1\ ;\ (m_1,\ m'_1)]$ + sign of b,

(6.36") $ZSk\ [j_i\ ;\ (\ell_1,\ \ell'_1)\ ;\ (m_1,\ m'_1)]$ + sign of b,

give a full identification of the spring area, including the communications between the sheets of the foliated structure represented in the $(a,b,\ |z|\ )$ space (fig. 6.31).

The cusp point is also surrounded by the curves s = constant (cf § 6.9.2) forming a family of loops, the set of these curves giving the shape of a cuspidal caustic [T1], singularity introduced in the theory of catastrophes. It results a superposition of sheets such that two distinct flip bifurcations occur simultaneously in different places in the phase plane, and two different attractive (when $|b| < 1$) cycles $(k;\ \ell_1)$ exist for the same parameters (a, b) value. In the parameter plane, a path passing through the cusp point and between its two branches gives rise to the following co-dimension two bifurcation ($|b| < 1$) :

(6.37) Attr.node cycle $(k;\ell_1) \to$ saddle cycle $(k;\ell_i;S_1>1)$+ 2 attr.node cycles $(k;\ell_1)$

It is an equivalent bifurcation which occurs when passing through a cusp point of a cross-road area.

In the (a, b, $|z|$) space, three sections (a), (b), (c), related to the partial foliated structure obtained from $\Lambda^{m_1}_{(k_1)_0}$, $\Lambda^{m'_1}_{k_1}$, are represented fig. 6.39. Below the part (2) of each sections one has a sheet of saddle (if $|b| < 1$) cycle (k ; ℓ_1 ; $S_1 > 1$) and a sheet of cycle (k; ℓ_1 ; $|S_1| < 1$) joining at $\Lambda^{\ell_1}_{(k)_0}$ (the section of these two sheets are not represented in the schema of sections).

Consider the representation in the (a, b) plane (fig. 6.31). Starting from the point *c*, an attractive (if $|b| < 1$) cycle (k; j; $|S_1| < 1$) is obtained following the path *cd*. From the point *d* this cycle becomes a saddle (k; j; $S_1 < -1$), and S_1 decreases down to a minimum value, before increasing and passing through the value $S_1 = -1$ at the point *e* located on $\Lambda^{\ell_1}_{k}$. Then this cycle becomes attractive inside the loop of the bifurcation curve. Following the paths *efg*, *efh* leads to two different attractive cycles of order k. So after a "subterranean" (repulsive) path *de* the cycle resurges as attractive inside the loop of $\Lambda^{\ell_1}_{k}$, which has given the name of"spring"(analogy with the resurgence of a subterranean river) to this communication. Inversely starting from the point *d*, along a path b = constant, S_1 decreases. However S_1 cannot continue to decrease when a decreases, because this cycle should exist for values a < a ($\Lambda_{(1)_0}$). Thus the multiplier S_1 must pass through a minimum, before increasing up to the point *a* on a fold bifurcation curve ($S_1 = +1$).

Fig. 6.32 - 6.34 represent situations of spring areas for T_b given by the diffeomorphism (6.26), which permits to complete the fig. 6.17. Fig. 6.32 and 6.33 (c) show the variation of S_1 along a path b = constant. Fig. 6.34 (personal communication of D. Fournier) represents the aspect of flip bifurcation curves related to embedded boxes inside Ω^1_4 . It appears that the organization of the set of these curves depends on the shape of the loop of Λ^1_4 . Fig. 6.34 represents ZS4 [1;2; (10,8)] (b > 0), i.e. k = 4, j = 1, $\ell_1 = 2$, $m_1 = 10$, $m'_1 = 8$, besides $j_1 = 1$, $\ell_{1,1} = 9$, $k_1 = 8$. Fig. 6.33 a corresponds to ZS6 [2; 3; (47,45)] (b < 0) i.e. k = 6, j = 2, $\ell_1 = 3$, $m_1 = 47$, $m'_1 = 45$, besides $j_1 = 14$, $\ell_{1,1} = 46$, $k_1 = 12$. Fig. 6.37b corresponds to ZS7 [6; 7; (303,301)] (b < 0), i.e. k = 7, j = 6, $\ell_1 = 7$, $m_1 = 303$, $m'_1 = 301$, besides $\ell_{1,1} = 302$, $j_1 = 206$, $k_1 = 14$.

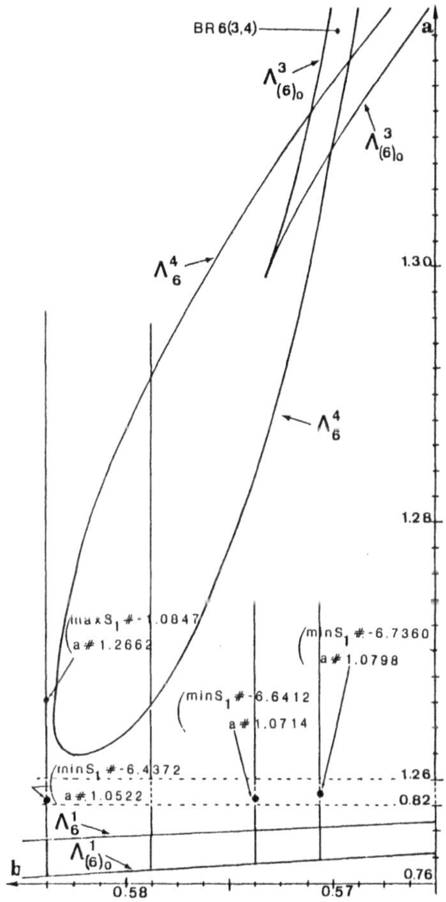

Non-embedded representation of cycles.

FIG.6.32 - Cusp and loop of a spring area .

Following a path a=constant on the left side of the loop, and starting from Λ^1_6, S_1 passes through a minimum and a maximum. When the path intersects the loop,before intersection S_1 passes through a minimum.

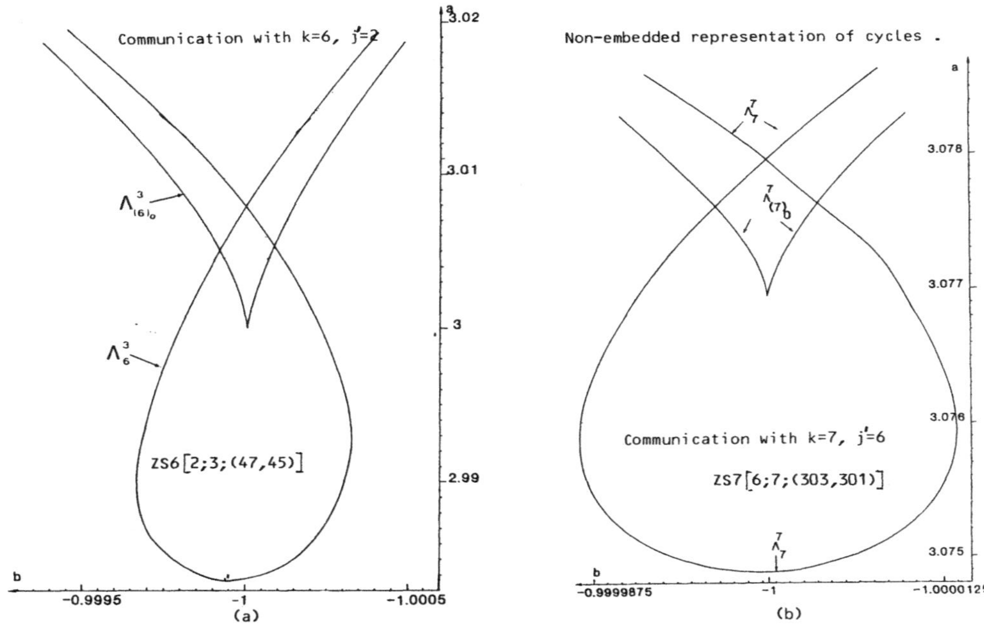

FIG.6.33 - (According to |H6|)

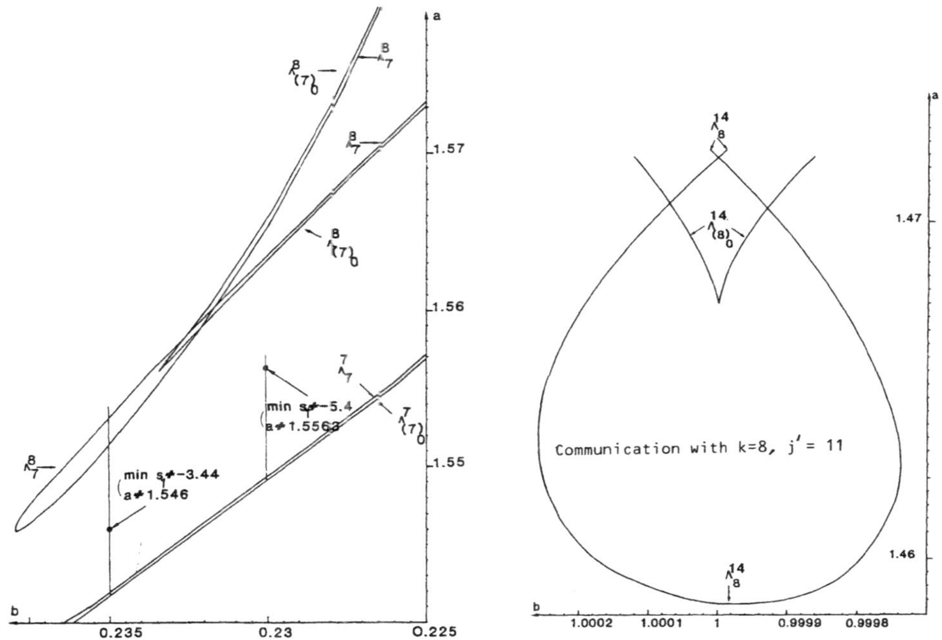

(c) (d)

FIG.6.33 - (Continuation) Non-embedded representation (according |H6|)

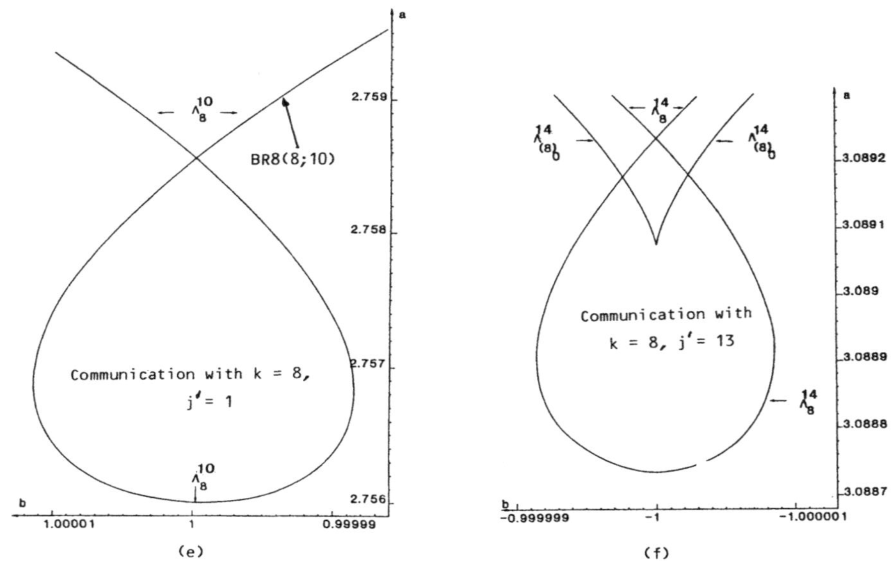

FIG.6.33 - (Continuation) . Non-embedded representation of cycles.

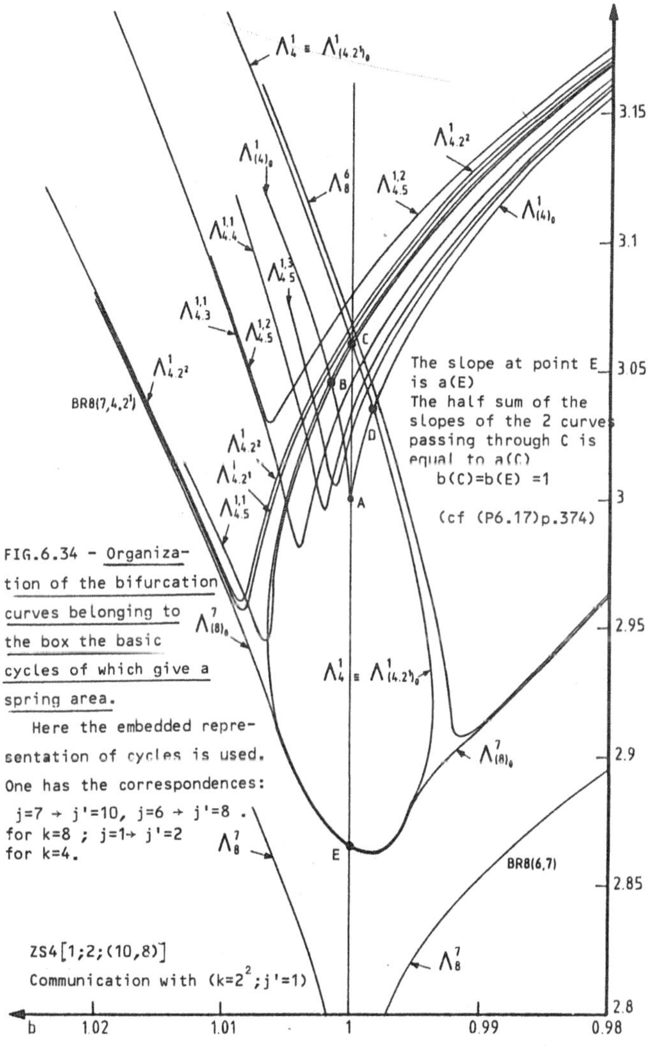

FIG.6.34 - Organization of the bifurcation curves belonging to the box the basic cycles of which give a spring area.

Here the embedded representation of cycles is used. One has the correspondences: $j=7 \rightarrow j'=10$, $j=6 \rightarrow j'=8$ for $k=8$; $j=1 \rightarrow j'=2$ for $k=4$.

For $b < 0$ it seems that one has the following relation between indexes of a spring area of (6.36) type :

$$(6.38) \quad m_1 = \ell_{1,1} + 1 = m'_1 + 2 \quad (b < 0)$$

Fig. 6.31 - 6.34 give only a local organization of communications related to spring areas. However from the parametric equations (6.6) of bifurcation curves, represented in the (a, b) plane, it appears that a spring area implies a set of more complicated structure of communications which concerns an important, $b < 0$, or $b > 0$, part of the (a, b) plane. Indeed in the symbolism (6.36'),(6.36"),a part is related to a branch BR k_1 (m_1, m'_1) of cross-road area ZEk_1 (m_1, m'_1), $k_1 = 2k$. Considering the second branch BR k_1 (m'_1, m_1), from parametric equations (6.6) it appears that this branch corresponds to a fold bifurcation curve with a cusp, and a flip bifurcation curve with a loop surrounding the cusp. The obtained situation implies the existence of a second spring area ZS $k\,2^1$ [j_1 ; (m'_1, m_1) ; (m_2, m'_2)] , $k_1 \equiv k2^1$. Here j_1 is the non-embedded index of the cycle ($k2^1$; j_1) born of the doubling order of the cycle (k; j) which communicates with the cycle (k; ℓ_1) (fig. 6.31). The representation of such a situation is given in the (a, b) plane in fig. 6.35, and in the (a, b, $|z|$) space in fig. 6.36. This process of communications formation is repeated for all the cycles ($k2^i$; j_i) of the Myrberg's spectrum ω_k^j , and one has [F8] :

(P6.15) Let ZSk [j; ℓ_1 ; (m_1, m'_1)] *be a spring area.*

(a) The Myrberg' spectrum ω_k^j, *and its* $\Lambda_{k2^i}^{ji}$ *flip bifurcation curves,* i = 0,1,2,...,*give rise to a cascade (or spectrum) of spring areas with doubling order (period) :*

$$ZSk2^i \; [j_i \; ; \; (m'_i \; , \; m_i) \; ; \; (m_{i+1}, \; m'_{i+1})] \; , \; i = 1,2,\ldots,(m'_0,m_0) \equiv \ell_1$$

(b) A cross-road area ZE$k2^i$(m_i, m'_i) *leads to a communication between the two spring areas :*

$$ZSk2^{i-1}[j_{i-1};(m'_{i-1},m_{i-1});(m_i,m'_i)] \text{ and } ZSk2^i[j_i;(m'_i,m_i);(m_{i+1},m'_{i+1})]$$

via the branches BR$k2^i$(m_i, m'_i), BR$k2^i$(m'_i , m_i) *belonging to this cross-road area. For* i = 1,2,3..., *these cross-road areas form a cascade by doubling order of* ZE $k2^i$(m_i, m'_i)

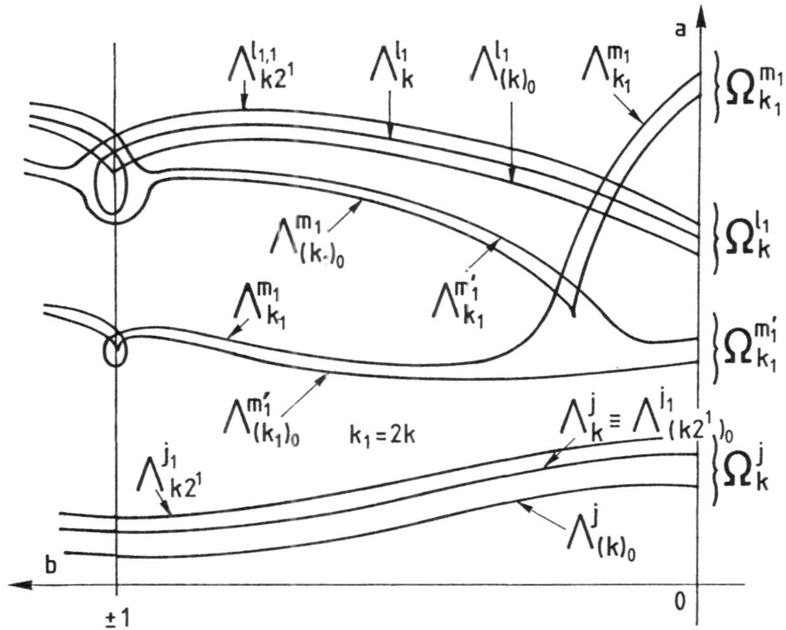

FIG.6.35 - Outset of a cascade of spring areas. 1st area ZSk$[j;l_1;(m_1,m_1')]$

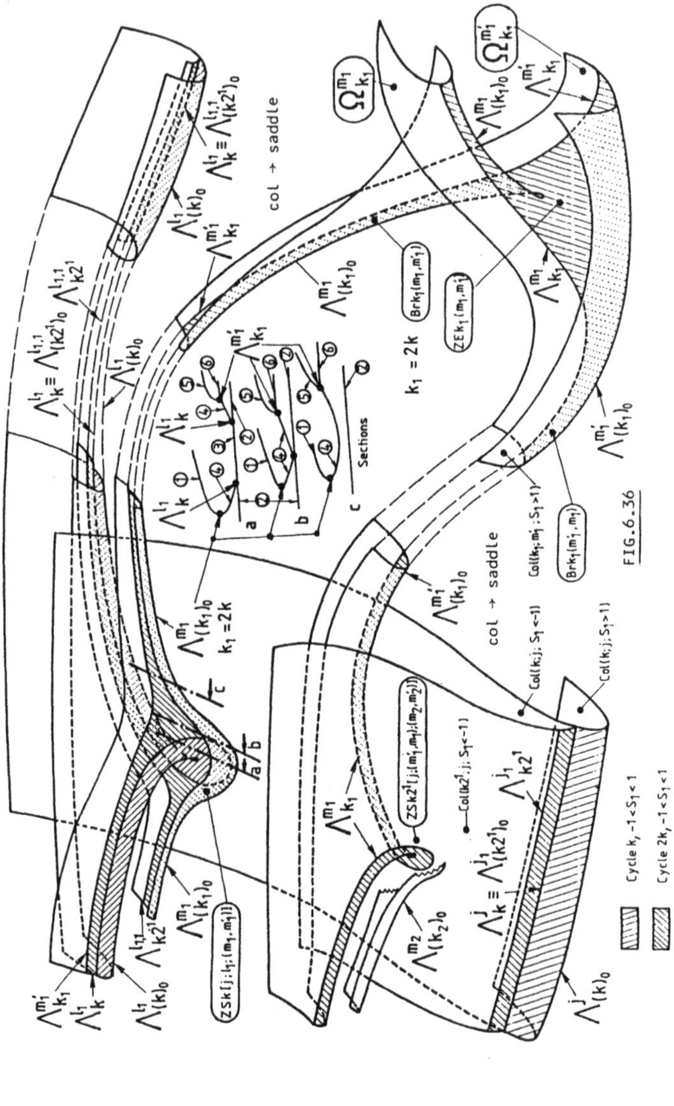
col → saddle
col → saddle
Sections
k1 = 2k
Cycle k1 -1<S1<1
Cycle 2k1 -1<S1<1

FIG. 6.36

(c) When $i \to \infty$*, the limit* $\Lambda^{j}_{(k)_s}$ *of the flip bifurcation curves* $\Lambda^{j_i}_{k2^i}$ *gives rise to the existence of an accumulation curve of the cascade of spring areas, and that of cross-road areas. This accumulation curve is the limit of* $\Lambda^{m'_i}_{(k2^i)_0}$*,* $i \to \infty$.

(d) This complex organization of communications is reproduced in a fractal way, related to the box-within-a-box bifurcations structure of the diffeomorphism T_b.

From the above considerations it appears the following property, which in general concerns the most part of dynamical systems :

(P6.16) For a dynamical system, the effective dimension (cf. § 1.5) of which is two, the local existence of a cusp point on a fold bifurcation curve may be related for a whole part of the (a, b) plane to a global complex structure of communications between sheets of a parameters plane.

In § 6.11 this complex structure is called communications cell.

6.10. CONTRACTED REPRESENTATION OF THE PARAMETERS PLANE. CONSEQUENCE.

Consider the expression (6.26) of T_b, and that (6.27) of T_b^{-1}. It appears that T_b^{-1} has an equivalent form to that of T_b, and leads to the same bifurcation curves with the parameters and variables correspondences [F9] :

$$a \to \alpha = a/b^2;\ b \to \beta = 1/b;\ x \text{ of } T_b \to y \text{ of } T_b^{-1},\ y \text{ of } T_b \to x \text{ of } T_b^{-1}.$$

Then $b = \pm\infty$ *corresponds to* $\beta = 0$*, and on* $|b| = 1$, $b = \beta$, $\alpha = a$. *Given the sign of* b*, it results that the part* $|b| < 1$ *of every bifurcation curve, represented with the coordinates* (a, b) *is the symmetrical, with respect to* $|b| = 1$ *, of the part* $|b| > 1$ *represented with the coordinates* (α, β) *(cf [F9], this property is due to H. Kawakami).*

The juxtaposition of the part $|b| < 1$ of the plane (a, b), and of the part $|\beta| < 1$ of the plane (α, β) on the both sides of $b = \beta = 1$

($b = \beta = 1$ being the same straight line) is called the *contracted representation of the parameters plane* (cf. the figures of § 6.11).

The property of symmetry in the contracted parameters plane gives rise to the proposition :

(P6.17) *At every point of* $|b| = 1$, $a = \bar{a}$, *through which only one bifurcation curve passes, the slope of this curve is* $(da/db) = b\,\bar{a}$.

At every point of $|b| = 1$, $a = \bar{a}$, *through which two bifurcation curves* (Γ_1) (Γ_2), *with slopes* $(da/db)_1$, $(da/db)_2$, *pass, the slopes are such that* : $2b\bar{a} = (da/db)_1 + (da/db)_2$.

In order to prove the first part of (P6.17) let $a = g(b)$ be the explicit (local) equation of an isolated branch of bifurcation curve passing through a point ($|b| = 1, a = g(b) = \bar{a}$). In the (α, β) plane this equation is $\alpha = h(\beta)$, or $a = b^2 h(1/b)$. So one has :

$$da/db = 2bh(1/b) - h'(1/b) \qquad h' = dh/db$$

When $b = \pm 1$, $2bh(\pm 1) - h'(\pm 1) = g'(\pm 1)$. In the contracted parameters plane the symmetry implies $h'(\pm 1) = g'(\pm 1)$ and $h(\pm 1) = \bar{a}$. Thus $b\bar{a} = h'(\pm 1) = g'(\pm 1)$

For the second part of (P6.17) consider $a = g_1(b)$, $a = g_2(b)$ the explicit (local) equations of two branches of bifurcation curves passing through a point ($|b| = 1$, $a = g_1(b) = g_2(b) = \bar{a}$). This point can be either point of intersection of a loop belonging to a spring area, or cusp point, or point of intersection of two bifurcation curves having the same nature (crossing area in § 6.11). Let $\alpha = h_1(\beta)$, $\alpha = h_2(\beta)$ the equations of these branches in the (α, β) plane. Then for $|b| = 1$ the symmetry leads to $2b\,h_1(\pm 1) - h_1'(\pm 1) = g_2(\pm 1)$, $2bh_2(\pm 1) - h'_2(\pm 1) = g'_1(\pm 1)$, $h_1(\pm 1) = h_2(\pm 1) = \bar{a}$, $h'_1(\pm 1) = g_1'(\pm 1)$, $h_2'(\pm 1) = g'_2(\pm 1)$ and

$$2b\bar{a} = g'_1(b) + g'_2(b) \quad \text{when } b = \pm 1$$

6.11. SYMBOLIC DYNAMICS OF COMMUNICATIONS IN THE FRACTAL FOLIATED BIFURCATIONS STRUCTURE. CONNECTIONS CHAIN.

6.11.1. NOTIONS OF CONNECTIONS CHAIN AND COMMUNICATIONS CELL.

The paragraph 6.11 again concerns the quadratic diffeomorphism T_b given by (6.26). It is founded on the concept of *connections chain* recently introduced in symbolic dynamics [F7] [F9], and on that of *communication cell* which is a schema of a complete set of communications in the parameters (a, b) plane when the sign of b is fixed. In this paragraph 6.11 *only the non-embedded representation of cycles is used.*

The notion of connections chain is defined via the representation of a *graph*, obtained as follows :

- Let E_k be the set of all the cycles having the same order k, constituting a list for which the adjoint relations are given, when the sign of b is fixed (cf. the table 1 giving the lists $6 \leqq k \leqslant 9$ for $b<0$).

- From E_k a sub-set called "connections chain" and designated by (CL) ("Chaîne de liaisons" in french) is extracted. This sub-set is constituted by cycles (kjdα) α = n, and p (§ 3.7) which are in adjoint relations with cycles (kj*dα). Thus self adjoint cycles do not belong to a connection chain.

- The graph of (CL) is such that :

- Two cycles of a (CL),(kjdn) (kjdp) born from the same fold bifurcation, are written on two lines and form a *double nucleus* represented by a closed curve surrounding the identification code of this pair (cf table 1, k = 8 the double nucleus j = 8 ; k = 9 the double nuclei j = 5, 12, 14, 16, 17, 19).

- When the fold bifurcation leads to a self adjoint cycle (kjsγ), γ = n (resp p), and a non-self adjoint cycle (kjdδ), δ = p (resp n), only this last cycle forms a nucleus with a single element, called *simple nucleus*. The non-self adjoint cycle (k1dn) born from a flip bifurcation also constitutes a simple nucleus. In table 1 a simple nucleus is indicated in the last column j' by a short horizontal segment on the right of j', which is not surrounded by a closed curve (example j' = 3,4,6,7,8,9,10,11,15,18,21,22,23,25,26 for k = 9).

- Starting from a given nucleus, each cycle is connected to its adjoint by a continuous line (adjoint connection). This connection attains a new nucleus, for which it is proceeded similarly, and this construction is repeated up to all the possible junctions may be achieved. So the described connections join a set of nuclei, and form the graph of a *connections chain of order* k (CL) or $(CL)_k$. Such connections chains (CL) are represented in table 1 ($b < 0$), for example when $k = 9$, the connections ($j = 3, 4$), ($j = 5, 12, 10, 11$), ($j = 6, 9$), ($j = 7, 8$), ($j = 14, 16, 17, 19$), ($j = 15, 18$), ($j = 21, 24$), ($j = 22, 23$), j = (25, 26) constitute 9 connections chains. Examples of connections chains also are given in fig. 6.37 to 6.45.

So the set E_k is made up of a set of connections chains and the set of the self adjoint cycles of order k. When $b < 0$, the results of § 3.6 lead to the following property

(P6.18) For $b < 0$, *the cycles belonging to the same connections chain have binary words with the same root. Moreover the turns number* γ *(§ 6.8) is the same for all these cycles.*

A *communications cell* (CC) is constituted by a set of fold and flip bifurcation curves, characterizing a complete set of communications between sheets of the foliated structure in the (a, b, $|z|$) space. These bifurcation curves are represented in the contracted parameters plane (§ 6.10). With the choice of axes (a, b) and (α, β) according to the considered region $|b| < 1$, $|b| > 1$, every bifurcation curve undergoes a slope discontinuity when passing through $|b| = 1$, which is due to the symmetries of (P6.17). Conventionally these discontinuities will not be represented in order to avoid ambiguities about the identification of the continuation of curves after crossing $|b| = 1$ (cf fig. 6.37 - 6.45). In all the fig. 6.37-6.45 it is always supposed that :

$$j_1 < j_2 < \ldots < j_i < j_{i+1} < \ldots$$

The basic elements of the communication cells are locally constituted by : spring areas (ZS), saddle areas (ZC), cross-road areas (ZC), and a new element called *crossing* (Cr) which plays an important role. A crossing is characterized by the intersections on $|b| = 1$ of two fold curves $\Lambda^{j_i}_{(k)_0}$, $i = 1, 2$ and their associated flip curves $\Lambda^{j'_i}_k$, $j_i = j'_i$

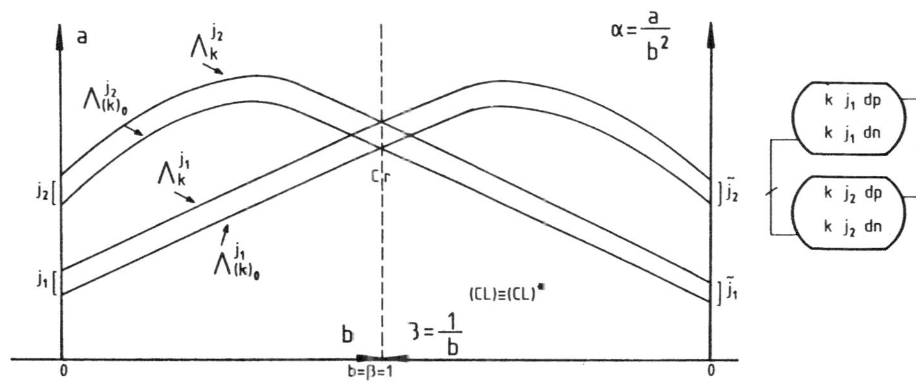

FIG.6.37 – The simple crossing and its connections chain.

if one, or two of these curves do not belong to a branch of a cross-road area (cf. fig. 6.37, 6.38).

6.11.2. CONNECTIONS CHAIN AND COMMUNICATIONS CELL OF SIMPLE SITUATIONS

When $|b| = 1$ it is important to note that the pair of cycles born of a spring area, or a saddle area,or a crossing have globally the symmetry $y = -x$ (if $b = -1$), or $y = x$ (if $b = +1$), i.e. the same symmetry as that of the symbolical phase plane (x', y'). For the spring area, and the saddle area this point results respectively from the co-dimension two bifurcations (6.20), (6.21) of the conservative case, which are (6.36), (6.37) for $|b| \neq 1$. These symmetries are those resulting from (P6.5) and (P6.6), i.e. adjoint relations are the symbolic images of the situations (ZS), (ZC), and (Cr). In other words the existence of two adjoint cycles $(k;j) = (k;j^*)$ corresponds either to a case for which two cycles can communicate via a cuspidal singularity of the (a, b) plane, located on $|b| = 1$, giving a zone of overlaps of sheets (fig. 6.29 and 6.31), or to the case of two pairs of cycles giving two folds of sheets (four sheets joining by pairs at $\Lambda^{j_i}_{(k)_0}$, $i = 1, 2$) intersecting $|b| = 1$ by crossing, or to more complex situations. Locally in the neighbourhood of $|b| = 1$ a crossing does not give rise to sheet communications with $j_1 \neq j_2$, but there exists cases such that communications occur outside this neighourhood (fig. 6.38).

A *saddle area* (ZC) with the cusp point located on $|b| = 1$ is related to a connection chain of order k, the simplest form of which is a connection between two simple nuclei (kjdp), (kj*dp) ($j = j_2$, $j^* = j_5$ in fig. 6.45 ; $b < 0$; $k = 7$, $j = 3$, $j^* = 4$; $k = 8$, $j = 10$, $j^* = 11$; $k = 9$, $j = 3$, $j^* = 4$; $k = 9$, $j = 6$, $j^* = 9$; $k = 9$, $j = 22$, $j^* = 23$; cf table 1). The connections chain of saddle area has a more complex structure in the case of fig. 6.43 with a pair of double nuclei and a pair of simple ones (for example $b < 0$, $k = 9$, $j = 5, 10, 11, 12$). In the simplest case the cycles ($S_1 > 1$) (kjdp), (kj*dp) belong to the same sheet as indicated in fig. 6.29.

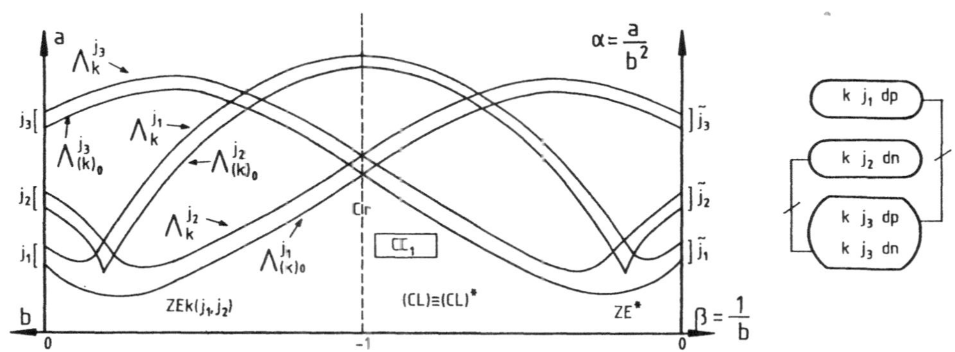

FIG.6.38 - Communications cell (CC_1) and its connections chain.

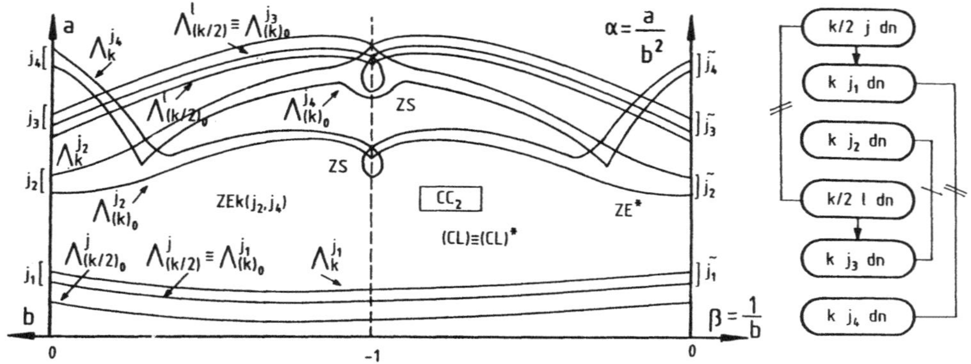

FIG.6.39 - Communications cell (CC_2) of a spring area. Connections chain.

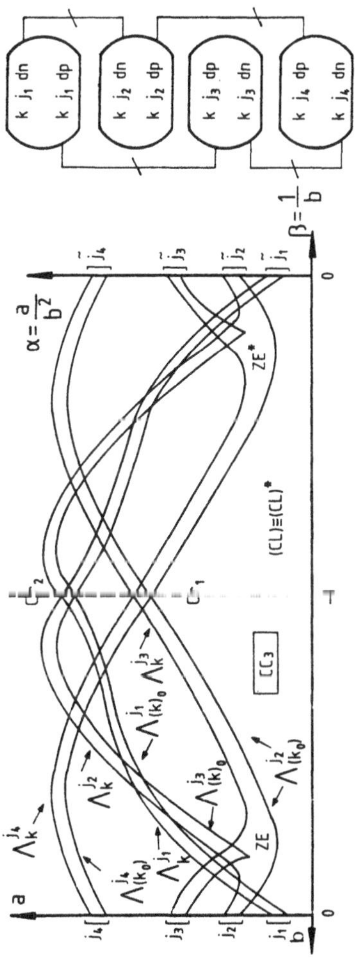

FIG.6.40 - Communications cell (CC_3) and its connections chain.

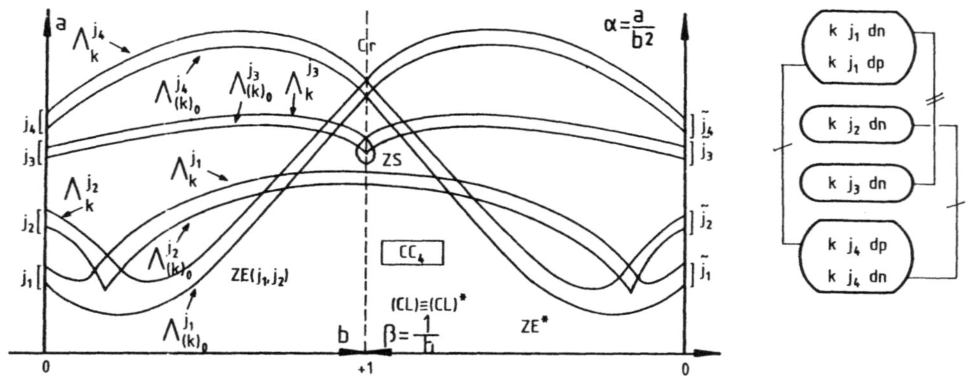

FIG.6.41 - Communications cell (CC_4) and its connections chain.

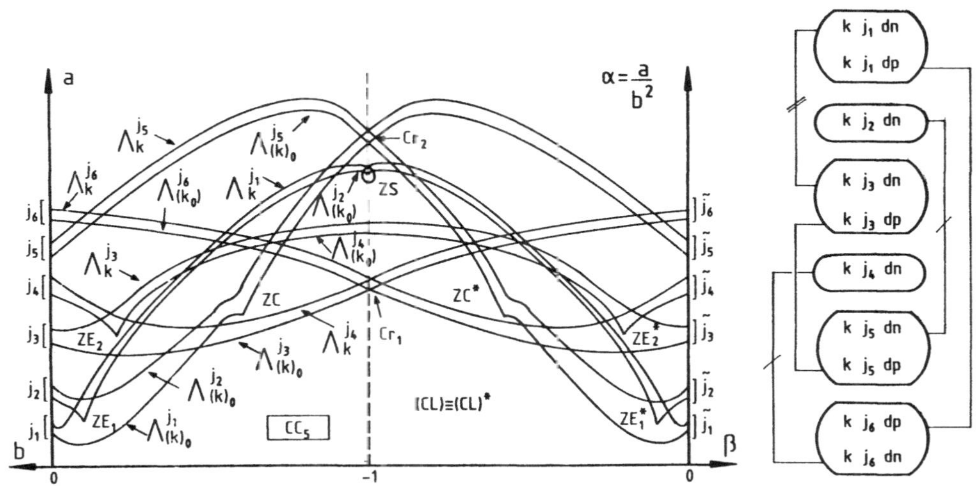

FIG.6.42 - Communications cell (CC_5) and its connections chain.

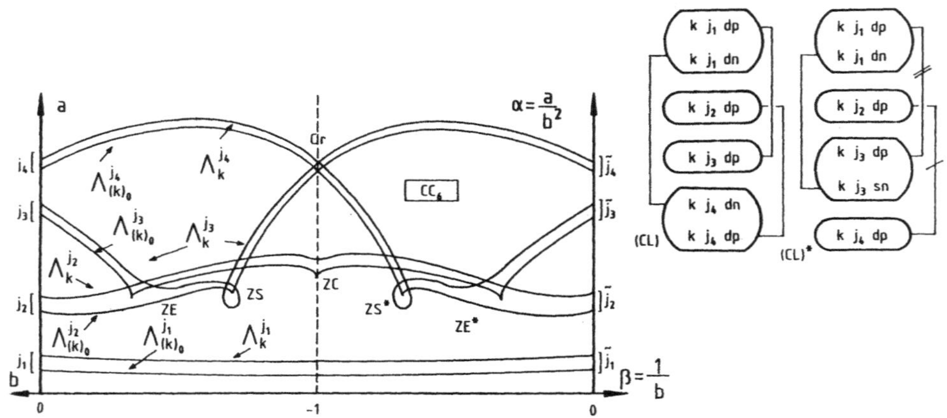

FIG.6.43 - Communications cell (CC_6) and its connections chain.

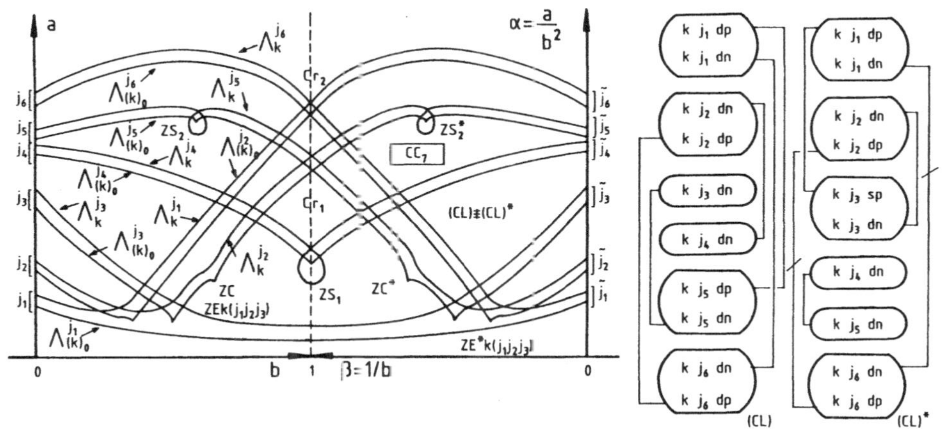

FIG.6.44 - Communications cell (CC_7) and its two adjoint connections chain.

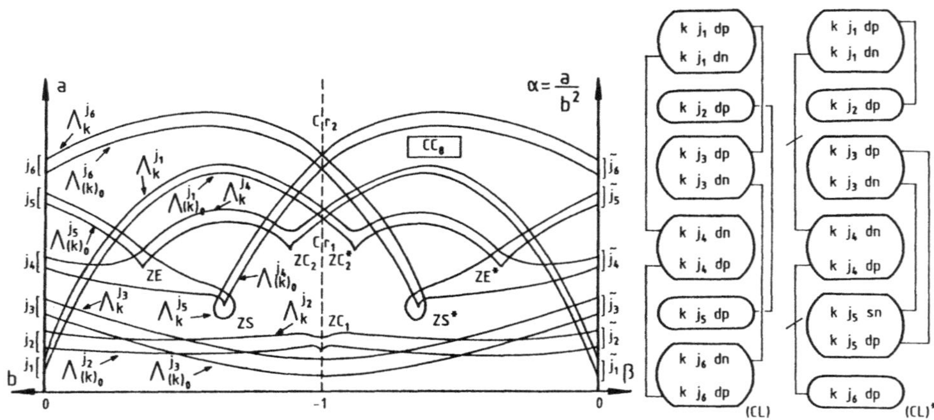

FIG.6.45 - Communications cell (CC_8) and its two adjoint connections chain.

A *spring area* (ZS) with a cusp point located on $|b| = 1$ is related to a connection chain of order k, the simplest form of which is a connection between two simple nuclei (kjdn), (kj*dn) [$b < 0$; $k = 6$, $j = 2$, $j^* = 3$; $k = 7$, $j = 6$, $j^* = 7$; $k = 8$, $j = 9$, $j^* = 12$; $k = 8$, $j = 13$, $j^* = 14$; $k = 9$, $j = 7$, $j^* = 8$; $k = 9$, $j = 21$, $j^* = 24$; $k = 9$, $j = 25$, $j^* = 26$]. However these connections chains can have a more complex structure as indicated in fig. 6.41, 6.42, 6.44. For spring areas of the simplest case the cycles ($S_1 < -1$) (kjdn) (kj*dn) belong to the same sheet as indicated in fig. 6.31.

From (P6.15) and fig. 6.31 it results that the single connections chain $(CL)_k$ of order k is not sufficient to give a global image of the communications implied by the existence of a spring area. To obtain this image, $(CL)_k$ must be associated with all the connections chains $(CL)_{k2^i}$ of order $k2^i$, $i = 1,2,\ldots$, resulting from (P6.15). Conventionally in the graph, the association is indicated by arrows, joining two nuclei of orders in the ratio of one to two, which figurate flip bifurcations related to two cycles belonging to each nucleus, as it appears in fig. 6.39. Then a cycle (k/2 jdn) of the list $E_{k/2}$ is the cycle (kj_1dp) of the list E_k, which leads to the existence of the cycle (kj_1dn). The corresponding connections chain, limited to the association of two orders k/2 and k, is represented fig. 6.39, and so represents the symbolic image of the communications cell of this figure

The simplest case of a *crossing* (Cr) is represented in fig. 6.37, with the corresponding connections chain. In this case, there is neither local communication of the cycles in the neighbourhood of $|b| = 1$, nor communication outside this neighbourhood. Two pairs of folded sheets overlap without communication, and fig. 6.37 does not represent a communication cell. It is not the case for fig. 6.38 where a communication of sheets (kj_1dn) (kj_3dn) occurs outside the neighbourhood of $|b| = 1$, via the existence of two cross-road areas (Fig. 6.46). Crossings, giving rise to communications outside the neighbourhood of $|b| = 1$, occur in more complex situations as indicated in fig. 6.40 - 6.45. It is worth of note that for $b > 0$, k being an odd integer, the fold and flip bifurcation curves are tangent when $b = 1$, and the crossing corresponds to a single point of $b = 1$.

6.11.3. CONNECTIONS CHAIN AND COMMUNICATIONS CELL OF MORE COMPLEX SITUATIONS

These situations are illustrated by fig. 6.40 - 6.45. In order to understand these ones, it is necessary to specify the numbering of indexes of the axis $\beta = 0$ in the contracted parameters plane. On $\beta = 0$, the non-embedded numbering of cycles is marked by indexes $\tilde{j}_i$, $i=1,2,\ldots$ These indexes follow the same order as that of cycles of $b = 0$, because T_b^{-1} given by (6.27) has a "symmetrical" form with respect to T_b. However, due to the presence of crossings (Cr) located on $|b| = 1$, a bifurcation curve of a (Cr) associated with an index j for $|b| < 1$, may continue in the region $|b| > 1$ with a different index $\tilde{j}$. Thus the same bifurcation curve may be associated with two different indexes $j \neq \tilde{j}$ in the whole part (with same sign of b) of the contracted parameters plane.

Then from a given communication cell (CC), it is possible to define a correspondence between indexes j_i and $\tilde{j}_i$, belonging to the same bifurcation curve, $i = 1, 2, \ldots, p$, p being the number of indexes of the same order k used for the definition of (CC). A part of these indexes are such that $j_i = \tilde{j}_i$. It is the case for a saddle area on $|b|=1$, a cusp point ($S_1 = +1$) and a loop ($S_1 = -1$) of a spring area on $|b| = 1$, the simple intersection of a bifurcation curve with $|b| = 1$ without any singularity, or even certain particular bifurcation curve of a crossing. Others indexes are such that $j_i \neq \tilde{j}_i$, the necessary condition of this situation being that j_i be the index of a bifurcation curve which defines a crossing (Cr).

Considering such a correspondence, a connections chain (CL), defined for $|b| < 1$, is transformed into an *adjoint connections chain* (CL)*, defined for $|b| > 1$. Then, when (CL) $\equiv$ (CL)* the connections chain (CL) is called *self-adjoint*. The different representations of fig. 6.37 to 6.45 show how (CL)* is deduced from (CL). In the graphs of these figures a double oblique mark, crossing through an adjoint connection between a cycle $(kj_a d\alpha)$ and a cycle $(kj_b d\alpha)$, $\alpha = n$ or p, indicates that $j_a = \tilde{j}_a$ and $j_b = \tilde{j}_b$. A single oblique mark indicates that $j_a = \tilde{j}_b$ and $j_b = \tilde{j}_a$. The absence of such a mark is related to the absence of such equalities between the considered indexes.

Every fig. 6.38 to 6.45 defines a typical communications cell (CC_i), $i = 1,2,\ldots,8$, the symbolical image of which is a typical connections chain (CL_i). For $i = 1,2,\ldots,5$, one has $(CL_i) \equiv (CL_i)^*$, and the corresponding connections chains are "connected", i.e. all the nuclei are continuously joined, and form a single connected set. For $i = 6, 7, 8$, $(CL_i) \not\equiv (CL_i)^*$, and the connections chains are not self adjoint.

When $i = 6, 7$, the connections chains (CL_i) are "connected", but it is not the case for the $(CL_i)^*$. Indeed for $(CL_i)^*$, $i = 6$, the simple nuclei j_2, j_4 are disjoint with respect other nuclei, and for $i = 7$ it is the case for the simple nuclei j_4, j_5. The case of the communications cell (CC_8) is more complex, because it is characterized by two disjoint partial connections chains, the union of which constitutes the "non-connected" connections chain (CL_8) (here the simple nuclei j_2, j_5 are without connection with other nuclei).

The cases $i = 6, 7, 8$ are such that $(CL_i)^*$ makes appear a self-adjoint cycle $(kjs\alpha)$, $\alpha = n$ or p, the points set of which has the symbolical symmetry $y' = -x'$ $(b < 0)$, $y' = x'$ $(b > 0)$. In all these situations, (CL_i) and $(CL_i)^*$ are necessary to identify the communications of (CC_i).

It is worth of note that in the case of (CC_8), from $(CL_8)^*$ the spring area ZS does not lead to symbolical symmetry for the cycles (kj_5sn) and (kj_3dn), and that one has a similar situation with (ZS) of (CC_6) and for the cycles (kj_1dn) (kj_3sn) of (CL_7) . Thus there exist two kinds of (ZS) and (ZC) : one having the symbolic symmetry and located on $|b| = 1$, the other without this symmetry and located outside a neighbourhood of $|b| = 1$.

The unicity of relation between a communication cell (CC_i), and the corresponding connections chain (CL_i) appears when $i < 7$, by consideration of the symbolical symmetries related to the (Cr), (ZC) and (ZS), which implies the existence of at least one cross-road area (ZE) for $|b| < 1$. When $i = 7, 8$ the situation is different, and this unicity is considered as a conjecture.

The described (CC_i) and (CL_i), $i = 1, 2, \ldots, 8$, are obtained for $k \leqq 10$, and are not the only possible. When k increases certainly more

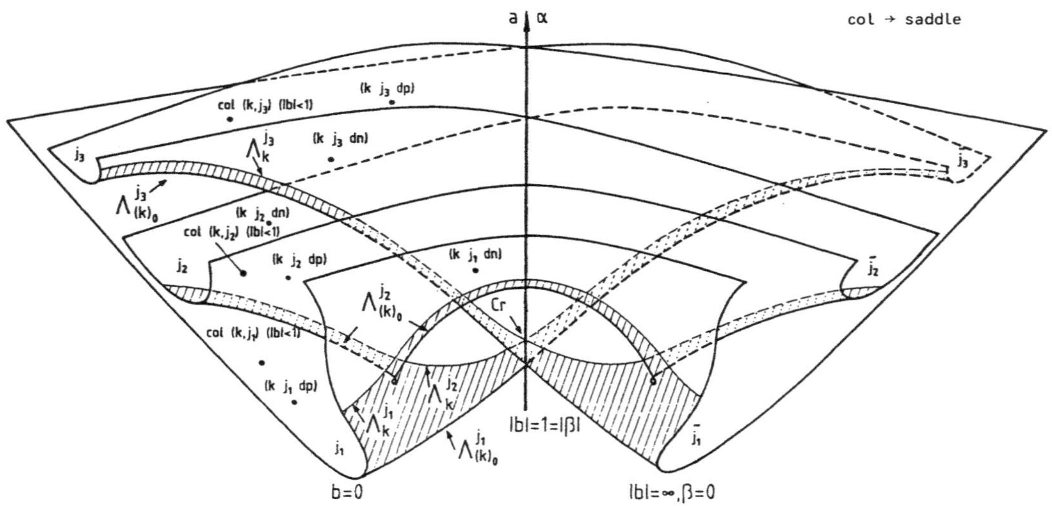

FIG.6.46 - Representation in the $(a,b,|z|)$ space of the communications cell (CC_1).

complicated structures occur. Moreover all the considered structures are reproduced in a fractal way according to the box-within-a-box bifurcations organization of T_b. The results of § 6.11 have been published in [F9].

Remark

(a) There exists a relation between the concept of adjoint cycles and that of *isotopic cycles* introduced in the interesting study of "knotted periodic orbits" [H11]. So adjoint cycles are always isotopic, but the inverse does not occur. Thus the cycles belonging to a connections chain are isotopic, but a complete set of isotopic cycles does not constitute a connections chain. In the same way the notion of *"putative braid"* of a knot introduced in [H11] is here related to that of turns (or winding) number obtained without using the knotted periodic orbits.

(b) Fig. 6.46 is given to illustrate the representation of the communications cell (CC_1) (fig. 6.38). (CC_2) is represented by fig. 6.36. Equivalent representations also can be obtained for (CC_i), i = 3,...,8, but the corresponding figures become too complex to lead to a sufficiently clear drawing.

6.11.4. ORGANIZATION OF CONNECTIONS CHAINS BELONGING TO A SAME LIST E_k (b < 0)

Considering the list E_k (b < 0) of all the cycles *with the non-embedded representation,* having the same order k, it appears that the connections chains (CL) can be divided into families. A *family is a set a connections chains the binary words of which have the same root* 10^{k-p} i.e the corresponding binary words of k bits are :

$$[q^{(1)j}_k] = [10^{k-p}\ 1\ a_{p-2}] ,$$

where a_{p-2} is a binary word of (p-2) bits, and the concatenation $1a_{p-2}$ is the stalk which is the variable part for the words of the family. A family is designed by its root.

The list $E_k(b < 0)$ is characterized by a first set of properties which appear from an examination of the lists with increasing k (γ is the turns (or winding) number). A part of these properties in given by fig. 6.47 - 6.52. Since only cycles of the same order k and in adjoint relations are in question, the complete representation ($kjd\alpha$), α = n or p, is shortened fig. 6.47-6.52 in the form ($j\alpha$).

(P6.19) *With* $b < 0$, *the two cycles* $j' = p_k$ *(ultimate number of* E_k *given by* $j_k = N_\lambda(k)$ *in (3.7)-(3.10), root* 10^{k-2}, $\gamma = 1$*), the two cycles* $j' = p_k - 1$ *(root* 10^{k-3}, $\gamma = 2$ *), and the cycle* ($k = 2^i$)*, or the two cycles* ($k \neq 2^i$), $j' = 1$*(root* 10^1) *, are always self adjoint.*

(P6.20) *(a) The cycles belonging to the same connections chain* (CL) *have the same turns (or winding) number* γ*, but connections chains belonging to the same family generally correspond to different turns numbers.*
(b) The connections chains belonging to the family 10^1*, with* k = *odd integer, have the same turns number* $\gamma = (k-1)/2$
(c) The family 10^1*, with* k = *even number, is constituted by two disjoint sub-sets: one belongs to the box of second kind* $\Omega_{2^1}(\lambda < \lambda^*_{2^1})$ *and corresponds to* $\gamma = k/2$*, the other* ($\lambda > \lambda^*_{2^1}$) *corresponds to* $\gamma = k/2 - 1$.

(P6.21) *(a) The family* 10^{k-4} *is constituted by only one connections chain, which is that of a spring area* (ZS) *for* $k \geqq 6$*, the corresponding graph is that of fig. 6.47.*
(b) The family 10^{k-5} *is constituted by two nested connections chains* (ZS), (ZC) *according to fig. 6.48, when* $k \geqq 8$.
(c) the family 10^{k-6} *is constituted by three connections chains according to fig. 6.49, when* $k \geqq \bar{k} = 10$*. When* $k = 9$*, the spring area* (ZS) *defined by* $j' = p_k-8$ *,* $j' = p_k-15$ *, disappears with respect to fig.6.49.*
(d) The family 10^{k-7} *is constituted by the connections chains of fig. 6.50 when* $k \geqq \bar{k} = 12$*. When* $k = 11$ *, this family looses the spring area* ($j' = p_k-16$, $j' = p_k-31$).
(e) The family 10^{k-8} *is constituted by the connections chains of fig. 6.52, when* $k \geqq \bar{k} = 14$*. When* $k = 13$*, this family looses the spring area* ($j' = p_k-32$, $j' = p_k-63$).
(f) Let ($j'_{max} - j'_{min}$ *)be the length of a connections chain,* j'_{max}*(resp*

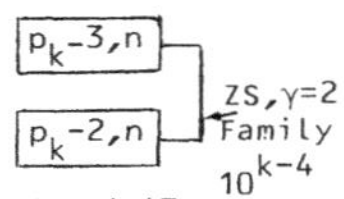

FIG.6.47

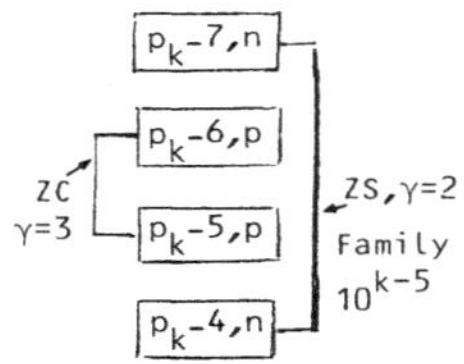

FIG.6.48

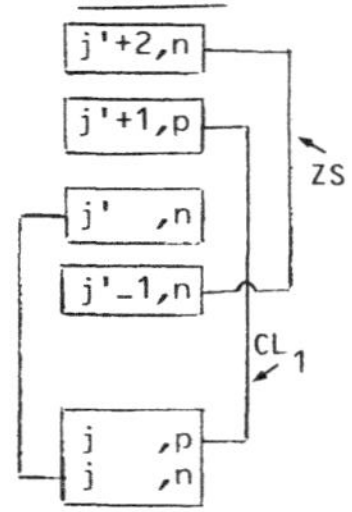

FIG.6.51

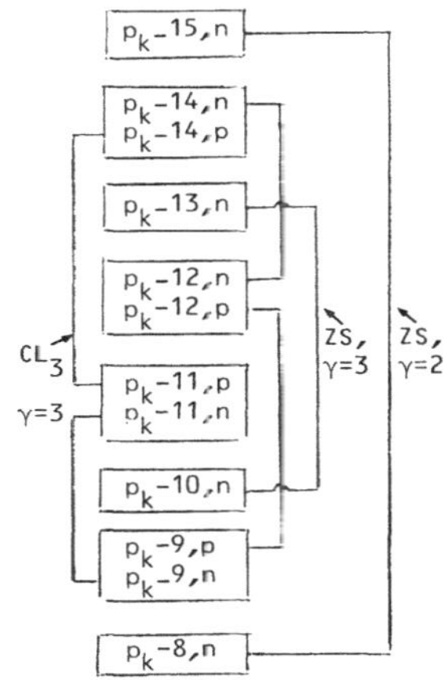

FIG.6.49
Family of connections chains 10^{K-6}.

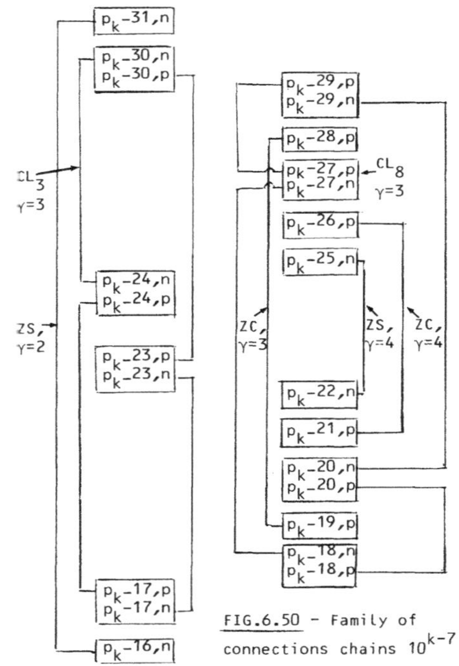

FIG.6.50 - Family of connections chains 10^{k-7}

FIG.6.50

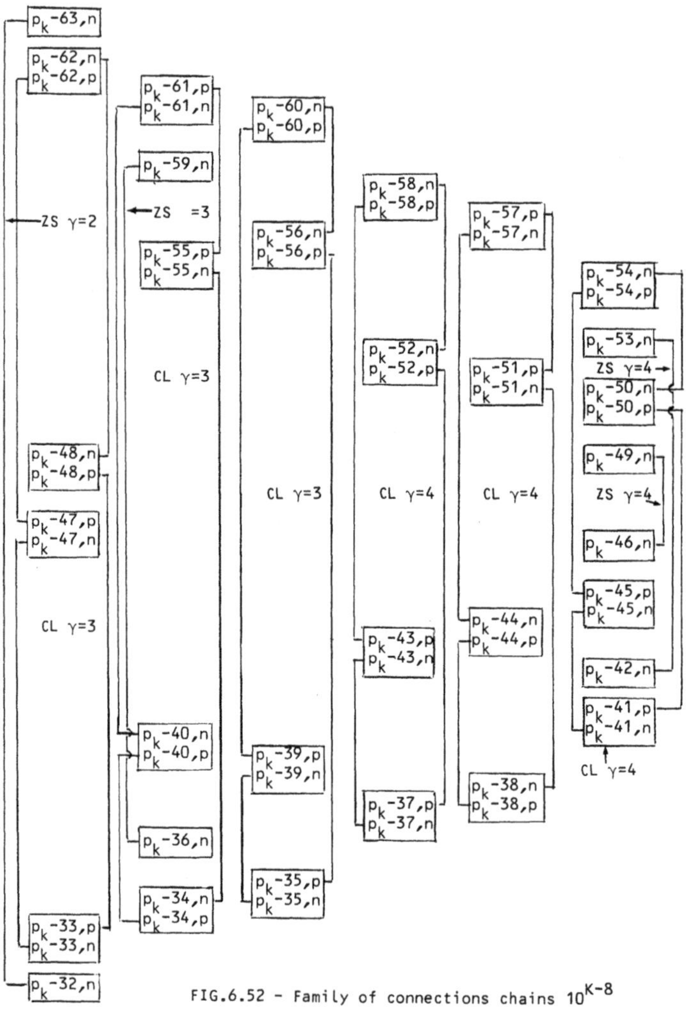

FIG.6.52 - Family of connections chains 10^{K-8}

j'_{min}) *being the highest (resp lowest) index of the cycles belonging to this chain. In the cases* (c) - (e) *the connections chain* (CL_3) *(fig. 6.40), having the maximum length, degenerates into the connections chain* (CL_1) *(fig. 6.38) when* $k = \bar{k} - 2$ *(cf. 6.51).*

(g) In the list E_k *, the presence of a pair of cycles the rotation sequence of which is related to a Poincaré's rotation number (self adjoint cycles) located between* $j = p_k-2$ *and* $j' = p_k-63$ *products a shift of one unity in the numbering of indexes of* (c) - (e).

Such an organization of the connections chains, inside a given family 10^{k-p}, $p > 8$, $k > 14$, expands so that the number of nested connections chains at first increases with p, passes through a maximum, and then decreases the last family being 10^1.

With increasing k and p the situation also becomes more and more complex with the presence of less simple connections chains than those considered in (P6.21) (for example (CL_i), i = 4-7, of fig. 6.41-44, and others more complex).

The given results are limited to the case $b < 0$. When $b > 0$, families of connections chains cannot be defined from roots of binary words, and the solution of the equivalent problem is less evident.

6.12. BIFURCATIONS STRUCTURE OF A TWO-DIMENSIONAL CUBIC DIFFEOMORPHISM

6.12.1. GENERALITIES

This paragraph considers some bifurcations properties of the cubic diffeomorphism T :

(6.39) $$x_{n+1} = ax_n(1-x_n) + y_n(1+\alpha x_n^2) \; ; \; y_{n+1} = bx_n, \; \alpha > 0, \; n=0,1,2..$$

x, y, being real variables, a, b, α real parameters. For $b = 0$, $n > 1$, T turns into the one-dimensional quadratic endomorphism T_0 :

(6.40) $x_{n+1} = ax_n (1 - x_n)$

The bifurcation structure of T_0 was identified in chapters 3 and 4. It is a *box-within-a-box structure*, with a fractal organization for the set of the bifurcations, in direct relation with the properties of $x_{n+1} = x_n^2 - \lambda$ (on account of the relation between coefficients of a quadratic endomorphism (§ 3.1) $4\lambda = a^2 - 2a$).

For $b \neq 0$, $\alpha = 0$, T is a quadratic diffeomorphism, with a constant jacobian $J = -b$, equivalent to (6.26). T ($b \neq 0$, $\alpha \neq 0$) has the same type of bifurcations structure as that of (6.26), but with an additional property due to the form of the jacobian J, $J = -b(1 + \alpha x^2)$, which is not constant. Then it results from this situation that a cycle (k;j) with multipliers (eigen-values) $S_{1,2} = \rho \exp (\pm i \varphi)$, $i = \sqrt{-1}$, can undergo the bifurcation (wrongly named "Hopf bifurcation") (P5.9) :

(6.41) attr. (or rep.) focus (k;j) $\rightleftarrows$ rep. (or att.) focus (k;j)+ $s(C_k^j)$

This occurs when ρ crosses through the critical value $\rho = 1$, for which the cycle is a *complex focus of multiplicity* $\ell \geq s$. In (6.41) "attr.", is written for "attractive", "rep." for "repulsive", (C_k^j) is constituted by k closed curves invariant by T^k, with a cyclic interchange by T characterized by j. Round each of the k points of the cycle (k;j), s such invariant closed curves are possible, and the most distant curve from one of these k points is attractive (resp. rep.). For (6.41), φ is supposed different from certain commensurable with 2π values, giving *exceptional cases* (§ 5.8 - 5.10). By "invariant closed curve", it is understood that there is either no cycle on the closed curve (the rotation number of this curve is irrational), or at least one saddle node (or saddle focus) pair of cycles of order qk on this curve (the rotation number is rational, equal to p/q, p,q = integers). T_0 and T_b ($b \neq 0$, $\alpha = 0$) can be considered as imbedded in the more general diffeomorphism $T_{b\alpha}$ given by (6.39).

The purpose of this paragraph is to describe in the (a, b) plane the bifurcations chains, equivalent to Myrberg's chains (doubling of the cycles order (period) for T_0, and corresponding to the doubling of the order for (C_k^j) (existence of chains of invariant closed curves $(C_{k2^i}^{j_i})$ $(i = 0,1,2,\ldots)$. In a sufficiently small neighbourhood of $b = 0$, the bifurcations structure of $T_{b\alpha}$ qualitatively is that of T_b (6.26). Results of this paragraph have been published [M40] [M58] (1983, 1984).

6.12.2. DESCRIPTION OF THE CHAINS OF BIFURCATIONS (6.41)

The curve $\rho = 1$ of the (a, b) plane is a segment of the curves described by :

$$(6.42)\qquad (-b)^k \prod_{i=1}^{k} (1 + \alpha x_i^2) = 1,\ X_{i+k} = X_i,\ X_{i+m} \neq X_i,\ m < k,$$

where X_i is the coordinates pair (x_i, y_i) of one of the k points of the cycle (k ; j). This cycle appears from a point (a, b) of the fold bifurcation curve $\Lambda^j_{(k)_0}$ $(S_1 = +1)$. Equation (6.42) says that the product of the two multipliers $S_1.S_2 = +1$. Then (6.42) represents three segments of curves :

Z_k^j with a φ-graduation, $0 \leqq \varphi \leqq \pi$, locus of the points such that $\rho = 1$;

Y_k^j having the point $\varphi = 0$ in common with Z_k^j, such that $S_1 = 1/S_2, S_{1,2} > 0$ real for the saddle (k ; j) born from $\Lambda^j_{(K)_0}$;

$\bar{Z}_k^j$ having the point $\varphi = \pi$ in common with Z_k^j, such that $S_1 = 1/S_2$, $S_{1,2} < 0$ real, for the saddle (k ; j) born from Λ_k^j , flip bifurcation curve on which one of the multipliers is -1.

The segment Z_k^j is located between $\Lambda^j_{(k)_0}$ $(\varphi = 0, S_1 = S_2 = 1)$, and Λ_k^j $(\varphi = \pi, S_1 = S_2 = -1)$. From (6.42) it results that cycles with odd k cannot be such that $S_1 S_2 = +1$ for $b > 0$, and so cannot give invariant closed curves from (6.41).

In § 6.1 it was indicated that the flip bifurcations (6.9) give rise to a Myrberg' spectrum ω_k^j . Then this spectrum gives rise to the chain of bifurcation curves

$$(6.43)\qquad Z_{k2^i}^{j_i} \rightarrow Z_{k2^{i+1}}^{j_{i+1}} + \bar{Z}_{k2^i}^{j_i} \qquad \text{when } \varphi(k2^i ; j_i) = \pi,\ j_0 \equiv j$$

$k = 2,3,4,\ldots,\ i = 0,1,2,\ldots,\ S_{1,2} = \exp(\pm i\varphi)$ for the cycle $(k2^i\ ;\ j_i)$ $Z^{j_i}_{k2^i}$ $(\varphi = 0)$ is on $\Lambda^{j_{i-1}}_{k2^{i-1}}$, $Z^{j_i}_{k2^i}$ $(\varphi = \pi)$ is on $\Lambda^{j_i}_{k2^i}$. *All the curves* $\Lambda^{j_i}_{k2^i}$, $Z^{j_i}_{k2^i}$, $\bar{Z}^{j_i}_{k2^i}$, $i = 0,1,2,\ldots$ *are located on the same Myrberg's spectrum* ω^j_k *in the* (a,b) *plane. For a different* k *or* j , *these curves belong to a different spectrum.*

Some curves $Z^{j_i}_{k2^i}$ (represented with continuous lines), $\bar{Z}^{j_i}_{k2^i}$ and Y^j_k (represented with broken lines) are given by fig. 6.53-6.57 (b < 0) and fig. 6.58, 6.59 (b > 0) when $\alpha = 1$.When $\alpha \to 0$ the set of these curves tends toward the straight lines $b = \pm 1$ of the (a, b) plane, and for $\alpha = 0$ bifurcations such (6.41) do not occur.

Fig. 6.53 makes appear Y_1 (b = -1) related to the saddle fixed point 0 (x = y = 0) and Z_1 ($k_1 = 1$, i = 0) related to the fixed point q_2 $(x = (a - \sqrt{\Delta})/(2b\alpha),\ \Delta = a^2 - 4\alpha b\,(a+b-1),\ y = bx)$. It appears that the curves $Z^{j_i}_{k2^i}$ either are always located on the same side of $\bar{Z}^{j_{i-1}}_{k2^{i-1}}$ for a given (k ; j) (fig. 6.54), or are located one one side of this curve when the index i is odd, and on the other side when i is even (fig. 6.57). *In all the cases two consecutive bifurcation curves* $Z^{j_i}_{k2^i}$ *have not a common tangent on the flip bifurcation curve* (for an equivalent remark concerning the non-autonomous Duffing-Van der Pol equation cf [K2] p. 258).

Crossing through $Z^{j_i}_{k2^i}$ by going from $\rho < 1$ toward $\rho > 1$, (the angle φ being different from values associated with exceptional cases) leads to three identified types of bifurcations :

- (a) The bifurcation (6.41) for an attractive ($\rho < 1$) focus cycle of order $k2^i$. For Z_1 (k = 1, i = 0), one has $\ell = s = 1$ and it seems that it is also the case for all the other curves $Z^{j_i}_{k2^i}$. In the region $\rho > 1$ of the (a, b) plane, the invariant closed curves $(C^{j_i}_{k2^i})$ disappear from the bifurcation described by fig. 5.39b, or 5.42b.
- (b) A saddle cycle (k ; j) is associated with a focus cycle (k ; j). Here (k ; j) can be a cycle $(k2^i\ ;\ j_i)$.In the region $\rho < 1$ of the (a,b) plane, there exists a segment V^j_k begining in $\varphi = 0$ of Z^j_k, such that the bifurcation described by fig. 5.37 (§ 5.15a) occurs on this segment.

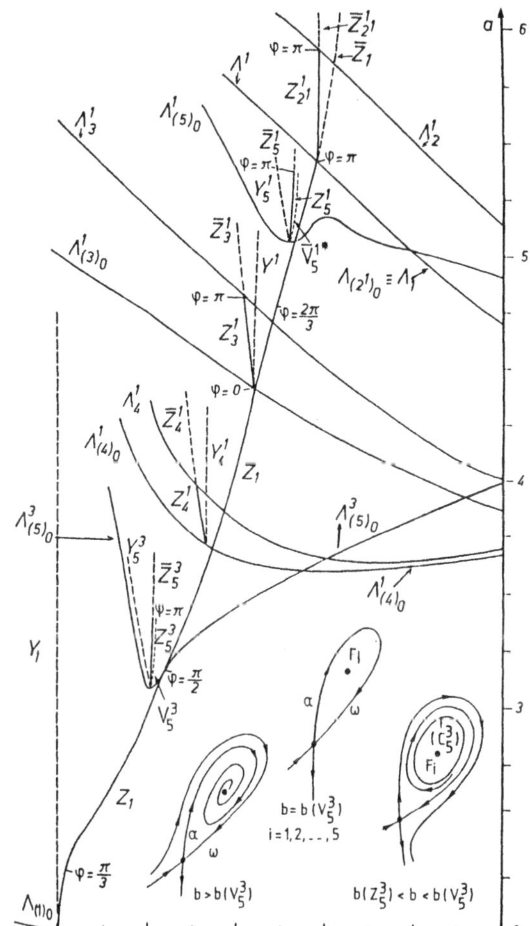

FIG.6.53 - Bifurcations structure of the cubic diffeomorphism (6.39) with α = 1. Passing through Z_k^j gives rise to an invariant closed curve

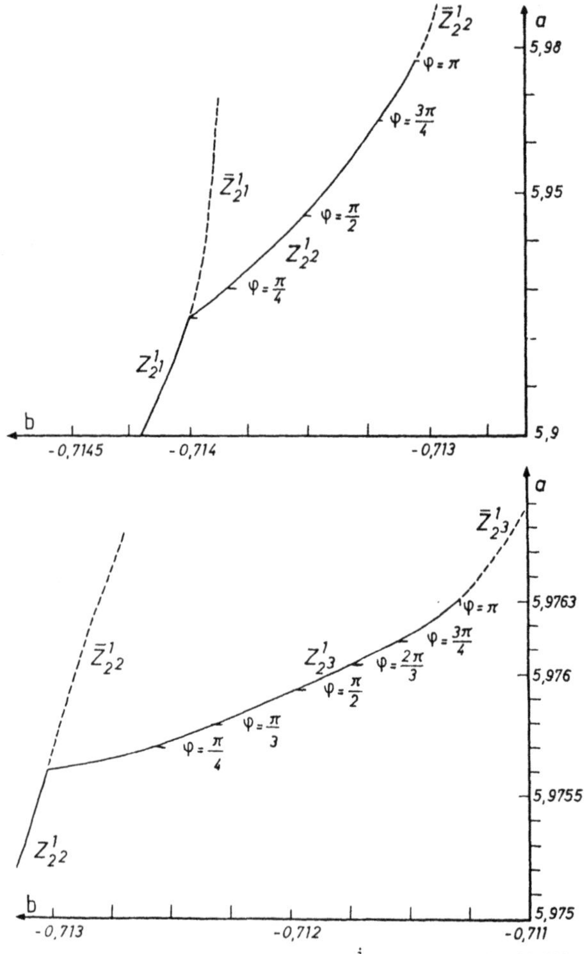

FIG.6.54 - Order doubling of the curves Z_k^j (diffeomorphism(6.39) with $\alpha = 1$). According to |M40||M58|.

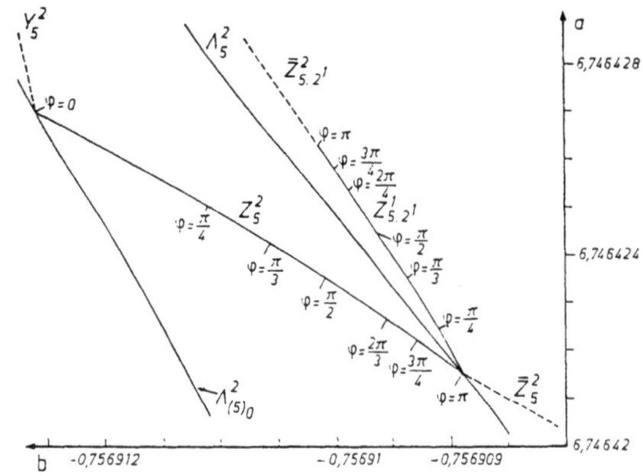

FIG.6.55 - Cubic diffeomorphism (6.39) α = 1 .

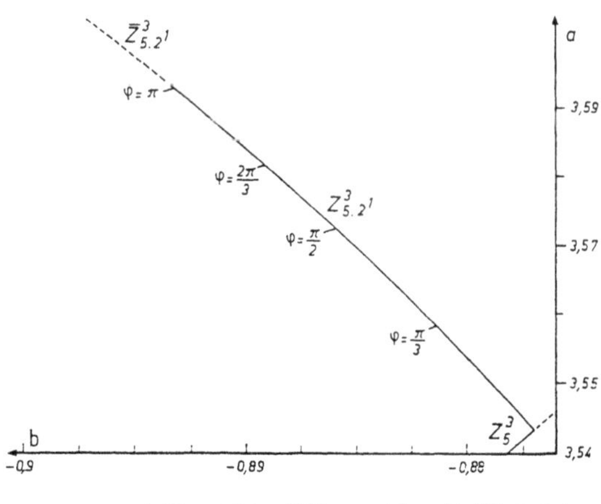

FIG.6.56 - Cubic diffeomorphism (6.39) α = 1 .
(According to |M40||M58|)

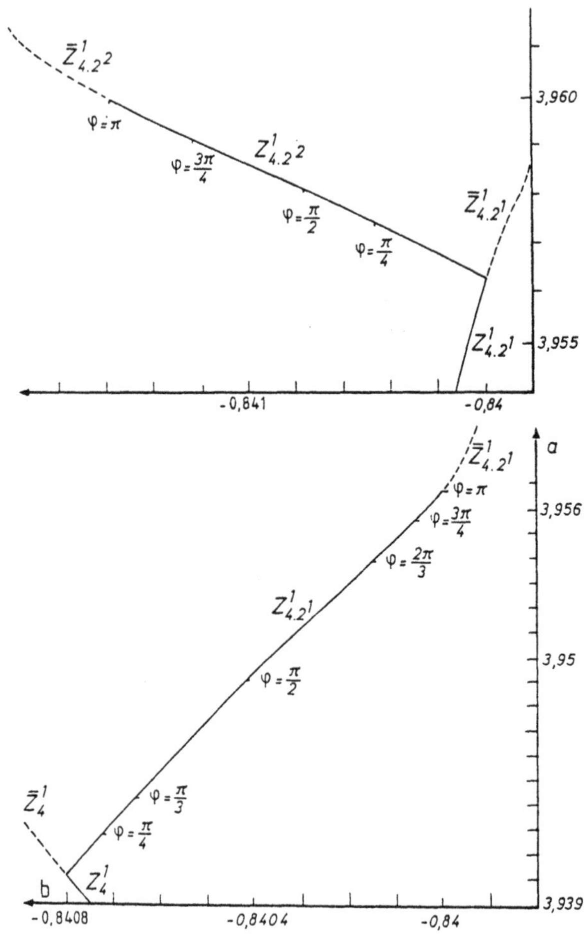

FIG.6.57 - Cubic diffeomorphism (6.39) $\alpha = 1$.
(According to |M40||M58|)

Let (D^j_k) be the region$(\rho < 1)$ of the (a, b) plane, which is bounded by Z^j_k, V^j_k, Λ^j_k .For parameters values located inside this region, every point of the attractive focus cycle (k ; j) is surrounded with one repulsive invariant closed curve by $T^k_{b\alpha}$. These k repulsive invariant closed curves constituting (C^j_k) (cyclic invariant closed curve) merge into the k points of the attractive focus cycle (k ; j) when (a, b) ϵ Z^j_k,and this cycle becomes repulsive for $\rho > 1$. Within the precision range of numerical tests, this situation was identified for k = 5, j = 3 (cf. fig. 6.53). It is worth of note that in a sufficiently small neighbourhood of Λ^j_k, or $\Lambda^j_{(k)_0}$, and sufficiently far from Z^j_k inside (D^j_k), the focus cycle becomes a node cycle.

- (c) In presence of homoclinic points related to the saddle cycle (k; j), the situation (b) is modified because the boundary V^j_k of (D^j_k) in (b) becomes the segment $\bar{V}^{j*}_k$ corresponding to the "first" (from Z^j_k) *homoclinic tangency* (fig. 5.39b). Inside (D^j_k), and in a sufficiently small neighbourhood of $\bar{V}^{j*}_k$, the repulsive cyclic invariant closed curve (C^j_k) folds upon itself in a very complicated way, practically giving a "chaotic annular area" which plays the role of a fuzzy basin boundary. When the point (a, b) approaches Z^j_k , (C^j_k) "unfolds itself" up to have an elliptical shape when it tends towards the attractive focus cycle (k ; j) before merging into the cycle, which then becomes repulsive.

In the case of a focus cycle $(k.2^i\ ;\ j_i)$ resulting from a Myrberg' spectrum, an equivalent bifurcation occurs. An area $(D^{j_i}_{k2^i})$ of the (a,b) plane, giving the existence of a $(C^{j_i}_{k2^i})$ cyclic invariant closed curve with the order $k2^i$, is defined. $Z^{j_i}_{k2^i}$ is one of the boundary of $(D^{j_i}_{k2^i})$. Another boundary is in general a segment $\bar{V}^{j_i*}_{k2^i}$ corresponding to the "first" *homoclinic tangency* (which takes place starting from (a, b) ϵ $Z^{j_i}_{k2^i}$), for the two invariant curves segments of the saddle $(k2^{i-1};\ j_i)$ For (a,b) ϵ $(D^{j_i}_{k2^i})$ sufficiently near $\bar{V}^{j_i}_{k2^i}$, $(C^{j_i*}_{k2^i})$ turns up on itself with a very complex way, following invariant curve loops of the saddle $(k2^{i-1};\ j_{i-1})$and becomes practically a *"chaotic annular area"* in the (x, y) plane. When (a, b) crosses through $\bar{V}^{j_i*}_{k2^i}$, this complex closed

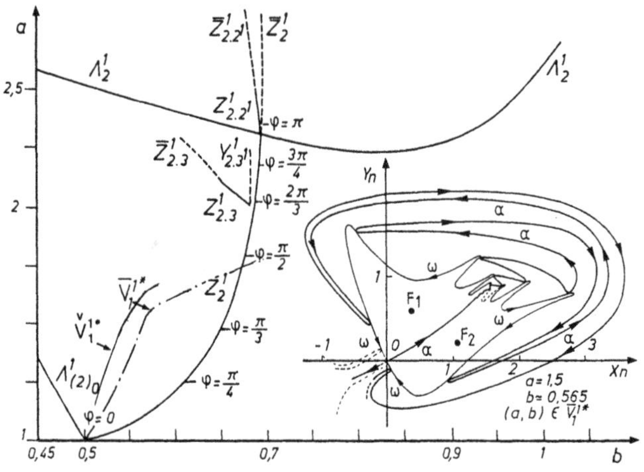

FIG.6.58 - Cubic diffeomorphism (6.39) α = 1

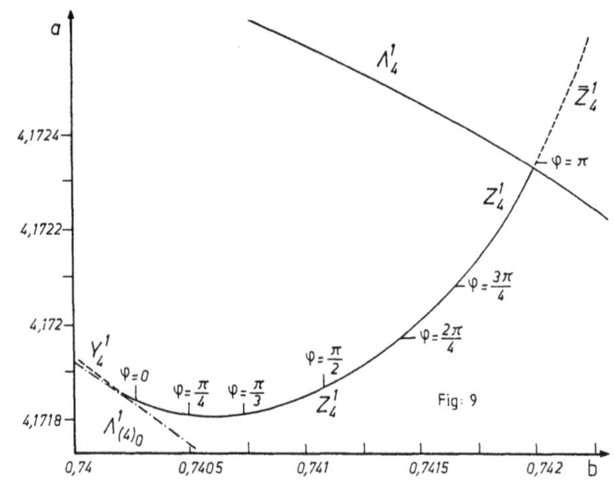

FIG.6.59 - Cubic diffeomorphism (6.39) α = 1 .

invariant curve is destroyed with the creation of homoclinic points. Such homoclinic points exist only for (a, b) located between the segments $\bar{V}^{j_i^*}_{k2^i}$, $\check{V}^{j_i^*}_{k2^i}$, this last segment (the farthest from Z^j_k with respect to $\bar{V}$) corresponding to a second homoclinic tangency, which is a second limit existence of homoclinic points. For certain values of (a, b), $\bar{V}^{j_i^*}_{k2^i}$ and $\check{V}^{j^*}_{k2^i}$ practically merge. Then $(C^{j_i}_{k2^i})$ appears practically from a loop of an invariant curve going from a saddle to other saddle (or the same saddle), which is a known bifurcation for two-dimensional autonomous differential equations. In general, the area $(D^{j_i}_{k2^i})$ does not stretch entirely from $\Lambda^{j_{i-1}}_{k.2^{i-1}}$ to $\Lambda^{j_i}_{k2^i}$, because $\bar{V}^{j_i^*}_{k2^i}$ crosses through $Z^{j_i}_{k2^i}$. Then for φ-values of the $(k2^i\ ;\ j_i)$ focus greater than a critical value, the $(C^{j_i}_{k2^i})$ disappear by the creation of homoclinic points for the invariant curves of the saddle $(k.2^{i-1}\ ;\ j_{i-1})$. So a segment, $\varphi_C < \varphi < \pi$, of $Z^{j_i}_{k2^i}$ can be such that crossing through it does not give rise to $(C^{j_i}_{k2^i})$.

Fig. 6.58 represent this last situation for Z^1_2 and the homoclinic tangency in the phase plane obtained for (a, b) $\in \bar{V}^{1*}_1$, and related to the saddle 0. The curve $\check{V}^{1*}_1$ is located on the left of $\bar{V}^{1*}_1$. For this curve $\check{V}^{1*}_1$, the α-invariant curve of fig. 6.58 tends toward a limit(L), and the ω-invariant curve of the saddle 0 is now tangent to (L) outside the domain limited by (L).

This bifurcation seems correspond to the birth of a saddle and node (q_2) fixed points merging for (a, b) $\in \check{V}^{1*}_1$.

Remarks

- (a) The (x_n, y_n) part of fig. 6.58 has the advantage to show a situation of homoclinic tangency when the multiplier associated with the ω-curve is negative. This produces an "oscillation" of the ω-curve on the both sides of an "average" line passing between the two points of the focus cycle F_1, F_2.
- (b) Except the existence of curves Z^j_k , qualitatively the bifurcations structure of (6.39) $\alpha \neq 0$ has numerous common points with that of (6.26). In particular the most part of cross-road areas, spring areas, saddle areas of (6.26) also are obtained with (6.39).

- (c) The diffeomorphism (6.39) $T_{b\alpha}$ *can be considered as the simplest model giving rise to a complete set of bifurcations with a fractal foliated structure, and with birth of cyclic invariant closed curves (impossible to obtain with (6.26), i.e.* $\alpha = 0$ *). From this point of view it gives the fundamental "mechanisms" occuring for example in non-autonomous differential equations such (1.12), with quasi periodic solutions, or more complex solutions. These "mechanisms" are obtained without the "parasitic effects" induced by more complicated models (cf. the preface).*

APPENDIX A

POINCARE'S INDEXES |B 3|

The notion of index associated with an equilibrium point of a two-dimensional autonomous differential equation is due to Poincaré. This notion was extended by Neimark to the singularities of recurrences |N 1|.

Let T be the two-dimensional recurrence given by :

(A.1) $x_{n+1} = f(x_n, y_n)$; $y_{n+1} = g(x_n, y_n)$

and the points $M_n(x_n, y_n)$, $M_{n+1}(x_{n+1}, y_{n+1})$, of an iterated sequence. Let (C) be a closed curve, without self-intersection, without passing through any fixed point of (A1), which encircles a fixed point P .

The Poincaré's index of the fixed point P is the number i such that the vector $\overrightarrow{M_n M_{n+1}}$ undergoes a rotation $2\pi i$, when M_n describes (C) with an angular variation equal to 2π .

From this definition, it results that the index of an attractive or repulsive node of type 1 is i=+1.

The index of an attractive node of type 2 is i=+1, that of a repulsive node of type 2 is i=-1

The index of an attractive, or repulsive, node of type 3 is i=+1.

The index of a saddle of type 1, or type 2 the multiplier of which having the higher absolute value being positive, is i=-1.

The index of a saddle of type 3, or type 2 the multiplier of which having the higher absolute value is negative, is i=+1 .

The index of an attractive, or repulsive, focus is i=+1 .

The index of a closed invariant curve (Γ) is i=+1.

As for the differential equations case one has the following theorem : " the sum of the indexes of the fixed points encircled by an invariant closed curve (Γ), without any fixed point, is equal to the index of this closed curve, i.e. i=+1 ."

This theorem gives a necessary condition of existence of an invariant closed curve in the phase plane . It is also a help to foresee the possible bifurcations occuring in (A.1).

In a critical case in the sense of Ljapunov, the index i=+1, or -1, or 0.

As example of illustration, the following recurrence T is considered

(A.2) $$x_{n+1} = y_n \ , \ y_{n+1} = (.2y_n - .5x_n)(x_n^2 + y_n^2 - 1.2x_n - \lambda y_n + .5) + x_n \ .$$

For λ = 1.3, T has two attractive fixed points $A_1=0$, A_3, which are two nodes of type 2, thus with indexes i=+1. The boundary separating the basins of these two attractors is the curve (F) passing through the saddle of type 2 A_2 having its multiplier of higher absolute value $S_1>0$, thus with the index i=-1. This curve (F) spiraling tends toward an invariant closed curve passing through a repulsive node cycle of order 5 and of type 1, N_i(i=1,...,5), and a saddle cycle of order 5 and type 1 C_i(i=1,...,5).

T has two other singularities located on (F) : a saddle cycle of type 1, order 2, B_1,B'_1, a repulsive focus cycle of order 2, B_2,B'_2 .

The sum of indexes of the fixed points A_1,A_2,A_3 is +1 which corresponds to the index of the invariant closed curve. Considering T^2 now one has seven fixed points with the four points of the two cycles. Their indexes have also a sum equal to +1, because the multiplier of higher absolute value of B_1,B'_1 is positive.

For λ = 1.1, A_1 is always an attractive node of type 2 (i=+1), but now B_1,B'_1 is an attractive node cycle of type 1, and A_3 is a saddle of of type 2 ($i = +1$). On the boundary (F) separating the basins of the two attractors are located the fixed point A_2 , repulsive node of type 2 ($i = -1$), and the cycle of order two, B_2, B'_2, saddle of type 1 ($i = -1$). Concerning the fixed points the sum of indexes is $i = +1$. Considering T^2 and its fixed points (points of the cycle of T) the sum of indexes is also +1 .

Figures (A_1), (A_2) show the two situations .

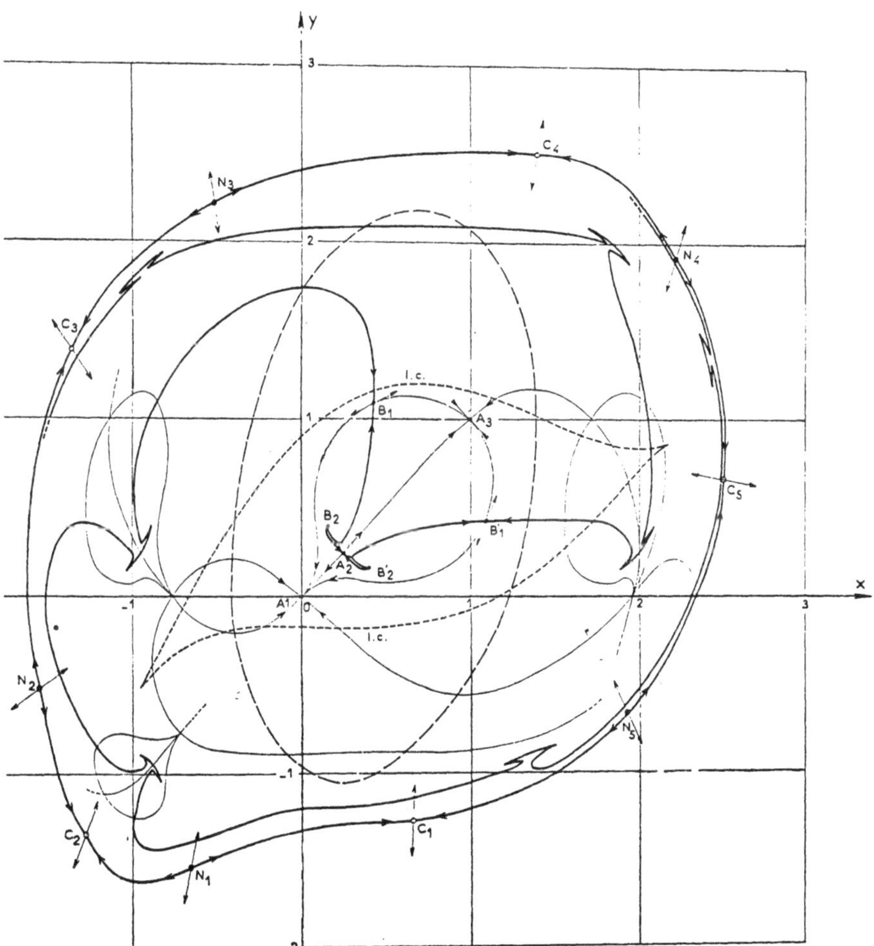

FIG. A1

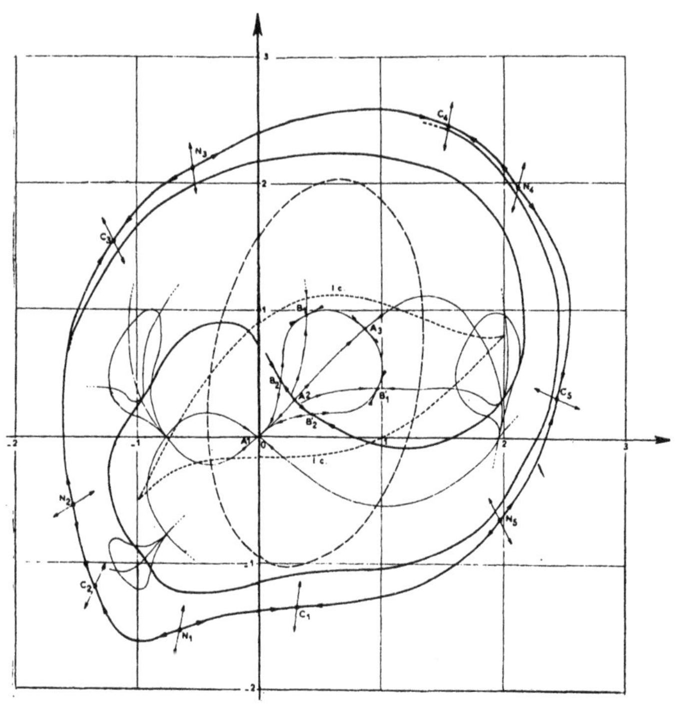

FIG. A2

APPENDIX B

EQUIVALENCE BETWEEN THE CANONICAL FORM OF A TWO-DIMENSIONAL RECURRENCE AND THE REDUCED CIGALA FORM

$$\begin{cases} r_{n+1} = r_n \cos\phi - s_n \sin\phi + \sum_{m=2}^{\infty} R_m(r_n, s_n) \\ s_{n+1} = r_n \sin\phi + s_n \cos\phi + \sum_{m=2}^{\infty} S_m(r_n, s_n) \end{cases} \qquad \begin{cases} x_{n+1} = x_n e^{j\phi} + \sum_{m=2}^{\infty} P_m(x_n, y_n) \\ y_{n+1} = y_n e^{-j\phi} + \sum_{m=2}^{\infty} Q_m(x_n, y_n) \end{cases}$$

$$r = (x+y)/2 \; ; \; s = -j(x-y)/2$$

$$P_m = R_m + j\,S_m \; ; \; Q_m = R_m - j\,S_m$$

Degree 2

$$R_2 = Ar^2 + Brs + Cs^2$$
$$S_2 = A'r^2 + B'rs + C's^2$$

$$C_{20} = [(A-C+B') + j(A'-B-C')]/4$$
$$C_{21} = [2(A+C) + j\,2(A'+C')]/4$$
$$C_{22} = [(A-B'-C) + j(A'+B-C')]/4$$

Degree 3

$$R_3 = Dr^3 + Er^2s + Frs^2 + Gs^3$$
$$S_3 = D'r^3 + E'r^2s + F'rs^2 + G's^3$$

$$C_{30} = [(D+E'-F-G') + j(D'-E-F'+G)]/8$$
$$C_{31} = [(3D+E'+F+3G') + j(3D'-E+F'-3G)]/8$$
$$C_{32} = [(3D-E'+F-3G') + j(3D'+E+F'+3G)]/8$$
$$C_{33} = [(D-E'-F+G') + j(D'+E-F'-G)]/8$$

Degree 4

$$R_4 = Hr^4 + Ir^3s + Jr^2s^2 + Krs^3 + Ls^4$$
$$S_4 = H'r^4 + I'r^3s + J'r^2s^2 + K'rs^3 + L's^4$$

$$C_{40} = [(H-J+L+I'-K') + j(-I+K+H'-J'+L')]/16$$
$$C_{41} = [(4H-4L+2I'+2K') + j(-2I-2K+4H'-4L')]/16$$
$$C_{42} = [(6H+6L+2J) + j(6H'+6L'+2J')]/16$$
$$C_{43} = [(4H-4L-2I'-2K') + j(2I+2K+4H'-4L')]/16$$
$$C_{44} = [(H-J+L-I'+K') + j(I-K+H'-J'+L')]/16$$

Degree 5

$$R_5 = Mr^5 + Nr^4s + Or^3s^2 + Pr^2s^3 + Qrs^4 + Rs^5$$
$$S_5 = M'r^5 + N'r^4s + O'r^3s^2 + P'r^2s^3 + Q'rs^4 + R's^5$$

$$C_{50} = [(M-O+Q+N'-P'+R') + j(-N+P-R+M'-O'+Q')]/32$$
$$C_{51} = [(5M-O-3Q+3N'+P'-5R') + j(-3N-P+5R+5M'-O'-3Q')]/32$$
$$C_{52} = [2(5M+O+Q+N'+P'+5R') + j\cdot 2(-N-P-5R+5M'+O'+Q')]/32$$
$$C_{53} = [2(5M+O+Q-N'-P'-5R') + j\cdot 2(N+P+5R+5M'+O'+Q')]/32$$
$$C_{54} = [(5M-O-3Q-3N'-P'+5R') + j(3N+P-5R+5M'-O'-3Q')]/32$$
$$C_{55} = [(M-O+Q-N'+P'-R') + j(N-P+R+M'-O'+Q')]/32$$

APPENDIX C

CRITICAL CASE WITH TWO PAIRS OF COMPLEX MULTIPLIERS IN A FOUR-DIMENSIONAL NON-LINEAR RECURRENCE |M 18|

1. NON EXCEPTIONAL CRTICAL CASE.

Let T be the four -dimensional recurrence, with real variables, having the fixed point 0 ($r_i=s_i=0$, $i=1,2$) the multipliers of which are $\exp(\pm j\phi_\alpha)$, $\alpha = 1,2$, $j = \sqrt{-1}$. By means of a linear transformation, T can be brought into the form:

$$(1) \begin{cases} \bar{R}_\alpha = r_\alpha \cos\phi_\alpha - s_\alpha \sin\phi_\alpha + \mathcal{R}_\alpha(r_1,r_2,s_1,s_2), \\ \bar{S}_\alpha = r_\alpha \sin\phi_\alpha + s_\alpha \cos\phi_\alpha + \mathcal{S}_\alpha(r_1,r_2,s_1,s_2), \quad \alpha = 1, 2 \end{cases}$$

It is supposed that the functions $\mathcal{R}_\alpha$, $\mathcal{S}_\alpha$ are continuous and differentiable till an order M,M being sufficiently large, and that their expansions begin with terms the degree of which is higher than one. In order to avoid a bulky notation, the indexes of iteration n, n+1, of the variables are not used . So the capital letters R_α, S_α correspond to the index n+1, and the small letters r_α, s_α correspond to the index n .The purpose of this appendix is the determination of the conditions of stability of the fixed point 0. In the neighbourhood of this point it is supposed the absence of homoclinic , or heteroclinic point.

In order to proceed as Cigala in the two-dimensional case,(1) is transformed,via the new variables defined by $r_\alpha = (x_\alpha + y_\alpha)/2$, $s_\alpha = j(y_\alpha - x_\alpha)/2$, into a recurrence with complex conjugate variables :

(1') $X_\alpha = e^{j\phi_\alpha} x_\alpha + \mathscr{X}_\alpha(x_1,y_1,x_2,y_2)$, $Y_\alpha = e^{-j\phi_\alpha} + \mathscr{Y}_\alpha(x_1,y_1,x_2,y_2)$

$\mathscr{X}_\alpha$, $\mathscr{Y}_\alpha$, $\alpha = 1, 2$, being made of complex conjugate terms, the lower degrees of which being higher than one. The set of the terms of degree m willbe designated by $\mathscr{X}_\alpha^{(m)}$ and $\mathscr{Y}_\alpha^{(m)}$:

$$\mathscr{X}_\alpha^{(m)} = \sum D_\alpha^{(m_1,m_2,n_1,n_2)} x_1^{m_1} y_1^{m_2} x_2^{n_1} y_2^{n_2}, \quad \mathscr{Y}_\alpha^{(m)} = \sum D_\alpha^{*(m_1,m_2,n_1,n_2)} y_1^{m_1} x_1^{m_2} y_2^{n_1} x_2^{n_2}$$

with $m_1 + m_2 + n_1 + n_2 = m$, D_α^* being the conjugate of D_α (below the index * indicates the complex conjugate of a letter without this index)

Let $m^{(\alpha)} = m_1^{(\alpha)} + m_2^{(\alpha)} + n_1^{(\alpha)} + n_2^{(\alpha)}$, $\alpha = 1,2$, the lowest degree of $\mathscr{X}_\alpha$. With the equivalent of the Cigala's transformation:

$$(2) \quad \left[\begin{array}{l} x_\alpha = u_\alpha + \sum B_\alpha\left[m_1^{(\alpha)},m_2^{(\alpha)},n_1^{(\alpha)},n_2^{(\alpha)}\right] u_1^{m_1(\alpha)} v_1^{m_2(\alpha)} u_2^{n_1(\alpha)} v_2^{n_2(\alpha)}, \\ y_\alpha = v_\alpha + \sum B_\alpha^*\left[m_1^{(\alpha)},m_2^{(\alpha)},n_1^{(\alpha)},n_2^{(\alpha)}\right] v_1^{m_1(\alpha)} u_1^{m_2(\alpha)} v_2^{n_1(\alpha)} u_2^{n_2(\alpha)}, \quad \alpha = 1, 2 \end{array} \right.$$

the coefficients B_α are determinated in order to eliminate the largest number of terms having the degree $m^{(\alpha)}$ in $\mathscr{X}_\alpha$, $\mathscr{Y}_\alpha$. Then these coefficients verify the relations :

$$(3) \quad \left[\begin{array}{l} B_\alpha\left[m_1^{(\alpha)},m_2^{(\alpha)},n_1^{(\alpha)},n_2^{(\alpha)}\right]\left[1 - e^{j\left[\phi_1(m_1^{(\alpha)}-m_2^{(\alpha)})+\phi_2(n_1^{(\alpha)}-n_2^{(\alpha)})-\phi_\alpha\right]}\right] = \\ = e^{-j\phi_\alpha} D_\alpha\left[m_1^{(\alpha)},m_2^{(\alpha)},n_1^{(\alpha)},n_2^{(\alpha)}\right] \end{array} \right.$$

For ϕ_1 and ϕ_2 it is supposed that :

$$(4) \quad \phi_1(m_1^{(\alpha)}-m_2^{(\alpha)}) + \phi_2(n_1^{(\alpha)}-n_2^{(\alpha)}) - \phi_\alpha \neq 2h_\alpha\pi, \quad \alpha = 1, 2,$$

h_α being an integer which is either positive, or negative, or equal to zero .

From (3), (4) it appears that all the terms of degree $m^{(\alpha)}$ can be eliminated via (2), when $m^{(\alpha)}$ is an even integer. This leads to a recurrence for which the two relations corresponding to the index α have their terms,of lowest degree different from one,with an odd parity. Now if $m^{(\alpha)}$ is odd, it is not possible to eliminate with (2) the terms having degrees such that :

$$(5) \quad \left\{ \begin{array}{lll} n_1^{(1)} - n_2^{(1)} = 0, & m_1^{(1)} - m_2^{(1)} - 1 = 0 & \text{if } \alpha = 1, \\ m_1^{(2)} - m_2^{(2)} = 0, & n_1^{(2)} - n_2^{(2)} - 1 = 0 & \text{if } \alpha = 2, \end{array} \right.$$

in (1'). Successive applications of the transformations (2) with increasing degrees in the general case permit to end at odd degrees m. Consider the particular case $m^{(1)} = m^{(2)} = 3$. After one transformation (2) which eliminates the terms of the second degree, eliminating the terms which do not verify (5), the following expression of the recurrence is obtained :

$$(6)\quad \begin{cases} U_1 = e^{j\phi_1} u_1 + A_1^{[2100]} u_1^2 v_1 + A_1^{[1011]} u_1 u_2 v_2 + \cdots, \\ U_2 = e^{j\phi_2} u_2 + A_2^{[0021]} u_2^2 v_2 + A_2^{[1110]} u_2 u_1 v_1 + \cdots, \\ \qquad V_1 = U_1^*, \quad V_2 = U_2^* \end{cases}$$

the dotted lines after the two signs + corresponding to terms having degrees higher than 3.

Let the polar representation $u_\alpha = \rho e^{j\phi_\alpha}$, $v_\alpha = \rho e^{-j\phi_\alpha}$. In a sufficiently small neighbourhood of the fixed point 0, (2) becomes the recurrence :

$$(7)\quad \begin{cases} P_i = \rho_i + E_3^{(i)} \rho_i^3 + E_1^{(i)} \rho_i \rho_k^2 + \cdots, \quad \boldsymbol{\theta}_i = \theta_i + \phi_i - |a_i| \rho_i^2 \sin(\phi_i + \mathscr{C}_i) - \\ \quad - |b_i| \rho_k^2 \sin(\phi_i + \mathscr{C}'_i) + \cdots, \quad i = 1,2 \ ; \ k = 1,2 \ ; \ i \neq k \\ E_3^{(i)} = a_i e^{-j\phi_i} + a_i^* e^{j\phi_i} \quad ; \quad E_1^{(i)} = b_i e^{-j\phi_i} + b_i^* e^{j\phi_i} \\ a_i = A_1^{[2100]},\ b_i = A_1^{[1011]} \text{ si } i = 1 \quad ; \quad a_i = A_1^{[0021]},\ b_i = A_1^{[1110]} \text{ si } i = 2 \\ \mathscr{C}_i = \arg a_i^*,\ \mathscr{C}'_i = \arg b_i^*,\ E_3^{(i)} = |a_i| \cos(\phi_i + \mathscr{C}_i),\ E_1^{(i)} = |b_i| \cos(\phi_i + \mathscr{C}'_i) \end{cases}$$

the variables and the coefficients of which are real. In this small neighbourhood of 0, the two equations in ρ_i can be considered as independant of the variables θ_i . They correspond to a critical case with two multiplers equal to +1 (case of §5.10). The invariant manifolds in the small neighbourhood of 0 defined by :

$$P_2 \,/\, P_1 = \rho_2 \,/\, \rho_1 = \operatorname{tg} \bar{\theta} ,$$

are the solutions of :

$$(8)\quad \rho_1 \rho_2 \left[(E_3^{(2)} - E_1^{(1)}) \rho_2^2 + (E_1^{(2)} - E_3^{(1)}) \rho_1^2 \right] = 0$$

In the (ρ_1, ρ_2) plane these solutions ($\rho_1 > 0$, $\rho_2 > 0$) are :

$$\rho_1 = 0, \quad \rho_2 = 0, \quad (E_3^2 - E_1^1) \rho_2^2 + (E_1^2 - E_3^1) \rho_1^2 = 0$$

the third solution exists only if the expression between square brackets of (8) has a real solution.

The necessary and sufficient condition in order that 0 be attractive is that ρ_1 and ρ_2 be decreasing functions when successive maps T are applied. On $\rho_2 = 0$ this corresponds to $E_3^{(1)} < 0$, on $\rho_1 = 0$ to $E_3^{(2)} < 0$. When the third solution exists, ρ_2^2 function of ρ_1 is put in the relation giving P_1 for obtaining the condition of stability of 0. It results the following proposition :

(P1) Let T be the recurrence given by (1). The necessary and sufficient condition of asymptotic stability of the fixed point 0 (0 is attractive) is that one has :

(a). $E_3^{(1)} < 0$, $E_3^{(2)} < 0$, when $(E_3^{(2)} - E_1^{(1)})(E_1^{(2)} - E_3^{(1)}) > 0$; or

(b). $E_3^{(1)} < 0$, $E_3^{(2)} < 0$, $(E_1^{(1)}E_1^{(2)} - E_3^{(1)}E_3^{(2)})(E_1^{(1)} - E_3^{(2)}) < 0$, when $(E_3^{(2)} - E_1^{(1)})(E_1^{(2)} - E_3^{(1)}) < 0$.

The fixed point becomes repulsive if one of the above inequalities is reversed .

2. EXCEPTIONAL CRITICAL CASES .

It is supposed now that one of the following relations :

(9) $\quad \phi_1(m_1^{(\alpha)} - m_2^{(\alpha)}) + \phi_2(n_1^{(\alpha)} - n_2^{(\alpha)}) - \phi_\alpha = 2h_\alpha\pi, \ \alpha = 1, 2,$

is satisfied.Then in this situation, or if the two relations are satisfied, some terms of $\mathscr{X}_\alpha^{(2)}$ cannot be eliminated with the transformation (2).

In order to analyse the possible situations it will be called "case $(m_1, m_2, n_1, n_2)^{(\alpha)}$" the case such that in $\mathscr{X}_\alpha^{(2)}$, after the elimination of all the possible terms of the second degree, it remains a monomial in $x_1\ y_1\ x_2\ y_2$, with respectively the degrees m_1, m_2, n_1, n_2

The angle $\mathscr{C}_\alpha^{(m_1, m_2, n_1, n_2)}$ represents below the argument of $D_\alpha^{*(m_1, m_2, n_1, n_2)}$

When some terms of the second degree cannot be eliminated, because (9) is satisfied, then in a sufficiently small neighbourhood (d) of the fixed point O the recurrence can be written in polar coordinates in one of the three following forms :

$$(10)\quad \begin{cases} P_i = \rho_i + \rho_i^2 |D_i^{(m_1,m_2,n_1,n_2)}| \cos\left[\beta\,\theta_i - a\,\phi_i - a\,\mathscr{C}_i^{(m_1,m_2,n_1,n_2)}\right] + \cdots \\ \boldsymbol{\theta}_i = \theta_i + \phi_i + a\,\rho_i |D_i^{(m_1,m_2,n_1,n_2)}| \sin\left[\beta\,\theta_i - a\,\phi_i - a\,\mathscr{C}_i^{(m_1,m_2,n_1,n_2)}\right] + \cdots \\ a = \pm 1\ ;\ i = 1,2\ ;\ \beta = 1,3 \end{cases}$$

$$(11)\quad \begin{cases} P_i = \rho_i + \rho_j^2 |D_i^{(m_1,m_2,n_1,n_2)}| \cos\left[\beta\,\theta_i + \gamma\,\theta_j - a\,\phi_i - a\,\mathscr{C}_i^{(m_1,m_2,n_1,n_2)}\right] + \cdots \\ \boldsymbol{\theta}_i = \theta_i + \phi_i + a |D_i^{(m_1,m_2,n_1,n_2)}| \rho_i^{-1} \rho_j^2 \sin\left[\beta\,\theta_i + \gamma\,\theta_j - a\,\phi_i - a\,\mathscr{C}_i^{(m_1,m_2,n_1,n_2)}\right] + \cdots \\ a = \pm 1\ ;\ i = 1,2\ ;\ j = 1,2\ ;\ i \neq j\ ;\ \beta = \pm 2, \pm 1\ ;\ \gamma = \pm 1, \pm 2\ ;\ \beta \neq \gamma\ ; \\ m_1 + m_2 + n_1 + n_2 = 2\ ;\ m_i \text{ or } n_i = 2 \end{cases}$$

$$(12)\quad \begin{cases} P_i = \rho_i + \rho_i \rho_j |D_i^{(m_1,m_2,n_1,n_2)}| \cos\left[\beta\,\theta_i + \gamma\,\theta_j - a\,\phi_i - a\,\mathscr{C}_i^{(m_1,m_2,n_1,n_2)}\right] + \cdots \\ \boldsymbol{\theta}_i = \theta_i + \phi_i + a\,\rho_j |D_i^{(m_1,m_2,n_1,n_2)}| \sin\left[\beta\,\theta_i + \gamma\,\theta_j - a\,\phi_i - a\,\mathscr{C}_i^{(m_1,m_2,n_1,n_2)}\right] + \cdots \\ a = \pm 1\ ;\ i,j = 1,2\ ;\ i \neq j\ ;\ \alpha,\beta = \pm 2, \pm 1\ ;\ \alpha \neq \beta\ ;\ m_1 + m_2 + n_1 + n_2 = 2\ ; \\ m_i = 1,\ n_i = 1 \end{cases}$$

Considering only the terms of the second degree, 8 exceptional critical cases are possible .

(A) <u>Case (2, 0, 0, 0)</u>$^{(\alpha)}$, the possible relations (9) are:

$$(13)\qquad \phi_1 = 2h_1\pi,\ (\alpha = 1) \quad ; \quad 2\phi_1 - \phi_2 = 2h_2\pi,\ (\alpha = 2)$$

For $\alpha = 1$, the considered recurrence is (10) with $i = \beta = a = 1$, associated with (7) the indexes i, k being $i = 2$, $k = 1$.

For $\alpha = 2$, the considered recurrence is (11) with $i = \beta = 2$, $j = a = 1$, $\gamma = -1$, associated with (7) , $i = 1$, $k = 2$.

For $\alpha = 2$, considering the auxiliary angle $\theta = 2\theta_1 - \theta_2$, it appears from (13) that inside (d) the manifold $\bar{\theta}$ passing through O , which is defined by :

$$\bar{\theta} - \phi_2 - \mathscr{C}_2^{(2000)} = \ell\pi, \quad \ell = 0 \text{ or } 1$$

is invariant because on this manifold one has $\dot{\boldsymbol{\theta}} = \theta + 2h_2\pi$. The consideration of the multiplier

$$\left[d\, \boldsymbol{\theta} / d\,\theta\right]_{\theta=\bar\theta} = 1 - |D_2^{(2000)}|\,\rho_2^{-1}\,\rho_1^2 \cos(\bar\theta - \phi_2 - \mathscr{C}_2^{(2000)}) + \cdots$$

related to the relation in θ , permits to show that $\bar\theta$ is attractive for $\ell = 0$. Moreover from (7) with i = 1, it appears that ρ_2 is decreasing by successive applications of T. For $\ell = 1$, $\bar\theta$ is repulsive, and ρ_2 is decreasing. A situation equivalent to that of a saddle is obtained.

For $\alpha = 1$, a manifold defined by $\bar\theta - \phi_1 - \mathscr{C}_1^{(2000)} = \ell\pi$, $\ell = 0$, or 1, is invariant and can be studied similarly.

(B) <u>Case (0200)$^{(\alpha)}$</u>, the possible relations (9) being

(14) $\quad -3\phi_1 = 2h_1\pi, \ (\alpha = 1) \ ; \ -2\phi_1 - \phi_2 = 2h_2\pi, \ (\alpha = 2)$

For $\alpha = 1$ the considerred recurrence is (10) with i = 1, a = -1, $\beta = 3$, and (7) with i = 2, k = 1. The projection of the solution on the plane (x_1, y_1) makes appear the invariant manifolds

$$3\bar\theta + \phi_1 + \mathscr{C}_1^{(0200)} = \ell\pi, \ \ell = 0,1,\ldots,5,$$

which are asymptotes. In this plane these asymptotes have the configuration of fig. 5.21, but the invariant curves are self intersecting.

For $\alpha = 2$ the considered recurrence is (11) with $i = \beta = 2$, $\gamma = j = 1$, a = -1, and (7) with i = 1, k = 2. Considering the auxiliary angle $\theta = 2\theta_1 + \theta_2$, and proceeding as in (A), the manifold $\bar\theta$ defined by

$$\bar\theta + \phi_2 + \mathscr{C}_2^{(0200)} = \ell\pi, \ \ell = 0,1$$

is invariant, which gives equivalent properties to those presented in (A) .

(C) <u>Case (0020)$^{(\alpha)}$</u> the possible relations (9) being

(15) $\quad -\phi_1 + 2\phi_2 = 2h_1\pi, \ (\alpha = 1) \ ; \ \phi_2 = 2h_2\pi, \ (\alpha = 2)$

For $\alpha = 1$, the considered recurrence is (11) with i = a = 1, $j = \gamma = 2$, $\beta = -1$, and (7) with i = 2, k = 1. The auxiliary angle $\theta = -\theta_1 + 2\theta_2$ permits to make appear the invariant manifolds $\bar\theta$ given by

$$\bar\theta - \phi_1 + \mathscr{C}_1^{(0020)} = \ell\pi$$

attractive with ρ_1 increasing for $\ell = 0$, repulsive with ρ_1 decreasing for $\ell = 1$. The case $\alpha = 2$ is similar to the corresponding case of (A)

(D) <u>Case (0002)$^{(\alpha)}$</u> the possible relations (9) being

(16) $\quad -\phi_1 - 2\phi_2 = 2h_1\pi, \ (\alpha = 1) \ ; \ -3\phi_2 = 2h_2\pi, \ (\alpha = 2)$

For $\alpha = 1$ the considered recurrence is (11) with $i = \beta = 1$, a = -1,

$j = \gamma = 2$, and (7) with $i = 2$, $k = 1$. The auxiliary angle $\theta = \theta_1 + 2\theta_2$, permits to make appear the invariant manifolds

$$\bar{\theta} + \phi_1 + \mathscr{C}_1^{(0002)} = \ell\pi,$$

having the properties of the case (C) with $\alpha = 1$.

(E) Cases $(\ 1100\ )^{(\alpha)}$, $(\ 0011\ )^{(\alpha)}$ and $(\ 1010\ )^{(\alpha)}$, the possible relations (9) being respectively

$-\phi_1 = 2h_1\pi$, $-\phi_2 = 2h_2\pi$ for the two first cases, and $\phi_2 = 2h_1\pi$ for the third case. The problem is solved as for (A) with $\alpha = 1$.

(F) Case $(\ 1001\ )^{(\alpha)}$ the possible relations (9) being

$$-\phi_1 = 2h_1\pi, \ (\alpha = 1) \quad ; \quad \phi_1 - 2\phi_2 = 2h_2\pi, \ (\alpha = 2)$$

The case $\alpha = 1$ is similar to that of (A) with $\alpha = 1$. The case $\alpha = 2$ corresponds to the recurrence (12) with $i = 2$, $j = a = \gamma = 1$, $\beta = -2$, and (7) with $i = 1$, $k = 2$. The auxiliary angle $\theta = \theta_1 - 2\theta_2$ permits to define the invariant manifolds

$$\bar{\theta} - \phi_2 - \mathscr{C}_2^{(1001)} = \ell\pi,$$

which have the properties of the case (A) with $\alpha = 2$.

(G) Case $(\ 0110\ )^{(\alpha)}$ the possible relations (9) being

$$-2\phi_1 + \phi_2 = 2h_1\pi, \ (\alpha = 1) \quad ; \quad \phi_1 = 2h_2\pi, \ (\alpha = 2)$$

The case $\alpha = 1$ corresponds to the recurrence (12) with $i = a = \gamma = 1$, $j = -\beta = 2$, and to (7) with $i = 2$, $k = 1$. From the auxiliary angle $\theta = -2\theta_1 + \theta_2$, one has the following invariant manifolds

$$\bar{\theta} - \phi_1 - \mathscr{C}_1^{(0110)} = \ell\pi,$$

which have the properties of the case (C) with $\alpha = 1$. The case $\alpha = 2$ corresponds to the case (A) with $\alpha = 1$.

(H) Case $(\ 0101\ )^{(\alpha)}$ the possible relations (9) being

$$-2\phi_1 - \phi_2 = 2h_1\pi, \ (\alpha = 1) \quad ; \quad -\phi_1 - 2\phi_2 = 2h_2\pi, \ (\alpha = 2)$$

The case $\alpha = 1$ corresponds to the recurrence (12) with $i = -a = \gamma = 1$, $j = \beta = 2$, and to (7) with $i = 2$, $k = 1$. From the auxiliary angle $\theta = 2\theta_1 + \theta_2$, one has the invariant manifolds

$$\bar{\theta},\ \bar{\theta} + \phi_1 + \mathscr{C}_1^{(0101)} = \ell\pi,$$

which have the properties of the case (C) with $\alpha = 2$. The case $\alpha = 1$ corresponds to the recurrence (12) with $i = \beta = 2$, $j = -a = 1$, and to (7) with $i = 1$, $k = 2$. Using the auxiliary angle $\theta = \theta_1 + 2\theta_2$, the invariant manifolds

$$\bar{\theta} + \phi_2 + \mathscr{C}_2^{(0101)} = \ell\pi$$

are obtained with the properties of the case (A) with $\alpha = 2$.

Consider $\phi_\alpha \neq 2h_\alpha\pi$ and $\phi_\alpha \neq 2h_\alpha\pi/3$, in the relations (13)-(19), one relation $\mu\phi_1 + \delta\phi_2 = 2h_\alpha\pi$, δ, $\mu = \pm 2$, or ± 1, (ϕ_1, ϕ_2 being commensurable or non-commensurable with 2π), occurs twice, on the one hand with $\alpha = 1$, onthe other hand with $\alpha = 2$. Thus if such a relation occurs, two terms $D_\alpha^{(m_1,m_2,n_1,n_2)}$, one with $\alpha = 1$, the other with $\alpha = 2$, cannot be eliminated from the transformation (2) , when the auxiliary angle $\theta = \mu\theta_1 + \delta\theta_2$ is used. Thus each of the two cases : $(2000)^{(2)}$ and $(1010)^{(1)}$, $(0200)^{(2)}$ and $(0101)^{(1)}$, $(0020)^{(1)}$ and $(1001)^{(2)}$, $(0002)^{(1)}$ and $(0101)^{(2)}$ corresponds to a same relation (9) . Then it is possible to define a family of invariant manifolds

$$\rho_2 = f(\theta)\rho_1$$

in asufficiently small neighbourhood of the fixed point O, where $f(\theta)$ is the solution of an one-dimensional differential equation. The manifolds of this family have in common the manifolds such that $\boldsymbol{\theta} = \theta$.

The elimination of the terms of degree 3 in (1') leads to the definition of 18 possible situations plus the situations resulting from the simultaneous presence of the two cases $\alpha = 1$ and $\alpha = 2$.

APPENDIX D

PARTICULAR CASE OF A TWO-DIMENSIONAL RECURRENCE WITH REAL VARIABLES REDUCED TO A ONE-DIMENSIONAL RECURRENCE WITH COMPLEX VARIABLE

Let T be the two-dimensional recurrence

(1) $x_{n+1} = f(x_n, y_n)$, $y_{n+1} = g(x_n, y_n)$

x, y being two real variables. In the particular case corresponding to real sufficiently smooth functions which satisfy the Cauchy-Riemann conditions :

(2) $(\partial f/\partial x) = (\partial g/\partial y)$, $(\partial f/\partial y) = -(\partial g/\partial x)$

the two-dimensional recurrence corresponds to an one-dimensional recurrence with the complex variable z , T' :

(3) $z_{n+1} = h(z_n)$, $z = x + jy$, $j = \sqrt{-1}$

As already indicated in the chapter 2, the first "global" studies of (3) are due to Julia and Fatou |J 5||F 1|. In (3) the critical points play a fundamental role in the definition of the properties of solutions .

The set of all the points such that

$z = h^k(z)$,with $|dh^k/dz| > 1$, k = 1, 2, 3, ...

i.e. the set of all the repulsive cycles, plays also such a role in relation to the first one . Julia called E this set, and proved that its derived E' contains E and is perfect. A first consequence was that E' constitutes the set of the eesential singularities for every limit function which is extracted from the sequence of the $h^k(z)$.

In page 49 of |J 5| it is said " *la structure de* E'*est la même dans toutes ses parties"* , i.e. with the present terminology the *structure of* E'*is fractal* .

Suppose the existence of an attractive fixed point, or an attractive cycle of order k. Julia and Fatou have shown that the basin of this attractor is bounded by the set E', this basin being in simple cases constituted by an all in one block part containing the attractor, called *immediate basin*, and which contains also a critical point. When the number of attractive fixed points is higher than one, or by consideration of T'^k when the attractor is a cycle (every point of this cycle being a fixed point of T'^k), the *total basin* is made up of the immediate basin and of infinitely many disjoint parts. When among the attractors are fixed points,at most one of these points has an all in one block immediate basin merging into the total basin, all the others have atotal basin made up of infinitely many disjoint parts.

Moreover the set E' may be either continuous, or discontinuous. When it is linear continuous,E' is a simple closed Jordan curve, and on this curve the points of E, everywhere dense, are generally points for which the tangent is not defined. This last property appears direcly by consideration of the two-dimensional recurrence (1). Indeed with the condition (2) fixed points ,and cycles of (1) are such that their multipliers are $S_1 = S_2$, the matrix of the linear approximation having a diagonal form, i.e. the corresponding singular points are *star nodes* (§ 5.2.4) for which the eigen directions are not defined |G21| |M12 |

The case of h(z) rational function of z was more particularly studied by Julia and Fatou. When this function is a polynomial the corresponding results can be related to those obtained with z= real variable in chapters 2-4. So the quadratic case:

(4) $$z_{n+1} = z_n^2 + p \ , \ p = a + jb \ ,$$

has the real two dimensional form:

(5) $$x_{n+1} = x_n^2 - y_n^2 + a \ , \ y_{n+1} = 2x_n y_n + b$$

The intersection of E' with the real axis gives the non-wandering set of (4) (z being real). In the parameters (a,b) plane the bifurcations structure on the a-axis is the box-within-a-box one described chapter 4. In the whole parameters plane a first study of this structure is due to |D 3|.

An interesting problem consists in the introduction of a parameter λ in (1) with the form:

$$(6) \qquad x_{n+1} = f(x_n, y_n, \lambda) , \quad y_{n+1} = g(x_n, y_n, \lambda) ,$$

such that (2) is satisfied only for $\lambda = \lambda_0$. Then (6) is a perturbation of (1). E' turns into a boundary for which the tangent now can be defined. The critical points turn into closed critical curves (cf. |G 35|) which tends toward these critical points when $\lambda \to \lambda_0$, and which play a fundamental role in the definition of the properties of the two-dimensional endomorphism (6) .

APPENDIX E

THE PROBLEM OF STRUCTURAL STABILITY IN THE CHAOTIC DYNAMIC SITUATIONS

Within the framework of the theory of bifurcations , A.A. Andronov and L.S. Pontrjagin introduced in 1937 the concept of *roughness*, now called *structural stability* . Roughly speaking a dynamical system is structurally stable, when the topological structure of its motions does not change for small changes of its parameters, or small C^1 perturbations of the structure of the equation describing these motions. Then it appears that structural stability is a necessary condition in order that a model be physically relevant.

It is worth of note also that after Andronov Pontrjagin this notion has been refined with those of *dangerous and non-dangerous stability boundaries* in the parameters space, or in the functional one, introduced by Bautin in 1949 |B5 | |G23| |G35| (cf. |M 45| for more details). Another refinement of the concept of structural stability is due to Shilnikov with the introduction of the notions of *accessible* and *non-accessible values of a parameter of bifurcation*, and also those of *dynamically defined dangerous boundary* and *dynamically non-defined dangerous boundary*. This last refinement is well adapted to the considered problem in presence of chaotic behaviours . Details about these notions can be found in |S 4 ||S 5 ||G 35|.

In |G18| Guckenheimer and Holmes show very justly that a narrow interpretation of the necessary condition of physical relevance from the original concept can lead to a difficulty in chaotic dynamic situtions. Indeed in the most part of examples of strange attractors obtained from classical models, these strange attractors are not structurally stable. To show this point it is sufficient to consider the non-classical bifurcation values $\tilde{\lambda}$, λ^*, ${}^*\lambda$, defined in chapters 3 and 4, which are accumulation values of boxes of the box-within-a-box structure, and which correspond to strange attractors called cyclic chaotic segments. In the original sense the corresponding equations are not structurally stable because the considered parameters values give bifurcations. Nevertheless, as shown in § 4.4.3, in a neighbourhood of these values, a certain average qualitative behaviour is preserved, i.e. they are <u>non-dangerous stability boundaries in the sense of Bautin</u>.

On the foundation of |M22|(1976) and the content of § 4.4.3, the mentioned difficulty of the necessary condition of physical relevance can be removed. Indeed |M22| deals with the double interpretation, determinist or of a statistical nature, which can be given to the considered complex bifurcations, and from this the things become clearer.

From the determinist point of view, which implies a "microscopic" consideration of the singularities set of the model, this one is not structurally stable in the original sense, because a parameter perturbation changes the micoscopic organization of these singularities.

From the statistical point of view, which leads to a "macroscopic" consideration of the singularities organization, the considered model is structurally stable in an extended sense . Indeed, as indicated in § 4.4.3, in a sufficiently small neighbourhood of the non-classical bifurcation, the iterated sequence tends in general toward an attractive cycle of order k', k' → ∞ when the parameter value tends toward the bifurcation value, but on an average the k' points of this cycle "reproduce" the shape of the cyclic chaotic segment of small order k, obtained for the bifurcation value. The corresponding behaviour was called "average value cyclic chaotic segment" ("segment stochastique cyclique en valeur moyenne " or " segment SVM ") in |M22| which gives the formulation of the double interpretation for four cases of complex bifurcations .

In order to solve the indicated difficulty it is proposed to refer to *microscopic structural stability*, or *macroscopic structural stability*, according to the emphasis on an aspect , or the other , this choice depending on the nature of the problem and its perturbations.

REFERENCES

A

A 1 A.A. ANDRONOV - "Les cycles limites de Poincaré et la théorie des oscillations auto entretenues" - C.R. Acad. Sc. Paris, 18 (1929), 559-562.

A 2 A.A. ANDRONOV, A.A. WITT, S. KHAIKIN - "Theory of oscillators" - Pergamon Press (1966).

A 3 A.A. ANDRONOV, E.A. LEONTOVITCH, I.I. GORDON, A.G. MAIER - "Theory of bifurcations of dynamic systems in the phase plane" - Nauka, Moscow (1967).

A 4 V.I. ARNOLD - "Small divisors. Mapping of the circumference onto itself" - Izv. Akad. Nauk SSSR Mat., 25, (2) (1961), 21-56.

A 5 V.I. ARNOLD, A. AVEZ - "Problèmes ergodiques de la mécanique" - Gauthier-Villars, Paris (1967).

B

B 1 P.C. BAAYEN, W. KUYK, M.A. MAURICE - "On the orbits of the hat-function, and on countable maximal semi-groups of continuous mappings of the unit interval into itself" - Stichting Mat. Centrum Report ZW 1 962-018, Amsterdam, 1-12.

B 2 J.P. BABARY, C. MIRA - "Sur un cas critique pour une récurrence autonome du deuxième ordre" - C.R. Acad. Sc. Paris, 268, sér. A (1969), 129-132.

B 3 J.P. BABARY, A. GIRAUD, C. MIRA - "Indice de Poincaré de points doubles d'une récurrence" - Electronic Letters, 4 (5) (1968), 90-91.

B 4 L.O. BARSUK, N.M. BELOSLUDSTIEV, Yu. I. NEIMARK, N.M. SALGANSKAJA - "Stability of a fixed point in a critical case. Bifurcations" - Radiofisica, 11, (11) (1968), 1632-1641.

B 5 N.N. BAUTIN - "Behaviour of dynamical systems close to their stability boundary" - Gostekhizdat, Moscow, 1949.

B 6 N.N. BAUTIN, E.A. LEONTOVITCH - "Methods of qualitative study in the plane of dynamical systems" - Nauka, Moscow, 1976.

B 7 J.BERNUSSOU, LIU HSU, C. MIRA - "Quelques exemples de solutions stochastiques bornées dans les récurrences autonomes du 2e ordre"- Proceedings of "Tranfsormations ponctuelles et Applications" (Colloque CNRS, sept. 1973, Toulouse), Editions CNRS Paris 1976), 195-220.

B 8 G.D. BIRKHOFF - "Surface transformations and their dynamical applications" - Acta Math., 43 (1922), 1-119.

B 9 G.D. BIRKHOFF - "Dynamical systems" - Am. Math. Soc. Colloquium Publications, 9 (1966) (new edition of the 1927 book).

B10 G.D. BIRKHOFF - "Nouvelles recherches sur les Systèmes Dynamiques"- Memoriae Pont. Acad. Sci. Novi Lyncae, S. 3, 1 (1935), 85-216.

B11 G.D. BIRKHOFF - "Dynamical systems with two degrees of freedom" - Trans. Amer. Math. Soc. 18 (1917), 199-300.

B12 G.D. BIRKHOFF - "Collected mathematical Papers" - Dover Pub. Inc. N.Y. (1968).

B13 G.D. BIRKHOFF, P.A. SMITH - "Structure analysis of surface transformations". Journal de Math. S. 9, vol. 7 (1926), 345-379.

B14 E.S. BOÏKO - "The Academician A.A.Andronov' school" - Ed. Nauka, Moscow (1983).

C

C 1 A.R. CIGALA - "Sopra un criterio di instabilita" - Annali di Matematica. Ser. 3, 11 (1905), 67-75.

C 2 P. COLLET, J.P. ECKMANN - "Iterated maps on the interval as dynamical systems" - Progress on Physics, Birkhäuser-Boston (1980).

C 3 M. COSNARD - "Sur les suites itératives oscillantes : démonstration d'un théorème de Pulkin" - Colloque sur les Processus Itératifs. La Garde Freinet, mai 1979, 16 pages.

C 4 J. COUOT - "Invariant measures and topological entropy of complex stationary states of dynamic systems" - Proceedings of 8th ICNO, Prague, 1978. Ed. Academia Prague (1979), 1, 205-210.

C 5 J. COUOT, C. GILLOT - "Equations fonctionnelles de densités de mesures invariantes par un endomorphisme, et simulations numériques".

Proceedings of EquaDiff 78, Firenze (1978), edited by R. Conti, G. Sestini, G. Villari, 139-147.

C 6 J. COUOT, C. MIRA - "Densités de mesures invariantes non classiques" - C.R. Acad. Sc. Paris, Ser. 1, 296 (1983), 233-236.

C 7 P. CUITANOVIC, J. MYRHEIM - "Complex universality" - Nordita preprint.

C 8 E. COTTON - "Sur la notion de nombre caractéristique de M.Liapunov". Ann. Ecole Norm. 3 (36) (1919), 128-185.

D

D 1 A. DENJOY - "Sur les courbes définies par des équations différentielles à la surface du tore" - Journ. de Math. 11 (4) (1932), 333-375.

D 2 B. DERRIDA, A. GERVOIS, Y. POMMEAU - "Universal metric properties of bifurcations of endomorphisms" - J. Phys. A 12, (1979), 269-296.

D 3 A. DOUADY, J.H. HUBBARD - "Iteration des polynômes quadratiques complexes" - C.R. Acad. Sc. Paris, 294, sér. A (1982), 123-126.

D 4 J. DREWNIAK - "Convergence géométrique des suites récurrentes" - Proceedings of "Transformations ponctuelles et Applications", Colloque CNRS, sept. 1973, Toulouse, Editions CNRS (1976), 55-65.

D 5 P.M. DIAMOND - "Analytic invariants of mappings of two variables". Journal of Math. Analysis and Appl. 27 (1969), 601-608.

F

F 1 P. FATOU - "Mémoire sur les équations fonctionnelles" - Bull. Soc. Math. France, 47 (1919), 161-271 ; 48 (1920), 33-94 ; 208-314.

F 2 P. FATOU - "Sur l'itération des fonctions transcendantes entiières". Acta Math. 47 (1926), 337-370.

F 3 M.J. FEIGENBAUM - "Quantitative universality for a class of non linear transformations" - J. Statist. Phys., 19 (1978), 25-52.

F 4 D. FOURNIER - "Structure de bifurcation d'un endomorphisme défini par un polynôme du 3e degré" - C.R. Acad. Sc. Paris, sér. 1, 294 (1982), 455-458.

F 5 D. FOURNIER - "Contribution à l'étude de la dynamique chaotique. Analyse de deux situations" - Thèse de 3e cycle (Math. Appl.), n° 2471, Université Paul Sabatier, Toulouse, (1981).

F 6 D. FOURNIER, H. KAWAKAMI, C. MIRA - "Sur les bifurcations d'un difféomorphisme quadratique bi-dimensionnel. Situations homoclines et hétéroclines. Zones Morse-Smale" - C.R. Acad. Sc. Paris, sér. 1, 298 (11) (1984), 253-256.

F 7 D. FOURNIER, H. KAWAKAMI, C. MIRA - "Séquences de Myrberg et communications entre feuillets du plan des bifurcations d'un difféomorphisme bi-dimensionnel" - C.R. Acad. Sc. Paris, sér. 1, 301 (6) (1985), 325-328.

F 8 D. FOURNIER, H. KAWAKAMI, C. MIRA - "Feuilletage du plan des bifurcations d'un difféomorphisme bi-dimensionnel. Doublement de l'ordre des zones sources et des zones échangeurs" - C.R. Acad. Sc. Paris, sér. 1, 301, (5), (1985), 223-226.

F 9 D. FOURNIER, H. KAWAKAMI, C. MIRA - "La chaîne de liaison : image symbolique d'un ensemble complet de communications entre feuillets du plan des bifurcations d'un difféomorphisme bi-dimensionnel" - C.R. Acad. Sc. Paris, sér. 1, 303 (7) (1986), 315-320.

G

G 1 T. GEISEL. J. NIERWETBERG - "Self generated diffusion and universal critical properties in chaotic systems" - Proceedings Lecture Notes in Physics, 179 (1983), Springer, 93-114.

G 2 C. GILLOT - "Caractérisation des suites de rotation d'un endomorphisme unimodal de (0.1)" - C.R. Acad. Sc. Paris, sér. 1, 293 (1981), 294-251.

G3 C. GILLOT - "Structure des itinéraires symboliques et de l'entropie topologique des transformations unimodales" - Thèse de Doctorat d'Etat (Math.), n° 1 134, Université Paul Sabatier, Toulouse (1984).

G 4 C. GILLOT, C. MIRA - "Monotonie de l'entropie topologique de la récurrence quadratique définie par $x^2 - \lambda$ " - Note interne "Syst. Dynam." INSA, 83-1 (4 fév. 1983).

G 5 A. GIRAUD - "Application des récurrences à l'étude de systèmes de commmande" - Thèse de Docteur-Ingénieur, n° 205, avril 1969, Toulouse.

G 6 C. GREBOGI, S.W. Mc DONALD, E. OTT, J.A. YORKE - "Final state sensitivity : an obstruction to predictability" - Physics Letters, 1983, 99 A n° 9, 415-418.

G 7 C. GREBOGI, E. OTT, J.A. YORKE - "Fractal basin boundaries, long-lived chaotic transients, and unstable-unstable pair bifurcations"- Physical Review Letters, 50 (13) (1983), 935-938.

G 8 C. GREBOGI, E. OTT - "Crises, sudden changes in chaotic attractors and transient chaos" - Physica 7D (1983), 181-200.

G 9 C. GREBOGI, E. OTT, J.A. YORKE - "Attractors on an N-torus : quasiperiodicity versus chaos" - Physica 15 D (1985), 354-373.

G10 C. GREBOGI, E. KOSTELICH, E. OTT, J.A. YORKE - "Multi-dimensioned interwined basin boundaries : basin structure of the kicked double rotor" - Plasma preprint UMLPF 86-033 (Feb. 1986), Physics Publication Number, 86-174.

G11 A. GREVY - "Sur les équations fonctionnelles" - Bull. Sci. Math. (2), 16 (1892), 311-313.

G12 A. GREVY - "Etude sur les équations fonctionnelles" - Ann. Sci. Ecole Norm. Sup. (3), 11 (1894), 249-323 ; 13 (1896), 295-338.

G13 A. GREVY - "Equations fonctionnelles avec second membre" - Bull. Soc. Math. France, 25 (1897), 57-63.

G14 J. GUCKENHEIMER - "The bifurcation of quadratic functions" - N.Y. Acad. Sc. 75/1 (1980), 343-347.

G15 J. GUCKENHEIMER - "On the bifurcation of maps of the interval" - Invent. Math. 39, 165-178.

G16 J. GUCKENHEIMER - "sensitive dependence on initial conditions for one-dimensional maps". Comm. Math. Phys. 70 (1979), 133-160.

G17 J. GUCKENHEIMER - "One-dimensional dynamics". N.Y. Acad. Sc. 75/1 (1980), 343-347.

G18 J. GUCKENHEIMER, Ph. HOLMES - "Non linear oscillations, dynamical systems, and bifurcations of vector fields" - Applied Mathematical Sciences, 42, Springer (1983).

G19 I. GUMOWSKI, C. MIRA - "Sur une solution particulière de l'équation de Schröder". C.R. Acad. Sc. Paris 259 (1964) 2952-2955.

G20 I. GUMOWSKI, C. MIRA - "Sur une solution particulière explicite de l'équation fonctionnelle de Schröder" - C.R. Acad. Sc. Paris, 259 (1964), 4476-4479.

G21 I. GUMOWSKI, C. MIRA - "Sur un algorithme de détermination du domaine de stabilité d'un point double d'une récurrence non linéaire du deuxième ordre à variables réelles" - C.R. Acad. Sc. Paris, 260 (1965), 6524-6527.

G22 I. GUMOWSKI, C. MIRA - "Sensitivity problems related to certain bifurcations in non linear recurrence relations" - Automatica, 5 (1969), 303-317.

G23 I. GUMOWSKI, C. MIRA - "L'Optimisation. La Théorie et ses Problèmes". Dunod, 1970.

G24 I. GUMOWSKI C MIRA - "Boundaries of stochasticity domains in Hamiltonian systems" - Proc. 8th Int. Conf. on High-Energy Accelerators. Cern, Genève (1971), 374-376.

G25 I. GUMOWSKI, C. MIRA - "Sur la distribution des cycles d'une récurrence ou transformation ponctuelle, conservative, du deuxième ordre" - C.R. Acad. Sc. Paris, sér. A, 274 (1972), 1271-1274.

G26 I. GUMOWSKI, C. MIRA - "Sur la structure des lignes invariantes d'une récurrence ou transformation ponctuelle conservative du deuxième ordre" - C.R. Acad. Sc. Paris, sér. A 275 (1972), 869-872.

G27 I. GUMOWSKI, C. MIRA - "Stochastic solutions in a conservative dynamic system". Proc. 6th Int. Conf. on Non Linear Oscillations, Poznan. (1972). PWN-Warzawa. (1973), 417-438.

G28 I. GUMOWSKI - "Some properties of large amplitude solutions of conservative dynamic systems. Part 1 : Quadratic and cubic non linearities. Part 2 : bounded non-linearities". Rapport CERN/SI/ Int. BR/72-1. Genève. 1972.

G29 I. GUMOWSKI - "Solution structure of a conservative second order recurrence with an unbounded non linearity". Colloque International CNRS. Transformations Ponctuelles et leurs Applications, Toulouse (1973). Editions CNRS. Paris 1976.

G30 I. GUMOWSKI, C. MIRA - "Bifurcation pour une récurrence du deuxième ordre, par traversée d'un cas critique avec deux multiplicateurs complexes conjugués". C.R. Acad. Sc. Paris, sér. A 278 (1974), 1591-1594.

G31 I. GUMOWSKI, C. MIRA - "Sur les récurrences ou transformations ponctuelles du premier ordre avec inverse non-unique". C.R. Acad. Sc. Paris, sér. A 280 (1975), 905-908.

G32 I. GUMOWSKI, C. MIRA - "Accumulation de bifurcations dans une récurrence". C.R. Acad. Paris, sér. A 281 (1975), 45-48.

G33 I. GUMOWSKI - "Contribution de l'automatique au problème de réversibilité microscopique - irreversibilité macroscopique". RAIRO de l'AFCET 10 (7) (1976), 7-42.

G34 I. GUMOWSKI, C. MIRA - "Box-within-a-box bifurcation structure and the phenomenon of chaos". Informatica-76, Bled (1976)

G35 I. GUMOWSKI, C. MIRA - "Dynamique Chaotique" Cépadues Editions Toulouse, 1980.

G36 I. GUMOWSKI, C. MIRA - "Recurrences and discrete dynamic systems". Lecture Notes in Mathematics 809 Springer 1980.

G 37 I. GUMOWSKI, C. MIRA - "Détermination des courbes invariantes fermées d'une récurrence ou d'une transformation ponctuelle non linaire, voisine d'une récurrence conservative linéaire". C.R. Acad. sc. Paris sér. A 276 (1973), 333-336.

G38 I. GUMOWSKI, C. MIRA - "Bifurcation destabilisant une solution chaotique d'un endomorphisme du 2e ordre". C.R. Acad. sc. Paris sér. A 286 (1978), 427-431.

G39 C. GILLOT - "Comportement local de l'entropie topologique pour une famille unimodale à un paramètre". Note interne du Service de Mathématiques. INSA Toulouse nov. 1982.

H

H 1 J. HADAMARD - "Sur les transformations ponctuelles". Bull. Soc. Math. France. 34 (1906), 349-363.

H 2 J. HADAMARD - "Les surfaces à courbures opposées et leurs lignes géodésiques". J. Math. 5 (4) (1898), 27-73.

H 3 J. HADAMARD - "Les transformations ponctuelles". Bull. Soc. Math. Fr. 34 (1906), 349-363.

H 4 H. El-HAMOULY, C. MIRA - "Lien entre les propriétés d'un endomorphisme et celles d'un difféomorphisme". C.R. Acad. Sc. Paris sér. 1 293 (1982), 525-528.

H 5 H. El-HAMOULY, C. MIRA - "Singularités dues au feuilletage du plan des bifurcations d'un difféomorphisme bi-dimensionnel". C.R. Acad. Sc. Paris, sér. 1 294 (1982), 387-390.

H 6 H. El-HAMOULY - "Sructure du plan des bifurcations d'un difféomorphisme bi-dimensionnel". Thèse de Docteur-Ingénieur (Math. Appl.) n° 799 Université Paul Sabatier Toulouse (1982).

H 7 M. HENON - "A two-dimensional mapping with a strange attractor". Commun. Math. Phys. 50 1976, 69-77.

H 8 M.R. HERMAN - "Sur la conjugaison différentiable des difféomorphismes du cercle à des rotations". Publ. Math. IMES 49, (1979).

H 9 J. HERSCHEL - "Considération of various points of analysis". Phil. Trans. R. Soc. London 2 (1814), 440-468.

H10 P. HOLMES, D. WHITLEY - "Bifurcations of one and two-dimensional maps". Phil. Trans. R. Soc. London A 311 (1984), 43-102.

H11 P. HOLMES, R.F. WILLIAMS - "Knotted periodic orbits in suspensions of Smale's horseshoe : Torus knots and bifurcation sequences". Arch. Rat. Mech. Anal. 90 (2) (1985), 115-194.

H12 E. HOPF - "Abzweigung einer periodischen lösung von einer stationaren Lösung". Berich. Stächs. Acad. Wiss. Leipzig. Math. Phys.

KI. 94 (19) 1942, 15-25.

H13 P. HOLMES - "Knotted orbits and bifurcation sequences in periodi-dically forced oscillations". Paper presented at ICTAM, Lyngby (1984).

H14 A. HALANAY, D. WEXLER - "Theoria calitativa a sistemelor cu impulsuri." Ed. Acad. Rep. Soc. România, Bucarest 1968.

I

I 1 ISAACS - "Iterates of fractional orders" - Canadian J. Math, 2 (1950), 409-416.

J

J 1 M.V. JAKOBSON - "Mapping of the circle on to itself" - Math. Sbornik, 85 (2), (1971).

J 2 M.V. JAKOBSON - "Absolutely continuous invariant measures for one-parameter families of one-dimensional maps" - Comm. Math. Phys., 81 (1981), 39-88.

J 3 L. JONCKER, D. RAND - "Bifurcations in one dimension. (1) The non-wandering set" - Invent. Math. 62 (1981), 347-365. (2) "A versal model for bifurcations" - Invent. Math. 63 (1981), 1-15.

J 4 G. JULIA - "Sur les substitutions rationnelles" - C.R. Acad. Sci. Paris, 164 (1917), 1098-1100.

J 5 G. JULIA - "Mémoire sur l'itération des fonctions rationnelles" - J. Math. Pures Appl., 4 (1) (1918), 7e série, 47-245.

J 6 G. JULIA -"Sur l'itération de fractions rationnelles" - C.R. Acad. Sci. Paris, 166 (1918), 61-64.

K

K 1 H. KAWAKAMI - "Algorithme numérique définissant la bifurcation d'un point homocline". C.R. Acad. Sc. Paris, sér. 1 293, (1981) 401-103.

K 2 H. KAWAKAMI - "Bifurcation of periodic responses in forced dynamic non linear circuits : computation of bifurcation values of the system parameters". IEEE Trans. Circ. Syst., CAS 31 (3), (1984) 248-260.

K 3 H. KAWAKAMI - "Algorithme optimal définissant les suites de rotation de $y_{n+1} = y_n^2 - \lambda$. Notion de cycle adjoint". C.R. Acad. Sc. Paris, sér. 1 301 (12), (1985), 643-648.

K 4 H. KAWAKAMI - "Le dictionnaire des suites de rotation binaires et décimales". Note Groupe d'Etudes des Systèmes Non Linéaires et Applications, INSA Toulouse (1985).

K 5 H. KAWAKAMI - "Table of rotation sequences of $x_{n+1} = x_n^2 - \lambda$". Preprint Summer Symposium ou Dynamical Systems in Japan (1986) Proc. to appear in World Scientific Publications.

K 6 G. KOENIGS - "Recherches sur les substitutions uniformes".Bull. Sci. Math. 7 (2) (1883), 340-358.

K 7 G. KOENIGS - "Recherches sur les intégrales de certaines équations fonctionnelles".Ann. Sci. Ecole Norm. Sup. (3) 1 (1884). Supplément 3-41.

K 8 G. KOENIGS - "Nouvelles recherches sur les équations fonctionnelles". Ann. Sc. Ecole Norm. Sup. 2 (3) (1885), 385-404.

K 9 H. KAWAKAMI, C. MIRA - "The rotation sequences of the unimodal one-dimensional endomorphisms". Publ. "Syst. Dyn". INSA 86 - 1 (26 pages).

L

L 1 S. LATTES - "Sur les équations fonctionnelles qui définissent une courbe ou une surface invariante par une transformation". Ann. di Matematica. (3) 13 (1906), 1-137.

L 2 S. LATTES - "Sur la convergence des relations de récurrence".C.R. Acad. Sci. Paris 150 (1910), 1106-1109.

L 3 S. LATTES - "Sur les formes réduites des transformations ponctuelles dans le domaine d'un point double". Bull. Soc. Math. France. 39 (1911), 309-345.

L 4 S. LATTES - "Sur les suites récurrentes non linéaires et sur les fonctions génératrices de ces suites". Ann. Fac. Sci. Toulouse (3) 3 (1912), 73-124.

L 5 L. LEAU - "Sur les équations fonctionnelles". C.R. Acad. Sci. Paris 120 (1895), 427-429.

L 6 L. LEAU - "Etude sur les équations fonctionnelles à une ou plusieurs variables".Ann. Fac. Sci. Toulouse 11 (1897), 1-110.

L 7 L. LEAU - "Sur un problème d'itération". Bull. Soc. Math. France 26 (1898), 5-9.

L 8 E.M. LEMERAY - "Un théorème sur les fonctions itératives". Bull. Soc. Math. France 23 (1895), 255-262.

L 9 E.M. LEMERAY - "Sur les fonctions itératives est sur une nouvelle fonction". Assoc. Française pour l'Avancement des Sci. C.R. 24e session (Bordeaux 1895) 24 (1896) 149-165.

L10 E.M. LEMERAY - "Sur la convergence des substitutions uniformes". Nouv. Ann. de Math. (3) 16 (1897) 306-319.

L11 E.M. LEMERAY - "Sur quelques algorithmes et sur l'itération". Bull. Soc. Math. France 26 (1898), 10-15.

L12 E.M. LEMERAY - "Sur la convergence des substitutions uniformes". Nouv. Ann. de Math. (3) 17 (1898), 75-80.

L13 N.N. LEONOV - "Map of the line on to itself". Radiofisica. 2 (6) (1959), 942-956.

L14 N.N. LEONOV - "Piecewise linear map". Radiofisica. 3 (3) (1960), 496-510.

L15 N.N. LEONOV - "Theory of discontinuous maps of the line". Radiofisica. 3 (5) (1960), 872-886.

L16 N.N. LEONOV - "Discontinuous map of the straight line". Dokl. Akad. Nauk. SSSR. 143 (5) (1962), 1038-1041.

L17 Li T.Y. YORKE J.A. - "Period three implies chaos". Amer. Math. Monthly. 82 (1975), 985-992.

L18 LIU HSU - "Contribution à l'étude des récurrences non linéaires. Application aux systèmes dynamiques conservatifs". Thèse Doctorat ès Sc. Phys. Toulouse 22-09-74, n° 597.

L19 LIU HSU - "Sur un pas critique à multiplicateurs unitaires pour une récurrence autonome du 2e ordre à variables réelles". C.R. Acad. Sc. Paris sér. A 275 (1972), 671-674.

L20 A.M. LJAPUNOV - "Problème général de la stabilité du mouvement". Ann. de la Fac. des Sc. de Toulouse 9 (1907), 203-391.

M

M 1 I.G. MALKIN - "Theory of motion stability". Gostekhizdat. Moscow 1952, 3352.

M 2 R.M. MAY - "Simple mathematical models with very complicated dynamics". Nature 261 (1976), 459-470.

M 3 S.W. Mc DONALD, GREBOGI C., OTT E., YORKE J.A. - "Structure and crises of fractal basin boundaries", Physics Letters, 1985, 107 A n° 2, 51-54.

M 4 S.W. Mc DONALD, GREBOGI C., OTT E., YORKE J.A. - "Fractal basin boundaries", Physica 17 D, 1985, 125-153.

M 5 N. METROPOLIS, M.L. STEIN, P.R. STEIN - "On finite limit sets for transformations on the unit interval". Journal of Combin. Th. (A) 15 (1973), 25-44.

M 6 J. MILNOR, R. Thurston - "On iterated maps of the interval". Unpublished notes (1977), Princeton University Press.

M 7 C. MIRA - "Extension des notions de points singuliers aux équations aux différences". C.R. Acad. Sc. Paris 226 (1963), 3809-3812.

M 8 C. MIRA, F. ROUBELLAT - "Cas où le domaine de stabilité d'un ensemble limite attractif d'une récurrence du deuxième ordre n'est pas simplement connexe". C.R. Acad. Sc. Paris 268 (A) (1969), 1657-1660.

M 9 C. MIRA, F. ROUBELLAT - "Etude d'un cas critique pour des récurrences du deuxième ordre". C.R. Acad. Sc. Paris 266 (A) (1968), 302-305.

M10 C. MIRA - "Etude de la frontière de stabilité d'un point double stable d'une récurrence non-linéaire autonome du deuxième ordre". Proceedings. International Pulse Symposium , Budapest (1968), D 43-7/11 1-28.

M11 C. MIRA, F. ROUBELLAT - "Etude de la traversée d'un cas critique pour une récurrence du deuxième ordre sous l'effet d'une variation de paramètre".C.R. Acad. Sc. Paris 267 (A) (1968), 969-972.

M12 C. MIRA - "Détermination pratique du domaine de stabilité d'un point d'équilibre d'une récurrence non-linéaire du deuxième ordre à variables réelles". C.R. Acad. Sc. Paris 261 (2), (1964), 5314-5317.

M13 C. MIRA - "Méthode de détermination du domaine de stabilité d'un point fixe d'une récurrence non linéaire". Proceedings Congrès d'Automatique Théorique. Paris (1965), Ed Dunod (1967), 281-292.

M14 C. MIRA - "Traversée d'un cas critique pour une récurrence du deuxième ordre, sous l'effet d'une variation de paramètre". C.R. Acad. Sc. Paris 268 (A) (1969), 621-624.

M15 C. MIRA - "Etude d'un premier cas d'exception pour une récurrence ou transformation ponctuelle du deuxième ordre". C.R. Acad. Sc. Paris 269 (A) (1969), 1006-1009.

M16 C. MIRA - "Etude d'un second cas d'exception pour une récurrence ou transformation ponctuelle du deuxième ordre". C.R. Acad. Sc. Paris 270 (A) (1970), 332-335.

M17 C. MIRA - "Sur les cas d'exception d'une récurrence ou transformation ponctuelle du deuxième ordre". C.R. Acad. Sc. Paris 270 (A) (1970), 466-469.

M18 C. MIRA - "Cas critique d'une récurrence ou d'une transformation ponctuelle du quatrième ordre avec multiplicateurs complexes". C.R. Acad. Sc. Paris 272 (A) (1971), 1727-1730.

M19 C. MIRA - "Sur un cas critique correspondant à deux multiplicateurs égaux à l'unité pour une récurrence ou transformation ponctuelle de deuxième ordre". C.R. Acad. Sc. Paris 275 (A) (1972), 85-88.

M20 C. MIRA - "Sur les courbes invariantes fermées des récurrences non-linéaires, voisines d'une récurrence linéaire conservative du 2e ordre". Proceedings. Transformations Ponctuelles et Applications. (Colloque CNRS sept 73 Toulouse). Editions CNRS Paris (1976), 177-194.

M21 C. MIRA - "Accumulations de bifurcations et structures boites-emboitées dans les récurrences et transformations ponctuelles". Proceedings of the VIIe International Conference on Non-Linear Oscillation. Berlin sept. 1975 (Akademic Verlag. Berlin 1977), Band I 2, 81-93.

M22 C. MIRA - "Sur la double interprétation, déterministe et statistique, de certaines bifurcations complexes". C.R. Acad. Sc. Paris 283 (A) (1976), 911-914.

M23 C. MIRA - "Sur la notion de frontière floue de stabilité". Proceedings of the third Brazilian Congress of Mechanical Engineering Rio de Janeiro dec. 1975, D 4, 905-918.

M24 C. MIRA - "Bifurcation boîtes-emboitées dans les endomorphismes définis par un polynôme du 3e degré". Note interne n° 2 du Groupe systèmes dynamiques Non-Linéaires et Applications. Avril 1977. Université Paul Sabatier, Toulouse.

M25 C. MIRA- "Systèmes à dynamique complexe et bifurcations boîtes-emboitées". 1re partie RAIRO Automatique. 12 (1) (1978), 63-74 ; 12 (2) (1978), 171-190.

M26 C. MIRA - "Dynamique complexe engendrée par une récurrence continue linéaire par morceaux du premier ordre". C.R. Acad. Sc. Paris 285 (A) (1977), 731-734.

M27 C. MIRA - "Sur la structure des bifurcations des difféomorphismes du cercle". C.R. Acad. Sc. Paris 287 (A) (1978), 883-886.

M28 C. MIRA - "Sur quelques problèmes de dynamique complexe". Colloque "Modèles mathématiques en biologie". Journées Math. de la Société Math. de France. Montpellier 22-24 nov. 1978. Proceedings. Lecture Notes in Biomathematics 41, 169-205.

M29 C. MIRA - "Dynamique complexe engendrée par une équation différentielle d'ordre 3". Proceedings. Equadiff 78. Firenze (1978), 25-36.

M30 C. MIRA - "Frontière floue séparant les domaines d'attraction de deux attracteurs". Exemples. C.R. Acad. Sc. Paris 288 (A) (1979), 591-594.

M31 C. MIRA - "Récurrences et bifurcations d'un domaine d'attraction avec modifications qualitatives importantes". Colloque sur les comportements généraux des processus itératifs. La Garde Freinet 21-23 mai 1979.

M32 C. MIRA - "Chaotic dynamics in point mapping, Relations between a 1-dimensional endomorphism and a 2-dimensional diffeomorphism". International conference on non linear oscillations (ICNO). Kiev 1981. Proceedings Kiev Naukova Dumka (1984) 2, 265-271.

M33 C. MIRA - "Ensembles rythmiques de suites de rotation pour un endomorphisme uni-bimensionnel et nombres de rotation d'un difféomorphisme conservatif bi-dimensionnel". C.R. Acad. Sc. Paris sér. 1 294 (1982), 689-693.

M34 C. MIRA - "Sur les points d'accumulation de boîtes appartenant à une structure de bifurcations boîtes emboitées d'un endomorphisme uni-dimensionnel". C.R. Acad. Sc. Paris sér. 1 295 (1982), 13-16.

M35 C. MIRA - "Equation de Schröder et solution des récurrences. Généralisation des ploynômes de Tchebycheff". Proceedings Colloque international du CNRS n° 332, "Théorie de l'itération et Applications", (Toulouse, 17-22 mai 1982), Editions du CNRS, 1982, 35-43.

M36 C. MIRA - "Imbedding of a one-dimensional endomorphism into a two-dimensional diffeomorphism. Implications". 7th International Sitgès Conference, 5-11 sept. 1982, "Dynamical Systems and chaos". Lecture Notes en Physics, n° 179, Springer Verlag, 180-187.

M37 C. MIRA, S. MULLENBACH - "Sur l'itération fractionnaire d'un endomorphisme quadratique. C.R. Acad. Paris sér. 1 297 (1983), 369-372.

M38 C. MIRA - "Coexistence de structures de bifurcations "boîtes emboitées" et "boîtes en files" dans un endomorphisme unidimensionnel". Publication "Systèmes Dynamiques". INSA, 83-2 (février 1983).

M39 C. MIRA - "Structure des bifurcations d'un endomorphisme quadratique uni-dimensionnel". Journées d'Analyse Non Linéaire avec Applications à la mécanique, (Lille, 17-19 mai 1983). Proceedings of Journées. Publication IRMA 1983, 5 (2), (MIRA 1-12).

M40 C. MIRA, J.F. MYOUPO - "Chaîne de bifurcations donnant naissance à des courbes fermées invariantes cycliques, pour un difféomorphisme cubique bi-dimensionnel". C.R. Acad. Sc. Paris sér. 1 299 (4) (1984), 109-112.

M41 C. MIRA - "Rotation sequences and bifurcations structure of one-dimensional endomorphisms". International Symposium on Iteration Theory and its Functional Equations. Schloss Hofen bei Lochau (Autriche), sept. 84, Lecture notes in Mathematics n° 1 163, SPRINGER VERLAG (1985), 115-124.

M42 C. MIRA - "La structure de bifurcation boîtes-emboîtées. Ses conséquences". Bulletin de l'Académie Royale de Belgique - classe des sciences, 5e série 71, (1-2), (1985), 92-125.

M43 C. MIRA - "Généralités sur l'outil récurrence". In ATP-CNRS : Traitement numérique des attracteurs étranges". ATP : Outils et Modèles Mathématiques pour l'automatique, l'analyse des systèmes et le traitement du signal. Grenoble 20 to 22-02-1985. Proceedings (1985), 9-33.

M44 C. MIRA - "Structures de bifurcation boîtes-emboîtées et systèmes dynamiques. Applications aux redresseurs à thyristors". Same ATP-CNRS as M43, 89-104.

M45 C. MIRA - "Some historical aspects concerning the theory of dynamical systems". In "Dynamical Systems. A renewal of mechanism". 250-261. World Scientific Publ. Co 1986.

M46 C. MIRA - "Modelling in chaotic dynamic conditions". Proceedings of Symposium IASTED, Identification and Pattern Recognition Toulouse (18-20 juin 86). Tome I, 31-56.

M47 C. MIRA - "Les problèmes de prédiction en Dynamique Chaotique". Colloque AFCET-Perspectives Systémiques. Cerisy-la-salle 19-20 sept. 86 (Proceedings to appear 21 p.).

M48 C. MIRA - "Quelques situations fondamentales dans les Systèmes Dynamiques non linéaires et Chaotiques. Exemples". Journées d'Etudes "Phénomènes non linéaires chaotiques". (CNET Issy/les Moulineaux 3-5 déc. 86). Proceedings 22-64.

M49 M. MISIUREWICZ - "Absolutely continuous measures for certain maps of the interval". Publ. IHES 53 (1981), 17-52.

M50 M. MISIUREWICZ - "The structure of mappings of an interval with zero entropy". Publ. IHES 53 (1981), 5-16.

M51 M. MIQIUREWICZ, W. SZELENK - "Entropy of piecewise monotone mappings". Asterisque 50 (1977), 299-310.

M52 M. MORSE, G.A. HEDLUND - "Symbolic dynamics". Am. J. Math. 60 (1938), 815-866.

M53 S. MULLENBACH - "Contribution à l'étude de l'itération fractionnaire des endomorphismes". Thèse 3e cycle (Math. Appliquées) n° 2 785, Université Paul Sabatier (1983).

M54 P.J. MYRBERG - "Iteration von Quadratwurzeloperationen". Ann. Acad. Sci. Fenn. sér. A 259 (1958), 1-10.

M55 P.J. MYRBERG - "Iteration der reellen Polynome zweiten Grades". I. II. III. Ann. Acad. Sci. Fenn. sér. A 256 (1958), 1-10 ; 268 (1959), 1-10 ; 336 (1963), 1-10.

M56 P.J. MYRBERG - "Sur l'itération des polynômes réels quadratiques". J. Math. Pures Appl. (9) 41 (1962), 339-351.

M57 P. MONTEL - "Leçons sur les récurrences et applications". Gauthier-Villars. 1957 Paris.

M58 J.F. MYOUPO - "Plan des bifurcations d'un difféomorphisme bidimensionnel cubique". Thèse de 3e cycle (Math. Appl.) n° 2 797 Université Paul Sabatier Toulouse 1983.

N

N 1 Yu. I. NEIMARK - "Method of maps in the non linear oscillations theory". Izv. vuis. uch. zav. Radiofisica 1 (1) (1958), 41-66 ; 1 (2) (1958), 95-117 ; 1 (5-6) (1958), 146-167.

N 2 Yu. I. NEIMARK - "Method of maps in the non linear oscillations theory". Proceedings of International Conference on Non linear Oscillations, kiev 1961 (ed. Ukr. Acad. Nauk) 1963, (2) 285-298.

N 3 Yu. I. NEIMARK - "Stability of a fixed point of a map for a critical case". Radiofisica 3 (2), (1960), 342-343.

N 4 Yu. I. NEIMARK - "Existence and structural stability of invariant manifolds of maps". Radiofisica 10 (3) (1967), 311-320.

N 5 Yu. I. NEIMARK - "Method of maps in the non linear oscillations theory". Ed. Nauka, Moscow (1972).

N 6 S.E. NEWHOUSE - "Diffeomorphisms with infinitely many sinks". Topology 13 (1974), 9-18.

N 7 S.E. NEWHOUSE - "Lecture on dynamical systems". In "Dynamical systems" CIME Lectures. Prog. Math. 8 Boston Birkhäuser.

N 8 V.V. NEMYTSKII, V.V. STEPANOV - "Qualitative theory of differential equations". Princeton University Press (1960).

P

P 1 E. PICARD - "Leçons sur quelques équations fonctionnelles". Gauthier Villars. Paris. 1950.

P 2 S. PINCHERLE - "On the complete iteration of $x^2 - 2$". Rend. della Reale Acad. dei Lincei, 29 (1920), 329-333.

P 3 H. POINCARE - "Sur les courbes définies par des équations différentielles". Journ. de Math. pures et appl. 1 (1885), 167-224.

P 4 Y. POMEAU, P. Manneville - "Intermittent transition to turbulence in dissipative dynamical systems". Commun. Math. Phys. 74 (1980), 189-197.

P 5 L.S. PONTRYAGIN - "Asymptotic behaviour of differential equations with small parameters as coefficients of highest derivatives". Izvestia Akad. Nauk. Mat. sér. 21 (1957), 605-620.

P 6 C.P. PULKIN - "Oscillating iterated sequences". Dokl. Akad. Nauk. SSSR. 73 (6) (1950), 1129-1132.

R

R 1 Y. RANJEVA - "Contribution à l'étude de la récurrence réelle $x_{n+1} = x_n^2 - \lambda$. Mémoire DEA Université Paul Sabatier Toulouse (juin 1985).

R 2 F. ROUBELLAT - "Contribution à l'étude des solutions des récurrences non-linéaires et applications aux systèmes à données échantillonnées". Thèse de Doctorat ès Sciences Physiques. Toulouse 26 sept 1969, n° 364.

R 3 F. ROUBELLAT - "Etude pour une récurrence du deuxième ordre des trajectoires au voisinnage d'un point double correspondant à un cas critique". C.R. Acad. Sc. Paris 268 (A) (1969), 194-197.

R 4 Rolland MARIBE - "Contribution à l'étude des difféomorphismes et des endomorphismes bi-dimensionnels. Comportements chaotiques". Thèse de 3e cycle (Math. Appl.) n° 2 608, Université Paul Sabatier Toulouse, 1982.

R 5 O.E. ROSSLER, C. KAHLERT - "Hyperchaos and Julia sets". Z. NATURFORSCH 41a (1986).

R 6 O.E. ROSSLER, J.L. HUDSON, J.A. YORKE - "CLoud attractors and time inverted Julia boundaries". Z. NATURFORSCH 41a (1986).

R 7 D. RUELLE - "Applications conservant une mesure absolument continue par rapport à dx sur [0,1]". Commun. Math. Phys. 55 (1977), 47-52.

S

S 1 A.N. SHARKOVSKII - "Coexistence of cycles of a continuous map of a line into itself". Ukraïnskii Mat. Jurnal. 16 (1) (1964), 61-71.

S 2 A.N. SHARKOVSKII - "Problem of the isomorphism of dynamical systems". Proceeding 5th. International Conference On Non Linear Oscillations, Kiev (1969) 2, 541-544.

S 3 A.N. SHARKOVSKII - "On some properties of discrete dynamical systems". Proceedings of "Theorie de l'itération et ses applications" (Toulouse 1982) Ed CNRS, 153-158.

S 4 L.P. SHILNIKOV, N.K. GAVRILOV - "3-dimensional dynamical systems close to systems with stucturally unstable homoclinic curve". Mat. Sbornik. 88 (130) n° 4 (1972), 476-492 and 90 (132) n° 1 (1973), 139-156.

S 5 L.P. SHILNIKOV - "Theory of bifurcations of dynamical systems and dangerous boundaries". Dokl. Akad. Nauk SSR. 224 (5) (1975), 1046-1049.

S 6 C. SIMO - "On the Henon - Pommeau attractor". J. Statist. Phys. 21 (4) (1979), 465-494.

S 7 D. SINGER - "Stable orbits and bifurcations of maps of the interval". SIAM J. Appl. Math. 35 (1978), 260-267.

S 8 S. SMALE - "Diffeomorphisms with many periodic points". In "Differential Combinatorial Topology". S.S. Cairns (ed.) Princeton University Press (1963), 63-80.

S 9 S. SMALE - "Differentiable dynamical systems". Bull. Amer. Math. Soc. 73 (1967), 747-817.

S10 J. SOTOMAYOR - "Generic one parameter families of vecor fields". Publications Mathématiques de l'I.H.E.S. 43 (1974), 5-46.

T

T 1 R. THOM - "Stabilité structurelle et morphogénèse". Ediscience Paris 1973.

U

U 1 S.M. ULAM, J. von NEWMAN - "On combinations of stochastic and deterministic processes". Bull. Am. Math. Soc. 53 (1947), 1120.

U 2 USHIKI - "Fine structure of bifurcation branches in Henon's family of Mappings". Unpublished work (1982).

U 3 B. ULEWICZ - "On conditions for attractive singular points". Proceedings of Transformations Ponctuelles et Applications. (Colloque CNRS sept. 73 Toulouse). Editions CNRS Paris (1976), 31-39.

U 4 B. ULEWICZ - "On fixed points of the continuous real functions". Ann. Polon. Math. 31 (1975), 55-63.

V

V 1 G. VALIRON - "Théorie des fonctions". Masson ed. Paris 1948.

Z

Z 1 C. ZORNING - "Personal communication"(1979) for correcting the erroneous expression of N_k given in [M 25] .

INDEX

www.ingramcontent.com/pod-product-compliance
Lightning Source LLC
Chambersburg PA
CBHW042030200926
51615CB00028B/3
9789971503246